Chemical Engineering

Chemical Engineering
Trends and Developments

Editors

Miguel A. Galán
Eva Martin del Valle

Department of Chemical Engineering,
University of Salamanca, Spain

John Wiley & Sons, Ltd

Telephone (+44) 1243 779777

Email (for orders and customer service enquiries): cs-books@wiley.co.uk
Visit our Home Page on www.wiley.com

Other Wiley Editorial Offices

John Wiley & Sons Inc., 111 River Street, Hoboken, NJ 07030, USA

Jossey-Bass, 989 Market Street, San Francisco, CA 94103-1741, USA

Wiley-VCH Verlag GmbH, Boschstr. 12, D-69469 Weinheim, Germany

John Wiley & Sons Australia Ltd, 33 Park Road, Milton, Queensland 4064, Australia

John Wiley & Sons (Asia) Pte Ltd, 2 Clementi Loop #02-01, Jin Xing Distripark, Singapore 129809

John Wiley & Sons Canada Ltd, 22 Worcester Road, Etobicoke, Ontario, Canada M9W 1L1

Wiley also publishes its books in a variety of electronic formats. Some content that appears in print may not be available in electronic books.

Library of Congress Cataloging-in-Publication Data

Chemical engineering : trends and developments / editors Miguel A. Galán, Eva Martin del Valle.
p. cm.
Includes bibliographical references and index.
ISBN-13 978-0-470-02498-0 (cloth : alk. paper)
ISBN-10 0-470-02498-4(cloth : alk. paper)
1. Chemical engineering. I. Galán, Miguel A., 1945– II. Martín del Valle, Eva, 1973–
TP155.C37 2005
660—dc22
2005005184

British Library Cataloguing in Publication Data

A catalogue record for this book is available from the British Library

ISBN-13 978-0-470-02498-0 (HB)
ISBN-10 0-470-02498-4 (HB)

Typeset in 10/12pt Times by Integra Software Services Pvt. Ltd, Pondicherry, India
Printed and bound in Great Britain by Antony Rowe Ltd, Chippenham, Wiltshire
This book is printed on acid-free paper responsibly manufactured from sustainable forestry in which at least two trees are planted for each one used for paper production.

Contents

List of Contributors

G. Akay (1) Process Intensification and Miniaturization Centre, School of Chemical Engineering and Advanced Materials, (2) Institute for Nanoscale Science and Technology, Newcastle University, Newcastle upon Tyne NE1 7RU, UK

Orlando M. Alfano INTEC (Universidad Nacional del Litoral and CONICET), Güemes 3450. (3000) Santa Fe, Argentina

Pedro E. Arce Department of Chemical Engineering, Tennessee Tech University, Cookeville, TN 38505, USA

Miguel J. Bagajewicz School of Chemical Engineering, University of Oklahoma, OK 73019-1004, USA

M.A. Bokhari (1) School of Surgical and Reproductive Sciences, The Medical School, (2) Process Intensification and Miniaturization Centre, School of Chemical Engineering and Advanced Materials, (3) Institute for Nanoscale Science and Technology, Newcastle University, Newcastle upon Tyne NE1 7RU, UK

V.J. Byron (1)School of Surgical and Reproductive Sciences, The Medical School, Newcastle University, Newcastle upon Tyne NE1 7RU, UK, (2)Process Intensification and Miniaturization Centre, School of Chemical Engineering and Advanced Materials

Ruben G. Carbonell Department of Chemical and Biomolecular Engineering, North Carolina State University, Raleigh, NC 27695-7905, USA

Alberto E. Cassano INTEC (Universidad Nacional del Litoral and CONICET), Güemes 3450. (3000) Santa Fe, Argentina

Ramon L. Cerro Department of Chemical and Materials Engineering, University of Alabama in Huntsville, Huntsville, AL 35899, USA

M.G. De Angelis Dipartimento di Ingegneria Chimica, Mineraria e delle Tecnologie Ambientali, Università di Bologna, viale Risorgimento 2, 40136 Bologna, Italy

E.M. Del Valle Department of Chemical Engineering, University of Salamanca, P/Los Caídos 1–5, 37008 Salamanca, Spain

M. Elena Diaz Martin Department of Chemical and Materials Engineering, University of Alabama in Huntsville, Huntsville, AL 35899, USA

F. Doghieri Dipartimento di Ingegneria Chimica, Mineraria e delle Tecnologie Ambientali, Università di Bologna, viale Risorgimento 2, 40136 Bologna, Italy

M. Dogru Process Intensification and Miniaturization Centre, School of Chemical Engineering and Advanced Materials, Newcastle University, Newcastle upon Tyne NE1 7RU, UK

M.A. Galán Department of Chemical Engineering, University of Salamanca, P/Los Caídos 1-5, 37008 Salamanca, Spain

M. Giacinti Baschetti Dipartimento di Ingegneria Chimica, Mineraria e delle Tecnologie Ambientali, Università di Bologna, viale Risorgimento 2, 40136 Bologna, Italy

Ignacio E. Grossmann Department of Chemical Engineering, Carnegie Mellon University, Pittsburgh, PA 15213, USA

Patrick V. Gurgel Department of Chemical and Biomolecular Engineering, North Carolina State University, Raleigh, NC 27695-7905, USA

R. Gutierrez Department of Chemical Engineering, University of Salamanca, P/Los Caidos 1-5, 37008, Salamanca, Spain

Janez Levec Department of Chemical Engineering, University of Ljubljana, and National Institute of Chemistry, PO Box 537 SI-1000 Ljubljana, Slovenia

Michel Quintard Institut de Mécanique des Fluides de Toulouse, Av. du Professeur Camille Soula, 31400 Toulouse, France

C.A. Ruiz Department of Chemical Engineering, University of Salamanca, P/Los Caídos 1–5, 37008 Salamanca, Spain

Jeffrey R. Salm Department of Chemical and Biomolecular Engineering, North Carolina State University, Raleigh, NC 27695-7905, USA

G.C. Sarti Dipartimento di Ingegneria Chimica, Mineraria e delle Tecnologie Ambientali, Università di Bologna, viale Risorgimento 2, 40136 Bologna, Italy

L.E. Skip Scriven Coating Process Fundamentals Program, Department of Chemical Engineering and Materials Science and Industrial Partnership for Research in Interfacial and Materials Engineering, University of Minnesota, 421 Washington Avenue S. E., Minneapolis, Minnesota 55455, USA

Guangquan Wang Department of Chemical and Biomolecular Engineering, North Carolina State University, Raleigh, NC 27695-7905, USA

Stephen Whitaker Department of Chemical Engineering and Material Science, University of California at Davis, Davis, CA 95459, USA

Preface

Usually the preface of any book is written by a recognized professional who describes the excellence of the book and the authors who are, of course, less well-known than himself. In this case, however, the task is made very difficult by the excellence of the authors, the large amount of topics treated in the book and the added difficulty of finding someone who is an expert in all of them. For these reasons, I decided to write the preface myself, acknowledging that I am really less than qualified to do so.

This book's genesis was two meetings, held in Salamanca (Spain), with the old student army of the University of California (Davis) from the late 1960s and early 1970s, together with professors who were very close to us. The idea was to exchange experiences about the topics in our research and discuss the future for each of them. In the end, conclusions were collected and we decided that many of the ideas and much of the research done could be of interest to the scientific community. The result is a tidy re-compilation of many of the topics relevant to chemical engineering, written by experts from academia and industry.

We are conscious that certain topics are not considered and some readers will find fault, but we ask them to bear in mind that in a single book it is impossible to include all experts and all topics connected to chemical engineering.

We are sure that this book is interesting because it provides a detailed perspective on technical innovations and the industrial application of each of the topics. This is due to the panel of experts who have broad experience as researchers and consultants for international industries.

The book is structured according to the suggestions of Professor Scriven. It starts by describing the scope and basic concepts of chemical engineering, and continues with several chapters that are related to separations processes, a bottleneck in many industrial processes. After that, applications are covered in fields such as reaction engineering, particle manufacture, and encapsulation and coating. The book finishes by covering process integration, showing the advances and opportunities in this field.

I would like to express my thanks to each one of the authors for their valuable suggestions and for the gift to being my friends. I am very proud and honoured by their friendship. Finally, a special mention for Professor Martín del Valle for her patience, tenacity and endurance throughout the preparation of this book; to say thanks perhaps is not enough.

For all of them and for you reader: thank you very much.

Miguel Angel Galán

1

The Art and Science of Upscaling

Pedro E. Arce, Michel Quintard and Stephen Whitaker

1.1 Introduction

The process of upscaling governing differential equations from one length scale to another is ubiquitous in many engineering disciplines and chemical engineering is no exception. The classic packed bed catalytic reactor is an example of a hierarchical system (Cushman, 1990) in which important phenomena occur at a variety of length scales. To design such a reactor, we need to predict the output conditions given the input conditions, and this prediction is generally based on knowledge of the rate of reaction *per unit volume of the reactor*. The rate of reaction per unit volume of the reactor is a quantity associated with the averaging volume $\mathscr{V}$ illustrated in Figure 1.1. In order to use information associated with the averaging volume to design successfully the reactor, the averaging volume must be *large enough* to provide a representative average and it must be *small enough* to capture accurately the variations of the rate of reaction that occur throughout the reactor.

To develop a qualitative idea about what is meant by *large enough* and *small enough*, we consider a detailed version of the averaging volume shown in Figure 1.2. Here we have identified the fluid as the γ-phase, the porous particles as the κ-phase, and ℓ_γ as the characteristic length associated with the γ-phase. In addition to the characteristic length associated with the fluid, we have identified the radius of the averaging volume as r_0. In order that the averaging volume be *large enough* to provide a representative average we require that $r_0 \gg \ell_\gamma$, and in order that the averaging volume be *small enough* to capture accurately the variations of the rate of reaction we require that $L, D \gg r_0$. Here the choice of the length of the reactor, L, or the diameter of the reactor, D, depends on the concentration gradients within the reactor. If the gradients in the radial direction are comparable to or larger than those in the axial direction, the appropriate constraint is $D \gg r_0$. On the other hand, if the reactor is adiabatic and the non-uniform flow near the walls of the reactor can be ignored, the gradients in the radial direction will be negligible

---

Chemical Engineering Edited by Miguel A. Galán and Eva Martin del Valle

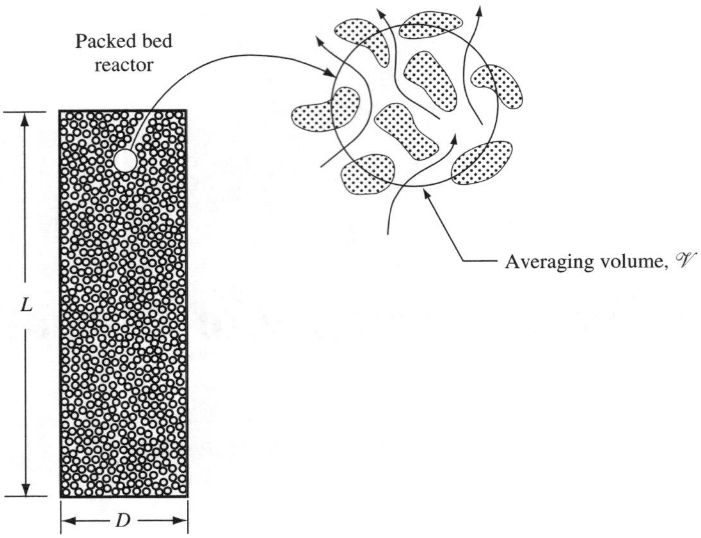

Figure 1.1 Design of a packed bed reactor

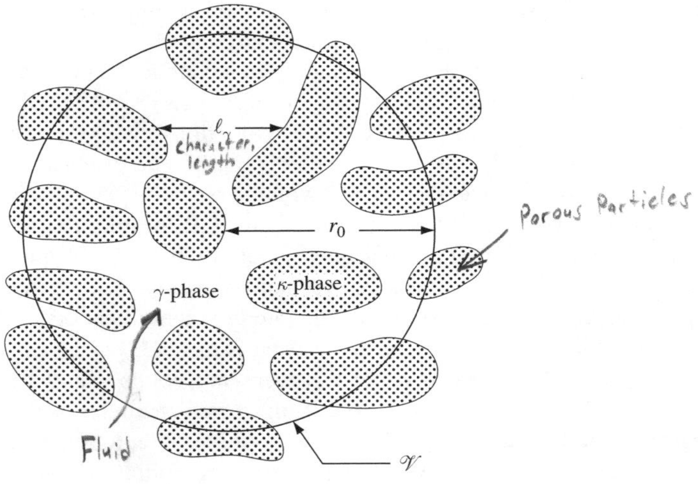

Figure 1.2 Averaging volume

and the appropriate constraint is $L \gg r_0$. These ideas suggest that the length scales must be *disparate* or *separated* according to

$$L, D \gg r_0 \gg \ell_\gamma \tag{1.1}$$

These constraints on the length scales are *purely intuitive*; however, they are characteristic of the type of results obtained by careful analysis (Whitaker, 1986a; Quintard and Whitaker, 1994a–e; Whitaker, 1999). It is important to understand that Figures 1.1 and 1.2

are not drawn to scale and thus are not consistent with the length scale constraints contained in equation 1.1.

In order to determine the average rate of reaction in the volume $\mathscr{V}$, one needs to determine the rate of reaction in the porous catalyst identified as the κ-phase in Figure 1.2. If the concentration gradients in both the γ-phase and the κ-phase are *small enough*, the concentrations of the reacting species can be treated as constants within the averaging volume. This allows one to specify the rate of reaction per unit volume of the reactor in terms of the concentrations associated with the averaging volume illustrated in Figure 1.1. A reactor in which this condition is valid is often referred to as an *ideal reactor* (Butt, 1980, Chapter 4) or, for the reactor illustrated in Figure 1.1, as a Plug-flow tubular reactor (PFTR) (Schmidt, 1998). In order to measure reaction rates and connect those rates to concentrations, one attempts to achieve the approximation of a uniform concentration within an averaging volume. However, the approximation of a uniform concentration is generally not valid in a *real reactor* (Butt, 1980, Chapter 5) and the concentration gradients in the porous catalyst phase need to be taken into account. This motivates the construction of a second, smaller averaging volume illustrated in Figure 1.3.

Porous catalysts are often manufactured by compacting microporous particles (Froment and Bischoff, 1979) and this leads to the micropore–macropore model of a porous catalyst illustrated at level II in Figure 1.3. In this case, diffusion occurs in the macropores, while diffusion and reaction take place in the micropores. Under these circumstances, it is reasonable to analyze the transport process in terms of a two-region model (Whitaker, 1983), one region being the macropores and the other being the micropores. These two regions make up the porous catalyst illustrated at level I in Figure 1.3. If the concentration gradients in both the macropore region and the micropore region are *small enough*, the concentrations of the reacting species can be treated as constants within this second averaging volume, and one can proceed to analyze the process of diffusion and reaction with

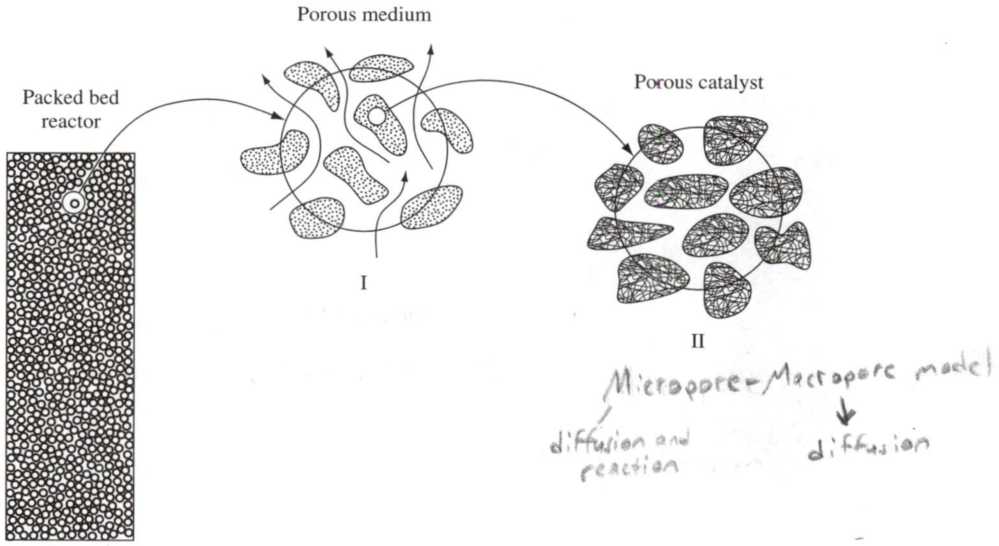

Figure 1.3 *Transport in a micropore–macropore model of a porous catalyst*

a one-equation model. This leads to the classic effectiveness factor analysis (Carberry, 1976) which provides information to be transported *up the hierarchy of length scales* to the porous medium (level I) illustrated in Figure 1.3. The constraints associated with the validity of a one-equation model for the micropore–macropore system are given by Whitaker (1983).

If the one-equation model of diffusion and reaction in a micropore–macropore system is not valid, one needs to proceed *down the hierarchy of length scales* to develop an analysis of the transport process in both the macropore region and the micropore region. This leads to yet another averaging volume that is illustrated as level III in Figure 1.4. Analysis at this level leads to a micropore effectiveness factor that is discussed by Carberry (1976, Sec. 9.2) and by Froment and Bischoff (1979, Sec. 3.9).

In the analysis of diffusion and reaction in the micropores, we are confronted with the *fact* that catalysts are not uniformly distributed on the surface of the solid phase; thus the so-called catalytic surface is highly non-uniform and spatial smoothing is required in order to achieve a complete analysis of the process. This leads to yet another averaging volume illustrated as level IV in Figure 1.5. The analysis at this level should make use of the method of area averaging (Ochoa-Tapia *et al.*, 1993; Wood *et al.*, 2000) in order to obtain a spatially smoothed jump condition associated with the non-uniform catalytic surface. It would appear that this aspect of the diffusion and reaction process has received little attention and the required information associated with level IV is always obtained by experiment based on the assumption that the experimental information can be used directly at level III.

The *train of information* associated with the design of a packed bed catalytic reactor is illustrated in Figure 1.6. There are several important observations that must be made

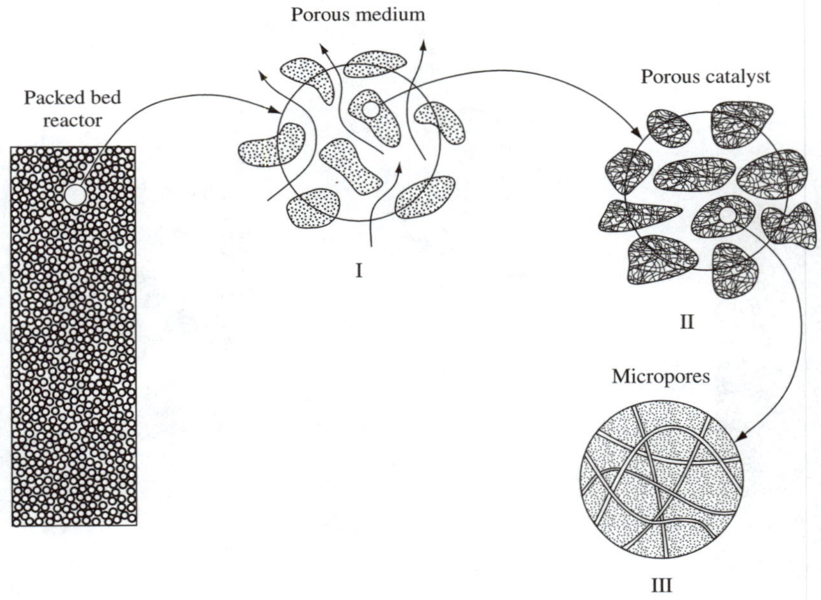

Figure 1.4 *Transport in the micropores*

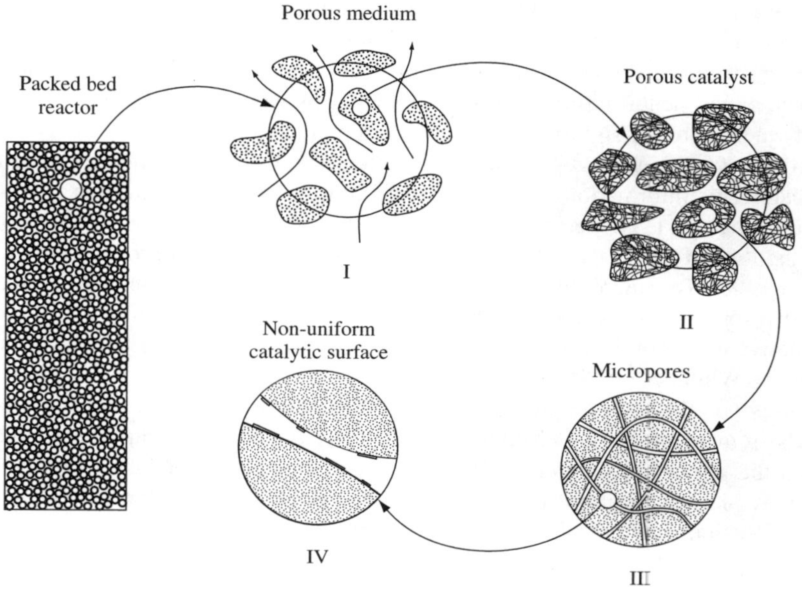

Figure 1.5 *Reaction at a non-uniform catalytic surface*

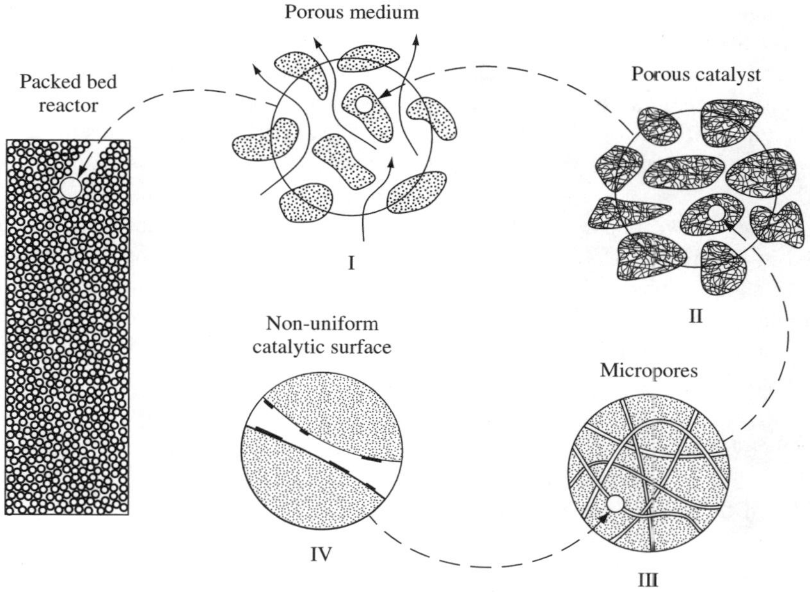

Figure 1.6 *Train of information.* Whitaker(1999), *The Method of Volume Averaging*, Figure 2, p.xiv; with kind permission of Kluwer Academic Publishers.

about this train. First, we note that the train can be continued in the direction of *decreasing length scales* in search for more fundamental information. Second, we note that one can board the train in the direction of *increasing length scales* at any level, provided that appropriate experimental information is available. This would be difficult to accomplish at level I when there are significant concentration gradients in the porous catalyst. Third, we note that *information is lost* when one uses the calculus of integration to move up the length scale. This information can be recovered in three ways: (1) intuition can provide the lost information; (2) experiment can provide the lost information; and (3) closure can provide the lost information. Finally, we note that information is *filtered* as we move up the length scales. By filtered we mean that not all the information available at one level is needed to provide a satisfactory description of the process at the next higher level. A quantitative theory of filtering does not yet exist; however, several examples have been discussed by Whitaker (1999).

In Figures 1.1–1.6 we have provided a qualitative description of the process of upscaling. In the remainder of this chapter we will focus our attention on level II with the restriction that the diffusion and reaction process in the porous catalyst is dominated by a single pore size. In addition, we will assume that the pore size is large enough so that Knudsen diffusion does not play an important role in the transport process.

1.2 Intuition

We begin our study of diffusion and reaction in a porous medium with a *classic, intuitive approach* to upscaling that often leads to confusion concerning homogeneous and heterogeneous reactions. We follow the intuitive approach with a rigorous upscaling of the problem of dilute solution diffusion and heterogeneous reaction in a model porous medium. We then direct our attention to the more complex problem of coupled, non-linear diffusion and reaction in a real porous catalyst. We show how the information lost in the upscaling process can be recovered by means of a closure problem that allows us to predict the tortuosity tensor in a rigorous manner. The analysis demonstrates the existence of a single tortuosity tensor for all N species involved in the process of diffusion and reaction.

We consider a two-phase system consisting of a fluid phase and a solid phase as illustrated in Figure 1.7. Here we have identified the fluid phase as the γ-phase and the solid phase as the κ-phase. The foundations for the analysis of diffusion and reaction in this two-phase system consist of the species continuity equation in the γ-phase and the species jump condition at the catalytic surface. The species continuity equation can be expressed as

$$\frac{\partial c_{A\gamma}}{\partial t} + \nabla \cdot (c_{A\gamma} \mathbf{v}_{A\gamma}) = R_{A\gamma}, \quad A = 1, 2, \ldots, N \tag{1.2a}$$

or in terms of the molar flux as given by (Bird *et al.*, 2002)

$$\frac{\partial c_{A\gamma}}{\partial t} + \nabla \cdot \mathbf{N}_A = R_{A\gamma}, \quad A = 1, 2, \ldots, N \tag{1.2b}$$

This latter form fails to identify the *species velocity* as a crucial part of the species transport equation, and this often leads to confusion about the mechanical aspects of

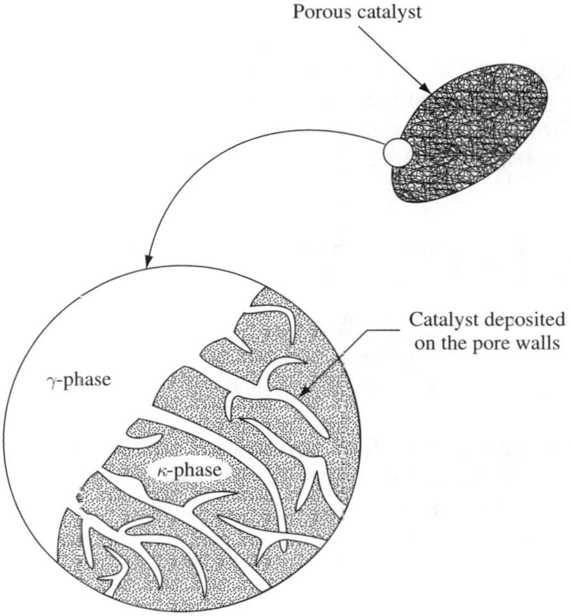

Figure 1.7 *Diffusion and reaction in a porous medium*

multi-component mass transfer. When surface transport (Ochoa-Tapia *et al.*, 1993) can be neglected, the jump condition takes the form

$$\frac{\partial c_{As}}{\partial t} = (c_{A\gamma}\mathbf{v}_{A\gamma}) \cdot \mathbf{n}_{\gamma\kappa} + R_{As}, \quad \text{at the } \gamma\text{–}\kappa \text{ interface}, \quad A = 1, 2, \ldots, N \tag{1.3a}$$

where $\mathbf{n}_{\gamma\kappa}$ represents the unit normal vector directed *from* the γ-phase *to* the κ-phase. In terms of the molar flux that appears in equation 1.2b, the jump condition is given by

$$\frac{\partial c_{As}}{\partial t} = \mathbf{N}_A \cdot \mathbf{n}_{\gamma\kappa} + R_{As}, \quad \text{at the } \gamma\text{–}\kappa \text{ interface}, \quad A = 1, 2, \ldots, N \tag{1.3b}$$

In equations 1.2a–1.3b, we have used $c_{A\gamma}$ to represent the *bulk concentration* of species A (moles per unit volume), and c_{As} to represent the *surface concentration* of species A (moles per unit area). The nomenclature for the homogeneous reaction rate, $R_{A\gamma}$, and heterogeneous reaction rate, R_{As}, follows the same pattern. The surface concentration is sometimes referred to as the *adsorbed concentration* or the *surface excess concentration*, and the derivation (Whitaker, 1992) of the jump condition essentially consists of a shell balance around the interfacial region. The jump condition can also be thought of as a surface transport equation (Slattery, 1990) and it forms the basis for various mass transfer boundary conditions that apply at a phase interface.

In addition to the continuity equation and the jump condition, we need a set of N momentum equations to determine the species velocities, and we need chemical kinetic constitutive equations for the homogeneous and heterogeneous reactions. We also need a method of *connecting* the surface concentration, c_{As}, to the bulk concentration, $c_{A\gamma}$.

Before exploring the general problem in some detail, we consider the typical *intuitive approach* commonly used in textbooks on reactor design (Carberry, 1976; Fogler, 1992; Froment and Bischoff, 1979; Levenspiel, 1999; Schmidt, 1998). In this approach, the analysis consists of the application of a *shell balance* based on the word statement given by

$$\begin{Bmatrix} \text{accumulation} \\ \text{of species } A \end{Bmatrix} = \begin{Bmatrix} \text{flow of species } A \text{ into} \\ \text{the control volume} \end{Bmatrix} - \begin{Bmatrix} \text{flow of species } A \text{ out} \\ \text{of the control volume} \end{Bmatrix} + \begin{Bmatrix} \text{rate of production} \\ \text{of species } A \text{ owing} \\ \text{to chemical reaction} \end{Bmatrix} \tag{1.4}$$

This result is applied to the cube illustrated in Figure 1.8 in order to obtain a balance equation associated with the accumulation, the flux, and the reaction rate. This balance equation is usually written with no regard to the *averaged or upscaled* quantities that are involved and thus takes the form

$$\frac{\Delta c_A}{\Delta t}\Delta x\Delta y\Delta z = \begin{matrix} (N_{Ax}|_x - N_{Ax}|_{x+\Delta x})\Delta y\Delta z \\ (N_{Ay}|_y - N_{Ay}|_{y+\Delta y})\Delta x\Delta z + R_A\Delta x\Delta y\Delta z \\ (N_{Az}|_z - N_{Az}|_{z+\Delta z})\Delta x\Delta y \end{matrix} \tag{1.5}$$

One divides this balance equation by $\Delta x\Delta y\Delta z$ and lets the cube shrink to zero to obtain

$$\frac{\partial c_A}{\partial t} = -\left(\frac{\partial N_{Ax}}{\partial x} + \frac{\partial N_{Ay}}{\partial y} + \frac{\partial N_{Az}}{\partial z}\right) + R_A \tag{1.6}$$

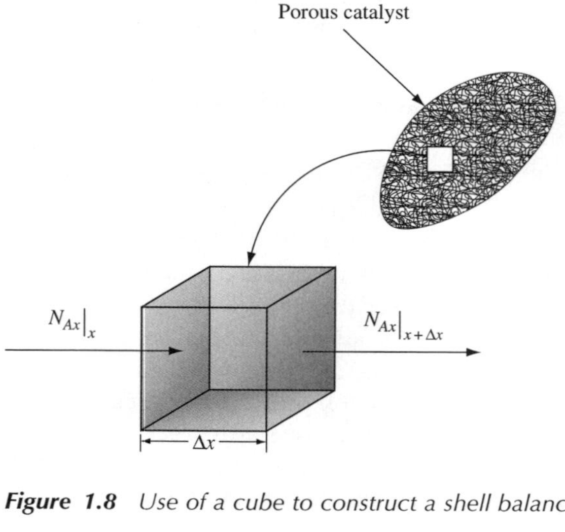

Figure 1.8 *Use of a cube to construct a shell balance*

In compact vector notation this takes a form

$$\frac{\partial c_A}{\partial t} = -\nabla \cdot \mathbf{N}_A + R_A \tag{1.7}$$

that can be *easily confused* with equation 1.2b. To be explicit about the confusion, we note that c_A in equation 1.7 represents a *volume averaged concentration*, $\mathbf{N}_A$ represents a *volume averaged molar flux*, and R_A represents a *heterogeneous rate of reaction*. Each one of the three terms in equation 1.7 represents something different than the analogous term in equation 1.2b and this leads to considerable confusion among chemical engineering students.

The diffusion and reaction process illustrated in Figure 1.7 is typically treated as one-dimensional (in the average sense) so that the transport equation given by equation 1.6 simplifies to

$$\frac{\partial c_A}{\partial t} = -\frac{\partial N_{Az}}{\partial z} + R_A \tag{1.8}$$

At this point, a vague reference to Fick's law is usually made in order to obtain

$$\frac{\partial c_A}{\partial t} = \frac{\partial}{\partial z}\left(\mathscr{D}_{\mathrm{e}} \frac{\partial c_A}{\partial z}\right) + R_A \tag{1.9}$$

where $\mathscr{D}_{\mathrm{e}}$ is identified as an *effective diffusivity*. Having dispensed with accumulation and diffusion, one often considers the first-order consumption of species A leading to

Heterogeneous reaction: $$\frac{\partial c_A}{\partial t} = \frac{\partial}{\partial z}\left(\mathscr{D}_{\mathrm{e}} \frac{\partial c_A}{\partial z}\right) - a_{\mathrm{v}} k c_A \tag{1.10}$$

where a_{v} represents the surface area per unit volume.

Students often encounter diffusion and *homogeneous reaction* in a form given by

Homogeneous reaction: $$\frac{\partial c_A}{\partial t} = \frac{\partial}{\partial z}\left(\mathscr{D} \frac{\partial c_A}{\partial z}\right) - k c_A \tag{1.11}$$

and it is not difficult to see why there is confusion about homogeneous and heterogeneous reactions. The essential difficulty results from the fact that the *upscaling* from an unstated point equation, such as equation 1.2, is carried out in a purely intuitive manner with no regard to the precise definition of the dependent variable, c_A. If the meaning of the dependent variable in a governing differential equation is not well understood, trouble is sure to follow.

1.3 Analysis

To eliminate the confusion between homogeneous and heterogeneous reactions, and to introduce the concept of upscaling in a rigorous manner, we need to illustrate the general features of the process without dealing directly with all the complexities. To do so, we

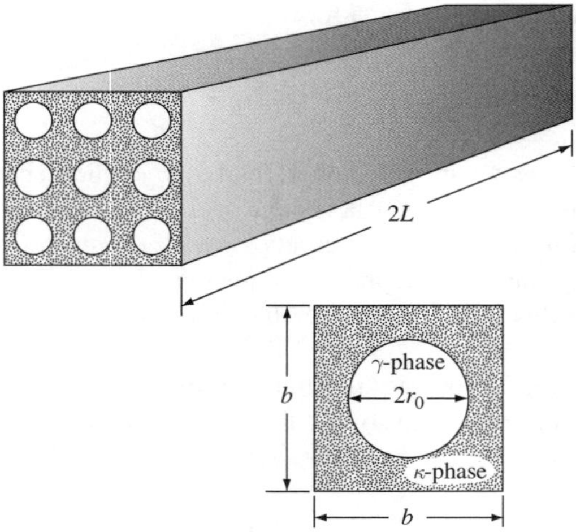

Figure 1.9 *Bundle of capillary tubes as a model porous medium*

consider a bundle of capillary tubes as a model of a porous medium. This model is illustrated in Figure 1.9 where we have shown a bundle of capillary tubes of length $2L$ and radius r_0. The fluid in the capillary tubes is identified as the γ-phase and the solid as the κ-phase. The porosity of this model porous medium is given by

$$\{\text{porosity}\} = \pi r_0^2 / b^2 \tag{1.12}$$

and we will use ε_γ to represent the porosity.

Our model of diffusion and heterogeneous reaction in one of the capillary tubes illustrated in Figure 1.9 is given by the following boundary value problem:

$$\frac{\partial c_{A\gamma}}{\partial t} = \mathscr{D}_\gamma \left[\frac{1}{r}\frac{\partial}{\partial r}\left(r \frac{\partial c_{A\gamma}}{\partial r} \right) + \frac{\partial^2 c_{A\gamma}}{\partial z^2} \right], \quad \text{in the } \gamma\text{-phase} \tag{1.13}$$

BC.1 : $$c_{A\gamma} = c_{A\gamma}, \quad z = 0 \tag{1.14}$$

BC.2 : $$-\mathscr{D}_\gamma \frac{\partial c_{A\gamma}}{\partial r} = k c_{A\gamma}, \quad r = r_0 \tag{1.15}$$

BC.3 : $$\frac{\partial c_{A\gamma}}{\partial z} = 0, \quad z = L \tag{1.16}$$

IC. : $$\text{unspecified} \tag{1.17}$$

Here we have assumed that the catalytic surface at $r = r_0$ is *quasi-steady* even though the diffusion process in the pore may be transient (Carbonell and Whitaker, 1984; Whitaker, 1986b). Equations 1.13–1.17 represent the physical situation in the *pore domain* and we need equations that represent the physical situation in the *porous medium domain*. This requires that we develop the area-averaged form of equation 1.13 and that we determine under what circumstances the concentration at $r = r_0$ can be replaced by the area-averaged concentration, $\langle c_{A\gamma}\rangle^\gamma$. The area-averaged concentration is defined by

$$\langle c_{A\gamma}\rangle^\gamma = \frac{1}{\pi r_0^2}\int_{r=0}^{r=r_0} 2\pi r c_{A\gamma}\,\mathrm{d}r \tag{1.18}$$

and in order to develop an area-averaged or *upscaled* diffusion equation, we form the intrinsic area average of equation 1.13 to obtain

$$\frac{1}{\pi r_0^2}\int_{r=0}^{r=r_0}\left(\frac{\partial c_{A\gamma}}{\partial t}\right)2\pi r\,\mathrm{d}r = \mathscr{D}_\gamma\left\{\frac{1}{\pi r_0^2}\int_{r=0}^{r=r_0}\frac{1}{r}\frac{\partial}{\partial r}\left(r\frac{\partial c_{A\gamma}}{\partial r}\right)2\pi r\,\mathrm{d}r + \frac{1}{\pi r_0^2}\int_{r=0}^{r=r_0}\left(\frac{\partial^2 c_{A\gamma}}{\partial z^2}\right)2\pi r\,\mathrm{d}r\right\} \tag{1.19}$$

The first and last terms in this result can be expressed as

$$\frac{1}{\pi r_0^2}\int_{r=0}^{r=r_0}\left(\frac{\partial c_{A\gamma}}{\partial t}\right)2\pi r\,\mathrm{d}r = \frac{\partial}{\partial t}\left\{\frac{1}{\pi r_0^2}\int_{r=0}^{r=r_0} c_{A\gamma}2\pi r\,\mathrm{d}r\right\} = \frac{\partial\langle c_{A\gamma}\rangle^\gamma}{\partial t} \tag{1.20}$$

$$\frac{1}{\pi r_0^2}\int_{r=0}^{r=r_0}\left(\frac{\partial^2 c_{A\gamma}}{\partial z^2}\right)2\pi r\,\mathrm{d}r = \frac{\partial^2}{\partial z^2}\left\{\frac{1}{\pi r_0^2}\int_{r=0}^{r=r_0} c_{A\gamma}2\pi r\,\mathrm{d}r\right\} = \frac{\partial^2\langle c_{A\gamma}\rangle^\gamma}{\partial z^2} \tag{1.21}$$

so that equation 1.19 takes the form

$$\frac{\partial\langle c_{A\gamma}\rangle^\gamma}{\partial t} = \mathscr{D}_\gamma\left\{\frac{2}{r_0^2}\int_{r=0}^{r=r_0}\frac{\partial}{\partial r}\left(r\frac{\partial c_{A\gamma}}{\partial r}\right)\mathrm{d}r\right\} + \mathscr{D}_\gamma\frac{\partial^2\langle c_{A\gamma}\rangle^\gamma}{\partial z^2} \tag{1.22}$$

Evaluation of the integral leads to

$$\frac{\partial\langle c_{A\gamma}\rangle^\gamma}{\partial t} = \mathscr{D}_\gamma\frac{\partial^2\langle c_{A\gamma}\rangle^\gamma}{\partial z^2} + \frac{2\mathscr{D}_\gamma}{r_0}\left.\frac{\partial c_{A\gamma}}{\partial r}\right|_{r=r_0} \tag{1.23}$$

and we can make use of the boundary condition given by equation 1.15 to incorporate the heterogeneous rate of reaction into the area-averaged diffusion equation. This gives

$$\frac{\partial\langle c_{A\gamma}\rangle^\gamma}{\partial t} = \mathscr{D}_\gamma\frac{\partial^2\langle c_{A\gamma}\rangle^\gamma}{\partial z^2} - \frac{2k}{r_0}\left.c_{A\gamma}\right|_{r=r_0} \tag{1.24}$$

Here we remark that the boundary condition is *joined* with the governing differential equation, and that means that the *heterogeneous* reaction rate in equation 1.15 is now beginning to 'look like' a *homogeneous* reaction rate in equation 1.24. This process, in which a boundary condition is joined to a governing differential equation, is inherent in all studies of multiphase transport processes. The failure to identify explicitly this process often leads to confusion concerning the difference between homogeneous and heterogeneous chemical reactions.

Equation 1.24 poses a problem in that it represents a *single* equation containing *two* concentrations. If we *cannot* express the concentration at the wall of the capillary tube in terms of the area-averaged concentration, the area-averaged transport equation will be of little use to us and we will be forced to return to equations 1.13–1.17 to solve the boundary value problem by classical methods. In other words, the upscaling procedure would fail without what is known as a *method of closure*. In order to complete the upscaling process in a simple manner, we need an estimate of the variation of the concentration across the tube. We obtain this by using the flux boundary condition to construct the following order of magnitude estimate:

$$\mathscr{D}_\gamma \left(\frac{c_{A\gamma}\big|_{r=0} - c_{A\gamma}\big|_{r=r_0}}{r_0} \right) = \mathbf{O}\left(k\, c_{A\gamma}\big|_{r=r_0} \right) \tag{1.25}$$

which can be arranged in the form

$$\frac{c_{A\gamma}\big|_{r=0} - c_{A\gamma}\big|_{r=r_0}}{c_{A\gamma}\big|_{r=r_0}} = \mathbf{O}\left(\frac{k r_0}{\mathscr{D}_\gamma} \right) \tag{1.26}$$

When $k r_0 / \mathscr{D}_\gamma \ll 1$ it should be clear that we can use the approximation

$$c_{A\gamma}\big|_{r=r_0} = \langle c_{A\gamma} \rangle^\gamma \tag{1.27}$$

which represents the *closure* for this particular process. This allows us to express equation 1.24 as

$$\frac{\partial \langle c_{A\gamma} \rangle^\gamma}{\partial t} = \mathscr{D}_\gamma \frac{\partial^2 \langle c_{A\gamma} \rangle^\gamma}{\partial z^2} - \frac{2k}{r_0} \langle c_{A\gamma} \rangle^\gamma, \quad \frac{k r_0}{\mathscr{D}_\gamma} \ll 1 \tag{1.28}$$

Here we see that the *heterogeneous reaction* rate expression that appears in the flux boundary condition given by equation 1.15 now appears as a *homogeneous reaction* rate expression in the area-averaged transport equation. It should be clear that the 'homogeneous reaction rate coefficient' contains the geometrical parameter, r_0, and this is a clear indication that $2k/r_0$ is something other than a true homogeneous reaction rate coefficient. When the constraint, $k r_0 / \mathscr{D}_\gamma \ll 1$, is not satisfied, the closure represented by equation 1.27 becomes more complex and this condition has been explored by Paine *et al.* (1983).

1.3.1 Porous Catalysts

When dealing with porous catalysts, one generally *does not* work with the intrinsic average transport equation given by

$$\underbrace{\frac{\partial \langle c_{A\gamma} \rangle^\gamma}{\partial t}}_{\substack{\text{accumulation}\\ \text{per unit volume}\\ \text{of fluid}}} = \underbrace{\mathscr{D}_\gamma \frac{\partial^2 \langle c_{A\gamma} \rangle^\gamma}{\partial z^2}}_{\substack{\text{diffusive flux}\\ \text{per unit volume}\\ \text{of fluid}}} - \underbrace{\frac{2k}{r_0} \langle c_{A\gamma} \rangle^\gamma}_{\substack{\text{reaction rate}\\ \text{per unit volume}\\ \text{of fluid}}} \tag{1.29}$$

Here we have emphasized the intrinsic nature of our area-averaged transport equation, and this is especially clear with respect to the last term which represents the rate of reaction *per unit volume of the fluid phase*. In the study of diffusion and reaction in real porous media (Whitaker, 1986a, 1987), it is traditional to work with the rate of reaction *per unit volume of the porous medium*. Since the ratio of the fluid volume to the volume of the porous medium is the porosity, i.e.

$$\varepsilon_\gamma = \{\text{porosity}\} = \frac{\left\{\begin{array}{c}\text{volume}\\ \text{of the fluid}\end{array}\right\}}{\left\{\begin{array}{c}\text{volume of the}\\ \text{porous medium}\end{array}\right\}} \tag{1.30}$$

the superficial averaged diffusion-reaction equation is expressed as

$$\underbrace{\varepsilon_\gamma \frac{\partial \langle c_{A\gamma}\rangle^\gamma}{\partial t}}_{\substack{\text{accumulation per}\\ \text{unit volume of}\\ \text{porous media}}} = \underbrace{\varepsilon_\gamma \mathscr{D}_\gamma \frac{\partial^2 \langle c_{A\gamma}\rangle^\gamma}{\partial z^2}}_{\substack{\text{diffusive flux per}\\ \text{unit volume of}\\ \text{porous media}}} - \underbrace{\frac{2\varepsilon_\gamma k}{r_0}\langle c_{A\gamma}\rangle^\gamma}_{\substack{\text{rate of reaction}\\ \text{per unit volume}\\ \text{of porous media}}} \tag{1.31}$$

Here we see that the last term represents the rate of reaction per unit *volume of the porous medium* and this is the traditional interpretation in reactor design literature. One can show that $2\varepsilon_\gamma/r_0$ represents the surface area per unit volume of the porous medium, and we denote this by a_v so that equation 1.31 takes the form

$$\varepsilon_\gamma \frac{\partial \langle c_{A\gamma}\rangle^\gamma}{\partial t} = \varepsilon_\gamma \mathscr{D}_\gamma \frac{\partial^2 \langle c_{A\gamma}\rangle^\gamma}{\partial z^2} - a_v k \langle c_{A\gamma}\rangle^\gamma \tag{1.32}$$

Sometimes the model illustrated in Figure 1.9 is extended to include tortuous pores such as shown in the two-dimensional illustration in Figure 1.10. Under these circumstances one often writes equation 1.32 in the form

$$\varepsilon_\gamma \frac{\partial \langle c_{A\gamma}\rangle^\gamma}{\partial t} = \frac{\varepsilon_\gamma \mathscr{D}_\gamma}{\tau} \frac{\partial^2 \langle c_{A\gamma}\rangle^\gamma}{\partial z^2} - a_v k \langle c_{A\gamma}\rangle^\gamma \tag{1.33}$$

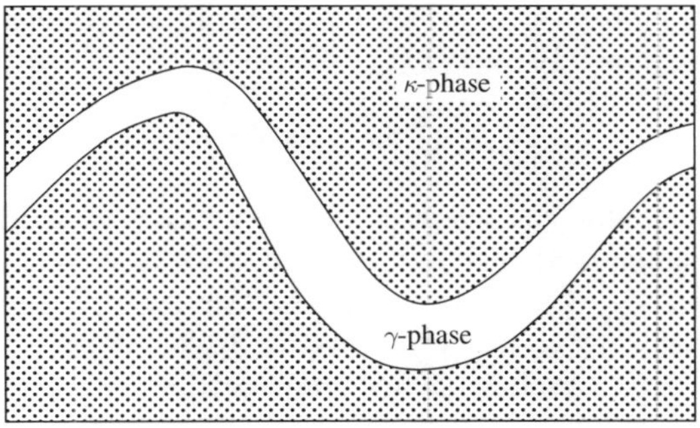

Figure 1.10 *Tortuous capillary tube as a model porous medium*

Here τ is a coefficient referred to as the *tortuosity* and the ratio, $\mathscr{D}_\gamma/\tau$, is called the effective diffusivity which is represented by D_{eff}. This allows us to express equation 1.33 in the traditional form given by

$$\varepsilon_\gamma \frac{\partial \langle c_{A\gamma}\rangle^\gamma}{\partial t} = \varepsilon_\gamma D_{\mathrm{eff}} \frac{\partial^2 \langle c_{A\gamma}\rangle^\gamma}{\partial z^2} - a_v k \langle c_{A\gamma}\rangle^\gamma \tag{1.34}$$

The step from equation 1.32 for a bundle of capillary tubes to equation 1.34 for a porous medium is *intuitive*, and for undergraduate courses in reactor design one might accept this level of intuition. However, the development leading from equations 1.13 through 1.17 to the upscaled result given by equation 1.32 is *analytical* and this level of analysis is *necessary* for an undergraduate course in reactor design. The more practical problem deals with *non-dilute* solution diffusion and reaction in porous catalysts, and a rigorous analysis of that case is given in the following sections.

1.4 Coupled, Non-linear Diffusion and Reaction

Problems of isothermal mass transfer and reaction are best represented in terms of the species continuity equation and the associated jump condition. We repeat these two equations here as

$$\frac{\partial c_{A\gamma}}{\partial t} + \nabla \cdot \left(c_{A\gamma}\mathbf{v}_{A\gamma}\right) = R_{A\gamma}, \quad A = 1, 2, \ldots, N \tag{1.35}$$

$$\frac{\partial c_{As}}{\partial t} = \left(c_{A\gamma}\mathbf{v}_{A\gamma}\right) \cdot \mathbf{n}_{\gamma\kappa} + R_{As}, \quad \text{at the } \gamma\text{–}\kappa \text{ interface}, \quad A = 1, 2, \ldots, N \tag{1.36}$$

A complete description of the mass transfer process requires a *connection* between the surface concentration, c_{As}, and the bulk concentration, $c_{A\gamma}$. One classic connection is based on *local mass equilibrium*, and for a linear equilibrium relation this concept takes the form

$$c_{As} = K_A c_{A\gamma}, \quad \text{at the } \gamma\text{–}\kappa \text{ interface}, \quad A = 1, 2, \ldots, N \tag{1.37a}$$

The condition of local mass equilibrium can exist even when adsorption and chemical reaction are taking place (Whitaker, 1999, Problem 1-3). When local mass equilibrium is not valid, one must propose an *interfacial flux constitutive equation*. The classic linear form is given by (Langmuir, 1916, 1917)

$$\left(c_{A\gamma}\mathbf{v}_{A\gamma}\right) \cdot \mathbf{n}_{\gamma\kappa} = k_{A1} c_{A\gamma} - k_{-A1} c_{As}, \quad \text{at the } \gamma\text{–}\kappa \text{ interface}, \quad A = 1, 2, \ldots, N \tag{1.37b}$$

where k_{A1} and k_{-A1} represent the adsorption and desorption rate coefficients for species A.

In addition to equations 1.35 and 1.36, we need N momentum equations (Whitaker, 1986a) that are used to determine the N species velocities represented by $\mathbf{v}_{A\gamma}$, $A = 1, 2, \ldots, N$. There are certain problems for which the N momentum equations consist of the total, or mass average, momentum equation

$$\frac{\partial}{\partial t}\left(\rho_\gamma \mathbf{v}_\gamma\right) + \nabla \cdot \left(\rho_\gamma \mathbf{v}_\gamma \mathbf{v}_\gamma\right) = \rho_\gamma \mathbf{b}_\gamma + \nabla \cdot \mathbf{T}_\gamma \tag{1.38}$$

along with $N-1$ Stefan–Maxwell equations that take the form

$$0 = -\nabla x_{A\gamma} + \sum_{\substack{E=1 \\ E \neq A}}^{E=N} \frac{x_{A\gamma} x_{E\gamma} (\mathbf{v}_{E\gamma} - \mathbf{v}_{A\gamma})}{\mathscr{D}_{AE}}, \quad A = 1, 2, \ldots, N-1 \tag{1.39}$$

This form of the species momentum equation is acceptable when molecule–molecule collisions are much more frequent than molecule–wall collisions; thus equation 1.39 is *inappropriate* when Knudsen diffusion must be taken into account. The species velocity in equation 1.39 can be *decomposed* into an average velocity and a diffusion velocity in more than one way (Taylor and Krishna, 1993; Slattery, 1999; Bird *et al.*, 2002), and arguments are often given to justify a particular choice. In this work we prefer a decomposition in terms of the *mass average velocity* because governing equations, such as the Navier–Stokes equations, are available to determine this velocity. The mass average velocity in equation 1.38 is defined by

$$\mathbf{v}_\gamma = \sum_{A=1}^{A=N} \omega_{A\gamma} \mathbf{v}_{A\gamma} \tag{1.40}$$

and the associated *mass diffusion velocity* is defined by the decomposition

$$\mathbf{v}_{A\gamma} = \mathbf{v}_\gamma + \mathbf{u}_{A\gamma} \tag{1.41}$$

The mass diffusive flux has the *attractive characteristic* that the sum of the fluxes is zero, i.e.

$$\sum_{A=1}^{A=N} \rho_{A\gamma} \mathbf{u}_{A\gamma} = 0 \tag{1.42}$$

As an alternative to equations 1.40–1.42, we can define a molar average velocity by

$$\mathbf{v}_\gamma^* = \sum_{A=1}^{A=N} x_{A\gamma} \mathbf{v}_{A\gamma} \tag{1.43}$$

and the associated molar diffusion velocity is given by

$$\mathbf{v}_{A\gamma} = \mathbf{v}_\gamma^* + \mathbf{u}_{A\gamma}^* \tag{1.44}$$

In this case, the molar diffusive flux also has the *attractive characteristic* given by

$$\sum_{A=1}^{A=N} c_{A\gamma} \mathbf{u}_{A\gamma}^* = 0 \tag{1.45}$$

However, the use of the molar average velocity defined by equation 1.43 presents problems when equation 1.38 must be used as one of the N momentum equations.

If we make use of the *mass* average velocity and the *mass* diffusion velocity as indicated by equations 1.40 and 1.41, the *molar* flux in equation 1.35 takes the form

$$\underbrace{c_{A\gamma} \mathbf{v}_{A\gamma}}_{\substack{\text{total molar} \\ \text{flux}}} = \underbrace{c_{A\gamma} \mathbf{v}_\gamma}_{\substack{\text{molar convective} \\ \text{flux}}} + \underbrace{c_{A\gamma} \mathbf{u}_{A\gamma}}_{\substack{\text{mixed-mode} \\ \text{diffusive flux}}} \tag{1.46}$$

Here we have decomposed the total molar flux into what we want, the molar convective flux, and what remains, i.e. a mixed-mode diffusive flux. Following Bird *et al.* (2002), we indicate the *mixed-mode diffusive flux* as

$$\mathbf{J}_{A\gamma} = c_{A\gamma}\mathbf{u}_{A\gamma}, \quad A = 1, 2, \ldots, N \tag{1.47}$$

so that equation 1.35 takes the form

$$\frac{\partial c_{A\gamma}}{\partial t} + \nabla \cdot (c_{A\gamma}\mathbf{v}_\gamma) = -\nabla \cdot \mathbf{J}_{A\gamma} + R_{A\gamma}, \quad A = 1, 2, \ldots, N \tag{1.48}$$

The single drawback to this mixed-mode diffusive flux is that it does not satisfy a simple relation such as that given by either equation 1.42 or equation 1.45. Instead, we find that the mixed-mode diffusive fluxes are constrained by

$$\sum_{A=1}^{A=N} \mathbf{J}_{A\gamma}(M_A/\overline{M}) = 0 \tag{1.49}$$

where M_A is the molecular mass of species A and $\overline{M}$ is the mean molecular mass defined by

$$\overline{M} = \sum_{A=1}^{A=N} x_{A\gamma} M_A \tag{1.50}$$

There are many problems for which we wish to know the concentration, $c_{A\gamma}$, and the normal component of the molar flux of species A at a phase interface. The normal component of the molar flux at an interface will be related to the adsorption process and the heterogeneous reaction by means of the jump condition given by equation 1.36 and relations of the type given by equation 1.37, and this flux will be influenced by the convective, $c_{A\gamma}\mathbf{v}_\gamma$, and diffusive, $\mathbf{J}_{A\gamma}$, fluxes.

The governing equations for $c_{A\gamma}$ and $\mathbf{v}_\gamma$ are available to us in terms of equations 1.38 and 1.48, and here we consider the matter of determining $\mathbf{J}_{A\gamma}$. To determine the mixed-mode diffusive flux, we return to the Stefan–Maxwell equations and make use of equation 1.41 to obtain

$$0 = -\nabla x_{A\gamma} + \sum_{\substack{E=1 \\ E \neq A}}^{E=N} \frac{x_{A\gamma} x_{E\gamma}(\mathbf{u}_{E\gamma} - \mathbf{u}_{A\gamma})}{\mathscr{D}_{AE}}, \quad A = 1, 2, \ldots, N-1 \tag{1.51}$$

This can be multiplied by the total molar concentration and rearranged in the form

$$0 = -c_\gamma \nabla x_{A\gamma} + x_{A\gamma} \sum_{\substack{E=1 \\ E \neq A}}^{E=N} \frac{c_{E\gamma}\mathbf{u}_{E\gamma}}{\mathscr{D}_{AE}} - \left\{ \sum_{\substack{E=1 \\ E \neq A}}^{E=N} \frac{x_{E\gamma}}{\mathscr{D}_{AE}} \right\} c_{A\gamma}\mathbf{u}_{A\gamma}, \quad A = 1, 2, \ldots, N-1 \tag{1.52}$$

which can then be expressed in terms of equation 1.47 to obtain

$$0 = -c_\gamma \nabla x_{A\gamma} + x_{A\gamma} \sum_{\substack{E=1 \\ E\neq A}}^{E=N} \frac{\mathbf{J}_{E\gamma}}{\mathscr{D}_{AE}} - \left\{ \sum_{\substack{E=1 \\ E\neq A}}^{E=N} \frac{x_{E\gamma}}{\mathscr{D}_{AE}} \right\} \mathbf{J}_{A\gamma}, \quad A = 1, 2, \ldots, N-1 \tag{1.53}$$

Here we can use the classic definition of the mixture diffusivity

$$\frac{1}{\mathscr{D}_{\mathrm{Am}}} = \sum_{\substack{E=1 \\ E\neq A}}^{E=N} \frac{x_{E\gamma}}{\mathscr{D}_{AE}} \tag{1.54}$$

in order to express equation 1.53 as

$$\mathbf{J}_{A\gamma} - x_{A\gamma} \sum_{\substack{E=1 \\ E\neq A}}^{E=N} \frac{\mathscr{D}_{\mathrm{Am}}}{\mathscr{D}_{AE}} \mathbf{J}_{E\gamma} = -c_\gamma \mathscr{D}_{\mathrm{Am}} \nabla x_{A\gamma}, \quad A = 1, 2, \ldots, N-1 \tag{1.55}$$

When the mole fraction of species A is small compared to 1, we obtain the *dilute solution* representation for the diffusive flux

$$\mathbf{J}_{A\gamma} = -c_\gamma \mathscr{D}_{\mathrm{Am}} \nabla x_{A\gamma}, \quad x_{A\gamma} \ll 1 \tag{1.56}$$

and the transport equation for species A takes the form

$$\frac{\partial c_{A\gamma}}{\partial t} + \nabla \cdot (c_{A\gamma} \mathbf{v}_\gamma) = \nabla \cdot (c_\gamma \mathscr{D}_{\mathrm{Am}} \nabla x_{A\gamma}) + R_{A\gamma}, \quad x_{A\gamma} \ll 1 \tag{1.57}$$

Given the condition $x_{A\gamma} \ll 1$, it is often plausible to impose the condition

$$x_{A\gamma} \nabla c_\gamma \ll c_\gamma \nabla x_{A\gamma} \tag{1.58}$$

and this leads to the following convective-diffusion equation that is ubiquitous in the reactor design literature:

$$\frac{\partial c_{A\gamma}}{\partial t} + \nabla \cdot (c_{A\gamma} \mathbf{v}_\gamma) = \nabla \cdot (\mathscr{D}_{\mathrm{Am}} \nabla c_{A\gamma}) + R_{A\gamma}, \quad x_{A\gamma} \ll 1 \tag{1.59}$$

When the mole fraction of species A is not small compared to 1, the *diffusive flux* in this *transport equation* will not be correct. If the diffusive flux plays an important role in the rate of heterogeneous reaction, equation 1.59 will not lead to a correct representation for the rate of reaction.

1.5 Diffusive Flux

We begin our analysis of the diffusive flux with equation 1.55 in the form

$$\mathbf{J}_{A\gamma} = -c_\gamma \mathscr{D}_{\mathrm{Am}} \nabla x_{A\gamma} + x_{A\gamma} \sum_{\substack{E=1 \\ E \neq A}}^{E=N} \frac{\mathscr{D}_{\mathrm{Am}}}{\mathscr{D}_{AE}} \mathbf{J}_{E\gamma}, \quad A = 1, 2, \ldots, N-1 \tag{1.60}$$

and make use of equation 1.49 in an alternate form

$$\sum_{A=1}^{A=N} \mathbf{J}_{A\gamma} \left(M_A / M_N \right) = 0 \tag{1.61}$$

in order to obtain N equations relating to the N diffusive fluxes. At this point we define a matrix $[R]$ according to

$$[R] = \begin{bmatrix} 1 & - & \dfrac{x_{A\gamma}\mathscr{D}_{\mathrm{Am}}}{\mathscr{D}_{AB}} & - & \dfrac{x_{A\gamma}\mathscr{D}_{\mathrm{Am}}}{\mathscr{D}_{AC}} & - & \cdots & - & \dfrac{x_{A\gamma}\mathscr{D}_{\mathrm{Am}}}{\mathscr{D}_{AN}} \\ -\dfrac{x_{B\gamma}\mathscr{D}_{\mathrm{Bm}}}{\mathscr{D}_{BA}} & + & 1 & - & \dfrac{x_{B\gamma}\mathscr{D}_{\mathrm{Bm}}}{\mathscr{D}_{BC}} & - & \cdots & - & \dfrac{x_{B\gamma}\mathscr{D}_{\mathrm{Bm}}}{\mathscr{D}_{BN}} \\ -\dfrac{x_{C\gamma}\mathscr{D}_{\mathrm{Cm}}}{\mathscr{D}_{CA}} & - & \dfrac{x_{C\gamma}\mathscr{D}_{\mathrm{Cm}}}{\mathscr{D}_{CB}} & + & 1 & - & \cdots & - & \dfrac{x_{C\gamma}\mathscr{D}_{\mathrm{Cm}}}{\mathscr{D}_{CN}} \\ \cdot & \cdot & \cdot & \cdot & \cdot & - & \cdots & - & \cdot \\ \cdot & \cdot & \cdot & \cdot & \cdot & - & \cdots & - & \cdot \\ \dfrac{M_A}{M_N} & + & \dfrac{M_B}{M_N} & + & \dfrac{M_C}{M_N} & + & \cdots & + & 1 \end{bmatrix} \tag{1.62}$$

and use equations 1.60 and 1.61 to express the N diffusive fluxes according to

$$[R] \begin{bmatrix} \mathbf{J}_{A\gamma} \\ \mathbf{J}_{B\gamma} \\ \mathbf{J}_{C\gamma} \\ \cdots \\ \cdots \\ \mathbf{J}_{N\gamma} \end{bmatrix} = -c_\gamma \begin{bmatrix} \mathscr{D}_{\mathrm{Am}} \nabla x_{A\gamma} \\ \mathscr{D}_{\mathrm{Bm}} \nabla x_{B\gamma} \\ \mathscr{D}_{\mathrm{Cm}} \nabla x_{C\gamma} \\ \cdots \\ \mathscr{D}_{N-1\mathrm{m}} \nabla x_{N-1\gamma} \\ 0 \end{bmatrix} \tag{1.63}$$

We assume that the inverse of $[R]$ exists in order to express the column matrix of diffusive flux vectors in the form

$$\begin{bmatrix} \mathbf{J}_{A\gamma} \\ \mathbf{J}_{B\gamma} \\ \mathbf{J}_{C\gamma} \\ \cdots \\ \cdots \\ \mathbf{J}_{N\gamma} \end{bmatrix} = -c_\gamma [R]^{-1} \begin{bmatrix} \mathscr{D}_{\mathrm{Am}} \nabla x_{A\gamma} \\ \mathscr{D}_{\mathrm{Bm}} \nabla x_{B\gamma} \\ \mathscr{D}_{\mathrm{Cm}} \nabla x_{C\gamma} \\ \cdots \\ \mathscr{D}_{N-1\mathrm{m}} \nabla x_{N-1\gamma} \\ 0 \end{bmatrix} \tag{1.64}$$

where the column matrix on the right-hand side of this result can be expressed as

$$\begin{bmatrix} \mathscr{D}_{Am}\nabla x_{A\gamma} \\ \mathscr{D}_{Bm}\nabla x_{B\gamma} \\ \mathscr{D}_{Cm}\nabla x_{C\gamma} \\ \cdots \\ \mathscr{D}_{N-1m}\nabla x_{N-1\gamma} \\ 0 \end{bmatrix} = \begin{bmatrix} \mathscr{D}_{Am} & 0 & 0 & \cdots & 0 & 0 \\ 0 & \mathscr{D}_{Bm} & 0 & \cdots & 0 & 0 \\ 0 & 0 & \mathscr{D}_{Cm} & \cdots & 0 & 0 \\ \cdot & \cdot & \cdot & \cdots & \cdot & \cdot \\ 0 & 0 & 0 & \cdots & \mathscr{D}_{N-1m} & 0 \\ 0 & 0 & 0 & \cdots & 0 & \mathscr{D}_{Nm} \end{bmatrix} \begin{bmatrix} \nabla x_{A\gamma} \\ \nabla x_{B\gamma} \\ \nabla x_{C\gamma} \\ \cdots \\ \nabla x_{N-1\gamma} \\ 0 \end{bmatrix} \tag{1.65}$$

The *diffusivity matrix* is now defined by

$$[D] = [R]^{-1} \begin{bmatrix} \mathscr{D}_{Am} & 0 & 0 & \cdots & 0 & 0 \\ 0 & \mathscr{D}_{Bm} & 0 & \cdots & 0 & 0 \\ 0 & 0 & \mathscr{D}_{Cm} & \cdots & 0 & 0 \\ \cdot & \cdot & \cdot & \cdots & \cdot & \cdot \\ 0 & 0 & 0 & \cdots & \mathscr{D}_{N-1m} & 0 \\ 0 & 0 & 0 & \cdots & 0 & \mathscr{D}_{Nm} \end{bmatrix} \tag{1.66}$$

so that equation 1.64 takes the form

$$\begin{bmatrix} \mathbf{J}_{A\gamma} \\ \mathbf{J}_{B\gamma} \\ \mathbf{J}_{C\gamma} \\ \cdots \\ \cdots \\ \mathbf{J}_{N\gamma} \end{bmatrix} = -c_\gamma [D] \begin{bmatrix} \nabla x_{A\gamma} \\ \nabla x_{B\gamma} \\ \nabla x_{C\gamma} \\ \cdots \\ \nabla x_{N-1\gamma} \\ 0 \end{bmatrix} \tag{1.67}$$

This result can be expressed in a form analogous to that given by equation 1.60 leading to

$$\mathbf{J}_{A\gamma} = -c_\gamma \sum_{E=1}^{E=N-1} D_{AE}\nabla x_{E\gamma}, \quad A = 1, 2, \ldots, N \tag{1.68}$$

In the general case, the elements of the diffusivity matrix, D_{AE}, will depend on the mole fractions in a non-trivial manner. When this result is used in equation 1.48 we obtain the non-linear, coupled governing differential equation for $c_{A\gamma}$ given by

$$\frac{\partial c_{A\gamma}}{\partial t} + \nabla \cdot \left(c_{A\gamma}\mathbf{v}_\gamma \right) = \nabla \cdot \left(c_\gamma \sum_{E=1}^{E=N-1} D_{AE}\nabla x_{E\gamma} \right) + R_{A\gamma}, \quad A = 1, 2, \ldots, N \tag{1.69}$$

We seek a solution to this equation subject to the jump condition given by equation 1.36 and this requires knowledge of the concentration dependence of the homogeneous and heterogeneous reaction rates and information concerning the equilibrium adsorption isotherm. In general, a solution of equation 1.69 for the system shown in Figure 1.7 requires *upscaling* from the point scale to the pore scale and this can be done by the method of volume averaging (Whitaker, 1999).

1.6 Volume Averaging

To obtain the volume-averaged form of equation 1.69, we first associate an averaging volume with every point in the $\gamma - \kappa$ system illustrated in Figure 1.7. One such averaging volume is illustrated in Figure 1.11, and it can be represented in terms of the volumes of the individual phases according to

$$\mathscr{V} = V_\gamma + V_\kappa \tag{1.70}$$

The radius of the averaging volume is r_0 and the characteristic length scale associated with the γ-phase is indicated by ℓ_γ as shown in Figure 1.11. In this figure we have also illustrated a length L that is associated with the distance over which significant changes in averaged quantities occur. Throughout this analysis we will assume that the length scales are disparate, i.e. the length scales are constrained by

$$L \gg r_0 \gg \ell_\gamma \tag{1.71}$$

Here the length scale, L, is a generic length scale (Whitaker, 1999, Sec. 1.3.2) determined by the gradient of the average concentration, and all three quantities in equation 1.71 are different to those listed in equation 1.1. We will use the averaging volume $\mathscr{V}$ to define two averages: the *superficial average* and the *intrinsic average*. Each of these averages is routinely used in the description of multiphase transport processes, and it is

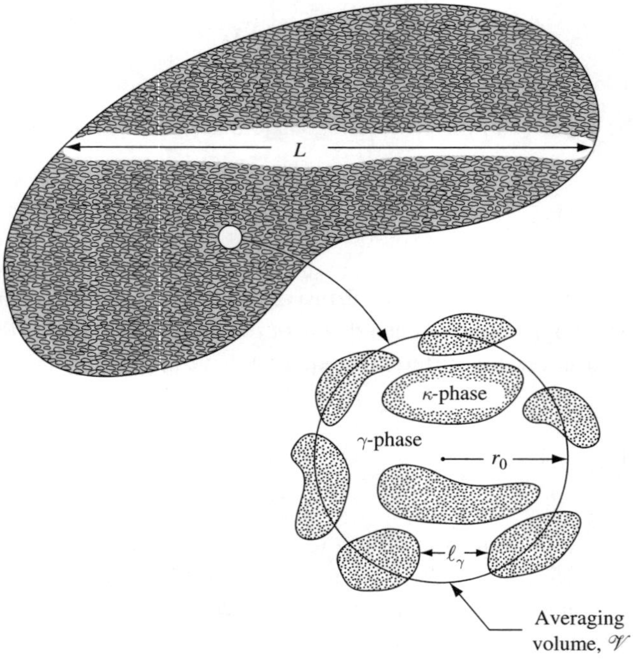

Figure 1.11 *Averaging volume for a porous catalyst*

important to define clearly each one. We define the *superficial average* of some function ψ_γ according to

$$\langle \psi_\gamma \rangle = \frac{1}{\mathscr{V}} \int_{V_\gamma} \psi_\gamma \, \mathrm{d}V \tag{1.72}$$

and we define the *intrinsic average* by

$$\langle \psi_\gamma \rangle^\gamma = \frac{1}{V_\gamma} \int_{V_\gamma} \psi_\gamma \, \mathrm{d}V \tag{1.73}$$

These two averages are related according to

$$\langle \psi_\gamma \rangle = \varepsilon_\gamma \langle \psi_\gamma \rangle^\gamma \tag{1.74}$$

where ε_γ is the volume fraction of the γ-phase defined explicitly as

$$\varepsilon_\gamma = V_\gamma / \mathscr{V} \tag{1.75}$$

In this notation for the volume averages, a *Greek subscript* is used to identify the particular phase under consideration, while a *Greek superscript* is used to identify an intrinsic average. Since the intrinsic and superficial averages differ by a factor of ε_γ, it is essential to make use of a notation that clearly distinguishes between the two averages.

When we form the volume average of any transport equation, we are immediately confronted with the average of a gradient (or divergence), and it is the gradient (or divergence) of the average that we are seeking. In order to interchange integration and differentiation, we will make use of the spatial averaging theorem (Anderson and Jackson, 1967; Marle, 1967; Slattery, 1967; Whitaker, 1967). For the two-phase system illustrated in Figure 1.11 this theorem can be expressed as

$$\langle \nabla \psi_\gamma \rangle = \nabla \langle \psi_\gamma \rangle + \frac{1}{\mathscr{V}} \int_{A_{\gamma\kappa}} \mathbf{n}_{\gamma\kappa} \psi_\gamma \, \mathrm{d}A \tag{1.76}$$

where ψ_γ is any function associated with the γ-phase. Here $A_{\gamma\kappa}$ represents the interfacial area contained within the averaging volume, and we have used $\mathbf{n}_{\gamma\kappa}$ to represent the unit normal vector pointing *from* the γ-phase *toward* the κ-phase.

Even though equation 1.69 is considered to be the preferred form of the species continuity equation, it is best to begin the averaging procedure with equation 1.35, and we express the superficial average of that form as

$$\left\langle \frac{\partial c_{A\gamma}}{\partial t} \right\rangle + \left\langle \nabla \cdot \left(c_{A\gamma} \mathbf{v}_{A\gamma} \right) \right\rangle = \langle R_{A\gamma} \rangle, \quad A = 1, 2, \ldots, N \tag{1.77}$$

For a rigid porous medium, one can use the transport theorem and the averaging theorem to express this result as

$$\frac{\partial \langle c_{A\gamma} \rangle}{\partial t} + \nabla \cdot \langle c_{A\gamma} \mathbf{v}_{A\gamma} \rangle + \frac{1}{\mathscr{V}} \int_{A_{\gamma\kappa}} \mathbf{n}_{\gamma\kappa} \cdot \left(c_{A\gamma} \mathbf{v}_{A\gamma} \right) \mathrm{d}A = \langle R_{A\gamma} \rangle \tag{1.78}$$

where it is understood that this applies to all N species. Since we seek a transport equation for the intrinsic average concentration, we make use of equation 1.74 to express equation 1.78 in the form

$$\varepsilon_\gamma \frac{\partial \langle c_{A\gamma}\rangle^\gamma}{\partial t} + \nabla\cdot\langle c_{A\gamma}\mathbf{v}_{A\gamma}\rangle + \frac{1}{\mathscr{V}}\int_{A_{\gamma\kappa}} \mathbf{n}_{\gamma\kappa}\cdot\left(c_{A\gamma}\mathbf{v}_{A\gamma}\right)\mathrm{d}A = \varepsilon_\gamma \langle R_{A\gamma}\rangle^\gamma \tag{1.79}$$

At this point, it is convenient to make use of the jump condition given by equation 1.36 in order to obtain

$$\varepsilon_\gamma \frac{\partial \langle c_{A\gamma}\rangle^\gamma}{\partial t} + \nabla\cdot\langle c_{A\gamma}\mathbf{v}_{A\gamma}\rangle = \langle R_{A\gamma}\rangle^\gamma - \frac{1}{\mathscr{V}}\int_{A_{\gamma\kappa}} \frac{\partial c_{\mathrm{As}}}{\partial t}\mathrm{d}A + \frac{1}{\mathscr{V}}\int_{A_{\gamma\kappa}} R_{\mathrm{As}}\,\mathrm{d}A \tag{1.80}$$

We now define the *intrinsic interfacial area average* according to

$$\langle \psi_\gamma\rangle_{\gamma\kappa} = \frac{1}{A_{\gamma\kappa}}\int_{A_{\gamma\kappa}} \psi_\gamma\,\mathrm{d}A \tag{1.81}$$

so that equation 1.80 takes the convenient form given by

$$\underbrace{\varepsilon_\gamma \frac{\partial \langle c_{A\gamma}\rangle^\gamma}{\partial t}}_{\text{accumulation}} + \underbrace{\nabla\cdot\langle c_{A\gamma}\mathbf{v}_{A\gamma}\rangle}_{\text{transport}} = \underbrace{\langle R_{A\gamma}\rangle^\gamma}_{\substack{\text{homogeneous}\\ \text{reaction}}} \underbrace{- a_{\mathrm{v}}\frac{\partial \langle c_{\mathrm{As}}\rangle_{\gamma\kappa}}{\partial t}}_{\text{adsorption}} + \underbrace{a_{\mathrm{v}}\langle R_{\mathrm{As}}\rangle_{\gamma\kappa}}_{\substack{\text{heterogeneous}\\ \text{reaction}}} \tag{1.82}$$

One must keep in mind that this is a *general result* based on equations 1.35 and 1.36; however, only the first term in equation 1.82 is in a form that is ready for applications.

1.7 Chemical Reactions

In general, the homogeneous reaction will be of no consequence in a porous catalyst and we need only direct our attention to the heterogeneous reaction represented by the last term in equation 1.82. The chemical kinetic constitutive equation for the *heterogeneous* rate of reaction can be expressed as

$$R_{\mathrm{As}} = R_{\mathrm{As}}\left(c_{\mathrm{As}}, c_{B\mathrm{s}}, \ldots, c_{N\mathrm{s}}\right) \tag{1.83}$$

and here we see the need to relate the surface concentrations, $c_{\mathrm{As}}, c_{B\mathrm{s}}, \ldots, c_{N\mathrm{s}}$, to the bulk concentrations, $c_{A\gamma}, c_{B\gamma}, \ldots, c_{N\gamma}$, and subsequently to the local volume-averaged concentrations, $\langle c_{A\gamma}\rangle^\gamma, \langle c_{B\gamma}\rangle^\gamma, \ldots, \langle c_{N\gamma}\rangle^\gamma$. In order for heterogeneous reaction to occur, adsorption at the catalytic surface must also occur. However, there are many transient processes of mass transfer with heterogeneous reaction for which the catalytic surface can be treated as *quasi-steady* (Carbonell and Whitaker, 1984; Whitaker, 1986b). When homogeneous reactions can be ignored and the catalytic surface can be treated as quasi-steady, the local volume-averaged transport equation simplifies to

$$\underbrace{\varepsilon_\gamma \frac{\partial \langle c_{A\gamma}\rangle^\gamma}{\partial t}}_{\text{accumulation}} + \underbrace{\nabla\cdot\langle c_{A\gamma}\mathbf{v}_{A\gamma}\rangle}_{\text{transport}} = \underbrace{a_{\mathrm{v}}\langle R_{\mathrm{As}}\rangle_{\gamma\kappa}}_{\substack{\text{heterogeneous}\\ \text{reaction}}} \tag{1.84}$$

and this result provides the basis for several special forms.

1.8 Convective and Diffusive Transport

Before examining the heterogeneous reaction rate in equation 1.84, we consider the transport term, $\langle c_{A\gamma}\mathbf{v}_{A\gamma}\rangle$. We begin with the mixed-mode decomposition given by equation 1.46 in order to obtain

$$\underbrace{\langle c_{A\gamma}\mathbf{v}_{A\gamma}\rangle}_{\substack{\text{total molar}\\ \text{flux}}} = \underbrace{\langle c_{A\gamma}\mathbf{v}_{\gamma}\rangle}_{\substack{\text{molar convective}\\ \text{flux}}} + \underbrace{\langle c_{A\gamma}\mathbf{u}_{A\gamma}\rangle}_{\substack{\text{mixed-mode}\\ \text{diffusive flux}}} \tag{1.85}$$

Here the convective flux is given in terms of the average of a product, and we want to express this flux in terms of the product of averages. As in the case of turbulent transport, this suggests the use of decompositions given by

$$c_{A\gamma} = \langle c_{A\gamma}\rangle^{\gamma} + \tilde{c}_{A\gamma}, \quad \mathbf{v}_{\gamma} = \langle \mathbf{v}_{\gamma}\rangle^{\gamma} + \tilde{\mathbf{v}}_{\gamma} \tag{1.86}$$

At this point one can follow a detailed analysis (Whitaker, 1999, Chapter 3) of the convective transport to arrive at

$$\underbrace{\langle c_{A\gamma}\mathbf{v}_{A\gamma}\rangle}_{\text{total flux}} = \underbrace{\varepsilon_{\gamma}\langle c_{A\gamma}\rangle^{\gamma}\langle \mathbf{v}_{\gamma}\rangle}_{\substack{\text{average convective}\\ \text{flux}}} + \underbrace{\langle \tilde{c}_{A\gamma}\tilde{\mathbf{v}}_{\gamma}\rangle}_{\substack{\text{dispersive}\\ \text{flux}}} + \underbrace{\langle \mathbf{J}_{A\gamma}\rangle}_{\substack{\text{mixed-mode}\\ \text{diffusive flux}}} \tag{1.87}$$

Here we have used the intrinsic average concentration since this is most closely related to the concentration in the fluid phase, and we have used the superficial average velocity since this is the quantity that normally appears in Darcy's law (Whitaker, 1999) or the Forchheimer equation (Whitaker, 1996). Use of equation 1.87 in equation 1.84 leads to

$$\varepsilon_{\gamma}\frac{\partial \langle c_{A\gamma}\rangle^{\gamma}}{\partial t} + \nabla\cdot\left(\varepsilon_{\gamma}\langle c_{A\gamma}\rangle^{\gamma}\langle \mathbf{v}_{\gamma}\rangle\right) = -\underbrace{\nabla\cdot\langle \mathbf{J}_{A\gamma}\rangle}_{\substack{\text{diffusive}\\ \text{transport}}} - \underbrace{\nabla\cdot\langle \tilde{c}_{A\gamma}\tilde{\mathbf{v}}_{\gamma}\rangle}_{\substack{\text{dispersive}\\ \text{transport}}} + \underbrace{a_{v}\langle R_{As}\rangle_{\gamma\kappa}}_{\substack{\text{heterogeneous}\\ \text{reaction}}} \tag{1.88}$$

If we treat the catalytic surface as *quasi-steady* and make use of a simple first-order, irreversible representation for the heterogeneous reaction, we can show that R_{As} is given by (Whitaker, 1999, Sec. 1.1)

$$R_{As} = -k_{As}c_{As} = -\left(\frac{k_{As}k_{A1}}{k_{As}+k_{-A1}}\right)c_{A\gamma}, \quad \text{at the } \gamma\text{–}\kappa \text{ interface} \tag{1.89}$$

when species A is consumed at the catalytic surface. Here we have used k_{As} to represent the intrinsic surface reaction rate coefficient, while k_{A1} and k_{-A1} are the adsorption and desorption rate coefficients that appear in equation 1.37b. Other more complex reaction mechanisms can be proposed; however, if a linear interfacial flux constitutive equation is valid, the heterogeneous reaction rates can be expressed in terms of the bulk concentration as indicated by equation 1.89. Under these circumstances the functional dependence indicated in equation 1.83 can be simplified to

$$R_{As} = R_{As}\left(c_{A\gamma}, c_{B\gamma}, \ldots, c_{N\gamma}\right), \quad \text{at the } \gamma\text{–}\kappa \text{ interface} \tag{1.90}$$

Given the type of constraints developed elsewhere (Wood and Whitaker, 1998, 2000), the interfacial area average of the heterogeneous rate of reaction can be expressed as

$$\langle R_{As}\rangle_{\gamma\kappa} = \langle R_{As}\rangle_{\gamma\kappa}\left(\langle c_{A\gamma}\rangle_{\gamma\kappa}, \langle c_{B\gamma}\rangle_{\gamma\kappa}, \ldots, \langle c_{N\gamma}\rangle_{\gamma\kappa}\right), \quad \text{at the } \gamma\text{–}\kappa \text{ interface} \tag{1.91}$$

Sometimes confusion exists concerning the idea of an *area-averaged bulk concentration*, and to clarify this idea we consider the averaging volume illustrated in Figure 1.12. In this figure we have shown an averaging volume with the centroid located (arbitrarily) in the κ-phase. In this case, the area average of the bulk concentration is given explicitly by

$$\langle c_{A\gamma}\rangle_{\gamma\kappa}\big|_{\mathbf{x}} = \frac{1}{A_{\gamma\kappa}(\mathbf{x})} \int_{A_{\gamma\kappa}(\mathbf{x})} c_{A\gamma}\big|_{\mathbf{x}+\mathbf{y}}\, dA \tag{1.92}$$

where $\mathbf{x}$ locates the centroid of the averaging volume and $\mathbf{y}$ locates points on the γ–κ interface. We have used $A_{\gamma\kappa}(\mathbf{x})$ to represent the area of the γ–κ interface contained within the averaging volume.

To complete our analysis of equation 1.91, we need to know how the area-averaged concentration, $\langle c_{A\gamma}\rangle_{\gamma\kappa}$, is related to the volume-averaged concentration, $\langle c_{A\gamma}\rangle^{\gamma}$. When convective transport is important, relating $\langle c_{A\gamma}\rangle_{\gamma\kappa}$ to $\langle c_{A\gamma}\rangle^{\gamma}$ requires some analysis; however, when *diffusive transport dominates* in a porous catalyst the area-averaged concentration is essentially equal to the volume-averaged concentration. This occurs because the *pore Thiele modulus* is generally small compared to one and the type of analysis indicated by equations 1.24–1.28 is applicable. Under these circumstances, equation 1.91 can be expressed as

$$\langle R_{As}\rangle_{\gamma\kappa} = \langle R_{As}\rangle_{\gamma\kappa}\left(\langle c_{A\gamma}\rangle^{\gamma}, \langle c_{B\gamma}\rangle^{\gamma}, \ldots, \langle c_{N\gamma}\rangle^{\gamma}\right), \quad \text{at the } \gamma\text{–}\kappa \text{ interface} \tag{1.93}$$

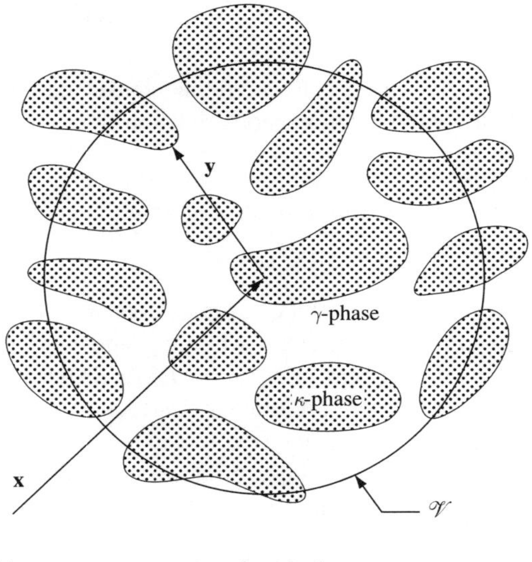

Figure 1.12 *Position vectors associated with the area average over the γ–κ interface*

and equation 1.88 takes the form

$$\varepsilon_\gamma \frac{\partial \langle c_{A\gamma}\rangle^\gamma}{\partial t} + \nabla \cdot \left(\varepsilon_\gamma \langle c_{A\gamma}\rangle^\gamma \langle \mathbf{v}_\gamma \rangle\right) = -\nabla \cdot \langle \mathbf{J}_{A\gamma}\rangle - \nabla \cdot \langle \tilde{c}_{A\gamma} \tilde{\mathbf{v}}_\gamma \rangle + a_v \langle R_{As}\rangle_{\gamma\kappa} \left(\langle c_{A\gamma}\rangle^\gamma, \langle c_{B\gamma}\rangle^\gamma, \ldots, \langle c_{N\gamma}\rangle^\gamma\right) \tag{1.94}$$

In porous catalysts one often neglects convective transport indicated by

$$\langle \tilde{c}_{A\gamma} \tilde{\mathbf{v}}_\gamma \rangle \ll \varepsilon_\gamma \langle c_{A\gamma}\rangle^\gamma \langle \mathbf{v}_\gamma \rangle \ll \langle \mathbf{J}_{A\gamma}\rangle \tag{1.95}$$

and this leads to a transport equation that takes the form

$$\varepsilon_\gamma \frac{\partial \langle c_{A\gamma}\rangle^\gamma}{\partial t} = -\nabla \cdot \langle \mathbf{J}_{A\gamma}\rangle + a_v \langle R_{As}\rangle_{\gamma\kappa} \left(\langle c_{A\gamma}\rangle^\gamma, \langle c_{B\gamma}\rangle^\gamma, \ldots, \langle c_{N\gamma}\rangle^\gamma\right), \quad A = 1, 2, \ldots, N \tag{1.96}$$

This result forms the basis for the classic problem of diffusion and reaction in a porous catalyst such as we have illustrated in Figure 1.5. It is extremely important to recognize that the *mathematical consequence* of equations 1.95 and 1.96 is that the mass average velocity has been set equal to zero; thus our substitute for equation 1.38 is given by the assumption

$$\mathbf{v}_\gamma = 0 \tag{1.97}$$

This assumption requires that we *discard* the momentum equation given by equation 1.38 and proceed to develop a solution to our mass transfer process in terms of the $N-1$ momentum equations represented by equation 1.39. The inequalities contained in equation 1.95 are quite appealing when one is dealing with a *diffusion process*; however, equation 1.97 is not satisfied by the Stefan diffusion tube process (Whitaker, 1991), nor is it satisfied by the Graham's law counter-diffusion process (Jackson, 1977). It should be clear that the constraints associated with the equalities given by equation 1.95 need to be developed. When convective transport is retained, some results are available from Quintard and Whitaker (2005); however, a detailed analysis of the coupled, non-linear process with convective transport remains to be done. At this point we leave those problems for a subsequent study and explore the diffusion and reaction process described by equation 1.96.

1.9 Non-dilute Diffusion

We begin this part of our study with the use of equation 1.68 in equation 1.96 to obtain

$$\varepsilon_\gamma \frac{\partial \langle c_{A\gamma}\rangle^\gamma}{\partial t} = \nabla \cdot \left\langle c_\gamma \sum_{E=1}^{E=N-1} D_{AE} \nabla x_{E\gamma} \right\rangle + a_v \langle R_{As}\rangle_{\gamma\kappa} \left(\langle c_{A\gamma}\rangle^\gamma, \langle c_{B\gamma}\rangle^\gamma, \ldots, \langle c_{N\gamma}\rangle^\gamma\right), \quad A = 1, 2, \ldots, N \tag{1.98}$$

where the diffusive flux is *non-linear* because D_{AE} depends on the $N-1$ mole fractions. This transport equation must be solved subject to the auxiliary conditions given by

$$c_\gamma = \sum_{A=1}^{A=N} c_{A\gamma}, \quad 1 = \sum_{A=1}^{A=N} x_{A\gamma} \tag{1.99}$$

and this suggests that numerical methods must be used. However, the diffusive flux must be arranged in terms of volume-averaged quantities before equation 1.98 can be solved, and any reasonable simplifications that can be made should be imposed on the analysis.

1.9.1 Constant Total Molar Concentration

Some non-dilute solutions can be treated as having a constant total molar concentration and this simplification allows us to express equation 1.98 as

$$\varepsilon_\gamma \frac{\partial \langle c_{A\gamma}\rangle^\gamma}{\partial t} = \nabla \cdot \left\langle \sum_{E=1}^{E=N-1} D_{AE} \nabla c_{E\gamma} \right\rangle + a_{\mathrm{v}} \langle R_{As}\rangle_{\gamma\kappa} \left(\langle c_{A\gamma}\rangle^\gamma, \langle c_{B\gamma}\rangle^\gamma, \ldots, \langle c_{N\gamma}\rangle^\gamma \right), \quad A = 1, 2, \ldots, N \tag{1.100}$$

The *restriction* associated with this simplification is given by

$$x_{A\gamma} \nabla c_\gamma \ll c_\gamma \nabla x_{A\gamma}, \quad A = 1, 2, \ldots, N \tag{1.101}$$

and it is important to understand that the *mathematical consequence* of this restriction is given by the assumption

$$c_\gamma = \langle c_\gamma \rangle^\gamma = \text{constant} \tag{1.102}$$

Imposition of this condition means that there are only $N-1$ independent transport equations of the form given by equation 1.100, and we shall impose this condition throughout the remainder of this study. The constraints associated with equation 1.102 need to be developed and the more general case represented by equations 1.98 and 1.99 should be explored.

At this point we decompose the elements of the diffusion matrix according to

$$D_{AE} = \langle D_{AE}\rangle^\gamma + \tilde{D}_{AE} \tag{1.103}$$

and if we can neglect $\tilde{D}_{AE}$ relative to $\langle D_{AE}\rangle^\gamma$, the transport equation given by equation 1.100 simplifies to

$$\varepsilon_\gamma \frac{\partial \langle c_{A\gamma}\rangle^\gamma}{\partial t} = \nabla \cdot \sum_{E=1}^{E=N-1} \langle D_{AE}\rangle^\gamma \langle \nabla c_{E\gamma} \rangle + a_{\mathrm{v}} \langle R_{As}\rangle_{\gamma\kappa} \left(\langle c_{A\gamma}\rangle^\gamma, \langle c_{B\gamma}\rangle^\gamma, \ldots, \langle c_{N-1\gamma}\rangle^\gamma \right), \quad A = 1, 2, \ldots, N-1 \tag{1.104}$$

We can represent this simplification as

$$\tilde{D}_{AE} \ll \langle D_{AE}\rangle^\gamma \tag{1.105}$$

and when it is *not satisfactory* it may be possible to develop a *correction* based on the retention of the spatial deviation, $\tilde{D}_{AE}$. However, it is not clear how this type of analysis would evolve and further study of this aspect of the diffusion process is in order.

1.9.2 Volume Average of the Diffusive Flux

The volume-averaging theorem can be used with the average of the gradient in equation 1.104 in order to obtain

$$\langle \nabla c_{E\gamma} \rangle = \nabla \langle c_{E\gamma} \rangle + \frac{1}{\mathscr{V}} \int_{A_{\gamma\kappa}} \mathbf{n}_{\gamma\kappa} c_{E\gamma} \, \mathrm{d}A \tag{1.106}$$

and one can follow an established analysis (Whitaker, 1999, Chapter 1) in order to express this result as

$$\langle \nabla c_{E\gamma} \rangle = \varepsilon_\gamma \nabla \langle c_{E\gamma} \rangle^\gamma + \frac{1}{\mathscr{V}} \int_{A_{\gamma\kappa}} \mathbf{n}_{\gamma\kappa} \tilde{c}_{E\gamma} \, \mathrm{d}A \tag{1.107}$$

Use of this result in equation 1.104 provides

$$\varepsilon_\gamma \frac{\partial \langle c_{A\gamma} \rangle^\gamma}{\partial t} = \nabla \cdot \left[\sum_{E=1}^{E=N-1} \langle D_{AE} \rangle^\gamma \left(\varepsilon_\gamma \nabla \langle c_{E\gamma} \rangle^\gamma + \underbrace{\frac{1}{\mathscr{V}} \int_{A_{\gamma\kappa}} \mathbf{n}_{\gamma\kappa} \tilde{c}_{E\gamma} \, \mathrm{d}A}_{\text{filter}} \right) \right] + a_\mathrm{v} \langle R_{As} \rangle_{\gamma\kappa} \tag{1.108}$$

where the area integral of $\mathbf{n}_{\gamma\kappa} \tilde{c}_{E\gamma}$ has been identified as a *filter*. Not all the information available at the length scale associated with $\tilde{c}_{E\gamma}$ will pass through this filter to influence the transport equation for $\langle c_{A\gamma} \rangle^\gamma$, and the existence of filters of this type is a recurring theme in the method of volume averaging (Whitaker, 1999).

1.10 Closure

In order to obtain a closed form of equation 1.108, we need a representation for the spatial deviation concentration, $\tilde{c}_{A\gamma}$, and this requires the development of the *closure problem*. When convective transport is negligible and homogeneous reactions are ignored as being a trivial part of the analysis, equation 1.48 takes the form

$$\frac{\partial c_{A\gamma}}{\partial t} = -\nabla \cdot \mathbf{J}_{A\gamma}, \quad A = 1, 2, \ldots, N-1 \tag{1.109}$$

Here one must remember that the total molar concentration is a specified constant; thus there are only $N-1$ independent species continuity equations. Use of equation 1.68 along with the restriction given by equation 1.101 allows us to express this result as

$$\frac{\partial c_{A\gamma}}{\partial t} = \nabla \cdot \sum_{E=1}^{E=N-1} D_{AE} \nabla c_{E\gamma}, \quad A = 1, 2, \ldots, N-1 \tag{1.110}$$

and on the basis of equations 1.103 and 1.105 this takes the form

$$\frac{\partial c_{A\gamma}}{\partial t} = \nabla \cdot \sum_{E=1}^{E=N-1} \langle D_{AE} \rangle^\gamma \nabla c_{E\gamma}, \quad A = 1, 2, \ldots, N-1 \tag{1.111}$$

If we ignore variations in ε_γ and subtract equation 1.108 from equation 1.111, we can arrange the result as

$$\frac{\partial \tilde{c}_{A\gamma}}{\partial t} = \nabla \cdot \left[\sum_{E=1}^{E=N-1} \langle D_{AE} \rangle^\gamma \nabla \tilde{c}_{E\gamma} \right] - \nabla \cdot \left[\sum_{E=1}^{E=N-1} \frac{\langle D_{AE} \rangle^\gamma}{\varepsilon_\gamma} \frac{1}{\mathscr{V}} \int_{A_{\gamma\kappa}} \mathbf{n}_{\gamma\kappa} \tilde{c}_{E\gamma} \, \mathrm{d}A \right] - \frac{a_\mathrm{v}}{\varepsilon_\gamma} \langle R_{\mathrm{As}} \rangle_{\gamma\kappa} \quad (1.112)$$

where it is understood that this result applies to all $N-1$ species. Equation 1.112 represents the governing differential equation for the spatial deviation concentration, and in order to keep the analysis relatively simple we consider only the first-order, irreversible reaction described by equation 1.89 and expressed here in the form

$$R_{\mathrm{As}} = -k_A c_{A\gamma}, \quad \text{at the } \gamma\text{–}\kappa \text{ interface} \quad (1.113)$$

One must remember that this is a severe restriction in terms of realistic systems and more general forms for the heterogeneous rate of reaction need to be examined. Use of equation 1.113 in equation 1.112 leads to the following form:

$$\frac{\partial \tilde{c}_{A\gamma}}{\partial t} = \nabla \cdot \left[\sum_{E=1}^{E=N-1} \langle D_{AE} \rangle^\gamma \nabla \tilde{c}_{E\gamma} \right] - \nabla \cdot \left[\sum_{E=1}^{E=N-1} \frac{\langle D_{AE} \rangle^\gamma}{\varepsilon_\gamma} \frac{1}{\mathscr{V}} \int_{A_{\gamma\kappa}} \mathbf{n}_{\gamma\kappa} \tilde{c}_{E\gamma} \, \mathrm{d}A \right] + \frac{a_\mathrm{v} k_A}{\varepsilon_\gamma} \langle c_{A\gamma} \rangle^\gamma \quad (1.114)$$

Here we have made use of the simplification

$$\langle c_{A\gamma} \rangle_{\gamma\kappa} = \langle c_{A\gamma} \rangle^\gamma \quad (1.115)$$

and the justification is given elsewhere (Whitaker, 1999, Sec. 1.3.3). In order to complete the problem statement for $\tilde{c}_{E\gamma}$, we need a boundary condition for $\tilde{c}_{E\gamma}$ at the γ–κ interface. To develop this boundary condition, we again make use of the *quasi-steady* form of equation 1.36 to obtain

$$\mathbf{J}_{A\gamma} \cdot \mathbf{n}_{\gamma\kappa} = -R_{\mathrm{As}}, \quad \text{at the } \gamma\text{–}\kappa \text{ interface} \quad (1.116)$$

where we have imposed the restriction given by

$$\mathbf{v}_\gamma \cdot \mathbf{n}_{\gamma\kappa} \ll \mathbf{u}_{A\gamma} \cdot \mathbf{n}_{\gamma\kappa}, \quad \text{at the } \gamma\text{–}\kappa \text{ interface} \quad (1.117)$$

This is certainly *consistent with* the inequalities given by equation 1.95; however, the neglect of $\mathbf{v}_\gamma \cdot \mathbf{n}_{\gamma\kappa}$ relative to $\mathbf{u}_{A\gamma} \cdot \mathbf{n}_{\gamma\kappa}$ is generally based on the *dilute solution* condition and the validity of equation 1.117 is another matter that needs to be carefully considered in a future study. On the basis of equations 1.68, 1.101, 1.103, and 1.105 along with equation 1.113, the jump condition takes the form

$$-\sum_{E=1}^{E=N-1} \mathbf{n}_{\gamma\kappa} \cdot \langle D_{AE} \rangle^\gamma \nabla c_{E\gamma} = k_A c_{A\gamma}, \quad \text{at the } \gamma\text{–}\kappa \text{ interface} \quad (1.118)$$

In order to express this boundary condition in terms of the spatial deviation concentration, we make use of the decomposition given by the first part of equation 1.86 to obtain

$$-\sum_{E=1}^{E=N-1} \mathbf{n}_{\gamma\kappa}\cdot\langle D_{AE}\rangle^{\gamma}\nabla\tilde{c}_{E\gamma} - k_A\tilde{c}_{A\gamma} = \sum_{E=1}^{E=N-1} \mathbf{n}_{\gamma\kappa}\cdot\langle D_{AE}\rangle^{\gamma}\nabla\langle c_{E\gamma}\rangle^{\gamma} + k_A\langle c_{A\gamma}\rangle^{\gamma}, \quad \text{at the } \gamma\text{–}\kappa \text{ interface} \tag{1.119}$$

With this result we can construct the following boundary value problem for $\tilde{c}_{A\gamma}$:

$$\underbrace{\frac{\partial \tilde{c}_{A\gamma}}{\partial t}}_{\text{accumulation}} = \underbrace{\nabla\cdot\left[\sum_{E=1}^{E=N-1}\langle D_{AE}\rangle^{\gamma}\nabla\tilde{c}_{E\gamma}\right]}_{\text{diffusion}} \underbrace{-\nabla\cdot\left[\sum_{E=1}^{E=N-1}\frac{\langle D_{AE}\rangle^{\gamma}}{\varepsilon_\gamma}\frac{1}{\mathscr{V}}\int_{A_{\gamma\kappa}}\mathbf{n}_{\gamma\kappa}\tilde{c}_{E\gamma}\,\mathrm{d}A\right]}_{\text{non-local diffusion}} + \underbrace{\frac{a_v k_A}{\varepsilon_\gamma}\langle c_{A\gamma}\rangle^{\gamma}}_{\text{reaction source}} \tag{1.120}$$

BC.1 :

$$\underbrace{-\sum_{E=1}^{E=N-1}\mathbf{n}_{\gamma\kappa}\cdot\langle D_{AE}\rangle^{\gamma}\nabla\tilde{c}_{E\gamma}}_{\text{diffusive flux}} - \underbrace{k_A\tilde{c}_{A\gamma}}_{\text{heterogeneous reaction}} = \underbrace{\sum_{E=1}^{E=N-1}\mathbf{n}_{\gamma\kappa}\cdot\langle D_{AE}\rangle^{\gamma}\nabla\langle c_{E\gamma}\rangle^{\gamma}}_{\text{diffusive source}} + \underbrace{k_A\langle c_{A\gamma}\rangle^{\gamma}}_{\text{reaction source}}, \quad \text{at the } \gamma\text{–}\kappa \text{ interface} \tag{1.121}$$

BC.2 :

$$\tilde{c}_{A\gamma} = F(\mathbf{r}, t), \quad \text{at } A_{\gamma e} \tag{1.122}$$

IC. :

$$\tilde{c}_{A\gamma} = F(\mathbf{r}), \quad \text{at } t = 0 \tag{1.123}$$

In addition to the flux boundary condition given by equation 1.121, we have added an unknown condition at the macroscopic boundary of the γ-phase, $A_{\gamma e}$, and an unknown initial condition. Neither of these is important when the separation of length scales indicated by equation 1.71 is valid. Under these circumstances, the boundary condition imposed at $A_{\gamma e}$ influences the $\tilde{c}_{A\gamma}$ field only over a negligibly small region, and the initial condition given by equation 1.123 can be discarded because the closure problem is quasi-steady. Under these circumstances, the closure problem can be solved in some representative, local region (Quintard and Whitaker, 1994a–e).

In the governing differential equation for $\tilde{c}_{A\gamma}$, we have identified the accumulation term, the diffusion term, the so-called non-local diffusion term, and the non-homogeneous term referred to as the *reaction source*. In the boundary condition imposed at the γ–κ interface, we have identified the diffusive flux, the reaction term, and two non-homogeneous terms that are referred to as the *diffusion source* and the *reaction source*. If the source terms in equations 1.120 and 1.121 were zero, the $\tilde{c}_{A\gamma}$-field would be generated only by the non-homogeneous terms that might appear in the boundary condition imposed at $A_{\gamma e}$ or in the initial condition given by equation 1.123. One can easily develop arguments indicating that the closure problem for $\tilde{c}_{A\gamma}$ is quasi-steady, thus the initial condition is of no importance (Whitaker, 1999, Chapter 1). In addition, one can develop arguments indicating that the boundary condition imposed at $A_{\gamma e}$ will influence the $\tilde{c}_{A\gamma}$ field over a negligibly small portion of the field of interest. Because of this, any useful solution to the closure problem must be developed for some *representative region* which is most often conveniently described in terms of a unit cell in a spatially periodic system. These ideas lead to a closure problem of the form

$$0 = \underbrace{\nabla \cdot \left[\sum_{E=1}^{E=N-1} \langle D_{AE} \rangle^\gamma \nabla \tilde{c}_{E\gamma} \right]}_{\text{diffusion}} - \underbrace{\nabla \cdot \left[\sum_{E=1}^{E=N-1} \frac{\langle D_{AE} \rangle^\gamma}{\varepsilon_\gamma \mathscr{V}} \int_{A_{\gamma\kappa}} \mathbf{n}_{\gamma\kappa} \tilde{c}_{E\gamma} \, dA \right]}_{\text{non-local diffusion}} + \underbrace{\frac{a_v k_A}{\varepsilon_\gamma} \langle c_{A\gamma} \rangle^\gamma}_{\substack{\text{reaction}\\ \text{source}}} \tag{1.124}$$

$$\text{BC.1}: \quad \underbrace{- \sum_{E=1}^{E=N-1} \mathbf{n}_{\gamma\kappa} \cdot \langle D_{AE} \rangle^\gamma \nabla \tilde{c}_{E\gamma}}_{\text{diffusive flux}} - \underbrace{k_A \tilde{c}_{A\gamma}}_{\substack{\text{heterogeneous}\\ \text{reaction}}} = \underbrace{\sum_{E=1}^{E=N-1} \mathbf{n}_{\gamma\kappa} \cdot \langle D_{AE} \rangle^\gamma \nabla \langle c_{E\gamma} \rangle^\gamma}_{\text{diffusive source}} + \underbrace{k_A \langle c_{A\gamma} \rangle^\gamma}_{\substack{\text{reaction}\\ \text{source}}}, \quad \text{at the } \gamma\text{–}\kappa \text{ interface} \tag{1.125}$$

$$\text{BC.2}: \quad \tilde{c}_{A\gamma}(\mathbf{r} + \ell_i) = \tilde{c}_{A\gamma}(\mathbf{r}), \quad i = 1, 2, 3 \tag{1.126}$$

Here we have used ℓ_i to represent the three base vectors needed to characterize a spatially periodic system. The use of a spatially periodic system does not limit this analysis to simple systems since a periodic system can be arbitrarily complex (Quintard and Whitaker, 1994a–e). However, the periodicity condition imposed by equation 1.126 can only be strictly justified when $\langle D_{AE} \rangle^\gamma$, $\langle c_{A\gamma} \rangle^\gamma$, and $\nabla \langle c_{A\gamma} \rangle^\gamma$ are constants and this does not occur for the types of systems under consideration. This matter has been examined elsewhere (Whitaker, 1986b) and the analysis suggests that the traditional separation of length scales allows one to treat $\langle D_{AE} \rangle^\gamma$, $\langle c_{A\gamma} \rangle^\gamma$, and $\nabla \langle c_{A\gamma} \rangle^\gamma$ as constants within the framework of the closure problem.

It is not obvious, but other studies (Ryan *et al.*, 1981) have shown that the reaction source in equations 1.124 and 1.125 makes a negligible contribution to $\tilde{c}_{A\gamma}$. In addition, one can demonstrate (Whitaker, 1999) that the heterogeneous reaction, $k_A \tilde{c}_{A\gamma}$, can be neglected for all practical problems of diffusion and reaction in porous catalysts. Furthermore, the non-local diffusion term is negligible for traditional systems, and under these circumstances the boundary value problem for the spatial deviation concentration takes the form

$$0 = \nabla \cdot \left[\sum_{E=1}^{E=N-1} \langle D_{AE} \rangle^\gamma \nabla \tilde{c}_{E\gamma} \right] \tag{1.127}$$

$$\text{BC.1}: \quad - \sum_{E=1}^{E=N-1} \mathbf{n}_{\gamma\kappa} \cdot \langle D_{AE} \rangle^\gamma \nabla \tilde{c}_{E\gamma} = \sum_{E=1}^{E=N-1} \mathbf{n}_{\gamma\kappa} \cdot \langle D_{AE} \rangle^\gamma \nabla \langle c_{E\gamma} \rangle^\gamma, \quad \text{at } A_{\gamma\kappa} \tag{1.128}$$

$$\text{BC.2}: \quad \tilde{c}_{A\gamma}(\mathbf{r} + \ell_i) = \tilde{c}_{A\gamma}(\mathbf{r}), \quad i = 1, 2, 3 \tag{1.129}$$

Here one must remember that the subscript A represents species A, B, C, . . . , $N-1$.

In this boundary value problem, there is only a single non-homogeneous term represented by $\nabla\langle c_{E\gamma}\rangle^\gamma$ in the boundary condition imposed at the γ–κ interface. If this source term were zero, the solution to this boundary value problem would be given by $\tilde{c}_{A\gamma} = \text{constant}$. Any constant associated with $\tilde{c}_{A\gamma}$ will not pass through the filter in equation 1.108, and this suggests that a solution can be expressed in terms of the gradients of the volume-averaged concentration. Since the system is linear in the $N-1$ independent gradients of the average concentration, we are led to a solution of the form

$$\tilde{c}_{E\gamma} = \mathbf{b}_{EA}\cdot\nabla\langle c_{A\gamma}\rangle^\gamma + \mathbf{b}_{EB}\cdot\nabla\langle c_{B\gamma}\rangle^\gamma + \mathbf{b}_{EC}\cdot\nabla\langle c_{C\gamma}\rangle^\gamma + \cdots + \mathbf{b}_{E,N-1}\cdot\nabla\langle c_{N-1\gamma}\rangle^\gamma \quad (1.130)$$

Here the vectors, $\mathbf{b}_{EA}$, $\mathbf{b}_{EB}$, etc., are referred to as the *closure variables* or the *mapping variables* since they map the gradients of the volume-averaged concentrations onto the spatial deviation concentrations. In this representation for $\tilde{c}_{A\gamma}$, we can ignore the spatial variations of $\nabla\langle c_{A\gamma}\rangle^\gamma$, $\nabla\langle c_{B\gamma}\rangle^\gamma$, etc. within the framework of a local closure problem, and we can use equation 1.130 in equation 1.127 to obtain

$$0 = \nabla\left[\sum_{E=1}^{E=N-1}\langle D_{AE}\rangle^\gamma \sum_{D=1}^{D=N-1}\nabla\mathbf{b}_{ED}\cdot\nabla\langle c_{D\gamma}\rangle^\gamma\right] \quad (1.131)$$

BC.1 :
$$-\sum_{E=1}^{E=N-1}\mathbf{n}_{\gamma\kappa}\cdot\langle D_{AE}\rangle^\gamma \sum_{D=1}^{D=N-1}\nabla\mathbf{b}_{ED}\cdot\nabla\langle c_{D\gamma}\rangle^\gamma = \sum_{E=1}^{E=N-1}\mathbf{n}_{\gamma\kappa}\cdot\langle D_{AE}\rangle^\gamma\nabla\langle c_{E\gamma}\rangle^\gamma, \quad \text{at } A_{\gamma\kappa} \quad (1.132)$$

BC.2 :
$$\mathbf{b}_{AE}(\mathbf{r}+\ell_i) = \mathbf{b}_{AE}(\mathbf{r}), \quad i=1,2,3, \quad A=1,2,\ldots,N-1 \quad (1.133)$$

The derivation of equations 1.131 and 1.132 requires the use of simplifications of the form

$$\nabla\left(\mathbf{b}_{EA}\cdot\nabla\langle c_{A\gamma}\rangle^\gamma\right) = \nabla\mathbf{b}_{EA}\cdot\nabla\langle c_{A\gamma}\rangle^\gamma \quad (1.134)$$

which result from the inequality

$$\mathbf{b}_{EA}\cdot\nabla\nabla\langle c_{A\gamma}\rangle^\gamma \ll \nabla\mathbf{b}_{EA}\cdot\nabla\langle c_{A\gamma}\rangle^\gamma \quad (1.135)$$

The basis for this inequality is the separation of length scales indicated by equation 1.71, and a detailed discussion is available elsewhere (Whitaker, 1999). One should keep in mind that the boundary value problem given by equations 1.131–1.133 applies to all $N-1$ species and that the $N-1$ concentration gradients are independent. This latter condition allows us to obtain

$$0 = \nabla\cdot\left[\sum_{E=1}^{E=N-1}\langle D_{AE}\rangle^\gamma\nabla\mathbf{b}_{ED}\right], \quad D=1,2,\ldots,N-1 \quad (1.136)$$

BC.1 :
$$-\sum_{E=1}^{E=N-1}\mathbf{n}_{\gamma\kappa}\cdot\langle D_{AE}\rangle^\gamma\nabla\mathbf{b}_{ED} = \mathbf{n}_{\gamma\kappa}\langle D_{AD}\rangle^\gamma, \quad D=1,2,\ldots,N-1, \quad \text{at } A_{\gamma\kappa} \quad (1.137)$$

Periodicity :
$$\mathbf{b}_{AD}(\mathbf{r}+\ell_i) = \mathbf{b}_{AD}(\mathbf{r}), \quad i=1,2,3, \quad D=1,2,\ldots,N-1 \quad (1.138)$$

At this point it is convenient to expand the closure problem for species A in order to obtain

First problem for species A

$$0=\nabla\cdot\Big\{\langle D_{AA}\rangle^{\gamma}\Big[\nabla\mathbf{b}_{AA}+(\langle D_{AA}\rangle^{\gamma})^{-1}\langle D_{AB}\rangle^{\gamma}\nabla\mathbf{b}_{BA}$$
$$+(\langle D_{AA}\rangle^{\gamma})^{-1}\langle D_{AC}\rangle^{\gamma}\nabla\mathbf{b}_{CA}+\cdots+(\langle D_{AA}\rangle^{\gamma})^{-1}\langle D_{A,N-1}\rangle^{\gamma}\nabla\mathbf{b}_{N-1,A}\Big]\Big\} \quad (1.139a)$$

$$\text{BC.}:\quad -\mathbf{n}_{\gamma\kappa}\cdot\nabla\mathbf{b}_{AA}-\mathbf{n}_{\gamma\kappa}\cdot(\langle D_{AA}\rangle^{\gamma})^{-1}\langle D_{AB}\rangle^{\gamma}\nabla\mathbf{b}_{BA}$$
$$-\mathbf{n}_{\gamma\kappa}\cdot(\langle D_{AA}\rangle^{\gamma})^{-1}\langle D_{AC}\rangle^{\gamma}\nabla\mathbf{b}_{CA}-\cdots \quad (1.139b)$$
$$-\mathbf{n}_{\gamma\kappa}\cdot(\langle D_{AA}\rangle^{\gamma})^{-1}\langle D_{A,N-1}\rangle^{\gamma}\nabla\mathbf{b}_{N-1,A}=\mathbf{n}_{\gamma\kappa},\quad \text{at } A_{\gamma\kappa}$$

$$\text{Periodicity}:\quad \mathbf{b}_{DA}(\mathbf{r}+\ell_i)=\mathbf{b}_{DA}(\mathbf{r}),\quad i=1,2,3,\quad D=1,2,\ldots,N-1 \quad (1.139c)$$

Second problem for species A

$$0=\nabla\cdot\Big\{\langle D_{AB}\rangle^{\gamma}\Big[(\langle D_{AB}\rangle^{\gamma})^{-1}\langle D_{AA}\rangle^{\gamma}\nabla\mathbf{b}_{AB}+\nabla\mathbf{b}_{BB}$$
$$+(\langle D_{AB}\rangle^{\gamma})^{-1}\langle D_{AC}\rangle^{\gamma}\nabla\mathbf{b}_{CB}+\cdots+(\langle D_{AB}\rangle^{\gamma})^{-1}\langle D_{A,N-1}\rangle^{\gamma}\nabla\mathbf{b}_{N-1,B}\Big]\Big\} \quad (1.140a)$$

$$\text{BC.}:\quad -\mathbf{n}_{\gamma\kappa}\cdot\nabla\mathbf{b}_{AA}-\mathbf{n}_{\gamma\kappa}\cdot(\langle D_{AA}\rangle^{\gamma})^{-1}\langle D_{AB}\rangle^{\gamma}\nabla\mathbf{b}_{BA}$$
$$-\mathbf{n}_{\gamma\kappa}\cdot(\langle D_{AA}\rangle^{\gamma})^{-1}\langle D_{AC}\rangle^{\gamma}\nabla\mathbf{b}_{CA}-\cdots \quad (1.140b)$$
$$-\mathbf{n}_{\gamma\kappa}\cdot(\langle D_{AA}\rangle^{\gamma})^{-1}\langle D_{A,N-1}\rangle^{\gamma}\nabla\mathbf{b}_{N-1,A}=\mathbf{n}_{\gamma\kappa},\quad \text{at } A_{\gamma\kappa}$$

$$\text{Periodicity}:\quad \mathbf{b}_{DB}(\mathbf{r}+\ell_i)=\mathbf{b}_{DB}(\mathbf{r}),\quad i=1,2,3,\quad D=1,2,\ldots,N-1 \quad (1.140c)$$

Third problem for species A

$$\text{etc.} \quad (1.141)$$

$N-1$ *problem for species A*

$$\text{etc.} \quad (1.142)$$

Here it is convenient to define a new set of *closure variables* or *mapping variables* according to

$$\mathbf{d}_{AA}=\mathbf{b}_{AA}+(\langle D_{AA}\rangle^{\gamma})^{-1}\langle D_{AB}\rangle^{\gamma}\mathbf{b}_{BA}+(\langle D_{AA}\rangle^{\gamma})^{-1}\langle D_{AC}\rangle^{\gamma}\mathbf{b}_{CA}$$
$$+\cdots+(\langle D_{AA}\rangle^{\gamma})^{-1}\langle D_{A,N-1}\rangle^{\gamma}\mathbf{b}_{N-1,A} \quad (1.143a)$$

$$\mathbf{d}_{AB}=(\langle D_{AB}\rangle^{\gamma})^{-1}\langle D_{AA}\rangle^{\gamma}\mathbf{b}_{AB}+\mathbf{b}_{BB}+(\langle D_{AB}\rangle^{\gamma})^{-1}\langle D_{AC}\rangle^{\gamma}\mathbf{b}_{CB}$$
$$+\cdots+(\langle D_{AB}\rangle^{\gamma})^{-1}\langle D_{A,N-1}\rangle^{\gamma}\mathbf{b}_{N-1,B} \quad (1.143b)$$

$$\mathbf{d}_{AC} = (\langle D_{AC}\rangle^{\gamma})^{-1}\langle D_{AA}\rangle^{\gamma}\mathbf{b}_{AC} + (\langle D_{AC}\rangle^{\gamma})^{-1}\langle D_{AB}\rangle^{\gamma}\mathbf{b}_{BC} + \mathbf{b}_{CC} + \cdots + (\langle D_{AC}\rangle^{\gamma})^{-1}\langle D_{A,N-1}\rangle^{\gamma}\mathbf{b}_{N-1,C} \tag{1.143c}$$

$$\text{etc.} \tag{1.143d}$$

With these definitions, the closure problems take the following simplified forms:

First problem for species A

$$0 = \nabla^2 \mathbf{d}_{AA} \tag{1.144a}$$

$$\text{BC.:} \quad -\mathbf{n}_{\gamma\kappa} \cdot \nabla \mathbf{d}_{AA} = \mathbf{n}_{\gamma\kappa}, \quad \text{at } A_{\gamma\kappa} \tag{1.144b}$$

$$\text{Periodicity:} \quad \mathbf{d}_{AA}(\mathbf{r} + \ell_i) = \mathbf{d}_{AA}(\mathbf{r}), \quad i = 1, 2, 3 \tag{1.144c}$$

Second problem for species A

$$0 = \nabla^2 \mathbf{d}_{AB} \tag{1.145a}$$

$$\text{BC.:} \quad -\mathbf{n}_{\gamma\kappa} \cdot \nabla \mathbf{d}_{AB} = \mathbf{n}_{\gamma\kappa}, \quad \text{at } A_{\gamma\kappa} \tag{1.145b}$$

$$\text{Periodicity:} \quad \mathbf{d}_{AB}(\mathbf{r} + \ell_i) = \mathbf{d}_{AB}(\mathbf{r}), \quad i = 1, 2, 3 \tag{1.145c}$$

Third problem for species A

$$\text{etc.} \tag{1.146}$$

N–1 problem for species A

$$\text{etc.} \tag{1.147}$$

To obtain these simplified forms, one must make repeated use of inequalities of the form given by equation 1.135. Each one of these closure problems is identical to that obtained by Ryan *et al.* (1981) and solutions have been developed by several workers (Chang, 1982, 1983; Ochoa-Tapia *et al.*, 1994; Quintard, 1993; Quintard and Whitaker, 1993a,b; Ryan *et al.*, 1981). In each case, the closure problem determines the *closure variable* to within an arbitrary constant, and this constant can be specified by imposing the condition

$$\langle \tilde{c}_{D\gamma}\rangle^{\gamma} = 0, \quad \text{or} \quad \langle \mathbf{d}_{GD}\rangle^{\gamma} = 0, \qquad \begin{cases} G = 1, 2, \ldots, N-1 \\ D = 1, 2, \ldots, N-1 \end{cases} \tag{1.148}$$

However, any constant associated with a closure variable will not pass through the filter in equation 1.108; thus this constraint on the average is not necessary.

1.10.1 Closed Form

The closed form of equation 1.108 can be obtained by use of the representation for $\tilde{c}_{E\gamma}$ given by equation 1.130, along with the definitions represented by equation 1.143. After some algebraic manipulation, one obtains

$$
\begin{aligned}
\varepsilon_\gamma \frac{\partial \langle c_{A\gamma}\rangle^\gamma}{\partial t} = \nabla \cdot \Bigg[& \varepsilon_\gamma \langle D_{AA}\rangle^\gamma \left(\mathbf{I} + \frac{1}{V_\gamma} \int_{A_{\gamma\kappa}} \mathbf{n}_{\gamma\kappa} \mathbf{d}_{AA} \, dA \right) \nabla \langle c_{A\gamma}\rangle^\gamma + \\
& + \varepsilon_\gamma \langle D_{AB}\rangle^\gamma \left(\mathbf{I} + \frac{1}{V_\gamma} \int_{A_{\gamma\kappa}} \mathbf{n}_{\gamma\kappa} \mathbf{d}_{AB} \, dA \right) \nabla \langle c_{B\gamma}\rangle^\gamma + \\
& + \varepsilon_\gamma \langle D_{AC}\rangle^\gamma \left(\mathbf{I} + \frac{1}{V_\gamma} \int_{A_{\gamma\kappa}} \mathbf{n}_{\gamma\kappa} \mathbf{d}_{AC} \, dA \right) \nabla \langle c_{C\gamma}\rangle^\gamma + \\
& + \varepsilon_\gamma \langle D_{A,N-1}\rangle^\gamma \left(\mathbf{I} + \frac{1}{V_\gamma} \int_{A_{\gamma\kappa}} \mathbf{n}_{\gamma\kappa} \mathbf{d}_{A,N-1} \, dA \right) \nabla \langle c_{N-1\gamma}\rangle^\gamma \Bigg] \\
& + a_v k_A \langle c_{A\gamma}\rangle^\gamma
\end{aligned}
\tag{1.149}
$$

Here one must remember that we have restricted the analysis to the simple linear reaction rate expression given by equation 1.113, and one normally must work with more complex representations for R_{As}.

On the basis of the closure problems given by equations 1.144a–1.147, we conclude that there is a single tensor that describes the tortuosity for species A. This means that equation 1.149 can be expressed as

$$
\begin{aligned}
\varepsilon_\gamma \frac{\partial \langle c_{A\gamma}\rangle^\gamma}{\partial t} &= \nabla \cdot \left[\varepsilon_\gamma \mathbf{D}_{AA}^{\text{eff}} \cdot \nabla \langle c_{A\gamma}\rangle^\gamma + \varepsilon_\gamma \mathbf{D}_{AB}^{\text{eff}} \cdot \nabla \langle c_{B\gamma}\rangle^\gamma \right. \\
& + \varepsilon_\gamma \mathbf{D}_{AC}^{\text{eff}} \cdot \nabla \langle c_{C\gamma}\rangle^\gamma \\
& \left. + \cdots + \varepsilon_\gamma \mathbf{D}_{A,N-1}^{\text{eff}} \cdot \nabla \langle c_{N-1\gamma}\rangle^\gamma \right] + a_v k_A \langle c_{A\gamma}\rangle^\gamma
\end{aligned}
\tag{1.150}
$$

where the effective diffusivity tensors are related according to

$$
\frac{\mathbf{D}_{AA}^{\text{eff}}}{\langle D_{AA}\rangle^\gamma} = \frac{\mathbf{D}_{AB}^{\text{eff}}}{\langle D_{AB}\rangle^\gamma} = \frac{\mathbf{D}_{AC}^{\text{eff}}}{\langle D_{AC}\rangle^\gamma} = \cdots = \frac{\mathbf{D}_{A,N-1}^{\text{eff}}}{\langle D_{A,N-1}\rangle^\gamma} \tag{1.151}
$$

The remaining diffusion equations for species $B, C, \ldots, N-1$ have precisely the same form as equation 1.150, and the various effective diffusivity tensors are related to each other in the manner indicated by equation 1.151. The generic closure problem can be expressed as

$$0 = \nabla^2 \mathbf{d} \tag{1.152a}$$

BC. : $$-\mathbf{n}_{\gamma\kappa} \cdot \nabla \mathbf{d} = \mathbf{n}_{\gamma\kappa}, \quad \text{at } A_{\gamma\kappa} \tag{1.152b}$$

Periodicity : $$\mathrm{d}(\mathbf{r} + \ell_i) = \mathbf{d}(\mathbf{r}), \quad i = 1, 2, 3 \tag{1.152c}$$

and solution of this boundary value problem is relatively straightforward. The existence of a single generic closure problem that allows for the determination of all the effective diffusivity tensors represents the main finding of this work. On the basis of this single closure problem, the tortuosity tensor is defined according to

$$\boldsymbol{\tau} = \mathbf{I} + \frac{1}{V_\gamma} \int_{A_{\gamma\kappa}} \mathbf{n}_{\gamma\kappa} \mathbf{d} \, dA \tag{1.153}$$

and we can express equation 1.151 in the form

$$\mathbf{D}^{\text{eff}}_{AA} = \boldsymbol{\tau} \langle D_{AA} \rangle^\gamma, \quad \mathbf{D}^{\text{eff}}_{AB} = \boldsymbol{\tau} \langle D_{AB} \rangle^\gamma, \ldots, \mathbf{D}^{\text{eff}}_{A,N-1} = \boldsymbol{\tau} \langle D_{A,N-1} \rangle^\gamma \tag{1.154}$$

Substitution of these results into equation 1.150 allows us to represent the local volume-averaged diffusion-reaction equations as

$$\varepsilon_\gamma \frac{\partial \langle c_{A\gamma} \rangle^\gamma}{\partial t} = \nabla \cdot \left[\sum_{E=1}^{E=N-1} \varepsilon_\gamma \boldsymbol{\tau} \langle D_{AE} \rangle^\gamma \cdot \nabla \langle c_{E\gamma} \rangle^\gamma \right] + a_v k_A \langle c_{A\gamma} \rangle^\gamma, \quad A = 1, 2, \ldots, N-1 \tag{1.155}$$

It is important to remember that this analysis has been simplified on the basis of equation 1.101 which is equivalent to treating c_γ as a constant as indicated in equation 1.102. For a porous medium that is isotropic in the volume-averaged sense, the tortuosity tensor takes the classical form

$$\boldsymbol{\tau} = \mathbf{I} \tau^{-1} \tag{1.156}$$

where $\mathbf{I}$ is the unit tensor and τ is *the* tortuosity. For isotropic porous media, we can express equation 1.155 as

$$\varepsilon_\gamma \frac{\partial \langle c_{A\gamma} \rangle^\gamma}{\partial t} = \nabla \cdot \left[\sum_{E=1}^{E=N-1} (\varepsilon_\gamma / \tau) \langle D_{AE} \rangle^\gamma \nabla \langle c_{E\gamma} \rangle^\gamma \right] + a_v k_A \langle c_{A\gamma} \rangle^\gamma, \quad A = 1, 2, \ldots, N-1 \tag{1.157}$$

Often ε_γ and τ can be treated as constants; however, the diffusion coefficients in this transport equation will be functions of the local volume-averaged mole fractions and we are faced with a coupled, non-linear diffusion and reaction problem.

1.11 Conclusions

In this chapter we have first shown how an intuitive upscaling procedure can lead to confusion regarding homogeneous and heterogeneous reactions, and in a more formal development we have shown how the coupled, non-linear diffusion problem can be analyzed to produce volume-averaged transport equations containing effective diffusivity tensors. The original diffusion-reaction problem is described by

$$\frac{\partial c_{A\gamma}}{\partial t} = \nabla \cdot \sum_{E=1}^{E=N-1} D_{AE} \nabla c_{E\gamma}, \quad A = 1, 2, \ldots, N-1 \tag{1.158a}$$

BC. :

$$- \sum_{E=1}^{E=N-1} \mathbf{n}_{\gamma\kappa} \cdot D_{AE} \nabla c_{E\gamma} = k_A c_{A\gamma}, \quad \text{at the } \gamma\text{–}\kappa \text{ interface} \tag{1.158b}$$

$$c_\gamma = \langle c_\gamma \rangle^\gamma = \text{constant} \tag{1.158c}$$

where the D_{AE} are functions of the mole fractions. For a porous medium that is isotropic in the volume-averaged sense, the upscaled version of the diffusion-reaction problem takes the form

$$\varepsilon_\gamma \frac{\partial \langle c_{A\gamma} \rangle^\gamma}{\partial t} = \nabla \cdot \left[\sum_{E=1}^{E=N-1} (\varepsilon_\gamma / \tau) \langle D_{AE} \rangle^\gamma \nabla \langle c_{E\gamma} \rangle^\gamma \right] + a_v k_A \langle c_{A\gamma} \rangle^\gamma, \quad A = 1, 2, \ldots, N-1 \tag{1.159}$$

Here we have used the approximation that D_{AE} can be replaced by $\langle D_{AE} \rangle^\gamma$ and that variations of $\langle D_{AE} \rangle^\gamma$ can be ignored within the averaging volume. The fact that only a single tortuosity needs to be determined by equations 1.152 and 1.153 represents the key contribution of this study. It is important to remember that this development is constrained by the linear chemical kinetic constitutive equation given by equation 1.113. The process of diffusion in porous catalysts is normally associated with *slow* reactions and equation 1.93 is satisfactory; however, the first-order, irreversible reaction represented by equation 1.113 is the exception rather than the rule, and this aspect of the analysis requires further investigation. The influence of a non-zero mass average velocity needs to be considered in future studies so that the constraint given by equation 1.97 can be removed. An analysis of that case is reserved for a future study which will also include a careful examination of the simplification indicated by equation 1.117.

Nomenclature

$A_{\gamma e}$	area of entrances and exits of the γ-phase contained in the macroscopic region, m^2
$A_{\gamma\kappa}$	area of the γ–κ interface contained within the averaging volume, m^2
a_v	$A_{\gamma\kappa}/\mathscr{V}$, area per unit volume, 1/m
$\mathbf{b}_\gamma$	body force vector, m/s^2
$c_{A\gamma}$	bulk concentration of species A in the γ-phase, moles/m^3
$\langle c_{A\gamma} \rangle$	superficial average bulk concentration of species A in the γ-phase, moles/m^3
$\langle c_{A\gamma} \rangle^\gamma$	intrinsic average bulk concentration of species A in the γ-phase, moles/m^3
$\langle c_{A\gamma} \rangle_{\gamma\kappa}$	intrinsic area average bulk concentration of species A at the γ–κ interface, moles/m^3
$\tilde{c}_{A\gamma}$	$c_{A\gamma} - \langle c_{A\gamma} \rangle^\gamma$, spatial deviation concentration of species A, moles/m^3

c_γ	$\sum_{A=1}^{A=N} c_{A\gamma}$, total molar concentration, moles/m^3
c_{As}	surface concentration of species A associated with the γ–κ interface, moles/m^2
$\mathscr{D}_{AB}$	binary diffusion coefficient for species A and B, m^2/s
$\mathscr{D}_{Am}$	$\mathscr{D}_{Am}^{-1} = \sum_{\substack{E=1 \\ E\neq A}}^{E=N} x_{E\gamma}/\mathscr{D}_{AE}$, mixture diffusivity, m^2/s
$[D]$	diffusivity matrix, m^2/s
D_{AE}	element of the diffusivity matrix, m^2/s
$\langle D_{AE}\rangle^\gamma$	intrinsic average element of the diffusivity matrix, m^2/s
$\tilde{D}_{AE}$	$D_{AE} - \langle D_{AE}\rangle^\gamma$, spatial deviation of an element of the diffusivity matrix, m^2/s
$\mathbf{J}_{A\gamma}$	$c_{A\gamma}\mathbf{u}_{A\gamma}$, mixed-mode diffusive flux, mole/m^{2}s
K_A	adsorption equilibrium coefficient for species A, m
k_{A1}	adsorption rate coefficient for species A, m/s
k_{-A1}	desorption rate coefficient for species A, 1/s
k_{As}	surface reaction rate coefficient, 1/s
ℓ_γ	small length scale associated with the γ-phase, m
r_0	radius of the averaging volume, m
L	large length scale associated with the porous medium, m
M_A	molecular mass of species A, kg/kg mole
$\overline{M}$	$\sum_{A=1}^{A=N} x_{A\gamma}M_A$, mean molecular mass, kg/kg mole
$\mathbf{n}_{\gamma\kappa}$	unit normal vector directed from the γ-phase to the κ-phase
$\mathbf{r}$	position vector, m
$R_{A\gamma}$	rate of homogeneous reaction in the γ- phase, moles/m^{3}s
R_{As}	rate of heterogeneous reaction associated with the γ–κ interface, moles/m^{2}s
$\langle R_{As}\rangle_{\gamma\kappa}$	area average heterogeneous reaction rate for species A, moles/m^{2}s
t	time, s
$\mathbf{T}_\gamma$	stress tensor for the γ-phase, N/m^2
$\mathbf{u}_{A\gamma}$	$\mathbf{v}_{A\gamma} - \mathbf{v}_\gamma$, mass diffusion velocity, m/s
$\mathbf{u}^*_{A\gamma}$	$\mathbf{v}_{A\gamma} - \mathbf{v}^*_\gamma$, molar diffusion velocity, m/s
$\mathbf{v}_{A\gamma}$	velocity of species A in the γ-phase, m/s
$\mathbf{v}_\gamma$	$\sum_{A=1}^{A=N} \omega_{A\gamma}\mathbf{v}_{A\gamma}$, mass average velocity in the γ-phase, m/s
$\mathbf{v}^*_\gamma$	$\sum_{A=1}^{A=N} x_{A\gamma}\mathbf{v}_{A\gamma}$, molar average velocity in the γ-phase, m/s
$\langle\mathbf{v}_\gamma\rangle^\gamma$	intrinsic mass average velocity in the γ-phase, m/s
$\langle\mathbf{v}_\gamma\rangle$	superficial mass average velocity in the γ-phase, m/s
$\tilde{\mathbf{v}}_\gamma$	$\mathbf{v}_\gamma - \langle\mathbf{v}_\gamma\rangle^\gamma$, spatial deviation velocity, m/s
$\mathscr{V}$	averaging volume, m^3
V_γ	volume of the γ-phase contained within the averaging volume, m^3
V_κ	volume of the κ-phase contained within the averaging volume, m^3
$x_{A\gamma}$	$c_{A\gamma}/c_\gamma$, mole fraction of species A in the γ-phase
$\mathbf{x}$	position vector locating the center of the averaging volume, m
$\mathbf{y}$	position vector locating points on the γ–κ interface relative to the center of the averaging volume, m

Greek letters

ε_γ volume fraction of the γ-phase (porosity)
$\rho_{A\gamma}$ mass density of species A in the γ-phase, kg/m^3
ρ_γ mass density for the γ-phase, kg/m^3
$\omega_{A\gamma}$ $\rho_{A\gamma}/\rho_\gamma$, mass fraction of species A in the γ-phase

References

Anderson T.B. and Jackson R. 1967. A fluid mechanical description of fluidized beds, Ind. Engng. Chem. Fundam., 6, 527–538.

Bird R.B., Steward W.E. and Lightfoot E.N. 2002. Transport Phenomena, 2nd edition. John Wiley & Sons, New York.

Birkhoff G. 1960. Hydrodynamics: A Study in Logic, Fact, and Similitude. Princeton University Press, Princeton, New Jersey.

Butt J.B. 1980. Reaction Kinetics and Reactor Design. Prentice-Hall, Englewood Cliffs, New Jersey.

Carberry J.J. 1976. Chemical and Catalytic Reaction Engineering. McGraw-Hill, New York.

Carbonell R.G. and Whitaker S. 1984. Adsorption and reaction at a catalytic surface: The quasi-steady condition, Chem. Eng. Sci., 39, 1319–1321.

Chang H-C. 1982. Multiscale analysis of effective transport in periodic heterogeneous media, Chem. Eng. Commun., 15, 83–91.

Chang H-C. 1983. Effective diffusion and conduction in two-phase media: A unified approach, AIChE J., 29, 846–853.

Cushman J.H. 1990. Dynamics of Fluids in Hierarchical Porous Media. Academic Press, London.

Fogler H.S. 1992. Elements of Chemical Reaction Engineering. Prentice Hall, Englewood Cliffs, New Jersey.

Froment G.F. and Bischoff K.B. 1979. Chemical Reactor Analysis and Design. John Wiley & Sons, New York.

Jackson R. 1977. Transport in Porous Catalysts. Elsevier, New York.

Langmuir I. 1916. The constitution and fundamental properties of solids and liquids I: Solids, J. Am. Chem. Soc., 38, 2221–2295.

Langmuir I. 1917. The constitution and fundamental properties of solids and liquids II: Liquids, J. Am. Chem. Soc., 39, 1848–1906.

Levenspiel O. 1999. Chemical Reaction Engineering, 3rd edition. John Wiley & Sons, New York.

Marle C.M. 1967. Écoulements monophasique en milieu poreux, Rev. Inst. Français du Pétrole, 22(10), 1471–1509.

Ochoa-Tapia J.A., del Río J.A. and Whitaker S. 1993. Bulk and surface diffusion in porous media: An application of the surface averaging theorem, Chem. Eng. Sci., 48, 2061–2082.

Ochoa-Tapia J.A., Stroeve P. and Whitaker S. 1994. Diffusive transport in two-phase media: Spatially periodic models and Maxwell's theory for isotropic and anisotropic systems, Chem. Eng. Sci., 49, 709–726.

Paine M.A., Carbonell R.G. and Whitaker S. 1983. Dispersion in pulsed systems I: Heterogeneous reaction and reversible adsorption in capillary tubes, Chem. Eng. Sci., 38, 1781–1793.

Quintard M. 1993. Diffusion in isotropic and anisotropic porous systems: Three-dimensional calculations, Transport Porous Med., 11, 187–199.

Quintard M. and Whitaker S. 1993a. Transport in ordered and disordered porous media: Volume averaged equations, closure problems, and comparison with experiment, Chem. Eng. Sci., 48, 2537–2564.

Quintard M. and Whitaker S. 1993b. One- and two-equation models for transient diffusion processes in two-phase systems. In, Advances in Heat Transfer, Harnett J.P., Irvine T.F. Jr and Cho Y.I. (Eds.), Vol. 23. Academic Press, New York, pp. 369–465.

Quintard M. and Whitaker S. 1994a. Transport in ordered and disordered porous media I: The cellular average and the use of weighting functions, Transport Porous Med., 14, 163–177.

Quintard M. and Whitaker S. 1994b. Transport in ordered and disordered porous media II: Generalized volume averaging, Transport Porous Med., 14, 179–206.

Quintard M. and Whitaker S. 1994c. Transport in ordered and disordered porous media III: Closure and comparison between theory and experiment, Transport Porous Med., 15, 31–49.

Quintard M. and Whitaker S. 1994d. Transport in ordered and disordered porous media IV: Computer generated porous media, Transport Porous Med., 15, 51–70.

Quintard M. and Whitaker S. 1994e. Transport in ordered and disordered porous media V: Geometrical results for two-dimensional systems, Transport Porous Med., 15, 183–196.

Quintard M. and Whitaker S. 2005. Coupled, non-linear mass transfer and heterogeneous reaction in porous media. In, Handbook of Porous Media, Vafai K. (Ed.), 2nd edition. Marcell Deckker, New York.

Ryan D., Carbonell R.G. and Whitaker S. 1981. A theory of diffusion and reaction in porous media, AIChE Symp. Ser., 202, 71, 46–62.

Schmidt L.D. 1998. The Engineering of Chemical Reactions. Oxford University Press, Oxford.

Slattery J.C. 1967. Flow of viscoelastic fluids through porous media, AIChE J., 13, 1066–1071.

Slattery J.C. 1990. Interfacial Transport Phenomena. Springer-Verlag, New York.

Slattery J.C. 1999. Advanced Transport Phenomena. Cambridge University Press, Cambridge.

Taylor R. and Krishna R. 1993. Multicomponent Mass Transfer. John Wiley & Sons, New York.

Whitaker S. 1967. Diffusion and dispersion in porous media, AIChE J., 13, 420–427.

Whitaker S. 1983. Diffusion and reaction in a micropore–macropore model of a porous medium, Lat. Am. J. Appl. Chem. Eng., 13, 143–183.

Whitaker S. 1986a. Transport processes with heterogeneous reaction. In, Concepts and Design of Chemical Reactors, Whitaker S. and Cassano A.E. (Eds.). Gordon and Breach Publishers, New York, pp. 1–94.

Whitaker S. 1986b. Transient diffusion, adsorption and reaction in porous catalysts: The reaction controlled, quasi-steady catalytic surface, Chem. Eng. Sci., 41, 3015–3022.

Whitaker S. 1987. Mass transport and reaction in catalyst pellets, Transport Porous Med., 2, 269–299.

Whitaker S. 1991. The role of the species momentum equation in the analysis of the Stefan diffusion tube, Ind. Eng. Chem. Res., 30, 978–983.

Whitaker S. 1992. The species mass jump condition at a singular surface, Chem. Eng. Sci., 47, 1677–1685.

Whitaker S. 1996. The Forchheimer equation: A theoretical development, Transport Porous Med., 25, 27–61.

Whitaker S. 1999. The Method of Volume Averaging. Kluwer Academic Press, Dordrecht, The Netherlands.

Wood B.D. and Whitaker S. 1998. Diffusion and reaction in biofilms, Chem. Eng. Sci., 53, 397–425.

Wood B.D. and Whitaker S. 2000. Multi-species diffusion and reaction in biofilms and cellular media, Chem. Eng. Sci., 55, 3397–3418.

Wood B.D., Quintard M. and Whitaker S. 2000. Jump conditions at non-uniform boundaries: The catalytic surface, Chem. Eng. Sci., 55, 5231–5245.

2

Solubility of Gases in Polymeric Membranes

M. Giacinti Baschetti, M.G. De Angelis, F. Doghieri and G.C. Sarti

2.1 Introduction

The solubility of gases and vapors in polymeric matrices is of significant importance in several applications, including membrane separations, development of barrier materials, and protective coatings. In membrane separations, the selectivity α_{ij} of component i versus component j is calculated as the corresponding permeability ratio, that is,

$$\alpha_{i,j} = \frac{P_i}{P_j} \tag{2.1}$$

where permeability P_k is defined as the ratio between mass flux J_k, through a membrane of thickness δ, and the partial pressure gradient across the membrane $(\Delta p_k/\delta)$:

$$P_i = \frac{J_i}{\Delta p_i/\delta} \tag{2.2}$$

In view of equation 2.2 and of Fick's law, the selectivity α_{ij} can be decomposed into its solubility and diffusivity contributions:

$$\alpha_{i,j} = \alpha_D \alpha_s = \frac{D_i}{D_j}\frac{S_{i,\mathrm{ave}}}{S_{J,\mathrm{ave}}} \tag{2.3}$$

The solubility factor $S_{i,\mathrm{ave}}/S_{j,\mathrm{ave}}$ is typically the leading term in determining selectivity α_{ij} for the case of rubbery polymers, and may be relevant also for the case of glassy

Chemical Engineering Edited by Miguel A. Galán and Eva Martin del Valle

membranes. The solubility isotherm for a given polymer penetrant system is clearly also a key parameter to determine the barrier properties of polymeric matrices. It is thus very important to rely either on direct experimental data or on thermodynamic relationships that allow solubility calculations based on the pure component properties.

In rubbery polymers, such relationships can be obtained in a rather straightforward way, since true thermodynamic equilibrium is reached locally immediately. In such cases, one simply has to choose the proper equilibrium thermodynamic constitutive equation to represent the penetrant chemical potential in the polymeric phase, selecting between the activity coefficient approach[1–5] or equation-of-state (EoS) method[6–12], using the most appropriate expression for the case under consideration. On the other hand, the case of glassy polymers is quite different insofar as the matrix is under non-equilibrium conditions and the usual thermodynamic results do not hold. For this case, a suitable non-equilibrium thermodynamic treatment must be used.

In the present work, we review reliable methods to evaluate solubility isotherms in polymeric phases, and examine the conditions needed for predictive calculations. The EoS approach will be used to calculate the chemical potential and the sorption isotherms of low molecular weight species in polymeric mixtures. Both the cases of equilibrium (e.g rubbery) and non-equilibrium (e.g. glassy) states will be treated, showing how the results of classical thermodynamics can be extended to the case of non-equilibrium states.

For the calculations, different EoS have been used: the lattice fluid (LF) model developed by Sanchez and Lacombe[8,13–15], as well as two recently developed equations of state – the statistical-associating-fluid theory (SAFT)[9,16–18] and the perturbed-hard-spheres-chain (PHSC) theory[10,11,19]. Such models have been considered due to their solid physical background and to their ability to represent the equilibrium properties of pure substances and fluid mixtures. As will be shown, they are also able to describe, if not to predict completely, the solubility isotherms of gases and vapors in polymeric phases, by using their original equilibrium version for rubbery mixtures, and their respective extensions to non-equilibrium phases (NELF, NE-SAFT, NE-PHSC) for glassy polymers.

2.2 Thermodynamic Models

In the present section, the general outline of the different EoS considered will be first recalled; then we report the basic results of the non-equilibrium analysis leading to their proper extension to glassy phases. It is not the aim of this section to offer an exhaustive presentation of the characteristics of the different models and of their detailed properties, but rather to point out the model relevant parameters and how they can be retrieved from pure component and mixture properties, independent of solubility isotherms. For further details, the reader is referred to the cited original papers.

2.2.1 Lattice Fluid Model

The Sanchez and Lacombe LF EoS[8,13–15] considers a compressible lattice for the representation of microstates of pure fluids and fluid mixtures. Such a lattice is made of cells, whose volume depends on mixture composition, which can be either empty or occupied by molecular segments of the components considered. The statistical analysis of the possible combinations of molecules in the lattice and the evaluation of the energetic interaction between adjacent occupied sites lead to expressions for the entropy, s, and internal energy, u, of the system from which an EoS can be built for the pure component or the mixture under consideration. In particular, the Helmholtz free energy density, a, is calculated as the sum of entropy and internal energy contribution, as usual:

$$a = u - Ts \tag{2.4}$$

where T represents the absolute temperature of the system. The pure component parameters in the LF model are the characteristic temperature T^*, the characteristic pressure p^*, and the characteristic density ρ^*, which somehow is related to the hypothetical density of the liquid phase at 0 K. For mixtures, the same characteristic parameters are calculated from those of the pure components by using well-established simple mixing rules. As it is often the case for mixtures, such rules contain adjustable binary parameters which are equal to the number of all possible pairs of different components in the mixture. Each binary parameter ψ_{ij} enters the definition of the quantity Δp^*_{ij}, appearing in the mixing rule for characteristic pressure p^*, and is related to the energetic interaction between dissimilar components in the mixtures:

$$\Delta p^*_{ij} = p^*_{ii} + p^*_{jj} - 2\psi_{ij}\sqrt{p^*_{ii}p^*_{jj}} \tag{2.5}$$

The default value $\psi_{ij} = 1$ can be used to recover the usual first-order approximation for the characteristic interaction energy, represented by a sort of geometric mean rule.

2.2.2 Tangent Hard Spheres Chain EoS

The tangent hard spheres chain models form a family of thermodynamic models that describe molecules as chains of spherical segments with an assigned mass and a temperature-dependent volume. Consecutive spheres are connected to each other to form a chain, and are able to interact energetically with segments of the same or of a different chain, according to a proper interaction potential. The two relevant models of this type considered in this chapter are known with the acronyms SAFT and PHSC. Among the different versions proposed in the literature for these models, use will be made of the SAFT model described in detail by Huang and Radosz (SAFT-HR)[9], and of the PHSC model proposed by Hino and Prausnitz[11] also known as PHSC square well (PHSC-SW).

The two models (SAFT and PHSC) differ substantially in the way they represent the different contributions to the expression of the residual Helmholtz free energy of a system. In the SAFT model, a^{res} is the sum of different contributions due to *hard spheres*, *dispersion*, *chain*, and *association*, respectively:

$$a^{\text{res}} = a^{\text{hd}} + a^{\text{disp}} + a^{\text{chain}} + a^{\text{assoc}} \tag{2.6}$$

The different terms represent segment–segment hard spheres interactions, a^{hd}, mean field contribution, a^{disp}, permanent bond energy between segments in the chain, a^{chain}, and free energy of specific hydrogen bond interactions between associating sites, a^{assoc}, if any.

The SAFT free energy expression for non-associating pure components contains only three parameters besides the molar mass (MW), i.e. the sphere radius, σ, the sphere mass, MW/m, and the characteristic energy of the interactions present in the dispersion contribution, u^0_{ii}.

Mixing rules are available to extend the models to multicomponent systems, with the use of adjustable binary parameters. In the absence of associating sites, we will consider only the binary interaction parameter k_{ij}, which enters the mixing rule for the characteristic interaction energy between pairs of unlike segments i and j:

$$u_{ij}^{0} = \left(1 - k_{ij}\right)\sqrt{u_{ii}{}^{0}u_{jj}{}^{0}} \tag{2.7}$$

The default value $k_{ij} = 0$ can be used to recover the typical first-order approximation for the characteristic interaction energy between unlike segments, given by the geometric mean rule.

In the PHSC EoS, the Helmholtz free energy is expressed as the sum of two different terms, one is a *reference* term accounting for chain connectivity and hard sphere interactions, and the other is a *perturbation* term, which represents the contributions of mean-field forces:

$$a^{\text{res}} = a^{\text{ref}} + a^{\text{pert}} \tag{2.8}$$

Following the notation of Song *et al.*[10], the pure component parameters involved in the expression for the Helmholtz free energy are, beyond the species molar mass, the sphere diameter σ, the mass per segment MW/m, the characteristic energy for the pair interaction potential ε, and, for the case of the PHSC-SW used in this chapter, the reduced well width λ, usually fixed to the value of 1.455 after Hino and Prausnitz[11]. The extension to mixtures can be obtained, as usual, through the introduction of appropriate mixing rules and of the binary parameters contained therein. The only adjustable binary parameters are the interaction parameters k_{ij} appearing in the expression of the characteristic energy for the interaction between pairs of unlike segments:

$$\varepsilon_{ij} = \left(1 - k_{ij}\right)\sqrt{\varepsilon_{ii}\varepsilon_{jj}} \tag{2.9}$$

Also in this case the default value of the binary parameters ($k_{ij} = 0$) represents the geometric mean approximation for the mixture interaction energy term.

2.2.3 Extension to the Non-Equilibrium Phases

The thermodynamic derivation of the NELF model has been reported in several publications[20–24]. From a more general point of view, such a model represents a special application of the non-equilibrium thermodynamics of glassy polymers (NET-GP) which indicates the relationships existing in general between the thermodynamic properties above and below the glass transition temperature; the NET-GP results hold for any thermodynamic model and are not limited to any particular EoS.

In the NET-GP analysis, the glassy polymer-penetrant phases are considered homogeneous, isotropic, and amorphous, and their state is characterized by the classical thermodynamic variables (i.e. composition, temperature, and pressure) with the addition of a single-order parameter, accounting for the departure from equilibrium. The specific volume of the polymer network, or, equivalently, the polymer density ρ_{pol}, is chosen as the proper order parameter. In other words, the hindered mobility of the glassy polymer chains freezes the material into a non-equilibrium state that can be labeled by the difference between actual polymer density ρ_{pol} and its equilibrium value at the given temperature, pressure, and mixture composition, ρ_{pol}^{EQ}.

The second key assumption in the NET-GP theory is related to the time evolution of the order parameter; in particular it is stated that the time rate of change of the polymer density depends only upon the state of the system:

$$\frac{d\rho_{pol}}{dt} = f(T, p, \Omega_{sol}, \rho_{pol}) \quad (2.10)$$

According to equation 2.10, the order parameter ρ_{pol} plays the role of an internal state variable[25] for the system, and basic thermodynamic relations of the NELF model and of NET-GP approach are derived by applying well-established thermodynamic results for systems endowed with internal state variables. In particular, it can be shown[21] that (i) the non-equilibrium Helmholtz free energy related to the glassy phase, a^{NE}, depends only on composition and polymer mass density and its value is not affected by the pressure of the system, and (ii) the non-equilibrium Helmholtz free energy, a^{NE}, coincides with the equilibrium value, a^{EQ}, calculated at the same temperature, composition, and polymer density. Once an expression for the equilibrium free energy a^{EQ} is found appropriate for the equilibrium polymer-penetrant mixture, the corresponding non-equilibrium equation is readily obtained through the simple relationship

$$a^{NE}(T, p, \Omega_{sol}, \rho_{pol}) = a^{EQ}(T, \Omega_{sol}, \rho_{pol}) \quad (2.11)$$

A corresponding relation can then be obtained for other thermodynamic properties and in particular for the non-equilibrium chemical potential μ_{sol}^{NE} in terms of the corresponding equilibrium function μ_{sol}^{EQ}, that is,

$$\mu_{sol}^{NE}(T, p, \Omega_{sol}, \rho_{pol}) = \mu_{sol}^{EQ}(T, \Omega_{sol}, \rho_{pol}) \quad (2.12)$$

It must be stated clearly that such results have been derived in a completely general manner and are thus independent from the particular EoS model used to describe the Helmholtz free energy or the penetrant chemical potential under equilibrium conditions. Non-equilibrium free energy functions can thus be obtained starting from different EoS such as LF, SAFT, PHSC, just to mention the relevant models considered in this chapter. The non-equilibrium information entering equations 2.11 and 2.12 is represented by the actual value of polymer density in the glassy phase, which must be known from a separate source of information, experimental data, or correlation, and cannot be calculated from the equilibrium EoS.

2.2.4 Determination of the Model Parameters

The pure component parameters of the models can be retrieved by using the volumetric data above the glass transition temperature, for the polymers, and using volumetric data and/or vapor pressure data for the penetrants. The binary interaction parameters can be obtained from gas–polymer equilibrium data in the rubbery phase, when available. In the absence of any direct experimental information, the first-order approximation can be used or, alternatively, they can be treated as adjustable parameters.

2.2.5 Solubility and Pseudo-Solubility Calculation

In the case of true thermodynamic phase equilibrium, in which the absolute minimum is attained for the system Gibbs free energy at given T and p, the solubility calculation is performed following the classical thermodynamic result which imposes the equality between the equilibrium chemical potential of the penetrant in the polymeric mixture ($\mu_{sol}^{EQ(s)}$) and in the external phase ($\mu_{sol}^{EQ(g)}$). The equilibrium solute content, Ω_{sol}^{EQ}, and polymer density, ρ_{pol}^{EQ}, can be calculated from the following conditions:

$$\begin{cases} \mu_{sol}^{EQ(s)}\left(T, \Omega_{sol}^{EQ}, \rho_{pol}^{EQ}\right) = \mu_{sol}^{EQ(g)}(T, p) \\ \left(\frac{\partial G^{(s)}}{\partial \rho_{pol}}\right)_{T,p,\Omega_{pol}} = 0 \end{cases} \tag{2.13}$$

The symbol $G^{(s)}$ represents the Gibbs free energy of the polymeric mixture per unit polymer mass.

For the solubility in glassy phases, the situation is substantially different since the polymer density does not match its equilibrium value ρ_{pol}^{EQ}, but it finally reaches an asymptotic value determined by the kinetic constraints acting on the glassy molecules, and is substantially dependent on the past history of the polymer sample. Thus the penetrant concentration in the polymeric phase reflects the pseudo-equilibrium state reached by the system. In view of the NET-GP results, such pseudo-equilibrium condition corresponds to the minimum Gibbs free energy for the system, under the constraint of a fixed value (the pseudo-equilibrium value) of the polymer density in the condensed phase:

$$\mu_{sol}^{NE}(s)\left(T, p, \Omega_{sol}, \rho_{pol}\right) = \mu_{sol}^{EQ(g)}(T, p) \tag{2.14}$$

In equation 2.14, the non-equilibrium solute chemical potential is calculated through the use of equation 2.12 and of an appropriate EoS for the polymer-penetrant system under consideration. The pseudo-equilibrium penetrant content in the polymer, Ω_{sol}, can be easily calculated whenever the value of the pseudo-equilibrium polymer density ρ_{pol} is known. Such a quantity represents, obviously, a crucial input for the non-equilibrium approach, since it labels the departure from equilibrium; it must be given as a separate independent information, and cannot be calculated simply from temperature and pressure since it depends also on the thermomechanical history of the sample.

Unfortunately, the polymer density value during sorption is not often readily available at all pressures and this limits the application of the NET-GP approach as a completely predictive tool. In several cases of practical interest, however, the pseudo-equilibrium density of the polymer can be easily known with negligible errors. One of these cases is encountered, for instance, in calculating the pseudo-solubility at low gas pressures, when the polymeric mixture is infinitely dilute and the volume of the polymer is not significantly affected by the presence of the solute. The density of the unpenetrated glass, ρ_{pol}^{0}, thus provides a very good estimate of the actual polymer density, ρ_{pol}, and the NET-GP approach can be applied in a straightforward and predictive way. Similar consideration also holds true when the solubility of non-swelling gases is to be determined at moderate pressures; under those conditions the pseudo-equilibrium problem can again be reduced to the following low-pressure approximation[24]:

$$\mu_{sol}^{NE(s)}\left(T, p, \Omega_{sol}, \rho_{pol}^{0}\right) = \mu_{sol}^{EQ(g)}(T, p) \tag{2.15}$$

When swelling agents or higher gas pressures are considered, practical application of the NET-GP approach needs some further observation. In particular, it can be noticed[26–28] that, generally, the polymer mass density during sorption varies linearly with gas pressure, in a relatively wide pressure range, at least for temperatures sufficiently below the glass transition, so that the following relationship is followed by polymer density[23]:

$$\rho_{\text{pol}}(p) = \rho_{\text{pol}}^{0}(1 - k_{\text{sw}}p) \tag{2.16}$$

where the swelling coefficient k_{sw} represents the effect of gas pressure on pseudo-equilibrium polymer density and is itself a non-equilibrium parameter, depending on thermomechanical and sorption history of the specific polymer sample. In view of equation 2.16, in the case of high-pressure gas sorption, the pseudo-equilibrium condition, equation 2.14, becomes

$$\mu_{\text{sol}}^{\text{NE(s)}}\left(T, p, \Omega_{\text{sol}}, \rho_{\text{pol}}^{0}(1 - k_{\text{sw}}p)\right) = \mu_{\text{sol}}^{\text{EQ(g)}}(T, p) \tag{2.17}$$

Through equation 2.17 the pseudo-equilibrium solubility can also be evaluated for the case of swelling penetrants, even in the cases in which polymer dilation is not known from direct experimental evidence. Indeed, the swelling coefficient, k_{sw}, can be treated as the only adjustable parameter in equation 2.17, and its value can be obtained, for example, from virtually a single experimental solubility datum at high pressure for the system under consideration[23].

In the following sections the different thermodynamic models presented in the preceding text will be applied in the calculation of the penetrant solubility for both the cases of rubbery and glassy polymeric systems. Binary as well as ternary systems will be considered to show the ability of the models to represent observed isotherms in rubbers as well as in glasses, based on their equilibrium versions and non-equilibrium extensions, respectively.

2.3 Comparison with Solubility Data

In this section, the predictive ability of the procedure presented above is examined, using the PHSC, SAFT, and LF EoS as equilibrium models. To this aim, gas–polymer solubility data taken from various literature sources are compared to the model predictions, performed in the pure predictive or correlative mode. For each gas–polymer mixture, we also report hereafter the pure component parameters for the polymer and the penetrant which were determined and used in the calculations. In the case of glassy mixtures, the value of the dry polymer density which has been used for the simulation is also specified. In order to test the behavior of the model in a variety of conditions, different mixtures and different types of miscibility data are examined, taking into account also ternary solutions formed by a single gas in a polymer blend or by mixed gases in a single polymer. The data relative to the glassy systems are classified on the basis of the swelling behavior of the penetrant, treating separately the non-swelling solvents, such as N_2, O_2, CH_4, and the swelling penetrants for which the calculation procedure is substantially different in the non-equilibrium case.

2.3.1 Infinite Dilution Solubility Coefficient Across T_g

In this section, we test the behavior of the NET-GP procedure for a gas–polymer mixture whose solubility has been characterized in both equilibrium and non-equilibrium conditions, that is, above and below the glass transition temperature T_g of the polymer. In Figure 2.1 we plot the value of the infinite dilution solubility coefficient of CO_2 in poly(bisphenol-A) carbonate (PC) as a function of the inverse absolute temperature, as measured by Wang and Kamiya[29]. The infinite dilution solubility coefficient, S_0, expressed in $cm^3(STP)/(cm^3 \cdot atm)$, is the slope of the solubility isotherm, in the limit of very low pressures:

$$S_0 = \lim_{p \to 0} \frac{C}{p} \tag{2.18}$$

where C is the penetrant concentration.

It is known from experience that CO_2 generally induces considerable swelling in polymeric matrices, but in this case, since we are exploring only the low-pressure range in which the polymer dilation is negligible, the solvent can be treated as a non-swelling one and the calculations were thus performed using a constant value of the polymer density, equal to ρ^0_{pol}. The experimental sorption data were taken both below and above the glassy transition temperature: when plotted in a semi-log scale, the solubility data lie on two lines characterized by different slopes, the glassy phase being characterized by a stronger temperature dependence of the gas solubility. In Figure 2.1, both the predictions of the SAFT EoS and of the corresponding NE-SAFT model are presented and compared

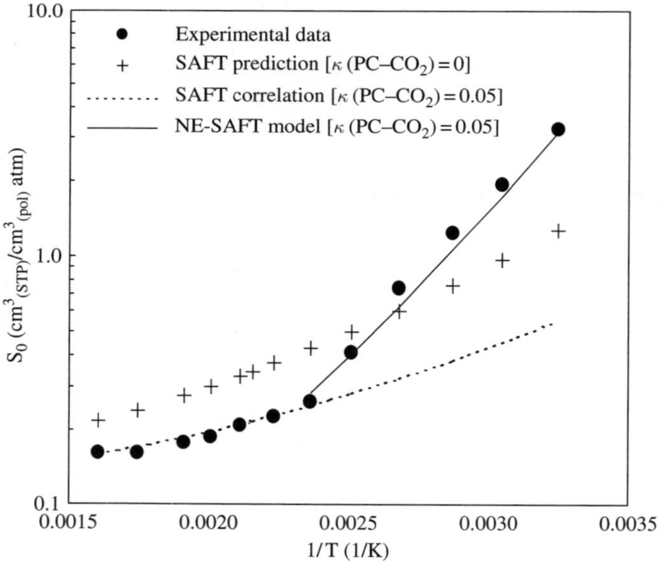

Figure 2.1 *Solubility coefficient of CO_2 in PC at infinite dilution reported as a function of inverse temperature. The SAFT and NE-SAFT calculations are also reported. The characteristic parameters used for the regression are listed in Table 2.1*

to the experimental data. As is evident, the equilibrium model cannot fit all the data, above and below T_g, with a unique value of the adjustable binary parameter k_{ij}.

Instead, the use of the NET-GP procedure allows the extension of the equilibrium SAFT model valid above T_g to temperatures below the glassy transition. The non-equilibrium model thus obtained satisfactorily represents the solubility coefficient in the glassy mixture, using the same value of the binary parameter ($k_{ij} = 0.05$) obtained from the equilibrium data, and the experimental value of the unpenetrated glassy polymer density. The value of the glassy polymer density has been calculated from the experimental value of 1.55 kg/L taken at $T_g(\approx 150°C)$ and adopting a cubic thermal expansion coefficient equal to 2.8×10^{-4}/K. The values of the pure component parameters for the SAFT EoS used are listed in Table 2.1. For the sake of brevity, only the results obtained with the SAFT and NE-SAFT models are shown in Figure 2.1, but similar results can be obtained of using LF and NELF, or PHSC-SW and NE-PHSC-SW, respectively.

This example clearly shows the potential of the NET-GP procedure in the description of the non-equilibrium states, when coupled to an EoS that adequately represents the equilibrium behavior.

2.3.2 Gas and Vapor Solubility in Rubbery Polymers

The ability of the EoS to predict the behavior of rubbery polymer solutions has been proved by various authors[30–34]. In this section we present two examples of the solubility of swelling penetrants in rubbery polymers, namely propane sorption in poly[dimethylsiloxane] (PDMS) at 35° C, and then toluene sorption in rubbery polymer blends formed by low-density poly[ethylene] (LDPE) and poly[vinylchloride] (PVC) at 30° C.

In Figure 2.2 the solubility of C_3H_8 in PDMS at 35° C is shown, expressed in grams per gram of polymer versus the gaseous phase pressure in megapascal[34]. The data are compared with the results obtained from three different EoS: LF, SAFT-PC, and PHSC-SW. The characteristic parameters used for the various EoS are listed in Table 2.2; for the case of the SAFT-PC and PHSC-SW the parameters of PDMS were evaluated in this work by best-fitting the EoS calculations to the pressure-volume-temperature (PVT) data of PDMS taken by Zoller and Walsh[35], relative to a high molecular weight polymer ($M_w = 1.5 \times 10^6$ g/mole). The experimental isotherm is slightly concave to the concentration axis, as is common for the case of swelling penetrants in rubbery polymers. While, for the case of the LF EoS, a slight adjustment of the binary parameter is needed in order to fit correctly the solubility isotherm ($k_{ij} = 1 - \psi_{ij} = 0.032$), the pure predictions obtained with SAFT-PC EoS and PHSC-SW EoS (i.e. using the default value $k_{ij} = 0$)

Table 2.1 *Characteristic parameters for the SAFT EoS for polycarbonate and CO_2*

	PC	CO_2
σ(Å)	3.043	3.171
MW/m (g/mol)	25.0	31.05
u_0/k (K)	371.0	216.08
Source	PVT data[35]	Ref.[9]

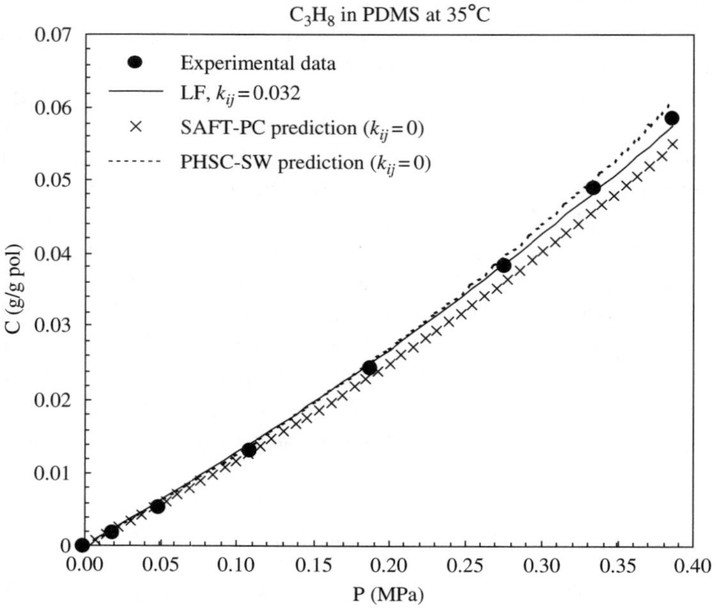

Figure 2.2 *Solubility isotherm of C_3H_8 in PDMS at 35° C (>T_g). The equilibrium models SAFT-PC (× symbols) and PHSC-SW (dashed line) provide good predictions of the solubility in a pure predictive way ($k_{ij} = 0$). A slight adjustment of the binary parameter ($k_{ij} = 1 - \psi_{ij} = 0.032$) is needed by the LF EoS (solid line). The characteristic parameters used for the calculations are listed in Table 2.2*

Table 2.2 *Characteristic pure component parameters for the various EoS for PDMS and C_3H_8*

	PDMS	C_3H_8
LF EoS		
ρ^*(kg/L)	1.0805	0.690
T^*(K)	498.0	375.0
p^*(MPa)	292.5	320.0
Source	Ref.[34]	Ref.[34]
SAFT-PC EoS		
σ(Å)	3.866	3.618
MW/m (g/mol)	39.86	22.03
u_0/k (K)	226.0	208.1
Source	PVT data[35]	Ref.[18]
PHSC-SW EoS		
σ(Å)	4.36	3.505
MW/m (g/mol)	54.2	20.63
u_0/k (K)	280.0	199.6
λ	1.3	1.455
Source	PVT data[35]	Ref.[11]

provide an extremely satisfactory representation of the sorption isotherm in the whole pressure range inspected.

The second example considers a blend formed by LDPE, with 30% crystallinity, and PVC. The polymer matrices examined are pure LDPE, the blends LDPE (80%)–PVC (20%) and LDPE (50%)–PVC (50%), and pure PVC, with toluene as the penetrant. Experimental data by Markevich *et al.*[36] report solubility of toluene in the above blends, at the temperature of 30° C, while toluene solubility in pure PVC was taken from Berens[37]. The glassy transition temperature is equal to −25° C for LDPE and to +75° C for pure PVC. Therefore, pure PVC is a glass at 30° C; however, due to the large swelling and plasticization of the polymer induced by toluene sorption, it can be seen that the sorption of toluene lowers the glass transition of PVC to temperatures below 30° C, already at relatively low toluene activities. That is also confirmed by the sorption isotherm which is concave to the concentration axis as is typical of rubbery polymers. The glass transition temperatures for the blends are estimated to be −10° C for the 80% LDPE blend and +17° C for the 50% LDPE blend, all below the temperature of the sorption experiment. The crystalline fraction of LDPE is assumed, as is usual, not to contribute to the sorption process, therefore we consider only the amorphous fraction of LDPE in the sorption calculations based on EoS. For the sake of simplicity, we present here only the results obtained with the LF equilibrium model.

The characteristic parameters for the blend–vapor mixture are calculated with the usual mixing rules valid for the LF EoS, by considering the blend–vapor mixture as a ternary mixture formed by the vapor and the two homopolymers. The characteristic parameters for the pure homopolymers are shown in Table 2.3. The values of the binary adjustable parameters in the LF EoS, ψ_{ij}, are reported in Table 2.4. In order to fit the solubility isotherm of toluene in pure LDPE, for this couple ψ_{ij} is adjusted to 0.961, while the default value of $\psi_{ij} = 1$ gives a good estimate of the solubility of toluene in pure PVC; the binary parameter associated with the LDPE–PVC pair has its default value. By using the above values of the three binary parameters, the model becomes predictive for the solubility in the polymer blend. Comparison between experimental data and predictive calculations is shown in Figure 2.3: as one can see, we obtain a good representation of the sorption isotherms, especially in the low-pressure range, and the dependence of solubility on the composition of the blend is definitely captured by the model.

2.3.3 Gas and Vapor Solubility in Glassy Polymers

Solubility isotherms of non-swelling penetrants in glassy polymers. The sorption isotherms of CH_4 in poly[phenilene oxide] (PPO) and poly[sulfone] (PSf) are now considered, as well as the sorption isotherm of N_2 in poly[ethylmethacrylate] (PEMA) at

Table 2.3 *Characteristic pure component parameters for the LF EoS for LDPE, PVC, and toluene*

Substance	ρ^*(kg/L)	T^*(K)	p^*(MPa)
LDPE[38]	0.883	693	400
PVC[38]	1.4577	736	415
Toluene[39]	0.966	543	402

Table 2.4 *Binary interaction parameters for the LDPE–PVC–toluene systems*

System	ψ_{ij}
LDPE–toluene	0.961
PVC–toluene	1.000
LDPE–PVC	1.000

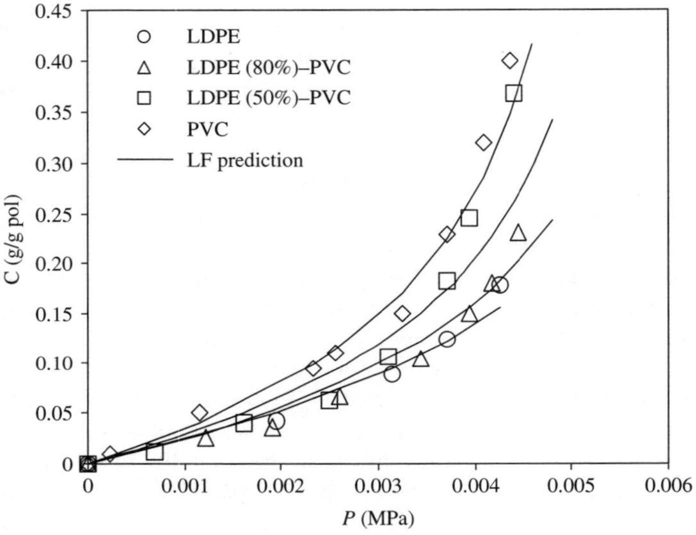

Figure 2.3 *Experimental sorption data for toluene in PVC, LDPE, and their blends at 30° C[36,37]. The solid lines are LF EoS correlations and predictions*

the temperature of 35° C [40–42], as typical examples of non-swelling penetrants. In the practical absence of polymer dilation, only the pure component parameters and the values of the unpenetrated polymer density are required for a complete description of the solubility isotherms through the NET-GP model. The pure component parameters for the different models used are listed in Table 2.5; they were taken from the literature or evaluated by best-fitting the EoS to the equilibrium volumetric data. The dry polymer density for PEMA was measured at the temperature of 25° C($\rho^0_{\text{pol}} = 1.124$ kg/L); the value at 35° C was thus extrapolated from volumetric data, obtaining $\rho^0_{\text{pol}} = 1.120$ kg/L; for PPO at 35° C, the experimental dry polymer density was taken from PVT data as 1.063 kg/L, while for PSf we obtained a value of 1.230 kg/L. The values of the binary parameters k_{ij} obtained for the various gas–polymer couples considered are listed in Table 2.6 and were adjusted on the low-pressure sorption data, according to the constant density assumption already recalled. The experimental data and the model calculations are shown in Figure 2.4.

Table 2.5 Pure component parameters for the different thermodynamic models used in the various gas–polymer systems

Substance	EoS	σ (A)	MW/m	u_0/κ (K)	References
PPO	SAFT-HR	3.043	24.01	320.0	This work
	PHSC-SW	3.600	37.73	293.0	This work
PSf	SAFT-HR	3.043	25.67	410	This work
	PHSC-SW	3.484	37.45	352	This work
PEMA	SAFT-HR	3.049	22.98	320.0	This work
	PHSC-SW	3.450	32.78	290.5	This work
CH_4	SAFT-HR	3.700	16.01	190.29	[9]
	PHSC-SW	3.672	16.01	164.9	[11]
N_2	SAFT-HR	3.575	28.01	123.53	[9]
	PHSC-SW	3.520	27.62	108	[1]
		T*	P*	ρ^*	
PPO		739	479	1.177	[38]
PSf		830	600	1.31	[24]
PEMA	LF	602	567.5	1.221	This work
N_2		145	160	0.943	[21]
CH_4		215	250	0.500	[21]

Table 2.6 Binary parameters for the different thermodynamic models used in the various gas–polymer systems

	Binary parameters k_{ij}		
System	SAFT-HR	PHSC-SW	SL$(1-\psi_{ij})$
PEMA–N_2	0.020	−0.018	0.030
PSf–CH_4	−0.015	−0.085	−0.030
PPO–CH_4	0.0	−0.085	−0.060

The solubility isotherms obtained from the non-equilibrium models for all these systems are always satisfactory and all the different models used give very similar results. One may notice that the worst case is represented by the PSf–CH_4 systems in which the NELF model slightly underestimates the experimental sorption data, especially at the higher pressure range, with an error, however, not exceeding 15%.

Solubility isotherms of swelling penetrants in glassy polymers. Application of the NET-GP results to the solubility of swelling penetrants in glassy polymers is analyzed by considering sorption of C_2H_4 in PC, of CO_2 in poly[methylmethacrylate] (PMMA) and in a perfluorinated matrix. The solubility of C_2H_4 in PC systems at 35° C has been experimentally studied by Jordan and Koros[43], who also measured polymer swelling during sorption. The experimental data of sorption and dilation, together with the predictions of the model, are shown in Figure 2.5. The dilation isotherm is inserted in the plot on the right-hand side of the figure, expressed in terms of polymer density versus

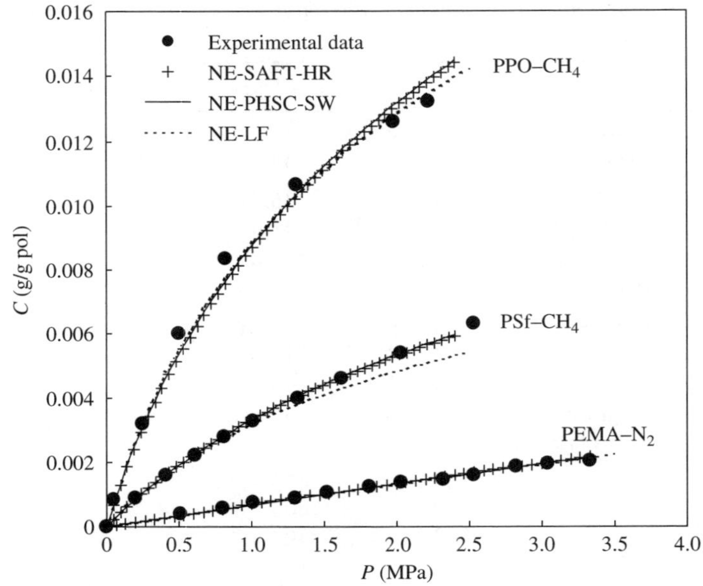

Figure 2.4 *Sorption isotherm for the systems CH_4–PPO, CH_4–PSf, and N_2–PEMA; the calculations of different non-equilibrium models (NELF, NE-SAFT, and NE-PHSC-SW) are also shown*

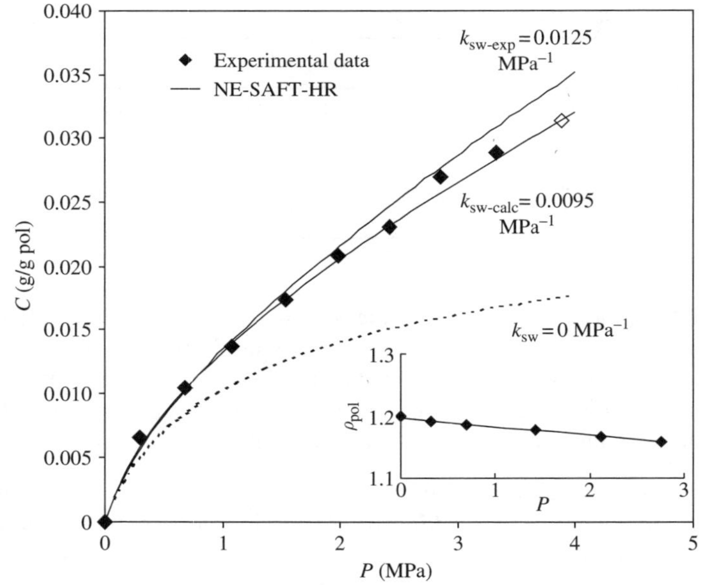

Figure 2.5 *Sorption isotherm for the system C_2H_4–PC; the polymer dilation is reported in the right frame in terms of polymer density versus pressure. Experimental data are from Ref.*[43]. *The results of the NE-SAFT-HR are also reported for two values of the swelling coefficient* k_{sw}*: experimental and calculated*

external penetrant pressure. In the present case, due to the penetrant characteristics, the swelling is not negligible and for a correct representation of the experimental data through the non-equilibrium models one needs to use the value of the polymer density at each pressure during sorption, which is given by polymer dilation data.

When the dilation data are readily available for the system under consideration, as in Figure 2.5, application of the NET-GP procedure is straightforward and the model can be used in an entirely predictive way, provided the value of the binary parameter is known. In Figure 2.5 the predictive calculations based on NE-SAFT are presented, using the first-order approximation of the binary parameter, $k_{ij} = 0.0$. The value of the swelling coefficient can be calculated from the original experimental dilation data ($k_{sw\text{-}exp} = 0.0125/\text{MPa}$) and such parameter is then used for the calculation of the solubility isotherm.

In Figure 2.5, we have also reported, for comparison, the results of the more general procedure, which must be followed in the case, actually very frequent, in which the experimental dilation data of the polymer are not available. According to the procedure illustrated in the previous section, the swelling coefficient is evaluated from one experimental solubility point. In the present case, the open symbol in the sorption curve in Figure 2.5 has been used to calculate the polymer volume and thus the swelling coefficient, obtaining $k_{sw\text{-}calc} = 0.0095/\text{MPa}$. The solubility isotherms obtained from NE-SAFT using the experimental and calculated values of the swelling coefficient are both shown in Figure 2.5. Clearly, the two different approaches to the solubility calculation give very good results. The dashed line in Figure 2.5 represents the prediction obtained by the same model neglecting swelling and assuming the polymer density is constant during sorption and equal to the density of the pure unpenetrated polymer. In that case, the low-pressure behavior is in good agreement with experimental data, while neglecting volume dilation leads to a huge underestimation of the solubility at higher penetrant pressures. The pure component characteristic parameters for the SAFT-HR EoS have been either calculated from experimental volumetric data or taken from the literature and are reported in Table 2.7.

As a further example of the applicability of the NET-GP results, we consider the solubility of CO_2 in PMMA at 33° C and in Teflon® AF1600 (poly[2,2-bistrifluoromethyl-4,5-difluoro-1,3-dioxole(87%)-*co*-tetrafluoroethylene]) at 35° C[44,45]. For both systems, the polymer dilation data are available from independent experimental measurements, enabling us to calculate the value $k_{sw\text{-}exp}$ of the experimental swelling coefficient. The swelling coefficient $k_{sw\text{-}calc}$ has also been estimated from the solubility data by using a single high-pressure solubility datum indicated in the figures as an open symbol.

Table 2.7 *Model parameters for the systems considered in the case of swelling penetrants*

Substance	ρ^0_{pol} (kg/L)	EoS	σ(Å)	MW/m (g/mol)	u_0/k (K)	References
PMMA	1.181	PHSC-SW	3.583	37.6	366.9	This work
CO_2		PHSC-SW	2.484	16.26	145.11	[6]
C_2H_4		SAFT-HR	3.49	19.16	216.06	[2]
			T*	P*	ρ*	
AF1600	1.840	SL	575	280	2.160	[9]
CO_2		SL	300	630	1.515	[11]

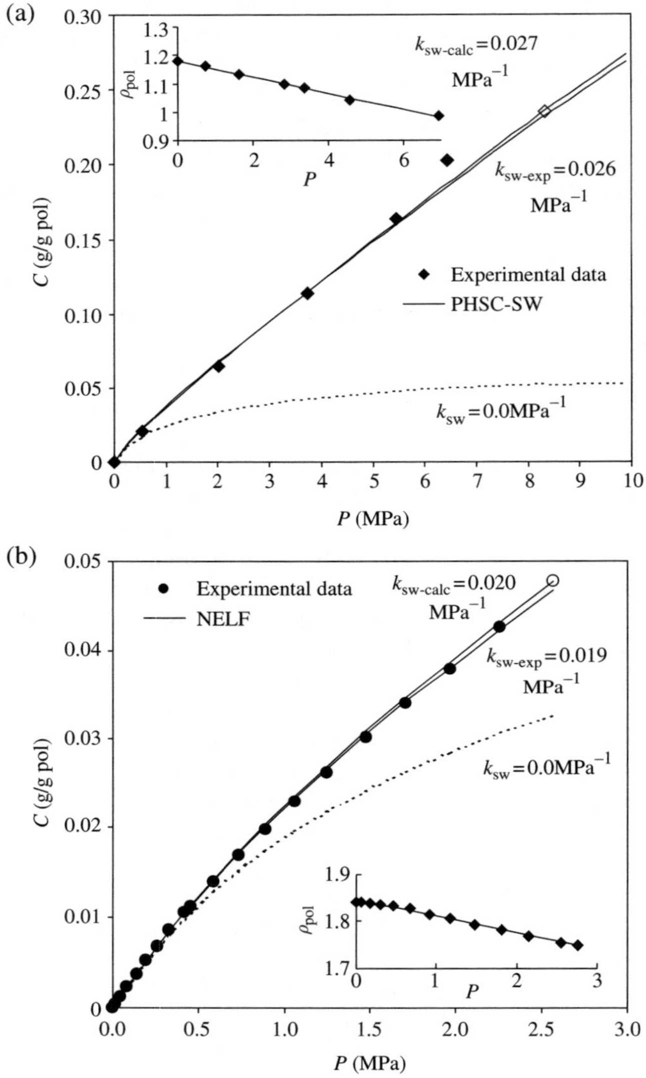

Figure 2.6 *CO_2 solubility in PMMA (a) at 33° C and in Teflon AF1600 (b) at 35° C, and the corresponding polymer dilation isotherms are reported as a function of pressure. Experimental data are from Refs.*[44,45]. *The results of the NE-PHSC-SW (case a) and NELF (case b) models are also reported for different values of the swelling coefficient* k_{sw}

The three models presented have been used for the calculations; however, for the sake of brevity the results explicitly presented in Figure 2.6a and b refer to NE-PHSC-SW for the PMMA–CO_2 system and NELF for the AF1600–CO_2 mixtures. As in the previous case, the model results obtained with a constant polymer density value have also been included in the two figures and are represented by a dashed line: once again, one can appreciate the importance of a correct estimation of volume dilation to account for the sorptive capacity of glassy polymers, especially at high pressure. The value of the pure unpenetrated polymer density as well as the pure component characteristic parameters used in the calculation are reported in Table 2.7, with the exception of polycarbonate whose data were already shown in Table 2.1. From Figure 2.6a and b, we conclude that the models used provide a very good fitting of the experimental data, regardless of the different procedure used for the estimation of the swelling coefficient and with the use of, at most, two data points. In particular, in this case we can notice the agreement between the values of the swelling coefficient obtained directly from the dilation data or estimated from the solubility datum: $k_{sw\text{-}exp}$ and $k_{sw\text{-}calc}$ are, respectively, equal to 0.026 and 0.027/MPa in the case of the CO_2–PMMA system, and to 0.019 and 0.020/MPa in the case of the CO_2–AF1600 mixtures. For CO_2–PMMA mixtures, the binary interaction parameter k_{ij} was adjusted using a low-pressure solubility datum and obtaining a value of 0.075, while in the case of the CO_2–AF1600 mixture k_{ij} was set to its default value $k_{ij} = 0$.

Gas solubility in glassy polymer blends and mixed gas solubility in glassy polymers. It is worthwhile to consider now some examples of more complex systems, as polymer blends and mixed gases, which are frequently encountered in gas separation with polymeric membranes or in barrier polymer applications. We first consider the solubility of a single gas in glassy polymer blends and then we turn to a case of mixed gas sorption, observing that reliable sorption data for such complex situations, in particular mixed gas sorption data in polymers, are rather rare in the open literature.

In Figure 2.7, we report the case of two glassy polymer blends: the solubility of CH_4 in PS–TMPC (tetramethyl polycarbonate) blends of different compositions (0–20–40–60–100% of PS) is shown in Figure 2.7a at 35° C[46] while the solubility isotherms of CO_2 in five different blends of (bisphenol-chloral) polycarbonate (BCPC) and PMMA (0–25–50–75% of PMMA)[47] at 35° C are shown in Figure 2.7b. The NELF estimation of the solubility is also reported, based only on the pure component characteristic parameters

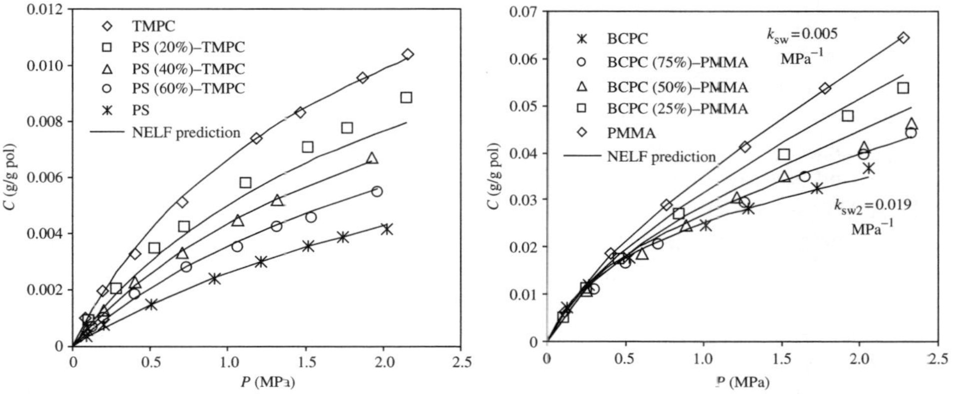

Figure 2.7 *(a) CH_4 solubility in PS–TMPC blends and (b) CO_2 solubility in PMMA–BCPC blends at 35° C, reported as a function of pressure. Experimental data are from Refs.*[46,47]*, respectively. The NELF model results are also reported as predicted on the basis of the binary systems data alone*

Table 2.8 *Pure polymer parameters and binary parameters for the different ternary solutions considered*

Substance	ρ^0_{pol}(kg/L)	EoS	T^*(K)	P^* (MPa)	ρ^* (kg/L)	References
TMPC	1.082		761.6	446.4	1.174	[23]
PMMA	1.188		695	560	1.27	[21]
PS	1.047	LF	750	360	1.099	[21]
BCPC	1.392		794	531.1	1.48	This work
C_2H_4			295	345	0.68	[21]

Binary parameters $k_{ij}(= 1 - \psi_{ij})$			
System	k_{12}	k_{23}	k_{13}
TMPC(1)–PS(2)–CH_4(3)	0.0	−0.059	−0.010
BCPC(1)–PMMA(2)–CO_2(3)	0.0	−0.028	−0.016
PMMA(1)–C_2H_4(2)–CO_2(3)	0.0	0.024	0.000

and on the pure polymer sorption isotherm, which allows for the estimation of the binary interaction parameters for the systems considered. In the case of CO_2 the swelling coefficient was also calculated. The parameters used are reported in Table 2.8, while the parameters for the blends have been calculated from those of the pure homopolymers through the appropriate mixing rules. The swelling coefficient of the blend is calculated as the volume average of the pure polymer swelling coefficients, based on the volume fractions in the unpenetrated blends[48]. By using the swelling and binary parameters obtained from the sorption isotherms in pure polymers, calculation of solubility in the blends is entirely predictive.

The two examples above show that the model allows us to predict accurately the solubility of the blends when the pure polymer sorption isotherm for the solvent under investigation is known and the binary parameter associated with the polymer–polymer pair is set to its default value, as in the present cases. The results are more than satisfactory, with average errors that seldom exceed the value of 10% in the case of PS–TMPC blends and are generally even lower for the other blends considered.

The reliability of the method can also be tested in the case of mixed gas sorption in a single glassy polymer. Also in this case a ternary mixture is present, formed by one polymer and two low molecular weight penetrants. An example is offered by the system PMMA–CO_2–C_2H_4 at 35° C, studied experimentally by Sanders *et al.* [49], whose data are here compared with the predictions of the NELF model. Here, the binary parameters for both polymer-penetrant pairs were set to the default values ($k_{12} = k_{13} = 0.0$), and swelling was neglected in view of the relatively low-pressure range inspected. Vapor–liquid equilibrium data for the penetrant mixture were used for the evaluation of the C_2H_4–CO_2 binary parameter, obtaining $k_{23} = 0.024$. Therefore, again in this case the extension of the NELF to the ternary system does not require any additional adjustable parameter and the results of the model are obtained in a completely predictive mode. In Figure 2.8a and b, the CO_2 and C_2H_4 concentration in the polymer are reported as a function of CO_2 partial pressure in the external gaseous phase, when the ethylene partial pressure is held constant at a fixed value of 2.06 ± 0.08 atm.

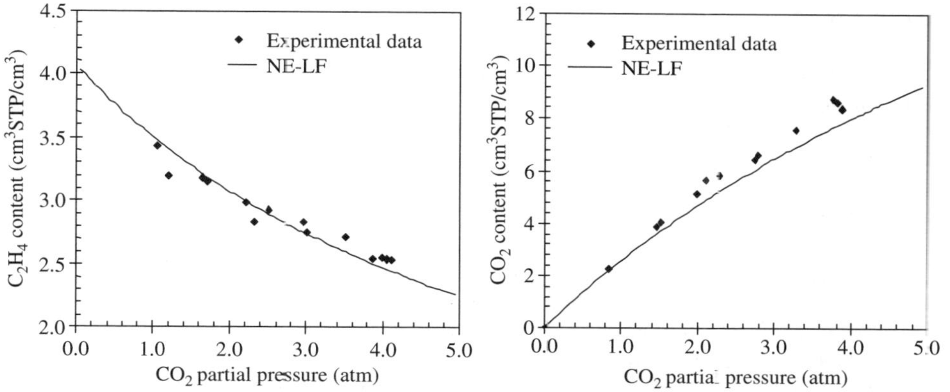

Figure 2.8 *(a) C_2H_4 and (b) CO_2 concentration in a ternary PMMA–C_2H_4–CO_2 mixture as a function of CO_2 partial pressure. The ethylene fugacity in the vapor phase was held constant during the measurements. The solid line represents the NELF prediction based on the binary mixtures data. Experimental data are from Ref.*[49]

As in the previous cases, the non-equilibrium model gives quite good results in predicting the experimental data; the ethylene content is in fact very well calculated and the slight underestimation of the CO_2 content at the higher penetrant partial pressure can be attributed to the polymer swelling that probably occurs in such a condition and which has been neglected in the calculation.

2.4 Conclusions

The solubility of low molecular weight penetrants in polymeric matrices can be satisfactorily calculated by using several EoS models such as LF, SAFT-HR, and PHSC-SW.

For rubbery phases, the models are used in their original equilibrium formulations, which requires knowledge of the pure component parameters and the binary interaction parameters entering the mixing rules associated with the models. The former can be retrieved from pure component volumetric data at different temperatures and pressures and, when applicable, from vapor pressure data; for each pair of substances the binary parameter is either retrieved from volumetric data or adjusted to the solubility data. In several cases, the default value offers a reasonable estimation of the solubility isotherms.

In the case of glassy polymers, on the other hand, the equilibrium approach is not applicable and a suitable non-equilibrium thermodynamic approach has been developed, NET-GP, which indicates how the selected free energy model can be extended to non-equilibrium glassy phases offering, in particular, explicit expressions for the penetrant chemical potential in glassy polymers. The departure from equilibrium is lumped into the glassy polymer density which is the only additional information that is needed, beyond the parameters used for the equilibrium rubbery phases. The model is fruitfully applied to the cases of non-swelling and swelling penetrants, as well as for calculating the solubility in polymer blends and of mixed gases.

Acknowledgements

This work has been partially supported by the University of Bologna (*progetto pluriennale 2004–06 and '60% funds'*).

References

[1] Flory P.J. 1941. J. Chem. Phys., 9, 660.
[2] Huggins M.L. 1941. J. Chem. Phys., 9, 440.
[3] Abrams M.M. and Prausnitz J.M. 1975. A.I.Ch.E J., 21, 116.
[4] Oishi T. and Prausnitz J.M. 1978. Ind. Eng. Chem. Res., 17, 333.
[5] Elbro H.S., Fredenslund A. and Rasmussen P. 1990. Macromolecules, 23, 4707.
[6] Patterson D. 1969. Macromolecules, 2, 672.
[7] Flory P.J. 1970. Disc. Faraday Soc., 49, 7.
[8] Sanchez I.C. and Lacombe R.H. 1978. Macromolecules, 11, 1145.
[9] Huang S.H. and Radosz M. 1990. Ind. Eng. Chem. Res., 29, 2284–2294.
[10] Song Y., Hino T., Lambert S.M. and Prausnitz J.M. 1996. Fluid Phase Equilib., 117, 69–76.
[11] Hino T. and Prausnitz J.M. 1997. Fluid Phase Equilib., 138, 105–130.
[12] Kang J.W., Lee J.H., Yoo K.P. and Lee C.S. 2002. Fluid Phase Equilib., 194, 77–86.
[13] Sanchez I.C. and Lacombe R.H. 1976. J. Phys. Chem., 80, 2352–2362.
[14] Lacombe R.H. and Sanchez I.C. 1976. J. Phys. Chem., 80, 2568–2580.
[15] Sanchez I.C. and Lacombe R.H. 1977. J. Polym. Sci.: Polym. Lett. Ed., 15, 71–75.
[16] Chapman W.G., Gubbins K.E., Jackson G. and Radosz M. 1989. Fluid Phase Equilib., 52, 31.
[17] Chapman W.G., Gubbins K.E., Jackson G. and Radosz M. 1990. Ind. Eng. Chem. Res., 29, 1709.
[18] Gross J. and Sadowsky G. 2001. Ind. Eng. Chem. Res., 40, 1244.
[19] Song Y., Lambert S.M. and Prausnitz J.M. 1994. Macromolecules, 27, 441.
[20] Doghieri F. and Sarti G.C. 1996. Macromolecules, 29, 7885.
[21] Sarti G.C. and Doghieri F. 1998. Chem. Eng. Sci., 53, 3435–3447.
[22] Doghieri F., Canova M. and Sarti G.C. 1999. Polymer membranes for gas and vapor separations, ACS Symp. Ser., 733, 179–193.
[23] Giacinti Baschetti M., Doghieri F. and Sarti G.C. 2001. Ind. Eng. Chem. Res., 40, 3027–3037.
[24] Doghieri F. and Sarti G.C. 1998. J. Membr. Sci., 147, 73.
[25] Coleman B.D. and Gurtin M.E. 1967. J. Chem. Phys., 47, 597.
[26] Koros W.J., Paul D.R. and Rocha A.A. 1976. J. Polym. Sci.: Polym. Phys. Ed., 14, 687.
[27] Koros W.J. and Paul D.R. 1978. J. Polym. Sci.: Polym. Phys. Ed., 16, 1947.
[28] Fleming G.K. and Koros W.J. 1990. Macromolecules, 23, 1353.
[29] Wang J.S. and Kamiya Y. 2000. J. Polym. Sci., Part B: Poly. Phys., 38, 883.
[30] Colina C.M., Hall C.K. and Gubbins K.E. 2002. Fluid phase equilib., 194–197, 553–565.
[31] Hariharan R., Freeman B.D., Carbonell R.G. and Sarti G.C. 1993. J. Appl. Polym. Sci., 50, 1781–1795.
[32] Kiszka M.B., Meilchen M.A. and McHugh M.A. 1988. J. Appl. Polym. Sci., 36, 583–597.
[33] Pope D.S., Sanchez I.C., Koros W.J. and Fleming G.K. 1991. Macromolecules, 24, 1779–1783.
[34] De Angelis M.G., Merkel T.C., Bondar V.I., Freeman B.D., Doghieri F. and Sarti G.C. 1999. J. Polym. Sci., Polym. Phys. Ed., 37, 3011–3026.
[35] Zoller P. and Walsh D. 1995. Standard Pressure–Volume–Temperature Data for Polymers. Technomic, Lancaster.
[36] Markevich M.A., Stogova V.N. and Gorenberg A.Ya. 1991. Polym. Sci. USSR, 1, 132–140.

[37] Berens A.R. 1985. J. Am. Water Works Assoc., 77(11), 57–64.
[38] Rodgers P.A. 1993. J. Appl. Polym. Sci., 48, 1061.
[39] Sanchez I.C. and Panayiotou C. 1994. In, Models for Thermodynamic and Phase Equilibria Calculations, Sandler S I. (Ed.). Dekker, New York, pp. 187–285.
[40] Chiou J.S. and Paul D.R. 1989. J. Membr. Sci., 45, 167.
[41] Davydova M.B. and Yampolskii Yu. P. 1991. J. Polym. Sci. USSR, 33, 495.
[42] McHattie J.S., Koros W.J. and Paul D.R. 1991. Polymer, 32, 840.
[43] Jordan S. and Koros W.J. 1995. Macromolecules, 28, 2228.
[44] Wissinger R.G. and Paulaitis M.E. 1991. Ind. Eng. Chem. Res., 30, 842.
[45] De Angelis M.G., Merkel T.C., Bondar V.I., Freeman B.D., Doghieri F. and Sarti G.C. 2002. Macromolecules, 35, 1276.
[46] Muruganandam N. and Paul D.R. 1987. J. Polym. Sci., Part B: Polym. Phys., 25, 2315.
[47] Raymond P.C. and Paul D.R. 1990. J. Polym. Sci., Part B: Polym. Phys., 28, 2103.
[48] Grassia F., Giacinti Baschetti M., Doghieri F. and Sarti G.C. 2004. Advanced materials for membrane separations, ACS Symp. Ser., 876, 55–73.
[49] Sanders E.S., Koros W.J., Hopfenberg H.B. and Stannett V.T. 1984. J. Membr. Sci., 18, 53.

3

Small Peptide Ligands for Affinity Separations of Biological Molecules

Guangquan Wang, Jeffrey R. Salm, Patrick V. Gurgel and Ruben G. Carbonell

3.1 Downstream Processing in Biopharmaceutical Production

The biotechnology industry faces serious challenges to profitability[1]. Since downstream processing accounts for 50–80% of the cost of manufacturing a therapeutic product, reductions in the number of steps in the purification train and increases in the yield and purity of the product in each step would effectively decrease production costs. In addition, there is a shift among regulatory agencies to what are called 'well-characterized biologics', which require that all manufactured biological products be essentially free of contaminants.

A typical industrial process for recovery and purification of therapeutic proteins or other biological molecules from fermentation broths or mammalian cell culture media often involves a complex series of steps. In the simplest case of a product excreted from a cell into a culture medium, the cells must first be separated from the broth by filtration or centrifugation. A rough separation step such as precipitation with ammonium sulfate followed by filtration can lead to a significant concentration of the desired protein in combination with other contaminants. A re-suspension of this precipitate can then be processed over a series of ion exchange columns for higher purification, followed by a polishing step using gel permeation chromatography. In a case where the product is bound to inclusion bodies or membranes within the cell, the purification is more complex, often involving the addition of chaotropic solvents and refolding of the protein into its active conformation. The large number of steps required to separate and purify biological molecules often results in significant losses of product yield and activity. For this reason, increasing attention is being given to affinity chromatography as a purification method

Chemical Engineering Edited by Miguel A. Galán and Eva Martin del Valle

that would result in both a reduction in the number of steps required for purification as well as a higher yield of product mass and activity.

3.2 Affinity Chromatography

In affinity chromatography, a ligand that is able to bind the target molecule specifically from a mixture is attached to a porous chromatographic support with suitable flow properties and binding capacity. As shown in Figure 3.1, the mixture containing the desired product is injected into the column, resulting in the adsorption of the target species on the chromatographic support. This adsorption step is continued until the column is saturated with the desired product and the solution begins to show a small amount of breakthrough of the product at the outlet. At this point, a washing buffer is introduced to remove from the resin any material that is bound non-specifically. Typically, this wash is carried out with a buffer with a slightly elevated ionic strength containing weak detergents. Once the contaminants are removed, an elution buffer of low or high pH or ionic strength, perhaps containing chaotropic solvents, is injected to remove the product from the chromatographic support in a concentrated and highly pure form. After several cycles of use, it is necessary to wash the column with acid or basic solutions in order to remove any pathogens or residual material that might contaminate the product.

To execute a successful affinity chromatography separation, the ligand must bind strongly enough to the product so that it will be able to selectively adsorb and concentrate it from the mixture. However, it cannot bind so strongly that harsh solvents are needed for elution thus reducing the product activity. Ligands with association constants equal to or higher than 10^6/M are generally preferred for this purpose. Finally, the capacity of the resin should be sufficiently high so that the column dimensions (and cost) are in a reasonable range. Capacities can range anywhere from 2 to 30 mg of product per

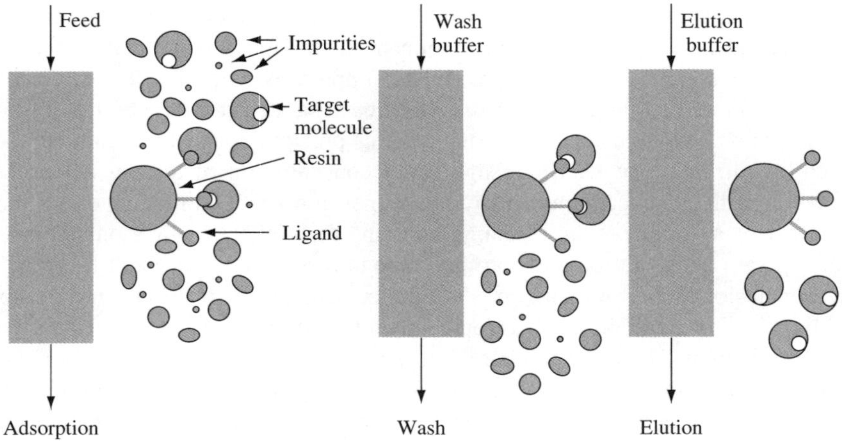

Figure 3.1 Schematic representation of affinity chromatography. The adsorption step is continued until the column is saturated. A wash step removes unwanted impurities bound to the resin. Elution of the product is executed by changing buffer conditions

milliliter of resin. There are many ligands for affinity chromatography that exploit nature's methods for molecular recognition. These include inhibitor–enzyme, receptor–protein, oligonucleotide–protein, and antigen–antibody pairs. Other naturally occurring biological molecules such as heparin and Protein A can do an excellent job of separating and purifying proteins from complex mixtures. Of these, the use of antibodies as affinity ligands for chromatography has been found to be the most flexible, especially in laboratory applications. Since either monoclonal or polyclonal antibodies can be generated against just about any biological molecule, literally thousands of purification processes rely on antibodies as the preferred ligands for affinity purification on a small scale.

Affinity chromatography is a very well established technique, whose roots can be traced to the work of Cuatrecasas, Wilchek, and Anfinsen in 1968[2]. Between the years 1994 and 1996 alone, over 150 patents per year were awarded using affinity chromatography as the method of choice for purification, and over 60% of the purification protocols published involve affinity chromatographic steps somewhere in the process[3]. Given this level of research activity, one would think that affinity chromatography steps would be common in the bioprocess industry, but this is not the case. There are many challenges to the establishment of affinity chromatography as a method of choice for industrial separations. The section that follows discusses some of these challenges and how small peptide ligands can play in a role in overcoming them[4–8].

3.3 Advantages of Small Peptide Ligands

The high cost of monoclonal or polyclonal antibodies often makes the use of affinity chromatography for any large-scale separation prohibitively expensive. For therapeutic products, the affinity ligand must be extremely well characterized, and it must have a purity that rivals that of the product itself. As a result, the mere production of the highly pure monoclonal or polyclonal antibody ligand results in a significant cost of generating the chromatographic support. In addition, all methods of immobilization or attachment of the antibody to the support result in some leakage of the ligand into the product with column use. In the case of production of therapeutics, this leakage needs to be carefully monitored and steps need to be introduced to remove small amounts of the antibody from the product solution. Failure to do this can result in immunogenic reactions in the patient caused by the presence of foreign animal proteins.

The elution and wash steps that are necessary in affinity chromatography can result in a reduction in the affinity of the antibody for the target molecule. Since antibodies are proteins that recognize other molecules as a result of their tertiary structure in solution, any solvent changes that result in small modifications of tertiary structure can cause reductions in affinity. Repeated numbers of adsorption, elution, and column cleaning cycles can deactivate what started as a highly effective but extremely expensive column. In addition, the binding of many antibodies to the product species is so specific and strong that rather extreme elution conditions are necessary for product recovery, resulting in loss of activity.

Regardless of the ligand type, chromatographic supports must be tested and validated against ligand leakage, binding of pathogens such as endotoxins and viruses, and for robust yield and purity results after being subjected to a large number of cycles of use. A great deal of work has been done on small organic ligand (including dyes) affinity chromatography and immobilized metal affinity chromatography (IMAC). Small organic molecule ligands and immobilized metal ions are more stable and less expensive than antibodies so that they can stand harsh operation conditions, but neither of them exhibit great selectivity when exposed to a complex mixture, unless the organic is specially designed to adsorb to a specific site on the surface of the protein. In addition, dye ligands and metal ions may be toxic and with metal ligands in particular, there is a problem with severe leakage from the support. There are small organic molecules currently in commercial use as ligands for the industrial purification of proteins, and some of the more successful ones involve triazine derivatives[3].

Peptides with anywhere from 3 to 25 amino acid residues can have affinities to molecules of interest that compare well with dyes, immobilized metal ions, and some antibodies. As opposed to monoclonal antibodies, small peptide ligands are much more stable because they do not require a specific tertiary structure to maintain their activity. Small peptides are also less likely than an antibody to cause immune response in case of leakage into the products. Peptides can be manufactured aseptically in large scale under GMP (good manufacturing practices) conditions at relatively low cost compared with antibodies, especially if they have less than 13 amino acids. The interactions between peptides and proteins are moderately strong so that the protein can be eluted under mild conditions without loss of protein activity. In addition to being good candidates as ligands in affinity chromatography, peptides are used widely to determine protein–protein interactions without *a priori* information on protein structure (e.g. in epitope mapping).

The identification of suitable peptide ligands for a given separation is often carried out using combinatorial peptide libraries generated with phage, yeast, ribosome, and other display vehicles. A great deal of work has been done with libraries of peptides containing 13–58 amino acids, multimeric or geometrically constrained peptides, Affibodies™, single-chain variable fragments of antibodies, etc. Because of the relatively large size of these peptides, the binding is often also determined by the three-dimensional configuration of the ligand, and as a result, these would be subject to the same stability difficulties experienced with whole antibodies. In addition, peptides with 13 or higher numbers of amino acid residues are very expensive to produce using synthetic routes, and must be produced by expression in recombinant organisms. As a result, little significant cost savings ensue from making a large peptide as opposed to a full antibody. Nevertheless, several groups have been actively pursuing this approach with some success, including Tecnogen in Milan, Affibody AB, Dyax, and Amersham Biosciences.

Our group has worked on an alternative approach based on the use of small peptides (three to eight amino acids) as ligands for affinity chromatography. For the case of peptides of this size it is possible to generate very large combinatorial peptide libraries on solid phase supports that are chromatographic resins. For example, over 34 million hexamers can be generated with 18 of 20 naturally occurring amino acids (excepting cysteine and methionine), each on a different resin bead, using only about 18 g of chromatographic resin. Using appropriate screening techniques, it is possible to isolate and sequence a very small number of leads that can then be studied further to verify binding to the target species. Since the peptides are already attached to the desired chromatographic support, they can then be made on a larger scale to generate the chromatographic column. It often happens that a peptide that can bind to a protein in solution or on the surface of a phage does not bind when immobilized on a solid support. The screening approach used in our work eliminates this possibility. Finally, when small peptides leak from a column they have a smaller chance of producing an immunogenic response, and since they do not exhibit a precise tertiary structure, their binding characteristics are not highly affected by solvents used in repeated cycles of adsorption, elution, and cleaning of the support. As opposed to antibodies, whose association constants with a target on a solid support can be 10^9–10^{10}/M, small peptides exhibit association constants around 10^6–10^8/M. These weaker association constants allow for relatively gentle elution conditions, higher product activity yields, and better column longevity.

Baumbach and Hammond first demonstrated the principle of using small peptide ligands from combinatorial libraries as affinity ligands in large-scale chromatography processes by using streptavidin as a target[4]. Since then, this technique has been successfully used to purify a variety of proteins, such as s-protein[6], von Willebrand factor[7], factor IX[9], factor VIII[10], trypsin[11], anti-MUC1 antibodies[12,13], α-1-proteinase inhibitor[5], monoclonal antibodies (IgG, IgA, IgE, IgM, IgY)[14], α-lactalbumin[15,16], fibrinogen[17], and staphylococcal enterotoxin B (SEB)[18].

In this chapter, we will try to summarize aspects of the construction and screening of combinatorial libraries of small peptides, the nature of the protein–ligand interactions on these supports, the effects of peptide density on performance, and the factors that determine the dynamic binding capacity of the resins. The concluding section will mention some of the current areas of investigation on novel applications for this promising technology.

3.4 Combinatorial Peptide Libraries

3.4.1 Phage-Displayed Libraries

One of the challenges for the use of peptide ligands in affinity chromatography is the identification of a sequence that shows affinity and specificity to the target protein. Examples in the literature show that designing a specific complementary peptide sequence is difficult even when the structure of the target protein is known[19,20]. The development of combinatorial libraries has allowed the screening of millions of peptide sequences to discover specific peptides that bind to the target protein. Peptide libraries can be generated either biologically or synthetically. Several combinatorial library methods have been described in the literature[21]. The most widely used biological libraries are phage-displayed libraries, while one-bead-one-peptide libraries are the dominant libraries obtained directly from chemical synthesis.

In phage-displayed peptide libraries, random oligonucleotides with a given length are generated and then inserted into bacterial phage gene III. The corresponding peptide coded by the inserted DNA is displayed at the N-terminal of the gene III protein (pIII) on the phage surface. Each phage displays one peptide sequence that is different from those on other phages. Affinity peptides on phage that bind to the target protein are selected through several rounds of affinity purification. Millions of phage particles are incubated with the target protein immobilized on a Petri dish or ELISA plates. Non-binding phages are washed out extensively, and then the bound phages are eluted under harsh conditions. The eluted phages are then amplified on agar medium and subjected to the next round of affinity purification. The tight-binding phages are then cloned and propagated in *Escherichia coli*. The amino acid sequence of the peptide on the phage is deduced by sequencing the coded DNA in the phage gene III[22–24].

Ligands identified from phage libraries frequently interact with natural binding sites on the target molecule and resemble the target's natural ligands. Thus phage-displayed random peptide libraries have been used to investigate protein–protein interactions in a variety of contexts. For example, phage-displayed random peptide libraries have been used to map the epitopes of monoclonal and polyclonal antibodies, and to identify peptide ligands for receptors, receptor ligands, and folded domains within larger proteins, such as several SH2, SH3 domains[25,26]. Recently, peptide ligands for some superantigens, for example, SEB and TSST-1, have been determined with phage-displayed random peptide libraries[27,28]. However, biopanning with phage-displayed libraries is slow and subject to non-specific binding.

Phage-displayed random peptide libraries have been constructed to display peptides of variable length ranging from 6 to 38 amino acids[25]. Phage display libraries have the advantage of allowing exposure of very large peptides as potential ligands. Once it is created, a phage library can be regenerated continuously and re-used unlike a synthetic library. Usually the diversity of the original phage library is on the order of 10^8 peptides. Selection must be avoided during the library expansion and propagation for phages with selective growth advantage[25]. Peptide synthesis on phages is limited to the 20 natural amino acids so that D-amino acids or other molecules cannot be used to increase the diversity of the library.

3.4.2 One-Bead-One-Peptide Libraries

Synthetic libraries can be created on solid supports through organic chemistry. There are several distinct combinatorial library methods[21]. The one-bead-one-peptide library method is used extensively in drug discovery processes due to its unique features[29]. Compared with other methods, the synthesis of a one-bead-one-peptide library is rapid with use of the 'split synthesis' approach. Because one bead has one unique peptide sequence, all of the beads can be tested concurrently but independently. Once positive beads have been identified, the chemical structure of the peptides on the beads can be directly determined by sequencing or by an encoding strategy. In addition, the libraries can be used either in the solid phase (i.e. peptides attached on solid) or in the solution phase (i.e. peptides cleaved from solid support). As in phage-displayed libraries, the screening of peptide ligands from one-bead-one-peptide libraries involves three steps[21]: construction of the library, screening the library with the target molecule, and determination of the peptide sequence.

Although peptide ligands from phage library have been presented on chromatographic support to purify proteins[4,6,7], it is possible that the microenvironment and the orientation of the peptides on the chromatographic support could be very different from that on phage. This can adversely affect the interactions between the peptide ligands and the target[9]. It has been found that some peptides derived from a phage library work only when the peptide is an integral part of the phage coat protein and not when isolated in free solution or on a solid support[1]. Thus one-bead-one-peptide libraries are uniquely suited to discover peptide ligands for protein purification.

3.4.3 Libraries on Chromatographic Resins

To avoid the possibility that a small peptide from a phage-displayed library will not bind adequately when immobilized on a solid support, Baumbach and Hammond suggested that combinatorial peptide libraries for protein purification be synthesized directly on resins that could be used as chromatographic supports on a large scale[4]. In this way, any ligand that is identified is already on a platform or format that would facilitate implementation in downstream processing. The one-bead-one-peptide solid phase library format is ideally suited for this purpose, if the library is built on chromatographic resins that can withstand the harsh solvent conditions used for peptide synthesis.

The first one-bead-one-peptide library was synthesized by Lam *et al.*[30] using the 'split synthesis' approach pioneered by Furka and Sebetyen[31]. Figure 3.2 shows how this approach is implemented in library construction. The resin beads are divided equally into separate reaction vessels of an automatic peptide synthesizer, each with a single amino acid. After the first amino acid is coupled to the resins, beads are repooled, mixed thoroughly, and redistributed into separate reaction vessels. The next coupling step is then performed. This divide–couple–recombine technique is repeated until the desired length of the peptide library is reached. There are X^n random sequences in the library, where X is the number of amino acids used for coupling and n is the length of the library. If X is the total number of natural amino acids (20) and $n = 6$ (hexamer library), the total number of different peptides generated is 64 million. Each resin bead displays multiple copies of only one peptide sequence. Thus libraries of this type are called 'one-bead-one-peptide' libraries[30]. Because other ligands besides naturally occurring amino acids, such as D-amino acids, oligonucleotides, synthetic oligomers, proteins, and small molecules, also can be coupled to solid resins, the idea of a one-bead-one-peptide library has been extended to one-bead-one-compound library[21]. The introduction of other compounds besides amino acids in combinatorial library construction increases the diversity of the library in comparison with a phage-displayed peptide library, in which phages only

- Synthetic peptide library constructed using 'divide, couple, recombine' technique
- Stepwise synthesis using F-moc and t-Boc solid phase chemistry and 18 of 20 naturally occurring L-amino acids (except Cys and Met)
- Method generates 34×10^6 unique peptide hexamers
- Peptide density ~100 μmoles/g

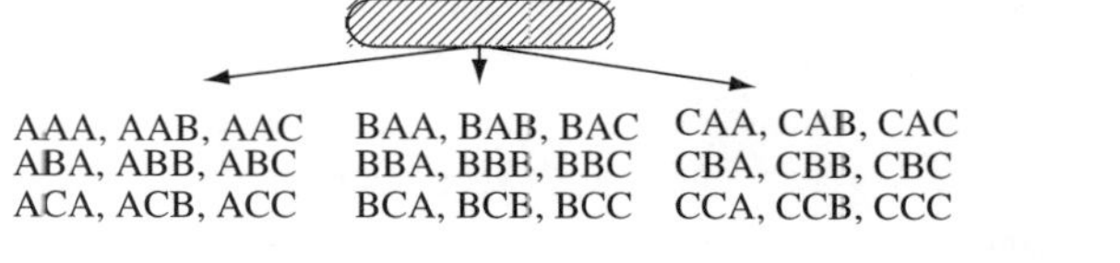

Figure 3.2 *Construction of a one-bead-one-peptide library. For N amino acids in the library, repetition of the split synthesis technique n times results in N^n different peptides in the library*

display peptides composed of natural amino acids. However, all synthetic methods have a practical limit on the size of the library as well as the length of the peptides on beads, while peptides on phage can be fairly large.

The choice of the solid support is critical for the library construction and the application of the library. The biological signal released from the peptides on a single bead depends quantitatively on the amount of the peptide on the bead. As a result, the size and substitution homogeneity is of the utmost importance. Meanwhile, the resin should resist the formation of clusters because clusters would prevent the statistical distribution of resin beads and lower the number of structures created. In addition, resins should be compatible with various organic and aqueous media. Solid beads with porous structure are preferred. The high surface area provided by porous resin results in a high ligand concentration, facilitating bead sequencing and providing high capacity for their use in chromatography. Moreover, the pores should be large enough to reduce diffusion resistance especially when using large proteins as targets. In order to avoid non-specific binding between the solid matrices and proteins, hydrophilic resins are preferred. If the peptide ligands will be used to purify protein in chromatography, the resins should have enough mechanical rigidity to withstand the high pressure used in liquid chromatography. A variety of polymer beads have been used to attach peptides in library construction, including polyhydroxylated methacrylate, polydimethylacrylamide, polyoxyethylene-grafted polystyrene, Tentagel, and so on[9,21].

The solid phase library used in the majority of the work done in our lab is constructed on Tosoh BioSciences, Inc. Toyopearl AF Amino 650 M resin as shown in Figure 3.3. Toyopearl resin is a functionalized hydroxylated polymethacrylate resin that works well with the F-moc and t-Boc chemistries used for peptide synthesis. The peptides are synthesized directly on the primary amines on the end of hydrophilic spacer arms that are attached to the resin surface. Toyopearl resin is also mechanically and chemically stable which allows it to be scaled up into a commercial setting. This resin has a maximum coupling density of approximately 400 μmoles of ligand per gram of dry resin, and in internal surface area of approximately 30 m^2/g. Libraries of these materials are normally

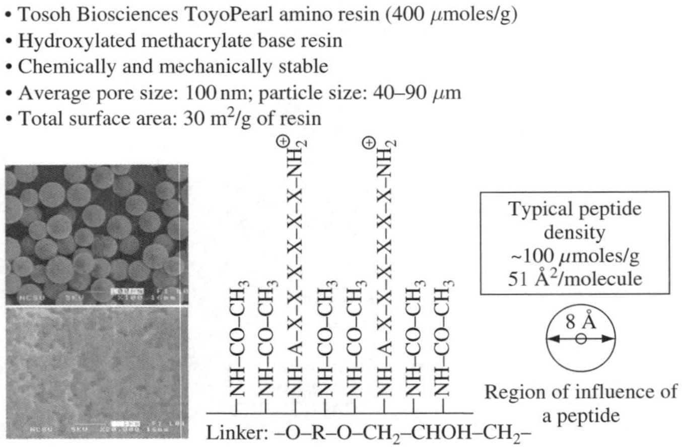

Figure 3.3 *Properties of Toyopearl library support and ligand structure*

synthesized with a peptide density of approximately 100 μmoles/g, and any uncoupled primary amines should be acetylated to reduce non-specific binding of target proteins. At this peptide density, the individual peptides are separated from each other by roughly 8 Å. The average pore size of 1000 Å and the large porosity of the resin make the pores accessible even to fairly large protein targets. These 65-μm-diameter resins are meant for large-scale chromatographic separations since they are rigid and exhibit excellent flow and chemical properties. As a result, the libraries created from these resins result in the identification of a peptide that binds to the target species and is already attached to a chromatographic support. This helps to eliminate any situations where a peptide might be identified using a phage-displayed library and not bind when attached to a chromatographic support. A hexamer library constructed on Toyopearl with 18 amino acids will weigh approximately 18 g of dry resin, which, when swollen in methanol, will occupy a volume of roughly 84.6 mL. Phage-displayed libraries can generate full diversity of very large peptides that cannot be achieved with the solid phase libraries since the amount of resin and amino acids required for chemical synthesis would make them prohibitively large and expensive.

3.5 Screening of One-Bead-One-Peptide Libraries

3.5.1 On-Bead Binding Screening

Both solid phase and solution phase methods have been used for screening one-bead-one-peptide combinatorial libraries. The most widely adopted method of screening is the 'on-bead' binding assay[21,32]. The target protein is incubated with the library, so that the protein can bind to beads that have specific peptide sequences favorable for adsorption. The binding of the target to the immobilized ligands is usually detected by using a reporter group such as an enzyme, a radionuclide, a fluorescent probe, or a color dye covalently attached to the target molecules. Alternatively, immunodetection schemes such as in enzyme-linked-immunosensitive assays (ELISA) can be used to detect binding to a resin bead. The signals generated from these reporter groups are proportional to the amount and density of peptides on the beads, as well as the size of the beads. Non-specific binding can result in high background with both immunodetection as well as direct detection methods, generating some difficulty in determining which beads contain true affinity ligands. This is usually eliminated by using blocking proteins (e.g. casein or bovine serum albumin), by washing with a high ionic strength buffer (e.g. 0.2–0.4 M NaCl) to reduce purely electrostatic binding, and by washing with nonionic detergents (e.g. 0.1% Tween 20).

One of the most convenient screening methods is the enzyme-linked colorimetric detection scheme. It has been used to discover the binding motifs for streptavidin[30,33], avidin[33], monoclonal antibodies[34], proteases[34], and MHC molecules[35]. The enzyme-linked colorimetric detection method is extremely rapid, taking a few hours to screen 10^7–10^8 beads. The problem with this method is that the enzyme molecule attached to the target can sterically affect the binding of the target to peptides on beads.

Radionuclide-labeled targets can be used to screen library beads to avoid this problem. The radionuclide probes, such as ^3H and ^{14}C, are particularly small compared with enzyme as reporter groups on the target, and it has been demonstrated that the labeled target shows almost the same biological properties as the natural target[36]. A schematic diagram

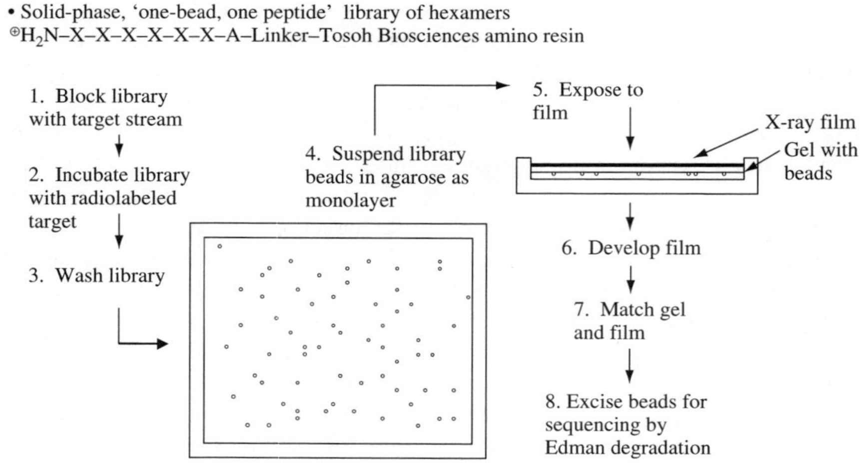

Figure 3.4 *Radiological screening of a hexamer peptide library by the method of Mondorf et al.*[38]

of the process is shown in Figure 3.4. The library is incubated with the radiolabeled target protein, washed, and then suspended in agarose gel. The slurry is poured onto a gel bond to form a monolayer so that all beads are spatially separated. Exposure of the gel to autoradiography film can locate the positive beads that are then isolated and sequenced. Several researchers have screened peptide libraries using radiolabeled targets[37–40]. The method developed by Mondorf *et al.*[38] using ^{14}C offers high resolution and sensitivity. It has been used to identify affinity peptide ligands for s-protein[38], fibrinogen[38], α-1-proteinase inhibitor[5], α-lactalbumin[15], recombinant factor VIII[41], and SEB[18].

Immunostaining schemes similar to ELISA also can be used to target the protein on beads. There are no modifications of the target using this method, so the bead-bound ligands bind directly to the native protein, and not to any adducts. However, the antibodies used in the detection system could bind to bead-bound ligands other than the targets to bring the possibility of interference and false positives. A two-color PEptide Library Immunostaining Chromatographic ANalysis (PELICAN) has been developed to determine beads specifically for the target from those beads resulting from antibody cross-reactivity[9]. It has been used to discover peptide leads for protease factor IX and fibrinogen[9,42]. Other on-bead screening schemes involve dye-labeled targets or fluorescently labeled target[43,44]. However, dyes always complicate the screening process by binding directly to many peptide ligands, and the autofluorescence of the library makes it unsuitable for this kind of screening process[21].

In order to minimize the number of false positive beads and make the screening more selective, two different screening methods can be sequentially used for one target. For example, a dual-color detection scheme that uses two sequential orthogonal probes in enzyme-linked colorimetric detection methods[45] and a cross-screening scheme that combines an enzyme-linked colorimetric method and a radiolabeled assay[46] have been developed. In this way, many of the initially determined positive beads are eliminated by the second screening method, and the chances of getting the true positive beads are greatly enhanced.

3.5.2 Screening of Soluble Combinatorial Peptide Libraries

One of the disadvantages of on-bead screening is the high peptide density required for peptide sequencing on beads. This can lead to multiple-point attachment of the target to the peptides so that non-specific interaction between target and peptides will be enhanced. Thus the selected peptide ligands may have less affinity and specificity to the target. Screening of soluble peptide libraries can make the affinity ligands more selective. The format of affinity chromatographic screening developed by Evans *et al.*[47] and Huang and Carbonell[8] is suitable for screening peptide libraries due to its rapidity. The targets are immobilized onto resins and then packed into a chromatographic column. The soluble peptide libraries are pumped into the column at a proper flow rate to ensure the peptides have enough time to bind to the immobilized targets. Then the column is washed thoroughly with binding buffer. The affinity peptide ligands bound to the targets are eluted and then isolated by reverse-phase chromatography. The fractions are then sequenced by Edman degradation or mass spectrometry. Huang and Carbonell have demonstrated this technique by showing that the identified sequence consensus, NFVE, is the same as that found from screening a phage-displayed library for s-protein[8]. Evans *et al.*[47] used a similar system to recover the known epitope, YGGFL, for a monoclonal antibody (3E-7), and then determine the affinity ligands for bacterial lipopolysaccharide (LPS, endotoxin). Although this technique is rapid and able to avoid false signals from non-specific binding, some hydrophobic leads could be missed due to their minimal solubility. The slow binding kinetics and the orientation of the immobilized targets may limit the contact between the immobilized targets and the free peptide ligands so that some potential leads could pass through the column. In addition, the methodologies used in this technique are more complex than those in on-bead screening[8].

3.5.3 Multi-Tiered Screening

A multi-tiered screening process was developed to help identify and eliminate false leads from the solid phase library screenings in our work[5,15,18] as shown in Figure 3.5.

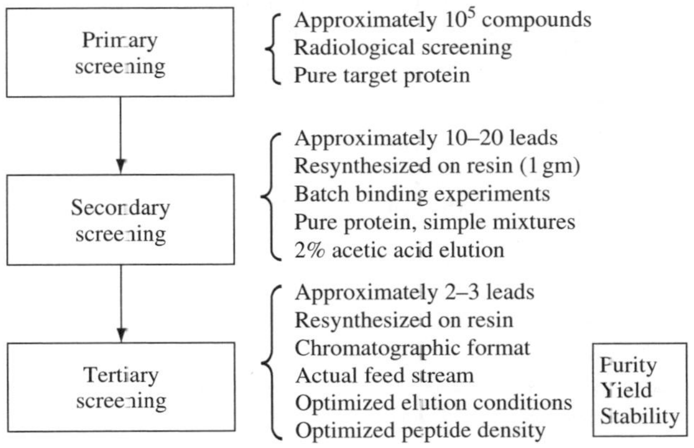

Figure 3.5 *Multi-tiered screening process. The path from primary to tertiary screening focuses the final sequence that gives the best combination of purity and yield in purification*

The 'primary screening' process involves identification of leads that are thought to bind specifically to the target protein. 'Secondary screening' is used to confirm the binding of the target protein to a larger mass of resin that contains the peptide ligands identified during the primary screening in a non-competitive format. Some peptides that bound weakly to the target protein are eliminated during secondary screening. The 'tertiary screening' process is performed in a column format to demonstrate the binding selectivity of the peptide ligands for the target protein from its native feed stream.

Since the largest amount of time in lead discovery is often spent on characterizing the one or two leads that are eventually shown to bind the target, considerable resources can be saved by decreasing the 20 or so leads identified during primary screening down to 4 or 5 leads for detailed characterization. Once leads from primary screening are sequenced, larger batches (1 g) of resin with each of the leads are re-synthesized to be used in secondary screening. Batch experiments with pure protein are carried out to verify binding of the target protein to the different peptide leads. From 20–30 beads in primary screening, only a few might show sufficient protein adsorption to move toward tertiary screening. In tertiary screening, issues such as the elution conditions and binding buffers used with the peptide are optimized to increase yields and purity during the separation. It is often the case that additional batches of the final one or two resins are synthesized to test the effects of peptide density on the adsorption and elution conditions. By using a multi-tiered screening process, it is possible to deduce a single lead that works well enough to achieve the desired purification.

The multi-tiered screening approach was applied by Bastek *et al.*[5] to screen approximately 6.5×10^5 leads from a 34×10^6 member hexameric peptide library against α-1-proteinase inhibitor (α_1-PI). Twenty-one leads were identified by primary screening as potential binders of α_1-PI. Though no true consensus sequence was identified, many of the sequences demonstrated a high degree of similarity in the types of amino acids present. In particular, two sequences differed by only one amino acid, IKRYYN and IKRYYL. Two other sequences demonstrating excellent homology were VIWLVR and IIWLYK.

Bastek *et al.*[5] then used secondary screening techniques to verify the binding of the leads identified during primary screening for α_1-PI. Nineteen of the sequences were found to bind α_1-PI in a non-competitive environment. Several of these peptides were able to elute α_1-PI using a 1 M NaCl wash. The rest of the leads eluted α_1-PI after washing with 2% acetic acid. During tertiary screening in human serum albumin (hSA), leads that eluted bound protein in 1 M NaCl were not able to purify α_1-PI from hSA as both proteins eluted at the same time. Leads that eluted α_1-PI after washing with acetic acid generated a clean peak of α_1-PI. Four of these leads achieved purities over 90% with yields ranging from 53% to 75% at 20° C. Three of these leads resulted in purities of 100% with yields ranging from 27% to 72% at 4° C.

Bastek *et al.*[5] also tested the ability of the identified affinity peptides to purify α_1-PI from effluent II + III, a process intermediate of the Cohn plasma fractionation process with eight major protein constituents. Several of the peptides achieved yields of 70–80% with purities ranging from 42% to 77%. Purification using these peptides matched or exceeded yields and purities reported in the literature using ion exchange chromatography.

Gurgel *et al.*[15] applied the multi-tiered screening approach to α-lactalbumin (α-La), a whey protein of significance to the food industry. Eighteen sequences were identified as potential α-La binders by screening approximately 2% of a solid phase hexameric peptide library. Unlike the leads identified by Bastek and co-workers, no obvious trends were observed and the distribution of amino acids seemed to be random. Secondary screening showed that 6 of the 18 leads bound more than 60% of the α-La. The remaining 12 leads bound less than 32% of the α-La, 10 of which bound less than 15%. Subsequent tertiary screening resulted in a final peptide, WHWRKR, which was effective in removing α-La from whey protein isolate (WHI).

Wang *et al.*[18] applied the multi-tiered screening approach to SEB, a primary toxin involved in food poisoning and leading to autoimmune diseases. Eleven leads from approximately 5% of a solid phase hexameric peptide library showed positive binding to ^{14}C-labeled SEB in the primary screening. Six leads had a tryptophan and a tyrosine at the N-terminus, and especially, two of them had a sequence consensus, YYW, at the N-terminus. It was also found that five of the six leads had at least two histidines at the C-terminus. Unlike the secondary screening used by Bastek and co-workers and Gurgel and co-workers, Wang and co-workers ran a second screening in both a competitive and non-competitive format. The non-competitive mode involved screening using only pure SEB. For the competitive mode, the peptide leads were tested using mixtures of casein and bovine serum albumin (BSA) spiked with SEB. Peptides that bound weakly to SEB or bound to other proteins were eliminated. There was only 1 lead, YYWLHH, that exhibited high specificity to SEB among those 12 lead candidates after the secondary screening. In the tertiary screening using a YYWLHH column, SEB was quantitatively recovered with high yield and purity from *E. coli* lysate, BSA, and its natural feed stream, *Staphylococcus aureus* fermentation broth. With most of the impurities passing through the column or washed out by 1 M salt, SEB was exclusively eluted by 2% acetic acid. The YYWLHH column also made it possible to obtain highly purified native SEB from a heterogeneous SEB preparation containing nicked or denatured protein. Wang and co-workers also showed that a peptide ligand derived from a phage-displayed peptide library by Goldman *et al.*[27] could not capture SEB when the ligand was immobilized on Toyopearl resin. This work confirmed that solid phase combinatorial peptide libraries are uniquely suited to discover peptide ligands for protein purification.

Both Bastek and co-workers and Gurgel and co-workers limited primary screening to less than 2% and Wang and co-workers to less than 5% of the overall library, showing that screening a complete library is impractical and, generally speaking, unnecessary. Even though a solid phase combinatorial library might contain 34 million different peptides, it has been found through experience with the radiological screening method that only about 1% of the library is necessary to have the primary screening process identify sufficient leads able to bind to a particular target protein. By using a protein that is not particularly hydrophobic and blocking agents that minimize non-specific binding, primary screening will identify only 10–100 beads from 350 000 beads that will bind the desired target protein in significant amounts to show up as radioactive leads.

In the following section we summarize some of what has been learned about the nature of the peptides identified as a result of the ligand screening process, and the types of interaction between the protein and the small peptide on the ligand surface. These aspects include the importance of electrostatic and hydrophobic interactions, the effect of peptide density on the magnitude of the adsorption equilibrium constant, and the rate of binding of the protein to the affinity support.

3.6 Characterization of Peptide Ligands

3.6.1 Single and Multipoint Attachment and the Effect of Peptide Density

Some peptide ligands identified by library screening are bioselective to their targets, while other peptide ligands behave as pseudo-affinity ligands. A study by Huang and carbonell showed that the peptide ligand YNFEVL is so specific to s-protein that randomization of the peptide sequence completely destroys the binding[6]. A similar study on the binding of von Willebrand factor (vWF) to the peptide RLRSFY shows that the randomization has little effect on the binding[7]. In the case of the s-protein, YNFEVL in solution could be used to elute the adsorbed protein on the resin surface, while RLRSFY, even at high concentrations, was unable to elute the adsorbed vWF. These results suggest that there is a binding cleft on the s-protein molecule that leads to specific interactions with the corresponding peptide ligand, while no such specific clefts exist on vWF. The pseudo-affinity peptide ligands are ideally suited for the capture or concentration of the target molecule at an early stage in purification[5]; other steps such as gel permeation chromatography could then be used to polish the products. The more bioselective peptide ligands can be efficient for purifying the protein in one step, but usually sample preparation prior to the affinity chromatography step is needed to protect the ligands and maximize the efficiency of the affinity column. As will be described in more detail in the following text, the type of ligand found by screening a combinatorial library depends in part on whether the target protein has a well-defined cleft or loop to which the peptide can bind, the size of the protein molecule, and the density of the peptides on the surface, as well as the types of interactions favored during binding.

Ligand density can affect the interactions between the peptide ligands and the target protein. If the protein has a cleft and the binding is attributed to single-point interactions between the protein and the ligand as shown in Figure 3.6, the capacity increases when increasing the ligand density, while the association constant may remain constant at low ligand density and decrease at high ligand density due to steric effects caused by the crowding of bound proteins around ligands on the surface. Thus there is an optimal peptide

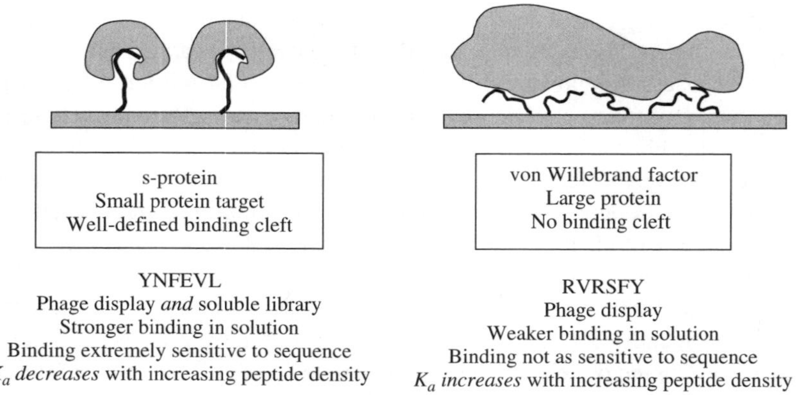

Figure 3.6 *Single-versus multi-point interaction. Both mechanisms of adsorption have been determined in peptide ligand affinity chromatography*

density that can provide a large capacity for binding without significantly affecting the magnitude of the adsorption equilibrium constant. If the binding is attributed to multipoint interactions, as is often the case with the binding of a large protein to a surface with high peptide density (Figure 3.6), increasing the ligand density typically increases both the capacity and the magnitude of the association constant. For highly specific ligands, increasing the ligand density may increase the steric hindrance at the surface and make the binding less efficient, thus decreasing the association constant and the utilization of the ligands. Small protein molecules with clearly defined binding clefts are much more likely to have monovalent interactions with the adsorbent. The binding of s-protein to the peptide sequence YNFEVL has been shown to have a 1:1 stoichiometry[48]. Adsorption isotherm measurements in a batch system have shown that the binding capacity increases from 0.0466 to 1.1650 μmol/mL, while the peptide utilization decreases from 96% to 40%, and the binding constant decreases from 1.2×10^5 to 5.6×10^4/M when the peptide density increases from 0.05 to 3.0 μmol/mL[6] (Figure 3.7a). The binding of large protein

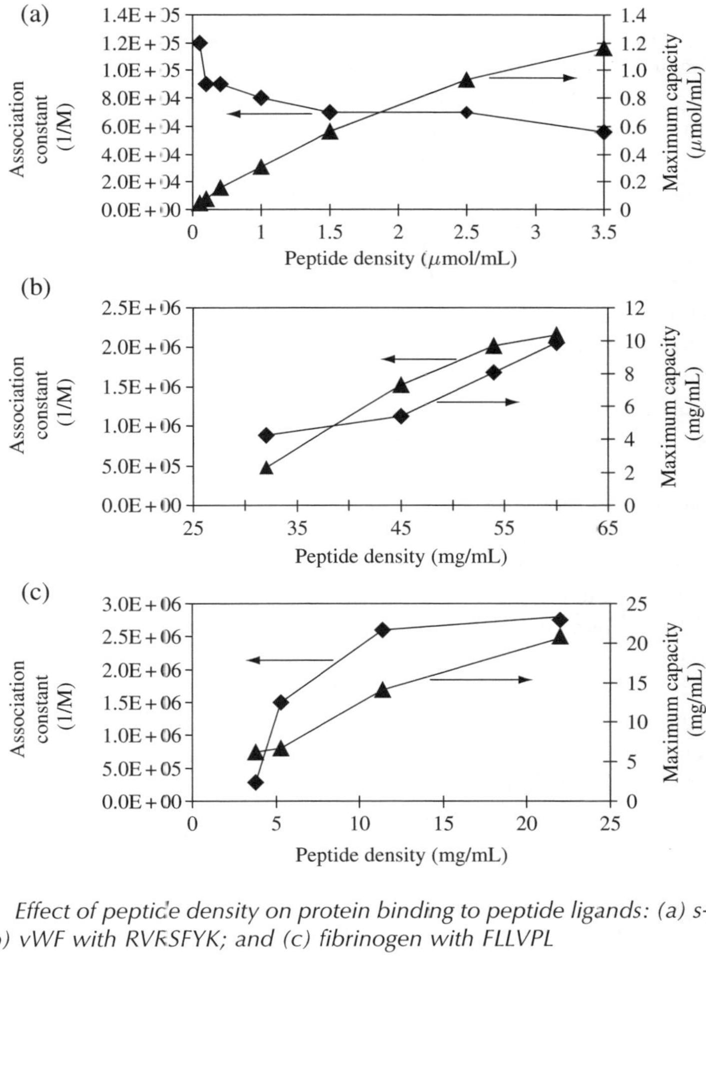

Figure 3.7 *Effect of peptide density on protein binding to peptide ligands: (a) s-protein with YNFEVL; (b) vWF with RVRSFYK; and (c) fibrinogen with FLLVPL*

molecules to peptide ligands is much more likely to be multivalent because the protein may cover many different peptide ligands upon adsorption. The results of binding of vWF to a small peptide ligand, Ac-RVRSFYK, immobilized on Toyopearl resin, show that the association constant increases from 8.82×10^5 to 2.06×10^6/M, and the maximum capacity from 2.32 to 10.33 mg/mL when the peptide density increases from 32 to 60 mg/mL[7] (Figure 3.7b). Kaufman *et al.*[17] also noted that the binding of fibrinogen to FLLVPL on Toyopearl amino resin was consistent with cooperative binding, either through multiple peptides attaching to the protein or changes in the structure of the peptide or the protein. The association constant increases from 2.9×10^5 to 2.8×10^6/M and the maximum capacity increases from 6.2 to 20.6 mg/mL when the FLLVPL density increases from 0.5 to 22 mg/mL resin (Figure 3.7c). Highly specific ligands require a much lower peptide density to obtain a high capacity, while the pseudo-affinity, multi-ligand mode of interaction is much less efficient in terms of ligand utilization, often requiring a ratio of over 100 peptides to bind each protein molecule.

3.6.2 Ligand–Target Interactions

As might be expected, the driving force for binding of the peptide ligands and the target protein molecule depends on the composition and orientation of the amino acids in the peptide ligand and on the protein surface. Any charged amino acids in the peptide ligand tend to form ionic interactions with the target molecule, while hydrophobic leads potentially contact hydrophobic patches on the target molecule driven by hydrophobic interactions. The terminal amine on the peptide is often positively charged and it can tend to attract negative-charged proteins to the surface. Once the protein is in the vicinity of the surface it can interact with hydrophobic and polar groups on the peptide, so that the net interaction tends to be much stronger than either pure ion exchange or pure hydrophobic interaction chromatography.

The differences between peptide affinity interactions and ion exchange chromatography are best reflected by the effect of salt concentration on adsorption. Traditional ion exchange resins adsorb at low salt concentrations, but elute proteins completely at salt concentrations between 0.25 and 0.5 M. Even though such weak binders that are predominantly driven by ion exchange are often identified through the peptide screening process from combinatorial libraries, they are not of real interest because they do not tend to be sufficiently specific to be classified as affinity binders. Peptides that are identified as a result of secondary and tertiary screening cannot be eluted simply by addition of 0.5 M NaCl because of the importance of other polar and non-polar interactions with the target.

Kaufman *et al.*[17] published a detailed characterization of FLLVPL, an affinity ligand for fibrinogen. The temperature dependence of the interaction between FLLVPL and fibrinogen was studied. As the temperature increased, the maximum fibrinogen binding capacity and the association constant increased. Changes in the association constant caused by increasing the system temperature were used to estimate the changes in enthalpy and entropy of the interaction of FLLVPL and fibrinogen. The Gibbs free energy was initially calculated using the thermodynamic equation $\Delta G = -RT \ln K$. The enthalpy of the interaction was estimated using the van't Hoff equation, $\partial \ln K/\partial T = \Delta H/RT^2$. The entropy of the interaction was then calculated using the relation between the Gibbs free energy change of adsorption and the related entropy and enthalpy changes, $\Delta G = \Delta H - T\Delta S$. For the interaction of FLLVPL and fibrinogen, the enthalpy and entropy changes were both positive while the free energy change was negative. In fact, the entropy change of adsorption is larger than the enthalpy change. Large changes in entropy are often associated with hydrophobic interactions since the orientation of water molecules at hydrophobic interfaces is one of the major contributors to the change in free energy of adsorption. The specific binding of fibrinogen to ligand FLLVPL is dominated by hydrophobic interactions with the peptide and ionic interactions with the free terminal amino group[17].

The binding of vWF to RVRSFY on Toyopearl resin is related to the pH of the binding buffer and the pI of the protein (pI = 5.8)[7]. RVRSFY has a net positive charge below pH 12, while vWF has a net negative charge above its pI. RVRSFY bound more than 90% of the loaded vWF at pH values between 5 and 12. Below pH 5, the binding of vWF fell to below 10%. Although NaCl at concentrations up to 2 M could not elute vWF from RVRSFY, $CaCl_2$ and $MgCl_2$ were able to elute approximately 80% of the bound vWF at concentrations larger than 0.3 M. It was also found that as the temperature increased, more vWF eluted when washed with $CaCl_2$. These results suggest that the interaction between vWF and RVRSFY has a large ionic component that is dependent on the charge difference of the protein and the peptide.

Gurgel *et al.*[49] presented a detailed characterization of the interaction between the peptide WHWRKR with α-La by looking at the effect of temperature on the α-La elution profile from a WHWRKR column. With increase in column temperature, more α-La was eluted in 2% acetic acid instead of in the salt wash. Intermittent chromatograms showed a gradual transition between the end point temperatures. It was also found that if the positive charge at the N-terminus of WHWRKR was acetylated, all of the α-La was eluted from the column after the acetic acid wash at high temperatures. Gurgel and co-workers suggested that at low temperatures, the interaction between WHWRKR and α-La is dominated by electrostatic interactions. As the temperature increases, hydrophobic interactions become the dominant binding mechanism. Chromatograms from AcWHWRKR column seem to support this hypothesis since removal of the positive terminal charge increases the overall interaction between the peptide and the protein. Thermodynamics also suggests that as the temperature is increased, the importance of electrostatic interactions will decrease. Instead, the entropic contributions of absorption dominate the interaction of the protein and the peptide.

Wang *et al.*[18] found that addition of 1M NaCl in the binding buffer favored somewhat the SEB binding to YYWLHH. Since the peptide ligand is positively charged and SEB is also positively charged at the pH of the binding buffer, adding salt tends to reduce electrostatic repulsion and enhance hydrophobic interactions with the aromatic residues on the peptide. It was also found that there was a significant reduction in binding of SEB to YYWLHH as a result of adding 0.05% Tween to the binding buffer. This indicates that hydrophobic interactions are the dominant driving force in binding to the peptide. It needs to be pointed out that the hydrophobic interactions between SEB and YYWLHH are apparently specific, as other peptides chosen from the primary screening with similar hydrophobicity cannot bind SEB even with the addition of 1 M NaCl. The positive charge at the N-terminus has no contribution to the SEB binding, but it could bind negatively charged impurities in the feed stream, e.g. DNA and RNA in the *E. coli* lysate, to block the binding sites for SEB. Wang and co-workers recommended adding salts in the binding buffer to eliminate the non-specific electrostatic binding. The blocked binding sites for SEB were completely recovered by adding 0.5 M NaCl in the binding buffer in the purification of spiked SEB from *E. coli* lysate.

3.6.3 Role of Peptide Amino Acid Sequence

Once a lead has been identified that binds the target protein in a competitive environment, the conditions and the peptide need to be optimized to meet the desired industrial application. Since it is rare to screen an entire solid phase library, mutations are often made to the identified sequences in an attempt to create a better binder. Understanding the role of each amino acid in the sequence can help to make mutations that will have a direct impact on the binding of the target molecule.

When the peptide is able to bind directly onto a cleft on the protein, as is the case with the s-protein and the peptide YNFEVL, even minor variations, or even exchanges of one amino acid for another in the same sequence, can disrupt the binding efficiency[6]. Randomizing the peptide sequence from the original can completely disrupt the ability of the peptide to bind to the cleft on the s-protein. This is the most precise and sensitive interaction described in the literature between a ligand from a peptide library and a target protein. However, in the case where the target molecule is large, and there is no clear binding cleft, small changes in peptide structure can result in some remarkable changes on interactions that affect both yield and purity of the active protein during the separation.

Buettner *et al.*[9] investigated the role each amino acid residue in the sequence YANKGY played in binding to factor IX plasma protein. Results showed that the core sequence KGY was essential to factor IX affinity. Truncations from the carboxy terminus resulted in significant reduction or elimination of all factor IX binding. Several peptides, such as YA, bound small amounts of factor IX. However, the peak area for such peptides did not grow proportionally to the amount of factor IX injected. For sequences that bound factor IX specifically, the peak area grew proportionally to the amount of factor IX injected. Buettner *et al.* observed that the truncated sequence NKGY bound more factor IX than ANKGY. However, the amount of bound factor IX was only proportional to the amount injected for the full sequence, demonstrating the importance of the entire sequence.

Huang *et al.*[7] demonstrated the potential benefits of point mutations on the sequence RLRSFY that was found to be specific to vWF. Twelve individual amino acid mutations were made to the sequence RLRSFY to try and increase yield and purity. Mutations were made so that the nature of the original amino acid was conserved: hydrophilic amino acids were replaced with hydrophilic amino acids and charged amino acids were replaced with charged amino acids. The mutated sequence RVRSFY achieved a 76% yield of vWF from Koate-HP versus the 50% yield obtained with the original RLRSFY sequence. The vWF was also almost pure without albumin contamination unlike the vWF recovered from RLRSFY. The mutations by Huang and co-workers also demonstrated the specificity of the overall amino acid sequence. Point mutations to the first amino acid in the sequence resulted in a significant reduction in vWF binding. The point mutation of RLRSFY to QLRSFY resulted in only a 2% recovery of vWF. Bastek *et al.*[5] observed a similar sensitivity to small mutations in the peptide structure on binding effectiveness. The sequence VIWLVR was identified using a multi-tiered screening process as a good binder for α_1-PI. Since Tryptophan is subject to degradation, the sequence was mutated to VIFLVR, and the effect on yield and purity was analyzed. Through this single-point mutation, the yield at 4° C increased from 67% to 83% while the purity dropped from 100% to 89%. At 20° C, the yield increased from 75% to 90% while the purity dropped from 100% to 84%.

3.6.4 Rates of Adsorption

One of the characteristics of affinity chromatography is that the rate of adsorption of a protein to a ligand on a resin tends to be a rate-limiting step. This is in sharp contrast to ion exchange or hydrophobic interaction chromatography, where the adsorption step can be considered to be essentially at equilibrium, and interparticle and intraparticle diffusion dominate the overall rate of adsorption.

Kaufman *et al.*[17] measured the effect of peptide density on the adsorption and desorption rate constants on the resin. Columns with varying peptide densities were challenged with a fibrinogen solution at a constant flow rate. The concentration of the exit stream was measured continuously as a function of time. The shapes of the breakthrough curves were modeled using a chromatography model that took into account axial dispersion, interparticle mass transfer, intraparticle diffusion, and the rates of adsorption and desorption of the protein to the surface. All the mass transfer parameters were estimated from correlations or measured directly, and the only remaining parameter in each run was the adsorption rate constant onto the resin. The rate constant for adsorption was obtained by finding the best fit to the breakthrough curve. The resulting analysis showed that the adsorption rates were indeed rate-limiting and that the rate of adsorption was relatively independent of peptide density.

Kaufman and co-workers also looked at the effect of flow rate on the column breakthrough experiments. Using a FLLVPL column with a peptide density of 11 mg/mL, fibrinogen was loaded onto the column at various flow rates (0.1, 0.5, and 1.0 mL/min). As the flow rate increased, the residence time inside the column decreased, resulting in a significant loss in dynamic binding capacity. Since adsorption is rate-liming, the reduced residence time resulted in significant losses in dynamic binding capacity at the faster flow rates and shorter residence times. Wang *et al.*[18] also found that the adsorption kinetics were rate-limiting in the adsorption of SEB to a peptide column.

3.6.5 Lifetime of Peptide Affinity Column

For an affinity resin to be cost-effective it must be reusable for an adequate number of cycles. Column lifetime is a crucial parameter to be determined in the validation process. Kaufman *et al.*[17] showed that the peptide ligand FLLVPE for fibrinogen purification could be subjected to 180 cycles of repeated loading of sample, washing, elution of fibrinogen, cleaning, and regeneration without either performance or peptide concentration loss. The column could be stored in 20% ethanol to maintain its full capacity and specificity to fibrinogen after regeneration.

Kelly *et al.*[50] presented a complete process validation study for a peptide ligand that was derived from a phage-displayed peptide library for the purification of recombinant B-Domain Deleted factor VIII (BDDrFVIII). The peptide column was reused 26 times without any loss in resin performance. The lifetime study was not extended further because there is no requirement for the process economics to go beyond this lifetime

3.7 Future Challenges and Opportunities

Affinity separation methods will play a significant role in the manufacturing of biologicals in the future. With the separation and purification of a product accounting for as much as 80% of the cost of production, technologies that are robust, inexpensive, and applicable to a wide range of targets will lead the continued development of affinity technologies, many of which will be based on ligands found from combinatorial libraries[51,52].

Combinatorial libraries can play an important role in the future development of affinity ligands for a wide range of potential separation applications, from therapeutics to pathogen removal and detection. This application complements other uses of combinatorial libraries, for drug development, organic and inorganic compound identification, and the development of new catalysts. For applications to protein therapeutics, the combinatorial libraries of peptides and other small ligands on chromatographic resins offer significant advantages. These include a library that is already on a platform that is ready for use on a larger scale, with ligands that are significantly more robust and less expensive than antibodies but more selective than simpler ion exchange or hydrophobic interaction chromatography. These libraries are likely to find an increasing number of applications in separations, sensing, diagnostics, and removal of a wide variety of different chemical species. Peptides from combinatorial libraries can also serve as templates for the design of small organic molecules based on triazine and other chemistries. Future work is likely to focus on ligands that bind to viruses, prion protein (for transmissible spongiform encephalopathies, TSE), toxins, and other harmful agents. These small, robust, and relatively inexpensive ligands may find a major role to play in large volume applications as surrogates for antibody functionality.

References

[1] Lowe C.R. 2001. Curr. Opin. Chem. Biol., 5, 248.

[2] Cuatrecasas P., Wilchek M. and Anfinsen C.B. 1968. Proc. Natl. Acad. Sci. USA, 61, 636.

[3] Lowe C.R. 1995. Chem. Soc. Rev., 24, 309.

[4] Baumbach G.A. and Hammond D.J. 1992. BioPharm, 24.

[5] Bastek P.D., Land J.M., Baumbach G.A., Hammond D.H. and Carbonell R.G. 2000. Separation Sci. Technol., 35, 1681.

[6] Huang P.Y. and Carbonell R.G. 1995. Biotechnol. Bioeng., 47, 288.

[7] Huang P.Y., Baumbach G.A., Dadd C.A., Buettner J.A., Masecar B.L., Hentsch M., Hammond D.J. and Carbonell R.G. 1996. Bioorg. Med. Chem., 4, 699.

[8] Huang P.Y. and Carbonell R.G. 1999. Biotechnol. Bioeng., 63, 633.

[9] Buettner J.A., Dadd C.A., Baumbach G.A., Masecar B.L. and Hammond D.J. 1996. Int. J. Pept. Protein Res., 47, 70.

[10] Amatschek K., Necina R., Hahn R., Schallaun E., Schwinn H., Josic D. and Jungbauer A. 2000. J. High Resol. Chromatogr., 23, 47.

[11] Makriyannis T. and Clonis Y.D. 1997. Biotechnol. Bioeng., 53, 49.

[12] Murray A., Sekowski M., Spencer D.I.R., Denton G. and Price M.R. 1997. J. Chromatogr. A, 782, 49.

[13] Murray A., Spencer D.I.R., Missailidis S., Denton G. and Price M.R. 1998. J. Pept. Res., 52, 375.

[14] Fassina G., Ruvo M., Palombo G., Verdoliva A. and Marino M. 2001. J. Biochem. Biophys. Methods, 49, 481.

[15] Gurgel P.V., Carbonell R.G. and Swaisgood H.E. 2001. Separation Sci. Technol., 36, 2411.

[16] Gurgel P.V., Carbonell R.G. and Swaisgood H.E. 2001. Bioseparation, 9, 385.

[17] Kaufman D.B., Hentsch M.E., Baumbach G.A., Buettner J.A., Dadd C.A., Huang P.Y., Hammond D.J. and Carbonell R.G. 2002. Biotechnol. Bioeng., 77, 278.

[18] Wang G., De J., Schoeniger J.S., Roe D.C. and Carbonell R.G. 2004. J. Pept. Res., 64, 51.

[19] Lawrence M.C. and Davis P.C. 1992. Proteins, 12, 31.
[20] Saragovi H.U., Greene M.I., Chrusciel R.A. and Kahn M. 1992. Biotechnology, 10, 773.
[21] Lam K.S., Lebl M. and Krchnak V. 1997. Chem. Rev., 97, 411.
[22] Cwirla S.E., Peters E.A., Barrett R.W. and Dower W.J. 1990. Proc. Natl. Acad. Sci. USA, 87, 6378.
[23] Devlin J.J., Panganiban L.C. and Delvin P.E. 1990. Science, 249, 404.
[24] Scott J.K. and Smith G.P. 1990. Science, 249, 386.
[25] Daniels D.A. and Lane D.P. 1996. Methods, 9, 494.
[26] Zwick M.B., Shen J. and Scott J.K. 1998. Curr. Opin. Biotechnol., 9, 427.
[27] Goldman E.R., Pazirandeh M.P., Mauro J.M., King K.D., Frey J.C. and Anderson G.P. 2000. J. Mol. Recognit., 13, 382.
[28] Sato A., Ida N., Fukuyama M., Miwa K., Kazami J. and Nakamura H. 1996. Biochemistry, 35, 10441.
[29] Lebl M., Krchnak V., Sepetov N.F., Seligmann B., Strop P., Felder S. and Lam K.S. 1995. Biopolymers (Peptide Science), 37, 177.
[30] Lam K.S., Salmon S.E., Hersh E.M., Hruby V.J., Kazmierski W.M. and Knapp R.J. 1991. Nature, 354, 82.
[31] Furka A. and Sebetyen F. 1991. Int. J. Pept. Protein Res., 37, 487.
[32] Lam K.S. and Lebl M. 1994. Methods, 6, 372.
[33] Lam K.S. and Lebl M. 1992. Immunol. Methods, 1, 11.
[34] Lam K.S., Lake D., Salmon S.E., Smith J., Chen M.L., Wade S., Abdul-Latif F., Knapp R.J., Leblova Z., Ferguson R.D., Krchnak V., Sepetov N.F. and Lebl M. 1996. Method. Enzymol., 9, 482.
[35] Smith M.H., Lam K.S., Hersh E.M., Lebl M. and Grimes W.J. 1994. Mol. Immunol., 31, 1431–1437.
[36] Jentoft N. and Dearborn D.G. 1983. Methods Enzymol., 91, 570.
[37] Kassarjian A., Schellenberger V. and Turck C.W. 1993. Pept. Res., 6, 129.
[38] Mondorf K., Kaufman D.B. and Carbonell R.G. 1998. J. Pept. Res., 52, 526.
[39] Nestler H.P., Wennemers H., Sherlock R. and Dong D.L.-Y. 1996. Bioorg. Med. Chem. Lett., 6, 1327.
[40] Turck C.W. 1994. Methods, 6, 394.
[41] Chen L.A., Buettner J.A. and Carbonell R.G. 2000. US Patent, No. 6,191,256.
[42] Buettner J.A., Dadd C.A., Baumbach G.A. and Hammond D.J. 1997. US Patent, No. 5,723,579.
[43] Chen J.K., Lane W.S., Brauer A.W., Tanaka A. and Schreiber S.L. 1993. J. Am. Chem. Soc., 115, 12591.
[44] Needels M.C., Jones D.G., Tate E.H., Heinkel G.L., Kochersperger L.M., Dower W.J., Barrett R.W. and Gallop M.A. 1993. Proc. Natl. Acad. Sci. USA, 90, 10700.
[45] Lam K.S., Wade S., Abdul-Latif F. and Lebl M. 1995. J. Immunol. Methods, 180, 219.
[46] Liu G. and Lam K.S. 2000. In, Combinatorial Chemistry, Fenniri H. (Ed.). Oxford University Press, New York, p. 33.
[47] Evans D.M., Williams K.P., McGuinness B., Tarr G., Regnier F., Afeyan N. and Jindal S. 1996. Nat. Biotechnol., 14, 504.
[48] Smith G.P., Schultz D.A. and Ladbury J.E. 1993. Gene, 128, 37.
[49] Gurgel P.V., Carbonell R.G. and Swaisgood H.E. 2001. J. Agric. Food Chem., 49, 5765.
[50] Kelly B.D., Tannat M., Magnusson R., Hagelberg S. and Booth J. 2004. Biotechnol. Bioeng., 87, 400.
[51] Labrou N.E. 2003. J. Chromatogr. B, 790, 67.
[52] Narayanan S.R. 1994. J. Chromatogr. A, 658, 237.

4

Bioprocess Scale-up: SMB as a Promising Technique for Industrial Separations Using IMAC

E.M. Del Valle, R. Gutierrez and M.A. Galán

4.1 Introduction

We would like to begin with a simple question: Where would biotechnologists and pharmacist be without liquid chromatography?

Column liquid chromatography can help in the separation of almost any mixture of components, to yield pure proteins, peptides or synthetical formulae for application. Potential applications are in agrochemicals, food, pharmaceutical industry, fine chemistry, etc. In these industries, the traditional separation processes (absorption, distillation column, liquid–liquid extraction, etc.) must be rejected either because of the thermal stability of the substances or for economic reasons. Consequently, separation by chromatographic methods appears to be competitive for very high purity separation.

The separation processes by preparative chromatography provide possibilities of separation with very high yield multi-component mixtures. This versatility is a result of establishing in many ways a difference in the affinity of components for a sorbent phase. The affinity can be based on size, charge or hydrophobocity, and can frequently moderate and frequently be modulated by the addition of solvents (in reversed phase chromatography) or salts (in ion exchange or hydrophobic interaction chromatography). Furthermore, there are many sorbents available, each with its own specific application area. Affinity chromatography is recognized, among the most selective chromatography separations, as a powerful technique for purifying enzymes and other biochemical materials.

Immobilized-metal affinity chromatography (IMAC) is a separation technique that uses covalently bound chelating compounds on solid chromatographic supports to entrap metal

Chemical Engineering Edited by Miguel A. Galán and Eva Martin del Valle

ions, which serve as affinity ligands for various proteins, making use of coordinative binding of some amino acid residues exposed on the surface. As with other forms of affinity chromatography, IMAC is used in cases where rapid purification and substantial purity of the product are necessary, although compared to other affinity separation technologies it cannot be classified as highly specific, but only moderately so. On the other hand, IMAC holds a number of advantages over biospecific affinity chromatographic techniques, which have a similar order of affinity constants and exploit affinities between enzymes and their cofactors or inhibitors, receptors and their ligands or between antigens and antibodies. The benefits of IMAC, ligand stability, high-protein loading, mild elution conditions, simple regeneration and low cost[1], are decisive when developing large-scale purification procedures for industrial applications.

Everson and Parker[2] were the first to adapt immobilization of chelating compounds to the separation of metalloproteins. The method became popular through the research work of Porath and co-workers[3–7] and Sulkowski[8–12] who laid the basis of the technique that is widely used today. It is applicable for a variety of purposes, including analytical and preparative purification of proteins, as well as being a valuable tool for studying surface accessibility of certain amino acid residues. Initially, IMAC techniques were used for separating proteins and peptides with naturally present, exposed histidine residues, which are primarily responsible for binding to immobilized metal ions. However, the work of Hochuli *et al.*[13,14] pioneered the efficient purification of recombinant proteins with engineered histidine affinity handles attached to the N- or C-terminus, especially in combination with the Ni(II)-nitrilotriacetic acid (Ni-NTA) matrix, which selectively binds adjacent histidines. Since numerous neighbouring histidine residues are uncommon among naturally occurring proteins, such oligo-histidine affinity handles form the basis for high selectivity and efficiency, often providing a one-step isolation of proteins at over 90% purity.

Another distinct advantage of this kind of IMAC over biospecific affinity techniques is its applicability under denaturing conditions. This is often necessary when recombinant proteins are highly expressed in *Escherichia coli* in the form of inclusion bodies. When appropriate cleavage sites are engineered between the affinity tags and proteins, with the purpose of enabling effective and precise tag removal after the main isolation step, IMAC seems to be an ideal solution for many applications. However, for the production of therapeutic proteins in substantial quantities, multiple operational cycles with high reproducibility are required as well as the minimal leaching of metal ion, exact terminals, and defined minimal values of host cell proteins, DNA, endotoxins, viruses, etc. To this end, the principles of the method have been studied intensively, and numerous modifications have been made for specific purposes. Because theoretical and practical issues of IMAC have already been widely reviewed by several authors[1,5,8,9,15,16], this short chapter will focus on novel uses and problems that have surfaced in recent years.

4.2 Purification of Proteins Using IMAC

Numerous natural proteins contain histidine residues in their amino acid sequence. However, histidines are mildly hydrophobic and only a few of them are located on the protein surface. For proteins with known 3D structure, data about the number and arrangement of surface histidine residues can be obtained from protein data banks. This can also serve as a basis for forecasting their behaviour in IMAC. In the coming years, with the development of proteomics, the number of proteins with known primary structure is bound to grow much faster than the number of 3D structures resolved, but structure modelling, based on the known primary amino acid sequence, will also become more useful and more accurate. However, until now, no data from systematic searches of the protein databases regarding surface histidines have been published.

For use in IMAC, protein-surface histidine residues must also be accessible to the metal ions and their bulky chelating compounds. However, the microenvironment of the binding residue, cooperation between neighbouring amino acid side groups, and local conformations play important roles in protein retention. In this way, IMAC can serve as a sensitive tool for revealing protein topography with respect to histidines and their surroundings[5,9]. Depending on proximity and orientation of histidines and density of the chelating groups and metal ions, as well as on spatial accessibilities between the support particles and the protein, multipoint binding of different histidines can be achieved[15]. In general, the protein shows the highest affinity for the metal surface arrangement which best matches its own distribution of functional histidines. Adjacent histidines can bind to the same or different chelating sites. Usually, one histidine is enough for weak binding to iminodiacetic acid-Cu(II) (in the following IDA-Cu(II)), while more proximal histidines are needed for efficient binding to Zn(II) and Co(II). Some interesting examples of using IMAC for proteins with naturally exposed histidine residues include human serum proteins[3,4,17], interferon[8], lactoferrin and myoglobin[11], tissue plasminogen activator[18], antibodies[19,20] and yeast alcohol dehydrogenase[21]. In general, a positive correlation is found between the number of accessible histidines and the strength of binding[8]. Separation of γ-chymotrypsin, a common contaminant in commercial α-chymotrypsin, was achieved by IMAC, due to various numbers of surface histidines[22], indicating possible industrial application. Interesting IMAC behaviour is exhibited by natural cytochrome C from different species, which differ in their histidine content[7]. Similarly, evolutionary variants of the lysozymes show varied affinities for IMAC matrices due to differences in the surface topography of histidines[23]. On the other hand, albumins contain up to 16 histidine residues in their structure but only one high-affinity binding site His3 at the N-terminus[8]. Recently, human serum proteins have been used for testing new IMAC affinity ligands[24,25].

4.2.1 Histidine Tags

Although the first demonstrations of IMAC were low-resolution group separations, the resolution has significantly improved with the use of genetically engineered affinity tags that can be attached to amino or carboxy terminals of the recombinant proteins. The first histidine-rich fusions were made on the basis of the high affinity of certain natural proteins containing histidine residues near the N-terminus. For instance, an octapeptide derived from angiotensine I was fused to the N-terminus of the TEM-β-lactamase and expressed in *E. coli* in the form of inclusion bodies. One-step purification of the recombinant protein from the resolubilized inclusion body material was achieved on IDA-Zn(II)[26]. Recently, a natural amino acid sequence, located on the N-terminus of chicken lactate dehydrogenase, has been described which is responsible for efficient binding to Co(II)-carboxymethylaspartate IMAC. The natural peptide contains six histidines, unevenly interleaved by other amino acid residues[27]. Its truncated version, designated as histidine affinity tag HAT, was fused to the N-terminals of three recombinant proteins to demonstrate its utility as a purification tag[28]. In the past, numerous histidine tags were employed, from very short ones, e.g. HisTrp, utilized for isolation of sulfitolized proinsulin[29] to rather long extensions, containing up to eight repeats of the peptide Ala-His-Gly-His-Arg-Pro, attached to various model proteins[30]. However, today by far the most widely used histidine tags consist of six consecutive histidine residues. After the appearance of the papers by Hochuli *et al.*[13,14], describing a new chelating matrix Ni-NTA and fusions with short peptides, containing two to six neighbouring histidines, these hexa-histidine tags have become very popular. Commercial expression vectors, containing nucleotides coding for His6, His10 and some other fusions, have been on the market for several years. However, His10 tags, even though efficient[31], have never received as much attention as His6. There are a very large number of papers on the use of His6 tag[13,14,32–35]. Recently, a versatile strategy using His6-GFP (green fluorescent protein), fused to the target protein, has been published, enabling simple fluorescence monitoring of the expression and localization, as well as easy purification of the fusion protein by IMAC[33,36]. The principle of polyhistidine tags is based on the premise that multiplicity of histidines may increase binding. On the other hand, very high affinity, which is an absolute requirement in some immobilized-metal-ion-based non-chromatographic technique single-stage processes, such as partitioning, is not always advantageous in chromatographic multi-stage processes[1]. An ideal affinity tag should enable effective but not too strong a binding, and allow elution of the desired protein under mild, non-destructive conditions. In the case of recombinant *E. coli*, many host proteins strongly adhere to the IMAC matrices, especially when charged with Cu(II) or Ni(II) ions, and are eluted with the target proteins. Therefore, new approaches for selecting improved histidine tags have focused on elution of the target protein in the 'contaminant-free' window. Interestingly, selection of an optimum tag by a phage-displayed library showed that tags with only two histidine residues possessed chromatographic characteristics superior to those of the most commonly used His6 tag[37,38]. Similarly, in many cases, IDA-Zn(II) may prove superior to either immobilized Cu(II) or Ni(II) ions, as a result of its relatively low binding affinity for host cell proteins[39]. Oligomeric proteins, as for example trimeric TNF-α, pose additional difficulties when one is searching for useful affinity tags, since interactions with the matrix are multiplied[40].

A different approach to achieving selective adsorption of engineered oligo-histidine-tagged proteins with minimal interference of host cell proteins involves 'tailor-made' chelating supports with very short spacer arms and low surface density of chelating groups[41].

Histidine tags seem to be compatible with all expression systems used today. Thus, His-tagged proteins can be successfully produced in procaryotic and eucaryotic organisms, intracellularly or as secreted proteins[42]. The use of long histidine tags in *E. coli* cells may reduce the accumulation level or induce the formation of inclusion bodies of otherwise soluble protein[43]. However, which position is preferable for the addition of His tag, N- or C-terminus, depends on the nature and intended use of the protein, and must be determined experimentally. Addition of His tag to the N-terminus of the protein appears to be more universal, if judged from the huge number of cases reported. Most likely, N-terminal tagging is more frequently used because several efficient endoproteases are available for precise cleavage of the tag after purification. Histidine tagging and IMAC have become a routine for easy first-time isolation of newly expressed proteins. In most cases, the histidine tags neither affect protein folding nor interfere significantly with the biological functionality.

4.2.2 Designed Histidine Patches and Motifs

In contrast to histidine tags, the possibility of engrafting new surface histidines for easy purification depends very much on the intended use of the protein. In therapeutic proteins there is a serious limitation of this approach because an authentic surface of the protein is usually required. Also 3D structure and active sites must be well characterized if one intends to design a protein with the desired affinity towards the chosen IMAC matrix. In many cases, a high enough affinity can be achieved when two or more surface histidines lie approximately in a plane. Thus, a concerted attachment of all exposed histidines is possible. Flexible loops are among the most attractive regions for the introduction of new histidines or for the replacement of existing amino acid residues. However, no universal rule exists and every protein and its 3D structure represent a special case. Therefore, we mention here just a few examples.

After recognizing that some high-affinity natural binding sites – such as His-X–His3, two histidines separated by a turn in α-helical structure, as in myoglobin or human fibroblast interferon – are most probably responsible for binding to IDA-Zn(II) and IDA-Co(II)[8,12], these sequences were engineered into cytochrome C and bovine somatotropin[1,44]. The mutant proteins actually demonstrated higher affinity. Similarly, the Zn(II)-binding site of human carbonic anhydrase, which includes three histidine residues, was successfully engineered onto the surface of the retinol-binding protein[45]. In general, the sites for introducing histidine residues must be exposed and separated structurally from the active site of the protein. Thus, their design is most easily accomplished when the biochemical properties and 3D structure are known. Another successful example of a newly introduced histidine cluster consists of mutants of glutathione transferase[45], constructed on the basis of the natural rat enzyme that contains two adjacent histidine residues forming a four-histidine cluster on the surface of the dimeric protein[47]. A similar effect was achieved by introducing one or two histidine residues into the flexible-loop region of the trimeric molecule of TNF-α[40,48], which resulted in planar surface clusters of three or six histidines and very good chromatographic characteristics in IMAC matrices. Although such newly designed histidine clusters can be very effective for rapid purification and can also be used for immobilization purposes, the engineered proteins are mutants which differ to a greater or a lesser extent from the authentic structures with respect to immunogenicity, biological activity, stability, etc. However, this approach could be very useful for many other groups of proteins not intended for human therapy, e.g. industrial enzymes, proteins for diagnostic purposes and enzymes for research.

4.2.3 Large-Scale Purification of Therapeutic Proteins

Many reports on IMAC used for purifying pharmaceutically interesting proteins, such as interferons, vaccines and antibodies, have been published but relatively few data exist on actual large-scale purifications of pharmaceutical proteins. However, IMAC offers possibilities for large-scale purification of many industrial enzymes as well as proteins for research in genetics, molecular biology and biochemistry[49–53].

Recently, some interesting reports on IMAC techniques used for purifying vaccines have appeared. For example, an efficient purification procedure for malaria vaccine candidates, expressed as His6-tagged proteins in *E. coli*, was described[54]. Addition of His6 tag to the hepatitis B virus core antigen (HBcAg), expressed in *E. coli*, enabled purification under milder denaturing conditions by Ni-NTA at high pH. Contaminating *E. coli* proteins and DNA were completely removed. This was otherwise impossible by standard sedimentation of virus-like particles in sucrose gradients[35]. Whole chimaeric virus-like particles of infectious bursal disease (young-chicken virus disease) were isolated on Ni(II) ProBond™ from insect cells, Sf-9, coinfected with two strains of baculoviruses. A His5 tag was added to one protein, which ensured sufficiently strong binding to the IMAC matrix and mild elution of particles. This approach avoided extensive centrifugation and led to simple and low-cost vaccine production[55].

A malaria-transmission-blocking vaccine candidate, based on the *Plasmodium falciparum* predominant surface protein Pfs25 with a His6-tag at the carboxyl terminus, was produced by secretion from *Saccharomyces cerevisiae* and purified on a large scale by Ni-NTA. Histidine-tagged protein exhibited higher potency and antigenicity than the original Pfs25[34]. This indicates that in some cases vaccination with His-tagged proteins may be advantageous.

His6 tag was also used for producing several clinical-grade single-chain Fv antibodies[32,56] and IMAC proved superior to traditional antigen affinity chromatography[32]. IMAC on Cu(II)-charged chelating sepharose has been used for large-scale preparation of clinical-grade factor IX[57].

There are many more reports on the application of His6 tag for IMAC isolation of potential therapeutics, but the majority of them describe preliminary procedures, and do not usually give details about histidine tag removal and final yields. However, IMAC technology should be further improved with respect to metal-ion leakage, dynamic capacity, reproducibility, etc. We can conclude that there are many attempts to use IMAC matrices for large-scale isolation of biopharmaceuticals, but many are still in the trial phase, or the data are not accessible to the public.

Expanded-bed adsorption (EBA) techniques constitute another broad field of IMAC application and require additional properties of column matrix, e.g. higher particle density and high resistance to harsh conditions during column cleaning or sanitization. Expanded-bed techniques are less attractive on a small, laboratory scale but potentially highly advantageous at an industrial scale. Downstream processing procedures from unclarified *E. coli* or yeast homogenates are being developed for native[21,58] as well as histidine-tagged proteins[59]. Generally, recoveries of over 80% of the protein were achieved in successful cases, but at least two major weak features must be further improved: low dynamic capacity and efficiency of clean in place (CIP) procedures for eliminating contaminants.

Elimination of centrifugation and filtration in large industrial-scale isolations is a major driving force for the introduction of EBA in the isolation of therapeutic proteins. Streamline chelating (Amersham Pharmacia Biotech) has been tried to purify two vaccine candidates for clinical studies: His6-tagged modified diphtheria toxin, expressed in *E. coli*, and malaria-transmission-blocking vaccine, secreted from *S. cerevisiae*[60].

The combination of IMAC and EBA techniques should provide a unique approach to simplifying the whole downstream process, reducing the number of steps and start-up investment, and thus making the purification more economical.

4.3 The Basis of the Problem

Although almost any separation is technically feasible, the efficiency of conventional fixed bed chromatography may be (too) low for industrial application. Current process scale chromatographic separations suffer from a few drawbacks:

- Mass transfer rates and pressure drop may be limiting. Both properties play a crucial role in chromatography performance, speed and scale-up.
- The sorbent inventory is high, which implies high costs, as the sorbents are costly.
- The use of (salty) buffers is large, which is very undesirable with respect to environmental as well as cost aspects.
- Products can only be harvested in a diluted form, which imposes the requirement for further processing.

These drawbacks are inherent to the current operation of chromatography.

Chromatography columns are operated in batches, which involves a small loading time, combined to a large time to elution. Meanwhile, mass transfer and equilibrium effects lead to the broadening of bands and the dilution of the separating fractions. This observation formed an important argument to begin the search for 'more efficient process chromatography'.

4.4 More Efficient Chromatographic Methods

In the literature, a few methods to improve chromatographic efficiency are described. Some promising examples are:

- displacement chromatography;
- two-way chromatography;
- recycle chromatography;
- use of ceramic monoliths as stationary phase;
- simulated moving bed chromatography.

In *displacement chromatography* a displacer is introduced after the feed injection[61–63]. The displacer has a high affinity towards the chromatography resin, which results in the development of an 'isotachic train', and an array of narrow, highly concentrated peaks of the pure components in order of their affinity.

The displacement train is a result of the roll-up effect that takes place when species interfere. Although displacement chromatography produces very pure and concentrated products, it has some severe disadvantages. One drawback is that the bands in the isotachic train are very narrow, which makes the harvest of the pure products a non-trivial task. A second drawback is the need to introduce an additional, strongly adsorbing species, which is undesirable as it is hard to remove from the resin.

In *two-way chromatography* displacement effects are exploited as well[64]. Here, there is no addition of a displacing species. By alternating the direction of flow, the more retained species in the feed serve as the displacer for the less retained species. Two-way chromatography may lead to an elevation of the concentration. However, a serious drawback is that it is far too complex for separation of a multi-component mixture.

The important feature of *recycle chromatography* is that products are harvested from the column before complete resolution has taken place[65]. Only the pure fronts and tails of the peaks are collected, whereas the 'unseparated' fraction leaving the column is mixed with fresh feed and resupplied to the column. This method of operation minimizes the losses of feed as the result of improper resolution; however, this is at the expense of column volume due to the reintroduction of the (diluted) recycle. A second drawback is the possible accumulation of undesired components, a major fear in pharmaceutical applications.

Use of ceramic monoliths as stationary phase in affinity chromatography: The search for faster and more efficient separation methods for the downstream processing of large, complex molecules resulted in the introduction of coated ceramic monoliths as active matrix supports in affinity chromatography[66]. Ceramic monoliths are a collection of square (or triangular) capillaries packed in a compact structure. The designation of a monolith is based on the number of cells per square inch of frontal area and on the thickness of the ceramic walls. The monolith used in the experiments described by del Valle *et al.*[66] were of the type 400/6, that is, a monolith with 400 cells/in^2 and a wall thickness of 6/1000 of an inch. The monoliths, made of cordierite ($Al_4Mg_2Si_5O_{18}$), a low thermal expansion material with outstanding mechanical resistance, were coated with agarose. The coating gel, which covers the inside walls of the capillaries with an average thickness between 10 and 50 μm, is thicker in the corners, leaving flow channels with near-circular cross-sections.

The macroporous structure of monoliths allows the overcoming of some of the disadvantages of conventional affinity chromatography[67]. Monoliths have lower mass transfer resistance and pressure drop than conventional random packed beds, and mass transfer within monolith channel rates can be substantially larger than mass transfer in packed beds used in conventional chromatography.

Whereas in packed bed chromatography, mass transfer rates and pressure drop may be limiting, in monoliths surface interactions determine the overall reaction rate[67]. The feasibility of using ceramic monoliths as support in affinity chromatography has been clearly established[66].

A ceramic monolith can be coated with an agarose gel and activated using the same procedure used to activate a bed of agarose beads. It is possible to increase or decrease the coating load in order to have a thicker or thinner coat, and it is possible to make a monolith bed as large as any commercially available monolith. There is no indication that any of the chemicals present in cordierite interfere with our separation process or that they even come in contact with the enzyme solutions[66].

There are many advantages in using a monolith for affinity chromatography separations:

1. There is very little pressure drop through the monolith[66]; thus pumping rate and superficial velocities are determined by mass transfer and adsorption needs and not by maximum pressure drop across the bed. Flow rates can be orders of magnitude larger in a monolith than in a bed of small agarose beads.
2. There is very little or no liquid trapped inside the monolith when there is no flow and the monolith is drained. Thus, back mixing with the tails of the adsorption or elution flows is very small.
3. Mass transfer effects on desorption rates are very small or negligible.

Numerical simulations have been made possible by the availability of experimental data on a well-characterized geometry and with accurate concentration measurements[67]. The simple geometry of ceramic monoliths is essential for accurate numerical modelling with no independent adjustable parameters such as tortuosities or effective diffusions within porous media. The only adjustable parameter, the peak diffusion at the moving contact line, will be eliminated when an acceptable, simple flow model of split-ejection streamlines is available.

Accurate numerical simulations, in turn, allow precise estimation of physical constants and point to areas or experimental conditions where data are needed in order to improve understanding. By dissecting the mass transfer problem into solvable, manageable mathematical expressions, one can explore in detail the values of relevant parameters and the sensibility of the overall solution to the actual values of these parameters. An example of how this can be accomplished is the discussion in the previous section of the determination of the inhibition constant, K_i. There are other concepts, however, that could greatly benefit from improved understanding.

The *footprint* of an enzyme becomes an important issue if adsorption is limited by the outside area of the coated channels, since the area of the footprint is inversely proportional to the amount of protein adsorbed. The mobility or overall diffusivity of the protein is also important when working under inertial or electrical fields that affect the mobility. The length of the spacer arm is a well-known fundamental issue in protein adsorption, but it will become determinant when proteins of large footprint must be adsorbed in scarce sites.

When monoliths are used as support, the scale-up problem is trivial. Every capillary channel behaves in the model, lab-scale and prototype chromatographic column exactly as it will behave in a large commercial column. As long as the fluid velocity inside the capillary is the same in the prototype and in the industrial unit, the results will be identical. If the diameter of the column is large, i.e. nearly a metre in diameter, problems of even flow distribution may develop at relatively low flow rates and a stochastic flow distribution model may have to be included. Thus, the ability to predict information and to use this information for the scale-up of separation/purification systems is a much-needed tool in the design of high throughput separation processes.

In *simulated moving bed (SMB) chromatography*, not only the liquid but also the resin is (simulated to be) in motion[68]. This countercurrent contact allows the continuous fractionation of a feed in two product fractions. The countercurrent operation assures a high driving force towards mass transfer. In the SMB, it is enough when the products exist in the pure form only at the product outlet ports. All these processes lead to a very efficient use of resin. As a result of the low dilution of products, the consumption of eluent can be reduced compared to fixed bed chromatography.

In the SMB system, the feed is continuously recycled in the system. This is advantageous from an efficiency point of view; however, it makes distinction of separate batches impossible. This was initially seen as a drawback of the technology in pharmaceutical applications.

In the SMB, it seems to be possible to reduce both resin inventory and eluent consumption and maintain a high product concentration at the same time. This is not possible using the other 'more efficient' options, which lead to improvement at only one of these points. Further advantages of SMB over the other options are that SMB is applicable to large-scale processes and does not involve any additional species that are hard to recover.

4.4.1 SMB History

By the early 1960s, SMB systems were being developed. The pioneering patent of Broughton of UOP[69] describes the setup of the SMB system and its application in the petrochemical industry. Since then, the technique found several large-scale applications, for instance in fractionation of saccharides (e.g. separation of glucose and fructose), xylenes, olefins and paraffins[70]. The SMB, according to the most commonly used Sorbex layout, is schematically depicted in Figure 4.1. It consists of four sections, which are numbered I through IV. The liquid and sorbent move countercurrent, as is indicated by arrows in the figure. Both liquid and sorbent are recycled. At the 'centre' of the system, a feed is introduced. The desorbent is introduced at the bottom, and delivers regeneration of the column.

The more retained components in the feed are predominantly transported in their sorbed form, and move downwards along with the resin. They are harvested in the extract product. The less retained components move upwards along with the liquid. They are harvested in the raffinate product. The direction of movement is determined by the ratio of the liquid to sorbent flow rate. A high ratio results in upward movement of the components.

In chromatography systems, the adsorbent cannot move, since this modify resolution. Because of this, the movement of the sorbent is stimulated instead. This is done by dividing the sorbent bed into small fractions, the size of one column. Thus, the system consists of (say) 12 interconnected columns. Per switch interval, the sorbent is moved one column in the opposite direction of liquid movement, thus simulating downward sorbent movement. Usually, three or four columns per section are sufficient to simulate the countercurrent movement[71]. After a number of cycles, the SMB is at a cyclic steady state: the profiles in the columns do not change when moving from one switch to another.

The SMB only functions properly when the ratio of liquid to sorbent flow rate has been chosen properly. Complete separation, that is, when there is a pure extract of the more retained component as well as a pure raffinate of the less retained component, can only then be achieved. The design of systems obeying linear isotherms, such as in separation of glucose and fructose, is relatively easy[72]. However, the design of a separation with nonlinear isotherms is much more complicated. That is why the new developments in SMB technology have only been initiated in the late 1980s, after the development of fast computers.

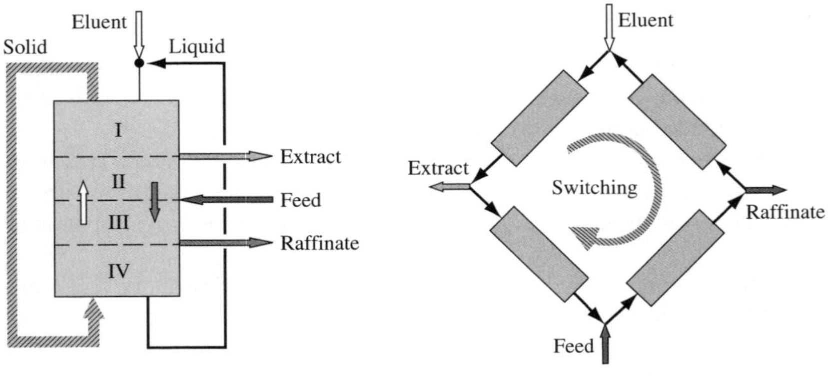

Figure 4.1 *Equivalent true countercurrent system*

Since then, robust design procedures have been developed for systems obeying Langmuir and stoichiometric isotherms. The basis of these design procedures lies in wave theory[73–75]. The most commonly used procedure for flow selection in SMB has been developed by Morbidelli and co-workers[76,77]. This method is also termed 'triangle theory', in reference to the triangularly shaped regions that form the 'working area'. Other procedures have been described as well[72,78,79]. Also, much attention has been paid to the computation of the profiles and performance of the SMB at given settings[72,80–82].

4.4.2 The Challenge

Looking at the literature, only a few applications of SMB in biotechnology were described[78,83–87].

Most of these considered a rather experimental setup, without the application of a robust design procedure. This defined the niche in which to position the current project. The aims of this project are as follows:

- demonstrate the possibility of fractionation of mixtures of proteins by SMB chromatography;
- develop methods for flow selection of the specific separations;
- optimize the fractionation processes.

The use of a gradient in salt concentration in the SMB was considered as a promising option for further improvement of ion exchange SMB processes.

4.4.3 Modelling

The moving bed simulation is carried out by connecting several chromatographic columns in series (Figure 4.2). The countercurrent movement is simulated by moving the feed stream and the input/output connections in a cyclic way, on the whole column sections.

The installation allows a continuous production by chromatographic separation by simulating the displacement of the countercurrent bed of the eluent phase. This simulation is done by sequenced displacement of the injection points, from one column to another, upstream to the eluent phase. The time interval between two displacements is called the switching time.

During this period of time, the chromatographic profile migrates in the same direction as the fluid inside the separator. The distribution of these points along the separator is selected according to the chromatographic profile. Desorbent is injected into a buffer zone. The mixture to be separated is injected into the column that has the richest mixture of an identical composition. The raffinate and the extract are collected at the outputs of the columns with the maximum purity and concentration (these last two will evolve during the sequence).

The starting of the installation is done in two stages:

- a transient mode, allowing the development of the required chromatographic profile in the separator;
- a pseudo-steady-state mode, allowing the continuous collection of the fractions of raffinate and extract at specified concentrations and purities.

In terms of operability, the optimal design of this type of process depends on various factors such as the number of sections of columns, their length and diameter, the flows,

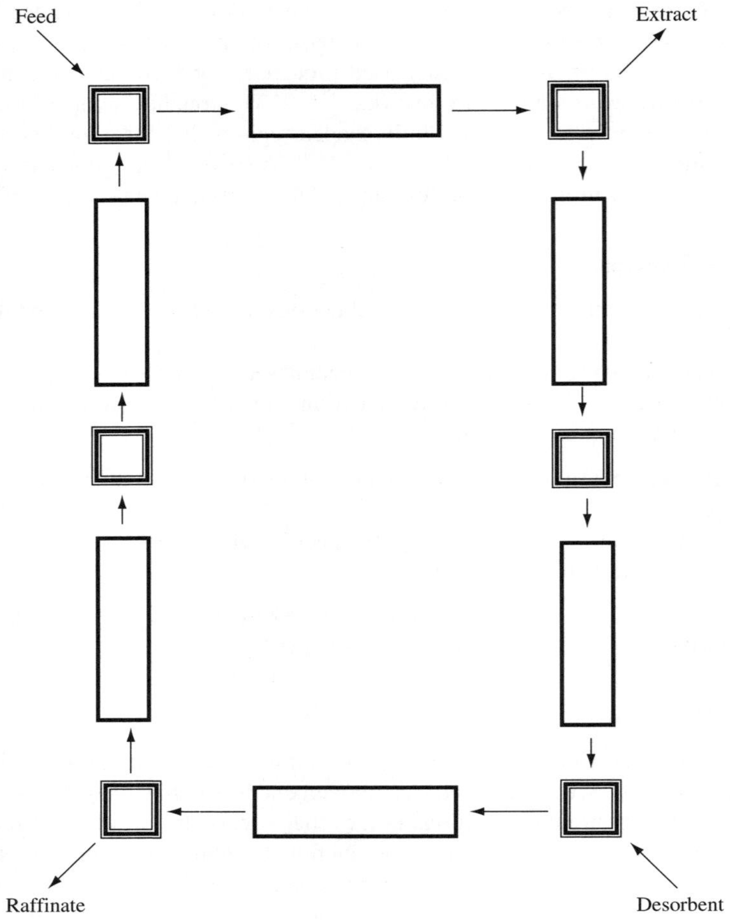

Figure 4.2 *SMB principle*

and the switching times between two points of product injection or collections. Because of the complex dynamic, the choice of the operational parameters is far from being easy. For this task, it is necessary to use a detailed and reliable dynamic model which will take into account the continuous dynamics of elementary columns as well as the management of the discrete state events resulting from the cyclic policy or from the production method retained for the various products to be split.

4.5 Representation of the Phenomena

In the case of a chromatographic column, several phenomena are involved (Figure 4.3):

- The aqueous solution transport and dispersion in the moving phase (A).
- The mass transfer between the solid phase and the moving phase (B).

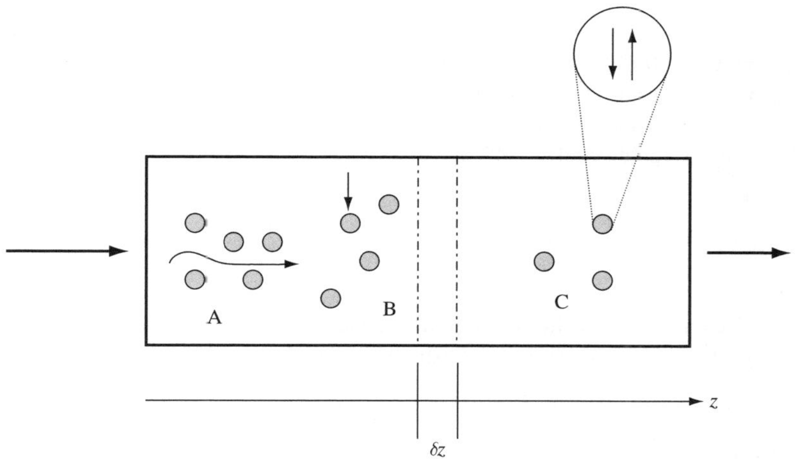

Figure 4.3 *Phenomena description in a chromatographic column*

- The aqueous solution transport and diffusion in the solid phase (C).
- The adsorption equilibrium in the solid phase (C).

Generally, the mass balance on an infinitely small column section z for each i component can be represented by the following equations.

4.5.1 In Moving Phase

Through a δz section (Figure 4.3) the relation that represents the evolution of the solute concentration in the moving phase is written:

$$-E_A \frac{\partial^2 C_i}{\partial z^2} + \frac{\partial(v \cdot C_i)}{\partial z} + \frac{\partial C_i}{\partial t} + \frac{1-\varepsilon}{\varepsilon}\frac{\partial C_{si}}{\partial t} = 0 \tag{4.1}$$

where

$\frac{1-\varepsilon}{\varepsilon}$ = the phases ratios (solid volume/liquid volume)
C_{si} = concentration for solute i in the solid phase
C_i = concentration of solute i in the liquid phase
ε = bead void fraction
E_A = axial dispersion coefficient
z = column length's discretization
t = time

4.5.2 In Solid Phase

The mass transfer between liquid and solid phases can be represented by considering the concentrations in identical solutes, on surface and inside pores, but supposing that there is a resistance to transfer between the solid phase and the liquid phase.

$$\frac{\partial C_{si}}{\partial t} = k(C^*_{si} - C_{si}) \tag{4.2}$$

where

C_{si} = concentration for solute i in the solid phase;
C_{si}^* = concentration for solute i in the solid phase when the thermodynamic equilibrium is reached;
k = mass transfer coefficient

and

$$\frac{1}{k} = \frac{1}{k_f} + \frac{1}{k_i} \tag{4.3}$$

Equation 4.3 clearly shows that the overall resistance to mass transfer is the sum of the resistance in the liquid film and the resistance in the pore fluid.

4.6 Conclusions

Simulating moving beds have been successfully and widely used in petrochemistry for almost 30 years. Clearly, this technology has a great potential for fine chemistry and pharmaceutical industry. More and more applications are described for the biochemical field (leading sometimes to 10 times lower eluent consumption compared with the usual chromatography). Since the small-scale units are already available, SMB can be used for very small-scale production (less than 1 kg) as well as for very large production (10 hundred tons per year), for very different enzymes.

SMB is basically a binary separator that presents three main advantages:

- It enables us to save significant amounts of eluent.
- It enables us to maximize productivity. The value of SMB with respect to batch chromatography is maximized for low selectivity problems or low efficiency systems.
- It is a continuous process that simplifies the operation and particularly the connection to associated equipments.

References

[1] Arnold F.H. 1991. Metal-affinity separations: a new dimension in protein processing, Biotechnology, 9, 150.
[2] Everson R.J. and Parker H.E. 1974. Zinc binding and synthesis eight-hydroxy-quinoline-agarose, Bioinorg. Chem., 4, 15.
[3] Porath J., Carlsson J., Olsson I. and Belfrage G. 1975. Metal chelate affinity chromatography: a new approach to protein fractionation, Nature, 258, 598.
[4] Porath J. and Olin B. 1983. Immobilized metal ion affinity adsorption and immobilized metal ion affinity chromatography of biomaterials: serum protein affinities for gel-immobilized iron and nickel ions, Biochemistry, 22, 1621.
[5] Porath, J. 1992. Immobilized metal ion affinity chromatography, Protein Expr. Purif., 3, 263.
[6] Ramadan N. and Porath J. 1985. Fe^{3+}-hydroxamate as immobilized metal affinity-adsorbent for protein chromatography, J. Chromatogr., 321, 93.
[7] Hemdan E.S., Zhao Y.J., Sulkowski E. and Porath, J. 1989. Surface topography of histidine residues: a facile probe by immobilized metal ion affinity chromatography, Proc. Natl. Acad. Sci. USA, 86, 1811.

[8] Sulkowski E. 1985. Purification of proteins by IMAC, Trends Biotechnol., 3, 1–7.
[9] Sulkowski E. 1989. The saga of IMAC and MIT, Bioassays, 10, 170–175.
[10] Sulkowski E. 1996. Immobilized metal-ion affinity chromatography: imidazole proton pump and chromatographic sequelae: I. Proton pump, J. Mol. Recognit., 9, 389–393.
[11] Sulkowski E. 1996. Immobilized metal-ion affinity chromatography: imidazole proton pump and chromatographic sequelae: II. Chromatographic sequelae, J. Mol. Recognit., 9, 494–498.
[12] Sulkowski E. 1987. Immobilized metal ion affinity chromatography of proteins. In, Protein purification, micro to macro, Burgess R. (Ed.). Allan R. Liss, New York, pp. 149–162.
[13] Hochuli E., Bannwarth W., Doebeli H., Gentz R. and Stueber D. 1988. Genetic approach to facilitate purification of recombinant proteins with a novel metal chelate adsorbent, Biotechnology, 6, 1321.
[14] Hochuli E., Doebeli H. and Schacher A. 1987. New metal chelate adsorbent selective for proteins and peptides containing neighbouring histidine residues. J. Chromatogr., 411, 177.
[15] Johnson R.D. and Arnold F.H. 1995. Multipoint binding and heterogeneity in immobilized metal affinity chromatography, Biotechnol. Bioeng., 48, 437.
[16] Yip T.T. and Hutchens T.W. 1994. Immobilized metal ion affinity chromatography, Mol. Biotechnol., 1, 151–164.
[17] Wu H.P. and Bruley D.F. 1999. Homologous human blood protein separation using immobilized metal affinity chromatography: protein C separation from prothrombin with application to the separation of factor IX and prothrombin, Biotechnol. Prog., 15, 928–931.
[18] Dodd I., Jalalpour S., Southwick W., Newsome P., Browne M.J. and Robinson J.H. 1986. Large scale, rapid purification of recombinant tissue-type plasminogen activator, Febs Lett., 209, 13.
[19] Boden V., Winzerling J.J., Vijayalakshmi M. and Porath J. 1995. Rapid one-step purification of goat immunoglobulins by immobilized metal ion affinity chromatography, J. Immunol. Methods, 181, 225.
[20] Freyre F.M., Vazquez J.E., Ayala M., Canaan-Haden L., Bell H., Rodriguez I., González A., Cintado A. and Gavilondo J.V. 2000.Very high expression of an anti-carcinoembryonic antigen single chain Fv antibody fragment in the yeast *Pichia pastoris*, J. Biotechnol., 76, 157.
[21] Willoughby N.A., Kirschner T., Smith M.P., Hjorth R. and Titchener-Hooker N.J. 1999. Immobilised metal ion affinity chromatography purification of alcohol dehydrogenase from baker's yeast using an expanded bed adsorption system, J. Chromatogr. A, 40, 195–204.
[22] Sagar S.L., Beitle R.R., Ataai M.M. and Domach M.M. 1992. Metal-based affinity separation of alpha- and gamma-chymotrypsin and thermal stability analysis of isolates, Bioseparation, 3, 291.
[23] Zhao Y.J., Sulkowski E. and Porath J. 1991. Surface topography of histidine residues in lysozymes. Eur. J. Biochem., 202, 1115–1119.
[24] Chaouk H. and Hearn M.T. 1999. New ligand, N-2-pyridylmethyl aminoacetate, for use in the immobilised metal ion affinity chromatographic separation of proteins, J. Chromatogr. A, 852, 105.
[25] Chaouk H. and Hearn M.T. 1999. Examination of the protein binding behaviour of immobilised copper(II)-2,6-diaminomethylpyridine and its application in the immobilised metal ion affinity chromatographic separation of several human serum proteins, J. Biochem. Biophys. Methods, 39, 161.
[26] Beitle R.R. and Ataai M.M. One-step purification of a model periplasmic protein from inclusion bodies by its fusion to an effective metal-binding peptide, Biotechnol. Prog., 9, 64.
[27] Chaga G., Hopp J. and Nelson P. 1999. Immobilized metal ion affinity chromatography on Co^{2+}-carboxymethylaspartate-agarose Superflow, as demonstrated by one-step purification of lactate dehydrogenase from chicken breast muscle, Biotechnol. Appl. Biochem., 29(Pt 1), 19.

[28] Chaga G., Bochkariov D.E., Jokhadze G.G., Hopp J. and Nelson P. Natural poly-histidine affinity tag for purification of recombinant proteins on cobalt(II)-carboxymethylaspartate crosslinked agarosa, J. Chromatogr. A, 864, 1999, 247.
[29] Smith M.C., Furman T.C., Ingolia T.D. and Pidgeon C. 1988. Chelating peptide-immobilized metal ion affinity chromatography: a new concept in affinity chromatography for recombinant proteins, J. Biol. Chem., 263, 7211.
[30] Ljungquist C., Breitholtz A., Brink-Nilsson H., Moks T., Uhlen M. and Nilsson B. 1989. Immobilization and affinity purification of recombinant proteins using histidine peptide fusions, Eur. J. Biochem., 186, 563.
[31] Grisshammer R. and Tucker J. 1997. Quantitative evaluation of neurotensin receptor purification by immobilized metal affinity chromatography, Protein Expr. Purif., 11, 53.
[32] Casey J.L., Keep P.A., Chester K.A., Robson L., Hawkins R.E. and Begent R.H. 1995. Purification of bacterially expressed single chain Fv antibodies for clinical applications using metal chelate chromatography, J. Immunol. Methods, 179, 105.
[33] Cha H.J., Wu C.F., Valdes J.J., Rao G. and Bentley W.E. 2000. Observations of green fluorescent protein as a fusion partner in genetically engineered *Escherichia coli*: monitoring protein expression and solubility, Biotechnol. Bioeng., 67, 565.
[34] Kaslow D.C. and Shiloach J. 1994. Production, purification and immunogenicity of a malaria transmission-blocking vaccine candidate: TBV25H expressed in yeast and purified using nickel–NTA agarosa, Biotechnology, 12, 494.
[35] Wizemann H. and von Brunn A. 1999. Purification of *E. coli*-expressed His-tagged hepatitis B core antigen by Ni^{2+}-chelate affinity chromatography. J. Virol. Methods, 77, 189–197.
[36] Wu C.F., Cha H.J., Rao G., Valdes J.J. and Bentley W.E. 2000. A green fluorescent protein fusion strategy for monitoring the expression, cellular location, and separation of biologically active organophosphorus hydrolase. Appl. Microbiol. Biotechnol., 54, 78–83.
[37] Goud G.N., Patwardhan A.V., Beckman E.J., Ataai M.M. and Koepsel R.R. 1997. Selection of specific peptide ligands for immobilised metals using a phage displayed library: application to protein separation using IMAC, IJBC, 2, 123.
[38] Patwardhan A.V., Goud G.N., Koepsel R.R. and Ataai M.M. 1997. Selection of optimum affinity tags from a phage-displayed peptide library: application to immobilized copper (II) affinity chromatography, J. Chromatogr. A, 787, 91.
[39] Pasquinelli R.S., Shepherd R.E., Koepsel R.R., Zhao A. and Ataai M.M. 2000. Design of affinity tags for one-step protein purification from immobilized zinc columns, Biotechnol. Prog., 16, 86.
[40] Gaberc-Porekar V., Menart V., Jevsevar S., Vidensek A. and Stalc A. 1999. Histidines in affinity tags and surface clusters for immobilized metal-ion affinity chromatography of trimeric tumor necrosis factor alpha, J. Chromatogr. A, 852, 117.
[41] Armisen P., Mateo C., Cortes E., Barredo J.L., Salto F., Diez B., Rodés L., García J.L., Fernández-Lafuente R. and Guisán J.M. 1999. Selective adsorption of poly-His tagged glutaryl acylase on tailor-made metal chelate supports, J. Chromatogr. A, 848, 61.
[42] Seidler A. 1994. Introduction of a histidine tail at the N-terminus of a secretory protein expressed in *Escherichia coli*, Protein Eng., 7, 1277.
[43] Gaberc-Porekar V. and Menart V. 2001. Review perspectives of immobilized-metal affinity chromatography, J Biochem. Biophys. Methods, 49, 335.
[44] Todd R.J., Van Dam M.E., Casimiro D., Haymore B.L. and Arnold F.H. 1991. Cu(II)-binding properties of a cytochrome c with a synthetic metal-binding site: His-X3-His in an alpha-helix, Proteins, 10, 156–161.
[45] Muller H.N. and Skerra A. 1994. Grafting of a high-affinity Zn(II)-binding site on the beta-barrel of retinol-binding protein results in enhanced folding stability and enables simplified purification, Biochemistry, 33, 14126.

[46] Yilmaz S., Widersten M., Emahazion T. and Mannervik B. 1995. Generation of a Ni(II) binding site by introduction of a histidine cluster in the structure of human glutathione transferase A1-1. Protein Eng., 8, 1163–1169.
[47] Chaga G., Widersten M., Andersson L., Porath J., Danielson U.H. and Mannervik B. 1994. Engineering of a metal coordinating site into human glutathione transferase M1-1 based on immobilized metal ion affinity chromatography of homologous rat enzymes, Protein Eng., 7, 1115.
[48] Menart V., Gaberc-Porekar V. and Harb V. 1994. Metal-affinity separation of model proteins having differently spaced clusters of histidine residues. In, Separations for biotechnology, Pyle D.L. (Ed.), Vol. 3. The Royal Society of Chemistry, Cambridge, pp. 308–313.
[49] Goodey A.R., Sleep D., van Urk H., Berenzenko S., Woodrow J.R., Johnson R.A., Wood P.C., Burton S.J. and Quirk A.V. 1996. Process of high purity albumin production, International patent WO 96/37515.
[50] de Hulster A.F. 1997. Development of a fed-batch fermentation protocol for high cell-density cultivation of recombinant Pichia pastoris, Human Serum Albumin production, Internal report ref. 9610, BIRD Engineering BV, Delft.
[51] Jacobs L. 1998. Large scale production of recombinant HSA with the yeast Pichia pastoris, Final report TwAiO – project, Delft University of Technology.
[52] Kerry-Wiliams S.M., Gilbert S.C., Evans L.R. and Ballance D.J. 1998. Disruption of the *Saccharomyces cerevisiae* YAP3 gene reduces the proteolytic degradation of secreted recombinant human serum albumin, Yeast, 14, 161.
[53] Kobayashi K., Tomomitsu K., Kuwae S., Ohya, T., Ohda T. and Omura T. 1996. Process for producing proteins, EP 0 736 605 A1.
[54] Takacs B.J. and Girard, M.F. 1991. Preparation of clinical grade proteins produced by recombinant DNA technologies, J. Immunol. Methods, 143, 231–240.
[55] Hu Y.C., Bentley W.E., Edwards G.H. and Vakharia V.N. 1999. Chimeric infectious bursal disease virus-like particles expressed in insect cells and purified by immobilized metal affinity chromatography, Biotechnol. Bioeng., 63, 721.
[56] Laroche-Traineau J., Clofent-Sanchez G. and Santarelli X. 2000. Three-step purification of bacterially expressed human single-chain Fv antibodies for clinical applications, J. Chromatogr. B Biomed. Sci. Appl., 737, 107.
[57] Feldman P.A., Bradbury P.I., Williams J.D., Sims G.E., Mcphee J.W., Pinnell M.A., Harris L., Crombie G.I. and Evans D.R. 1994. Large-scale preparation and biochemical characterization of a new high purity factor IX concentrate prepared by metal chelate affinity chromatography, Blood Coagul. Fibrin., 5, 939.
[58] Clemmitt R.H. and Chase H.A. 2000. Immobilised metal affinity chromatography of beta-galactosidase from unclarified *Escherichia coli* homogenates using expanded bed adsorption, J. Chromatogr. A, 874, 27.
[59] Clemmitt R.H. and Chase H.A. 2000. Facilitated downstream processing of a histidine-tagged protein from unclarified *E. coli* homogenates using immobilized metal affinity expanded-bed adsorption, Biotechnol. Bioeng., 67, 206.
[60] Noronha S., Kaufman J. and Shiloach J. 1999. Use of streamline chelating for capture and purification of poly-His-tagged recombinant proteins. Bioseparation, 8, 145.
[61] Brooks C.A. and Cramer S.M. 1992. Steric mass-action ion-exchange: displacement profiles and induced salt gradients, AIChE. J., 38, 1969.
[62] Horváth C., Nahum A. and Frenz J.H. 1981. High performance displacement chromatography, J. Chromatogr., 218, 365.
[63] Subramanian G., Phillips M.W. and Cramer S.M. 1988. Displacement chromatography of biomolecules, J. Chromatogr., 493, 341.
[64] Bailly M. and Tondeur D. 1981. Two-way chromatography: flow reversal in nonlinear preparative liquid chromatography. Chem. Eng. Sci., 36, 455.

[65] Bailly M. and Tondeur D. 1982. Recycle optimization in non-linear productive chromatography I mixing recycle with fresh feed, Chem. Eng. Sci., 37, 1199.

[66] Martin del Valle E.M., Galán M.A. and Cerro R.L. 2003. Use of ceramic monoliths as stationary phases in affinity chromatography, Biotechnol. Prog., 19, 921.

[67] Montes Sanchez F.J., Martin del Valle E.M., Galán M.A. and Cerro R.L. 2004. Modeling of monolith-supported affinity chromatography, Biotechnol. Prog., 20, 811.

[68] Ballanec B. and Hotier G. 1993. From batch to simulated countercurrent chromatography. In, Preparative and production scale chromatography, Ganetsos G. and Barker P.E. (Eds.), Marcel Dekker, New York, pp. 301–357.

[69] Broughton D.B. 1961. US patent 02985589.

[70] Johnson J.A. and Kabza R.G. 1993. Sorbex: industrial-scale adsorptive separation. In, Preparative and production scale chromatography, Ganetsos G. and Barker P.E. (Eds.), Marcel Dekker, New York, pp. 5–12.

[71] Hidajat K., Ching C.B. and Ruthven D.M. 1986. Simulated countercurrent adsorption processes: a theoretical analysis of the effect of subdividing the adsorbent bed, Chem. Eng. Sci., 41, 2953.

[72] Ruthven D.M. and C.B. Ching. 1989. Countercurrent and simulated countercurrent adsorption separation processes, Chem. Eng. Sci., 44, 1011.

[73] Helfferich F.G. and Klein G. 1970. Multicomponent chromatography: theory of interference, Marcel Dekker, New York.

[74] Rhee H.-K., Aris R. and Amundson N.R. 1970. On the theory of multicomponent chromatography, Philos. Trans. R. Soc. Lond. A, 267, 419.

[75] Rhee H.-K., Aris R. and Amundson N.R. 1971. Multicomponent adsorption in continuous countercurrent exchangers, Philos. Trans. R. Soc. Lond. A., 269, 187.

[76] Storti G., Masi M., Carrà S. and Morbidelli M. 1989. Optimal design of multicomponent countercurrent adsorption separation processes involving non-linear equilibria, Chem. Eng. Sci., 44, 1329.

[77] Storti G., Mazzotti M., Morbidelli M. and Carrà S. 1993. Robust design of binary countercurrent adsorption separation processes, AIChE. J., 39, 471.

[78] Hashimoto K., Adachi S. and Shirai Y. 1988. Continuous desalting of proteins with a simulated moving bed adsorber, Agric. Biol. Chem., 52, 2161.

[79] Ma Z. and Wang N.-H.L. 1997. Standing wave analysis of SMB chromatography linear systems, AIChE. J., 43, 2488.

[80] Yun T., Zhong G. and Guiochon G. 1997. Experimental study of the influence of the flow rates in SMB chromatography. A.I.Ch.E. J., 41, 2970.

[81] Zhong G.M. and Guiochon G. 1994. Theoretical analysis of band profiles in nonlinear ideal countercurrent chromatography. J. Chromatogr., 688, 1.

[82] Zhong, G.M. and Guiochon G. 1996. Analytical solution for the linear ideal model of simulated moving bed chromatography. Chem. Eng. Sci., 51, 4307.

[83] Adachi S. 1994. Simulated moving bed chromatography for continuous separation of two components and its application to bioreactors, J. Chromatogr., 658, 271.

[84] Gottschlich N., Weidgen S. and Kasche V. 1996. Continuous biospecific affinity purification of enzymes by simulated moving bed chromatography. Theoretical description and experimental results, J. Chromatogr., 719, 267.

[85] Huang S.H., Lee W.S. and Lin C.K. 1988. Enzyme separation and purification using improved simulated moving bed chromatography. In, Horizontals of biochemical engineering, Aiba S. (Ed.), Oxford University Press, pp. 58–72.

[86] Maki H., Fukuda H. and Morikawa H. 1987. The separation of glutathione and glutamic acid using a simulated moving bed adsorber system. J. Ferment. Technol., 65, 61.

[87] Van Walsem H.J. and Thompson M.C. 1997. Simulated moving bed in the production of lysine, J. Biotechnol., 59, 127.

5

Opportunities in Catalytic Reaction Engineering. Examples of Heterogeneous Catalysis in Water Remediation and Preferential CO Oxidation

Janez Levec

5.1 Introduction

The development of new catalysts during the last two decades has introduced more environmentally accepted processes into the production of commodities. The industrial solid catalysts that once played a major role in bulk chemicals manufacture are nowadays distributed among the industrial sectors so that about 25% of produced catalysts are used in the chemical industry, 40% in the petroleum industry, 30% in environmental protection, and 5% in the production of pharmaceuticals[1]. Environmental catalysis accounts for (i) waste minimization by providing alternative catalytic synthesis of important compounds without the formation of environmentally unacceptable by-products, and (ii) emission reduction by decomposing environmentally unacceptable compounds by using catalysts. Waste minimization is linked with the reaction(s) selectivity and therefore a proper choice of catalyst plays a decisive role. Emission reduction usually refers to end-of-the-pipe treatment processes where the selectivity of catalyst, if used, is not an important issue. Because it is almost impossible to transform the raw materials into the desired products without any by-product(s), one must take account of the necessity of providing a production process with an end-of-the-pipe treatment unit. Only then can

Chemical Engineering Edited by Miguel A. Galán and Eva Martin del Valle

such production be considered benign and harmless to the environment. In this chapter, three examples of environmental catalysis are presented: in the first two, the use of catalysts in water remediation technology is discussed, while in the third example the use of catalysts for CO cleanup of hydrogen for fuel cells is briefly presented.

In a great majority of industrial processes, water is used as a solvent, reaction or transport medium, therefore it is not surprising that many efforts in the last two decades have been made concerning the abatement of pollutants from industrial aqueous waste streams. The increasing demand for the reuse of water and increasingly stringent water quality regulations calls for the treatment of all kinds of wastewaters. The incapability of conventional methods to remove effectively many organic pollutants has made it evident that new, compact, and more efficient systems are needed. Therefore the interest in innovative methods of wastewater treatment based on catalytic oxidation has been growing rapidly.

Another example that concerns public health is groundwater polluted with nitrates and nitrites, which are found in agriculture areas worldwide. Besides biological digestion, heterogeneous liquid-phase hydrogenation over a solid catalyst is another promising technique for removal of these pervasive contaminants. Here the catalyst must exhibit a very high selectivity toward the nitrogen production and must strongly suppress the formation of highly unwanted ammonia. Although the catalytic liquid-phase hydrogenation of nitrate-polluted drinkable water is still in development, more research efforts on new selective catalysts may speed up this process and make it commercially feasible.

Energy production by fuel cells is undoubtedly a process that minimizes the waste. Fuel cells are unique devices for converting the energy of chemical systems into electric power. Proton exchange membrane fuel cells (PEMFC) seem to be the most attractive because they operate at low temperatures. They use hydrogen, which is generated by conventional processes such as steam reforming, partial oxidation, and a combination of both. These are catalytic by nature but result in H_2-reach gas with concentrations of CO too high to be directly used in PEMFC. In order to meet the requirements of PEMFC-grade hydrogen, additional processes must be employed to further reduce the level of CO concentration, since CO poisons the Pt-containing gas diffusion electrode in PEMFC. The selective oxidation of CO in a stream of high hydrogen excess over a solid catalyst, also called preferential CO oxidation (PROX), is considered as one of the most promising methods for tracing CO cleanup. In order to further developing the PROX process many research groups intensively seek new, more selective and reliable catalysts.

5.2 Catalytic Oxidation of Wastewaters

Oxidation processes may be classified into two main groups termed *advanced oxidation processes* (AOPs) and *thermal liquid-phase* or *wet oxidation* (WO), depending on the conditions in which high energy intermediates responsible for the destruction of organic compounds dissolved in water are generated. In AOPs, the generation of active oxygen species, such as hydroxyl radicals, takes place near the ambient temperature and pressure, whereas in the thermal processes these intermediates are formed by thermal reactions at high temperatures and pressures. The use of solid a catalyst to reduce reaction conditions was been proposed in the mid-1970s[2,3,4] and immediately attracted many researchers in the area of heterogeneous catalysis. Contrary to the severe reaction conditions in conventional WO (200–310° C; 20–150 bar), the catalytic process employs milder conditions: temperatures are typically in the range of 130–250° C and pressures of 20–50 bar, with residence times of about 1 h.

5.2.1 Catalysts

Although a number of catalysts are known to have the ability to promote oxidation of organics in aqueous phase, not all catalysts have been found to be suitable. The conditions (temperature in the range of 130–250° C) under which small amounts of organics dissolved in a large amount of water are oxidized force a severe demand on the physical and chemical properties of catalysts. In the last two decades, many studies have revealed that some catalysts sustain these severe conditions and exhibit a life long enough to be considered economically feasible. Literature surveys of the catalyst used in WO are given elsewhere [5–10]. In general, some metal-oxide-compounded catalysts (Cu, Zn, Co, Mn, Bi) are reported to exhibit high activity but they all suffer metal leaching and consequently lose the activity[11,12]. The catalyst deactivation may also occur during the oxidation of aromatic compounds due to the formation of polymeric products[13]. Catalysts based on precious metals deposited on stable supports, such as titanium and cerium oxides, are less prone to deactivation and have already exhibited good results in commercial application[14–19].

Catalysts that have exhibited a reasonably long lifetime consist of rather expensive metals, which is a drawback for any end-of-the-pipe treatment process. It is unlikely that any one type of catalyst can be successfully used for treating many varieties of aqueous waste streams, therefore many different catalyst systems are needed. Investigators should look for systems with less expensive but catalytically active compounds, e.g. manganese and copper, and decrease their solubility by incorporating them into a lattice of catalyst support to accomplish the task.

5.2.2 Oxidation Kinetics

Unfortunately, a vast portion of the WO works reported in the literature deals with the non-catalyzed oxidation kinetics for single compounds. In a review by Matatov-Meytal and Sheintuch[7], it was found that pure compounds such as phenol, benzene, dichlorobenzene, and acetic acid obey a first-order rate law with respect to the substrates and mainly half order with respect to the oxygen concentration. A thorough kinetic investigation in an isothermal, differentially operated fixed bed reactor with the oxygen pre-saturated aqueous solutions has revealed that the catalytic oxidation of acetic acid, phenol, chlorophenol, and nitro-phenol can be well expressed by means of the Langmuir–Hinshelwood kinetic formulation[4,20–22], namely

$$-r_{\mathrm{poll}} = \frac{k_{\mathrm{app}} K_{\mathrm{poll}} C_{\mathrm{poll}} C_{\mathrm{O_2}}^{1/2}}{1 + K_{\mathrm{poll}} C_{\mathrm{poll}}} \tag{5.1}$$

where K_{poll} stands for equilibrium adsorption constant and C_{poll} for the concentration of pollutant. Hamoudi *et al.*[23] introduced a rather complex kinetic model for the catalytic oxidation of aqueous phenolic solutions. Their model is based on the Langmuir–Hinshelwood–Hougen–Watson approach and also accounts for the catalyst deactivation.

However, kinetic models that solely predict the disappearance rate of pure compounds are not sufficient for design purposes. What is needed is a tool capable of predicting complete conversion of all organic species present in wastewater, regardless of whether they are originally present or formed as intermediate products. Therefore the rate law has to be expressed by means of a lumped parameter such as total organic carbon (TOC), which accounts for all organic species present in wastewater, or chemical oxygen demand (COD), which also takes into account oxidizable inorganics. For non-catalytic oxidation Li *et al.*[24] proposed a generalized lumped kinetic model, which is based on a simplified reaction scheme with acetic acid as the rate-limiting intermediate as shown in Figure 5.1.

Here the rates of all steps are of the first-order. This type of lumped kinetic model was recently successfully employed by Pintar *et al.*[25] for the prediction of TOC reduction in Kraft bleach plant effluents over Ru/TiO_2 catalyst. These authors assumed the second-order kinetic behavior of all reaction steps and ended up with the following equations for a slurry system

$$\frac{\mathrm{d}\,[\mathrm{TOC}-2\cdot\mathrm{Ac}]}{\mathrm{d}t} = -k_1\,[\mathrm{TOC}-2\cdot\mathrm{Ac}]^2 - k_2\,[\mathrm{TOC}-2\cdot\mathrm{Ac}]^2 \tag{5.2}$$

$$\frac{\mathrm{d}\,[\mathrm{Ac}]}{\mathrm{d}t} = 0.5\cdot k_2\,[\mathrm{TOC}-2\cdot\mathrm{Ac}]^2 - k_3\,[\mathrm{Ac}]^2 \tag{5.3}$$

with $[\mathrm{TOC}] = [\mathrm{TOC}]_0$, $[\mathrm{Ac}] = [\mathrm{Ac}]_0$, and $t = 0$, where brackets stand for concentration of TOC and acetic acid (Ac). The capability of predicting the TOC decay and accumulation of acetic acid by this model is illustrated in Figures 5.2 and 5.3. It is interesting to note that the second-order rate law is in agreement with the lumped kinetics found by Donlagic and Levec[26] for WO of an azo dye compound in batch reactor.

Belkacemi *et al.*[27] proposed an inhibition-deactivation reaction scheme (Figure 5.4) for the catalytic removal of TOC from raw high-strength alcohol distillery waste liquors. The kinetic model involving the rates for three carbon lumps, namely carbon in liquid (TOC), carbon deposited on solid catalyst (C_S), and carbon in gas phase (C_G), consists of the following set of equations:

$$-\frac{\mathrm{d}[\mathrm{TOC}]}{\mathrm{d}t} = k_T\,[\mathrm{TOC}]\,\tau + k_{WO}\,[\mathrm{TOC}]\,\tau + m_{cat}k_{CWO}\,[O_2]^{0.45}\,[\mathrm{TOC}]\left(1-\frac{C_S}{C_{S\infty}}\right) \tag{5.4}$$

$$-\frac{\mathrm{d}C_S}{\mathrm{d}t} = k_f C_{S\infty}\,[\mathrm{TOC}]\left(1-\frac{C_S}{C_{S\infty}}\right) \tag{5.5}$$

with $[\mathrm{TOC}_0] = [\mathrm{TOC}] + C_G + C_S$, $[\mathrm{TOC}] = [\mathrm{TOC}_0]$, $C_G = C_S = 0$, and $t = 0$.

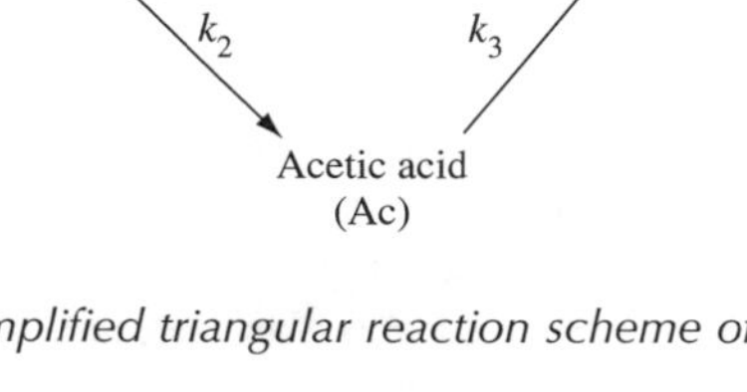

Figure 5.1 *Simplified triangular reaction scheme of wet oxidation*

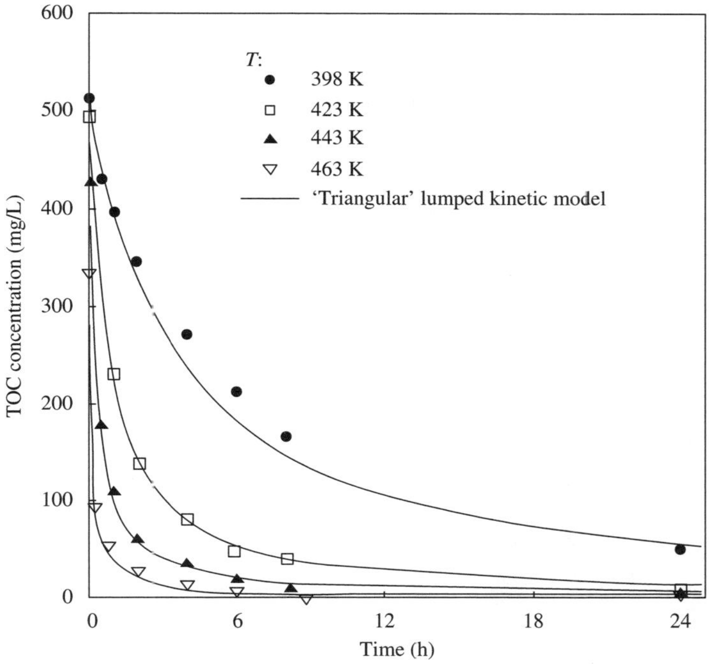

Figure 5.2 *Experimental and predicted concentration profiles for total organic carbon during oxidation of Kraft bleach plant effluent over Ru/TiO$_2$ catalyst. Adopted from Ref.*[25]

In equation 5.4, k_T, k_{WO}, and k_{CWO} are the rate constants for thermolysis, homogeneous oxidation, and catalytic oxidation, respectively, whereas τ represents a simple inhibition function expressed as

$$\tau = 1 - \frac{[\mathrm{TOC}_\infty]}{[\mathrm{TOC}]}.$$

In the above expression, $[\mathrm{TOC}_\infty]$ is the asymptotic residual organic carbon, which cannot be oxidized (R_{CL}) further. Agreement between the experimental data for combined thermolysis, catalytic, and non-catalytic WO and the model prediction is shown in Figure 5.5. From this plot one can conclude that the oxidation progress is terminated once the catalyst is deactivated due to the adsorption of carbonaceous intermediates on its surface. However, for practical design purposes one should use the lump kinetic approach based on the triangular reaction scheme such as depicted in Figure 5.1. It is believed that the rate laws can be expressed mostly by a simple power function.

5.2.3 Oxidation Process

It seems that Katzer *et al.*[28] were the first to evaluate the catalytic WO as a potential wastewater treatment technology. They have shown that catalytic oxidation may not be economically feasible for waste streams containing small amounts of organic material, as is the case with the conventional WO process. For dilute or very dilute wastes the

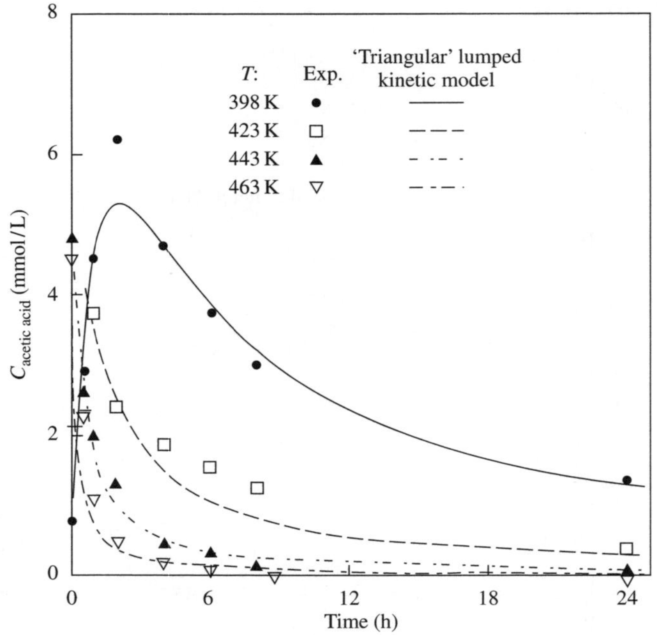

Figure 5.3 *Experimental and predicted concentration profiles for acetic acid during oxidation of Kraft bleach plant effluent over Ru/TiO$_2$ catalyst. Adopted from Ref.*[25]

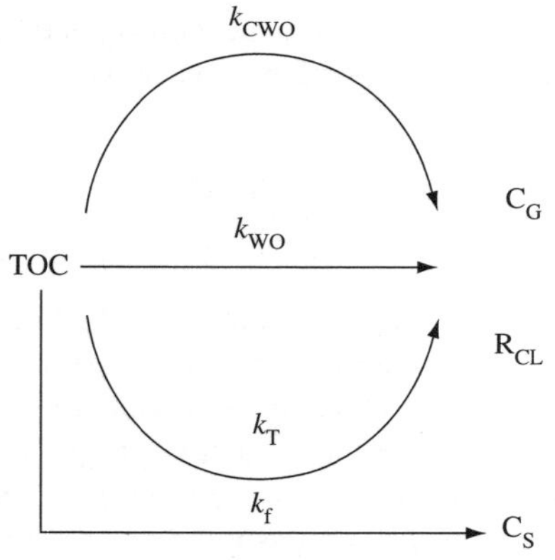

Figure 5.4 *Inhibition–deactivation reaction scheme proposed by Belkacemi* et al.[27]

adiabatic temperature rise is too small, therefore additional fuel is needed. To achieve economic throughputs and conversions, the oxidation has to be carried out autothermally, which requires preheating of the feed stream by the stream leaving the oxidation reactor. Because energy costs preclude vaporization, wet reactors must operate at pressures above

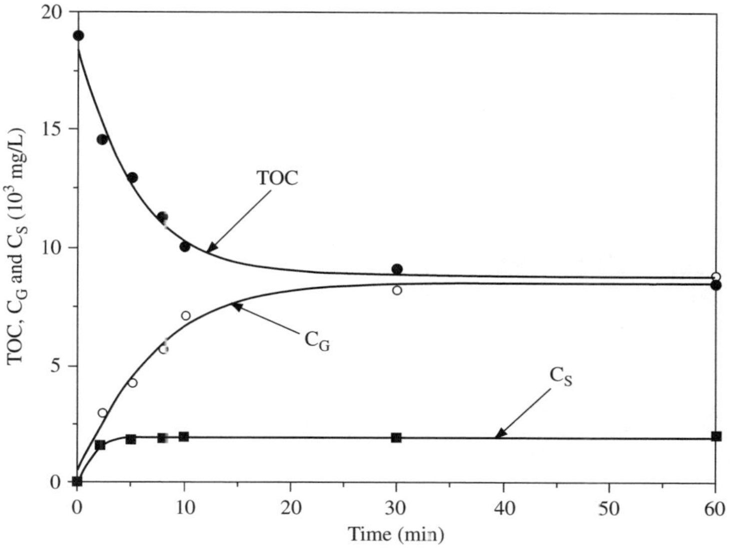

Figure 5.5 *Experimental and predicted concentration profiles for combined thermolysis, catalytic, and non-catalytic wet oxidation of high-strength alcohol distillery liquor over MnO_2/CeO_2 catalyst. Adopted from Ref.*[27]

the vapor pressure of water. Equipment to achieve intimate contacting of the three phases has been predominantly in the form of slurry or fixed bed reactors in which the two fluid phases flow through a stationary bed of catalyst concurrently upwards (bubble-column fixed bed) or downwards (trickle bed). Trickle bed reactors avoid the disadvantages of separating small catalyst particles from the liquid stream associated with slurry reactors and also avoid the limitation of flow rates encountered with up-flow through fixed beds. A reactor with a high liquid to catalyst ratio (e.g. slurry reactor) should not be used for wastewaters containing pollutants that tend to polymerize[13]. A block diagram of the process employing a trickle bed reactor is shown in Figure 5.6.

Several catalytic WO processes were commercialized in the mid-1980s in Japan. They are all based on heterogeneous catalysts containing precious metals deposited on titania or titania–zirconia carriers. In comparison to conventional WO units, some of these processes are able to oxidize recalcitrant acetic acid and ammonia. Wastewater treated in these units can either be discharged directly into an open body of water or reused as process water. The catalytic wet oxidation system (NS-LC) of Nippon-Shokubai, for example, which operates at temperature of 220° C and total pressure of 40 bars, is capable of achieving a 99% reduction of TOC at a liquid-hourly-space velocity (LHSV) of 2. It employs a $Pt–Pd/TiO_2–ZrO_2$ catalyst in the form of honeycomb or particles. For high COD value wastewaters, it consists of a shell-and-tube reactor with catalyst-filled tubes. The Osaka Gas catalytic process uses a catalyst composed of a mixture of precious and base metals on titania or titania–zirconia carriers (honeycomb or spheres). The catalyst lifetime is reported to be longer than eight years. It treats efficiently a variety of municipal waste streams as well as industrial wastewaters.

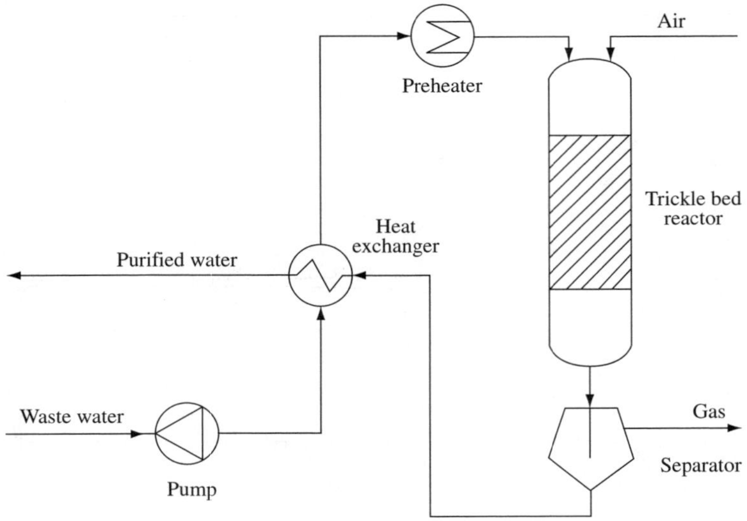

Figure 5.6 *Schematic drawing of a CWO process with single-pass trickle bed reactor*

When a wastewater contains relatively low concentrations of organic material the driving force for the chemical oxidation is very low, therefore some kind of a pre-concentration may be needed[29]. A process which uses activated carbon as an adsorption–pre-concentration step is shown in Figure 5.7 in combination with a trickle bed reactor.

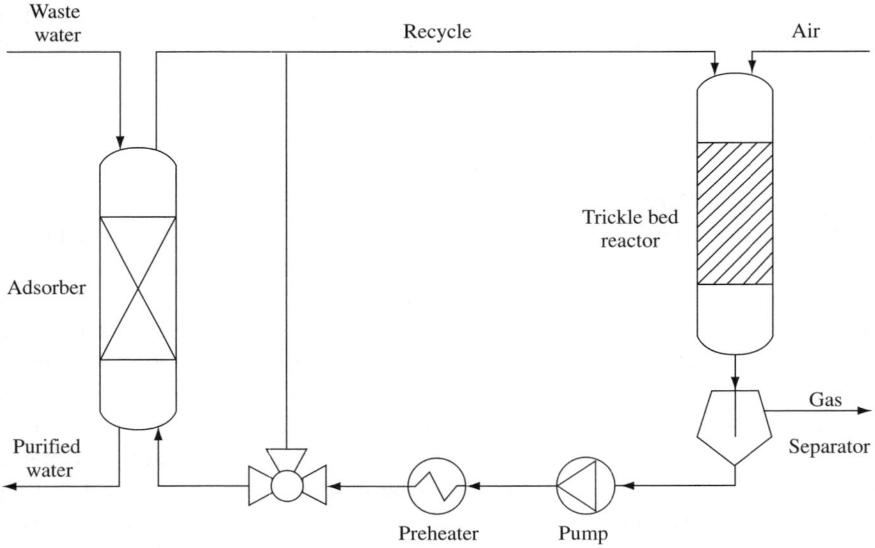

Figure 5.7 *Schematic drawing of a CWO process with adsorber for pre-concentration and recycled trickle bed reactor*

Once the carbon bed is saturated with organics, hot water at temperatures up to 180° C and elevated pressure is recycled through the adsorber where most of the organics are desorbed, and through the reactor where the organics are subsequently catalytically destructed. Polaert *et al.*[30] recently proposed a similar two-step adsorption–desorption process. They used a single bi-functional reactor where activated carbon is first used as an adsorbent and then as a catalyst. These two combined adsorption–oxidation processes offer good potential for treating diluted wastewaters at moderate flow rates. In the proposed two processes, it is advantageous that eventually leached metal catalysts also play an active catalytic role[12] within the close loop and thus they do not pollute the environment.

It should be emphasized that catalytic WO processes are primarily designed to oxidize organic pollutants into intermediates more amenable to biological treatment, since the complete oxidation may be prohibitively expensive. Therefore the catalytic WO units are installed at the very source of water pollution and are usually used as a pre-treatment for cheaper classical biological systems.

5.3 Catalytic Denitrification of Drinkable Water

Nitrates and nitrites are ubiquitous groundwater contaminants particular in the areas of extensive agriculture fertilizing. The toxicity of nitrates to humans is due to the body's reduction of nitrate to nitrite. The content of nitrates in groundwater that exceed the maximum admissible concentration (e.g. 50 mg/L as set by the European Water Directive) must be reduced in order to avoid health risk. Therefore the removal of nitrates from drinkable water is an emerging technology, which is going to keep busy many researchers in the coming years. Kapoor and Viraraghavan[31] presented in their state-of-the-art review all treatment methods that are currently available. According to the capital and operational cost, the ion-exchange technique is the most favorable but disposing the spent brine for regeneration poses a serious problem in non-coastal locations. However, the most promising techniques for nitrate removal without creating secondary waste are biological digestion and catalytic denitrification by employing noble metal catalysts. The main reasons for slow transfer of biological digestion into a practice are concerns of possible bacterial contamination and presence of residual organics in treated water, which additionally increase the chlorine demand. As an alternative to the biological digestion, Vorlop and Tacke[32] introduced the reduction of nitrates in drinkable water by hydrogen over a solid catalyst at mild reaction conditions: temperatures between 5° C and 25° C, and hydrogen partial pressures up to 7 bar. Pintar[10] has recently provided a thorough review on the catalytic treatment of drinkable water.

5.3.1 Catalysts

In this process, nitrates are selectively reduced via intermediates into nitrogen, therefore the electro-neutrality of the aqueous solution is sustained by replacing nitrates with hydroxide ions. Supported bimetallic catalysts such as Pd–Cu, Pd–Sn, Pd–In, and Pt–Cu are known to have great potential for the reduction of nitrates. Unfortunately, these catalysts are not suppressing sufficiently the side reaction toward the formation of ammonia, which is highly undesirable in water for drinking (below 0.5 mg/L). Hörold *et al.*[33] have shown that Pd hydrogenation catalyst doped with Cu is a selective bimetallic catalyst for the transformation of nitrates to nitrogen. At the initial nitrate concentration level of 100 mg/L they achieved the selectivity of 82 mol%, which can be even further increased by employing a mixture of supported Pd–Cu and Pd catalysts and lowering the hydrogen partial pressures. It is believed that the key intermediate product in the process of catalytic nitrate reduction is NO[34]. Pintar and Kajiuchi[35] and later Deganello *et al.*[36], who studied the same reaction over various Pd–Cu bimetallic catalysts, reported that the nitrate to nitrite reduction undergoes a structure-insensitive reaction. It was demonstrated that the reaction selectivity strongly depends on the spatial distribution of Pd and Cu metallic phases; the highest selectivity was obtained with a catalyst sample in which the very first sub-layers were enriched by palladium atoms[37,38]. Similar behavior with respect to the minimum accumulation of nitrite ions was obtained with Pd–Cu bimetallic catalysts prepared by the sol–gel preparation technique[39]. A reason for the low reaction selectivity found in some cases (less than 70 mol%) might be the inappropriateness of the textural properties of the catalyst surface and due to the ratio of the two metals[33,36]. Besides Pd–Cu bimetallic catalysts, new alumina-supported Pd–Sn and Pd–In catalysts have been used for the efficient treatment of nitrate-polluted drinkable water[40]. It seems, however, that the latter two types of catalyst prepared by a deposition–precipitated method are more active as well as more selective for nitrate removal. According to Prüsse *et al.*[40] the selectivity may also increase if formic acid is used as a source of hydrogen. Recently a Pd/SnO_2 catalyst[41], palladium–tin catalysts on acrylic resin catalyst[42], titania-supported Pd–Cu catalysts[43], and palladium- and platinum-based catalysts doped with copper, silver, or gold[44] were employed successfully. Pd–Cu/Mg/Al–hydrotalcites based catalysts have also experienced a good activity/selectivity[45]. Matatov-Meytal *et al.*[46] used woven cloths made of glass fibers impregnated with Pd and reported about the same activity as found with the conventional powdered catalysts. However, knowledge about the selective catalytic reduction of nitrates in drinkable water is still far from complete, therefore more mechanistic studies with different catalytic systems would be appreciated.

5.3.2 Reduction Kinetics

Quantitative rate data on the catalytic reduction of nitrates in drinkable water are relatively scarce. One of the first works concerning kinetics is that of Tacke and Vorlop[47] who employed a Pd–Cu bimetallic catalyst containing 5 wt.% of Pt and 1.25 wt.% of Cu in a slurry reactor. Measurements of the initial rates resulted in a power-law rate expression. They reported a power of 0.7 with respect to the nitrate concentration, and an independency on the hydrogen partial pressure providing this pressure exceeded 1 bar. Pintar *et al.*[48] reported a complete kinetic model of the Langmuir–Hinshelwood type written in the form

$$-r_{NO_3^-} = -\frac{dC_{NO_3^-}}{dt} = \frac{k_{app.}K_{NO_3^-}K_{H_2}^{1/2}C_{NO_3^-}p_{H_2}^{1/2}}{(1+K_{NO_3^-}C_{NO_3^-})(1+K_{H_2}^{1/2}p_{H_2}^{1/2})} \tag{5.6}$$

which accounts for both the non-competitive equilibrium nitrate and the dissociative hydrogen adsorption steps. Here the irreversible bimolecular reaction between adsorbed nitrate ion and hydrogen is considered as the rate-limiting step. In slurry reactor, the concentration of nitrite as an intermediate product showed no retardation on the rate of nitrate reduction. Studying the rate of nitrate reduction by using various nitrate salts as sources of nitrate ions it was demonstrated that the apparent rate constant increases in the order $K^+ < Na^+ < Ca^{2-} < Mg^{2+} < Al^{3+}$ and changes proportionally with the ionization potential of the cations present in the aqueous solution[49]. The permanent hardness of drinkable water exhibited no retardation either on the rate of nitrate removal or on the reaction selectivity. On the other hand, the nitrate disappearance rate and the nitrogen production rate decreased appreciably in the presence of hydrogen carbonates.

5.3.3 Denitrification Process

For the time being, most of the processes of catalytic nitrate removal have been carried out in a batch mode. In the early 1990s Sell *et al.*[50] performed pilot studies in continuous fixed and expanded bed reactors using Pd–Cu bimetallic catalyst at ambient temperature and hydrogen partial pressures up to 6 bar. Only water with pre-dissolved hydrogen was fed into these reactors (liquid-full operation). They found higher reduction rates in the expanded bed reactor compared with the fixed bed, most probably due to the mass transfer limitation in the latter. However, they reported the rates of nitrate disappearance up to 2.5 g/h per kilogram of catalyst without any accumulation of nitrite ions or the formation of ammonia. Pintar and Batista[51] also demonstrated catalytic denitrification in an isothermal bubble-column fixed bed reactor to be a feasible process in which the maximum permissible concentration of ammonium ion was not exceeded. When drinking water was used as the reaction medium instead of distilled water, the nitrate disappearance rate as well as the reaction selectivity decreased appreciably and, consequently, the nitrite concentration exceeded the maximum admissible value (0.02 mg/L). In the bubble-column fixed bed reactor, the reaction rate was controlled by hydrogen transport from the gas to the liquid phase. In a trickle bed reactor, Pintar and Batista[51] experienced lower conversions of nitrates at the same LHSV and higher ammonia production. The latter was mainly due to incomplete catalyst wetting, which caused a high flux of hydrogen to the dry surface and therefore higher surface hydrogen concentrations. However, a typical decay of nitrate and appearance of nitrite and ammonia during denitrification process is illustrated in Figure 5.8.

The above investigations[51] undoubtedly demonstrated that bimetallic catalysts available today might not be appropriate for the direct treatment of contaminated waters, in particular if they contain higher amounts of hydrogen carbonates. To overcome the disadvantage of ammonia production, many other approaches have been proposed, e.g. membranes[52,53], structured membranes[54,55], and hollow fibers[55]. Most of these techniques were found to be promising; however, a technological breakthrough in the area may be considered by introducing the integrated process, which combines the conventional ion-exchange process with the denitrification reactor[51,57,58]. In the integrated process, the disadvantages of each standalone process are completely eliminated. Thus, there is no production of a secondary waste stream containing high amounts of nitrate, sulfate, and chlorine ions as in the ion-exchange process. On the other hand, already purified water cannot be contaminated with ammonia. The process is schematically represented in Figure 5.9.

In the ion-exchange part, the column is packed with nitrate-selective resin in a chloride form, which prevents treating waters with a concentration of nitrates exceeding 50 mg/L.

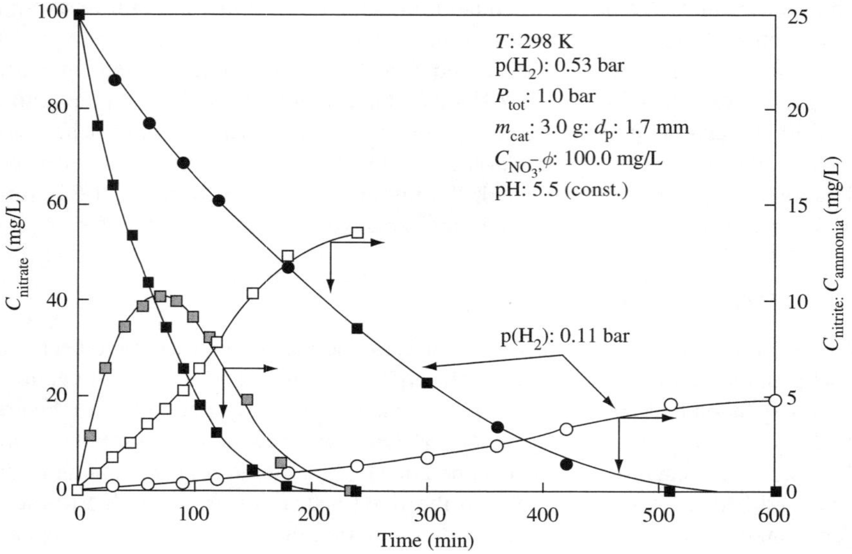

Figure 5.8 *Typical concentration profiles for nitrate, nitrite, and ammonia (○ and □) during hydrogenation of a model drinking water at two different hydrogen pressures over a bimetallic Pd(1%)–Cu(0.3%)/Al_2O_3 catalyst*

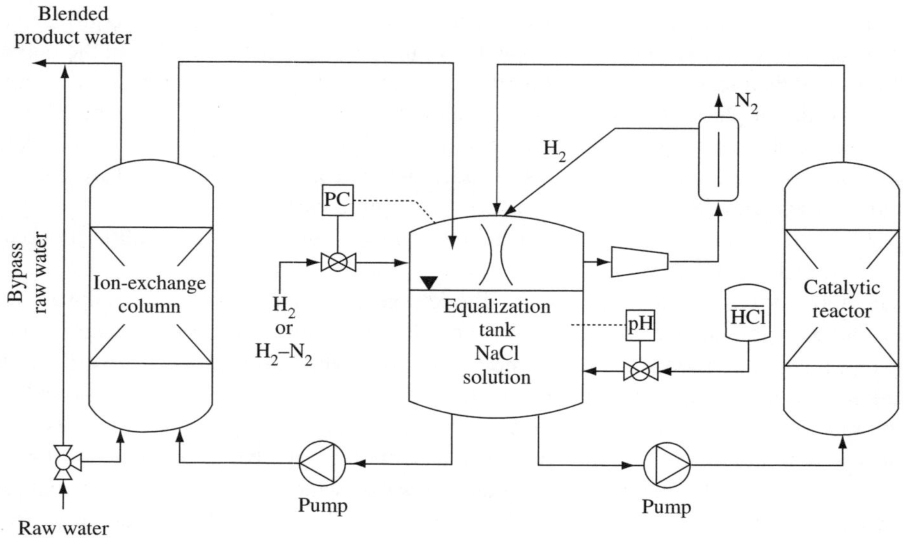

Figure 5.9 *Schematic drawing of a combined ion exchange–catalytic denitrification process*

Depending on the field capacity, contaminated level, and water demands the nitrate-free effluent from the ion-exchange column can be arbitrarily blended with the bypassed non-treated groundwater in order to meet the water quality requirements. The spent resin is regenerated in close circuit with sodium chloride solution, the concentration of which can be approximately an order of magnitude lower compared with the conventional ion-exchange regeneration process. The regenerant solution containing high amounts of nitrates circulates through the fixed bed reactor packed with a proprietary Pd–Cu/Al_2O_3 bimetallic catalyst where nitrates are reduced by hydrogen. Nitrite and ammonia formed in the reactor cannot be transported into the ion-exchange-treated effluent and therefore do not affect the water quality. The fixed bed reactor operates at ambient temperature and pressure. The pH of the liquid phase is controlled and kept at a value of 5 simply by adding HCl to the NaCl solution. The regenerant continues recycling through the ion-exchange column and the reactor until the resin reaches sufficient chloride loading. The integrated process is capable of destroying about 95% of waste brine, which is a tremendous improvement over the conventional ion-exchange process. Accumulation of sulfate ions in the closed circuit was found harmless to the nitrate capacity of the chloride-type resin. The use of HCl for pH control is also advantageous because there is no need for the regenerant makeup with NaCl. Namely, the stoichiometrically required amount of chloride ions for the subsequent regeneration cycle is provided by the neutralization of hydroxide ions produced in the nitrate reduction step. In a continuous and efficiently designed operation, one fixed bed denitrification reactor may serve at least three ion-exchange columns.

5.4 Preferential CO Oxidation

Selective oxidation of carbon monoxide in excess of the hydrogen or PROX process has become a renewed interest for many researchers due to its potential of making hydrogen acceptable for low-temperature PEMFC. In order to avoid problems associated with hydrogen storage (in automotive application in particular), hydrogen should be produced by the autothermal reforming of hydrocarbon fuels at a place of utilization. Indeed, the catalytic decomposition of ammonia appears to be another appealing process for CO-free hydrogen fuel. However, the composition of gas leaving the reformer and water gas-shift (WGS) reactor does not fit the requirements for PEMFC application. Typically, it contains 45–75 vol% of H_2, 15–25 vol% of CO_2, a few vol% of H_2O, traces of unconverted fuel, and 0.5–2 vol% of CO. CO is a catalyst poison for the Pt gas diffusion anode of PEMFC, which can only tolerate CO in concentrations below 50 ppm[59], in some cases even below 10 ppm[60]. In order to reduce the CO concentration in reformate, several approaches are currently under investigation: preferential oxidation (PROX), catalytic methanation, and Pd-membrane separation. Among these, PROX offers the lowest cost process for reducing CO content to the desired level without excessive hydrogen consumption.

5.4.1 Catalysts

A PROX catalyst needs to be active and selective; it should oxidize CO without oxidizing a large amount of hydrogen. It should prevent side reactions such as reverse WGS and methanation and be capable of operating under high steam and CO_2 content. Oxidation of hydrogen reduces the overall fuel efficiency on one side, but depending on the catalyst nature it may also affect the catalyst activity via produced water. The lower the catalyst activity, the higher is the O_2/CO ratio that has to be used in order to completely oxidize CO to CO_2. Catalysts employed in PROX reactors mostly involve high platinum group metals (PGM) loading on high surface area supports and are operated in a temperature range of 80–200° C. Catalyst formulation comprises Pt or promoted Pt, Ru, Pd, alloys of Pt–Sn or Pt–Ru, or Rh on alumina or on molecular sieves[61] or more recently, Au catalysts[62]. Among these, Pt catalysts have been the most studied and appear to offer the best activity and selectivity over a wide range of temperatures. Copper catalysts on alternative supports such as ceria, ceria-samaria, or other ceria-promoted supports have also been developed in an attempt to provide selective surface oxygen for CO oxidation at low temperatures[63,64]. Ghenciu[59], Shore and Farrauto[60], and Choudhary and Goodman[65] have recently provided a review on fuel processing catalysts for hydrogen production for PEMFC technology.

Oh and Sinkevitch[66] compared the efficiency of several PGM loading (0.5%) catalysts for the PROX reaction and found that the CO conversion decreases in the following order: $Ru/Al_2O_3 > Rh/Al_2O_3 > Pt/Al_2O_3 > Pd/Al_2O_3$. Although the Pd catalyst exhibited a similar activity as Ru and Rh catalysts at low temperatures, its activity at higher temperatures was found to be significantly poorer. The authors attributed this effect to the change in oxidation state of Pd with increasing reaction temperature. Despite Ru and Rh catalysts demonstrating high efficiency, these catalysts have not been explored in full detail. Nevertheless, there are two drawbacks of a ruthenium catalyst: first, its operating temperature is 140–200° C, thus well above the operating temperature of PEMFC, and second, it also acts as a methanation catalyst[60].

Gold-based catalysts have been investigated with a great interest although metal gold is known to be inefficient in CO oxidation. On the other hand, supported nano-gold clusters are found highly active and therefore promising in the PROX process. For example, on manganese oxide-supported gold catalyst, over 95% conversions of CO were reported in a temperature range of 50–80° C, which fits the operating temperature of PEMFC[67]. In spite of the fact that supported gold catalysts are almost insensitive to CO_2 and that their activity even enhances with moisture, Pt catalysts have an advantage over Au catalysts since the latter undergo relatively fast deactivation induced by oxygen[65]. The stability of nano-Au catalyst can be substantially increased by a temperature-programmed reduction–oxidation treatment of an Au–phosphine complex on TiO_2[68]. It is also worth mentioning here that a bimetallic carbon-supported PtSn system shows some superiority over Pt/Al_2O_3 catalyst[69].

Recently mixed oxides of Cu and Ce have been reported as very promising catalysts for PROX. In a comparative study of $Pt/\gamma\text{–}Al_2O_3$, $Au/\alpha\text{–}Fe_2O_3$, and $CuO\text{–}CeO_2$ catalysts for the selective oxidation of CO in excess of hydrogen, Avgouropoulos *et al.* [70] have undoubtedly demonstrated that the Au catalyst is the most active at low temperature, while the selectivity of $CuO\text{–}CeO_2$ is remarkably higher than that of both Au and Pt systems. Platinum on alumina was found to be the most resistant to water and CO_2. At temperatures between 45° C and 90° C, an inexpensive nano-structured $Cu_{0.1}Ce_{0.9}O_{2-y}$ catalyst (prepared by a sol–gel technique) was found provide 100% selectivity with CO conversions up to about 60% as depicted in Figure 5.10[64]. At higher temperatures much higher conversions can be attained but at the expense of selectivity. Kandoi *et al.*[71] have recently provided a theoretical basis for why the catalysts based on Au and Cu are superior to Pt-based catalyst for the oxidation of trace CO in reformate gases at low temperatures, i.e. close to the operational temperature of PEMFC.

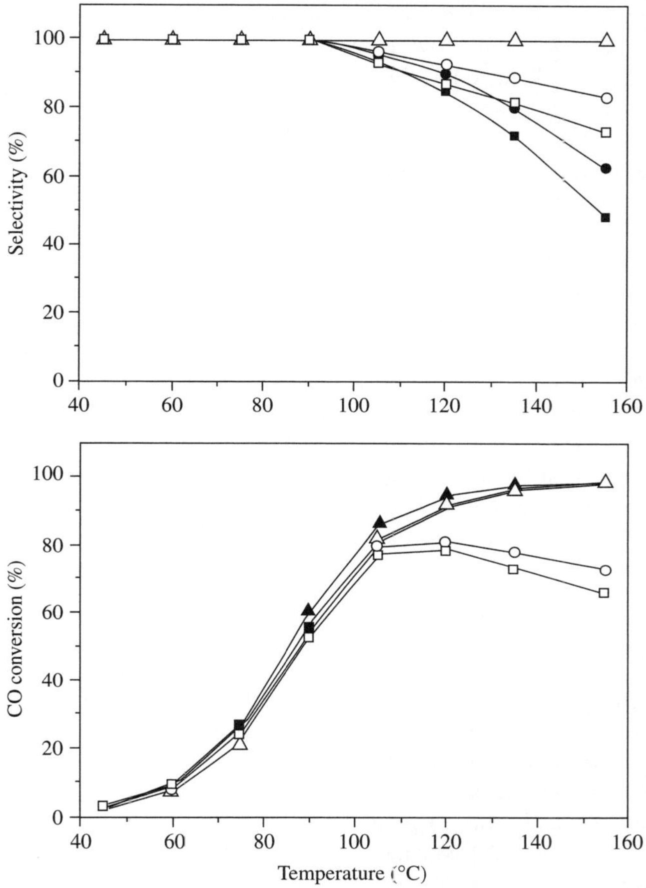

Figure 5.10 *Selectivity and CO conversion as a function of temperature obtained over nanostructured copper-cerium oxide at different values of λ (full and empty symbols) and hydrogen concentration in the reactor feed (▲ and △ no hydrogen in feed)*

5.4.2 PROX Kinetics

Initial conclusions on the kinetics of selective oxidation of CO in hydrogen reach gas were drawn from what is known about CO oxidation (in the absence of H_2) on single-crystalline and supported PGM catalysts. Namely, some early investigations under ultra-high vacuum conditions[72] and more recently in high pressure region[73] have found two distinct reaction regimes: (i) *high rate regime*, which occurs at high temperatures on the surface covered with very small amounts of adsorbed CO, and (ii) *low rate regime*, which takes place at low temperatures on the catalyst surface predominantly covered with adsorbed CO. Both regimes were modeled by the Langmuir–Hinshelwood mechanism, which is associated with the reaction order approaching -1 with respect to the CO partial pressure and close to $+1$ for the oxygen partial pressure for the *low rate regime*, and $+1$ for both the components in the *high rate regime*. Assuming the addition of hydrogen does not change the oxidation mechanism, one would expect that the reaction takes place in the *low rate regime* since the low temperature operation of PROX (below 250° C, and $\lambda \leq 2(=2p_{O_2}/p_{CO})$) is dictated by PEMFC. Kahlich *et al.*[74] have published an extensive kinetic study of the selective oxidation of CO in hydrogen reach gas on Pt/Al_2O_3. In a wide range of CO partial pressures and temperatures between 150° C and 250° C, and the process parameter λ values relevant to PROX, they have shown that the oxidation rate can be well represented by the power-law kinetics: the order of −0.4 with respect to the CO partial pressure and order of +0.8 with respect to the oxygen partial pressure. The reaction orders and the activation energy of 71 kJ/mol was found consistent with the reaction occurring on the surface predominantly covered by adsorbed CO, which blocks the oxygen adsorption. An interesting kinetics is found in the work of Han *et al.*[75] who oxidized CO in a methanol reformate over a $Ru/\gamma\text{–}Al_2O_3$ catalyst. In a temperature range of 80–120° C, these authors proposed the power-law kinetics with the temperature-dependent orders between −0.29 and −0.66 for carbon monoxide and +0.30 and +0.80 for oxygen, respectively, whereas the activation energy was reported to be 48 kJ/mol.

Oxidation kinetics on transition metal mixed oxide catalysts was also first interpreted by means of the Langmuir–Hinshelwood mechanism and in terms of a synergistic effect resulting from the interaction of different materials[76]. The rate equation in the following form was proposed:

$$r_{CO} = \frac{k_L K_L P_{CO} P_{O_2}^m}{1 + K_L P_{CO}} \tag{5.7}$$

where the parameters k_L and K_L represent the surface reaction rate and CO adsorption equilibrium constants, respectively, and $P =$ partial pressure. Both parameters are well correlated by the Arrhenius law. The reaction order with respect to oxygen was reported to be a very small number, close to 0, whereas the activation energy was found in a range of 73–94 kJ/mol. Sedmak *et al.*[64] modeled the kinetics of selective CO oxidation over the $Cu_xCe_{1-x}O_{2-y}$ nanostructured catalysts by means of the Mars–van Krevelen type of kinetics, which is based on a redox mechanism, thus

$$r_{CO} = \frac{k_{CO} k_{O_2} P_{CO} P_{O_2}^n}{0.5 k_{CO} P_{CO} + k_{O_2} P_{O_2}^n} \tag{5.8}$$

The parameters k_{CO} and k_{O_2} are taken to be the reaction rate constants for the reduction of surface by CO and the surface re-oxidation by O_2, respectively, and are subjected to the Arrhenius law. The order with respect to oxygen (n) takes a value of 0.2. It is interesting to note that the experimental data of Sedmak *et al.*[64] can also be well correlated by equation 5.7 with $m = 0.15$. However, the transient experiments with a step change of CO concentration in the reactor feed have revealed the involvement of lattice oxygen even at low temperatures, thus confirming the appropriateness of using the Mars–van Krevelen kinetic formulation[77,78]. Figure 5.11 shows a comparison between the experimental responses and the predictions made by a fixed bed reactor model accounting for the rate law given by equation 5.8.

5.4.3 PROX Process

From a process point of view (reformer→WGS reactor→PROX reactor→fuel cell) the PROX reactor should be advantageously operated between the outlet temperature of

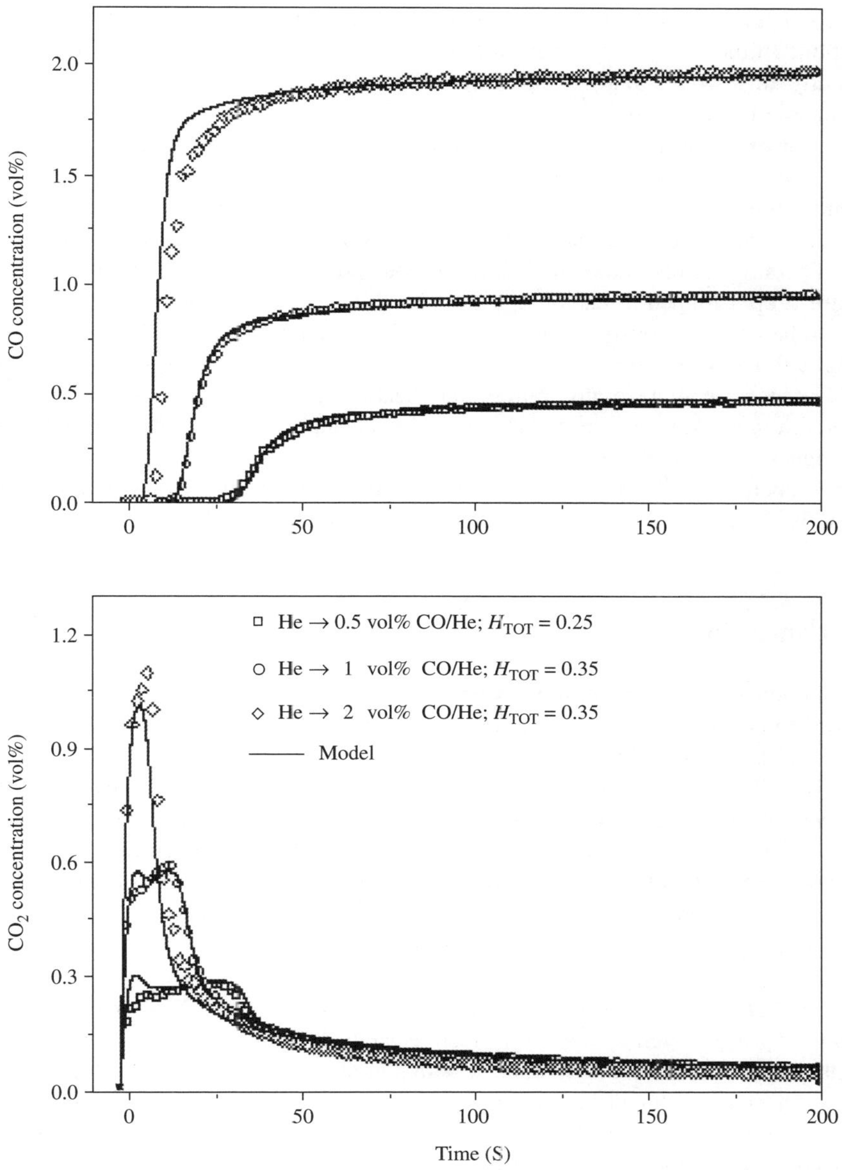

__Figure 5.11__ Experimental and predicted responses in CO and CO_2 concentrations in the reactor outlet stream after step change in the reactor feed stream from helium to a different CO concentration on fully oxidized copper-cerium oxide

the WGS reactor and the inlet temperature of fuel cell (~80° C). The process must be primarily designed for highly selective oxidation of CO since any loss of hydrogen, the primary electrochemical fuel, reduces the competitiveness and efficiency of the process. Bearing in mind that the process is installed as the very last stage of the hydrogen production line, it must provide hydrogen fuel over a wide range of PEMFC output conditions (turndown ratios). Therefore the PROX catalyst must also be able to operate efficiently in a wide range of space velocities, under severe transient conditions. It must handle the CO content (e.g. 10 ppm) at high as well as at low throughputs. It is known, for example, that at very low space velocities, a Pt-based catalyst may produce CO by the reverse WGS reaction. The inlet temperature has to be compatible with the outlet temperature of the upstream WGS unit, and because the oxidation of CO and H_2 is a highly exothermic reaction, the PROX exit stream must be cooled down to the operating temperature of PEMFC (about 80° C). Efficient cooling and minimal pressure drop within the reactor in particular dictate the catalyst shape design, which seems to be advantageously in the form of washcoat monoliths. In order to achieve high thermal efficiency, a fuel processor should consist of steam reformer, CO shift converter, PROX reactor, steam generator, burner, and heat exchanger in one package[78]. Since the temperature control in the secondary hydrogen cleaning system is crucial, it should be advantageous to employ microstructured reactors, which are known for their feasibility of dynamic operation[79,80]. Some design considerations of the PROX process and catalysts are briefly discussed in a recent article by Shore and Farrauto[60].

5.5 Conclusions

The common key issue in all three examples is a catalyst. While a catalyst for WO does not need to be selective, denitrification of drinkable water and preferential oxidation of CO call for a very selective catalyst.

Catalytic WO can be considered nowadays as a mature technology. Nevertheless, due to the variety of wastewater that has to be treated, one type of catalyst cannot fulfill all the needs. Therefore the catalyst must be tailored for each particular application and made of inexpensive materials. In order to reduce leaching, the catalytically active compounds have to be incorporated into a lattice of catalyst support. It would be advantageous to design a catalyst for treatment in single-pass reactors that had a minimum lifetime of over 500 h.

Removal of nitrate from drinkable water is still far from maturity. A direct treatment or single-pass process even with a very selective catalyst is not likely to be feasible because drinkable water should not be in contact with a noble metal catalyst. At this moment one can consider the combined ion exchange and denitrification process to be advantageous over the conventional ion-exchange technology, the main drawback of which is a harmful disposing of large amounts of spent brine. In the combined process, a Pd–Cu bimetallic catalyst seems to be efficient enough but more work in a pilot scheme would positively help to push the technology forward.

While more CO-tolerant fuel cells are being developed, efforts in developing more selective catalysts to remove higher amounts of CO (0.5–1.0%) from the hydrogen-reach reformate prior to entering the cell are continuing. These efforts are accompanied by cost reduction. Monolithic types of catalysts, especially those containing Pt, have already been successfully demonstrated in the PROX process. Nevertheless, some other inexpensive catalytic systems, e.g. copper-cerium, also remain attractive for low-temperature operation.

References

[1] Kochloefl K. 2001. Development of industrial solid catalyst, Chem. Eng. Technol., 24, 229–234.
[2] Sadana A. and Katzer J.R. 1974. Catalytic oxidation of phenol in aqueous solutions over copper oxide, Ind. Eng. Chem. Fundam., 13, 127–134.
[3] Baldi G., Goto S., Chow C.K. and Smith J.M. 1974. Catalytic oxidation of formic acid in water, Ind. Eng. Chem. Proc. Des. Dev., 13, 447–452.
[4] Levec J. and Smith J.M. 1976. Oxidation of acetic acid solutions in a trickle-bed reactor, AIChE J., 22, 159–168.
[5] Mishra V.S., Mahajani V.V. and Joshi J.B. 1995. Wet air oxidation, Ind. Eng. Chem. Res., 34, 2–48.
[6] Levec J. and Pintar A. 1995. Catalytic oxidation of aqueous solutions of organics. An alternative method for removal of toxic pollutants from wastewaters, Catal. Today, 24, 51–58.
[7] Matatov-Meytal Y.I. and Sheintuch M. 1998. Catalytic abatement of water pollutants, Ind. Eng. Chem. Res., 37, 309–326.
[8] Luck F. 1999. Wet air oxidation: past, present and future, Catal. Today, 53, 81–91.
[9] Imamura S. 1999. Catalytic and noncatalytic wet oxidation, Ind. Eng. Chem. Res., 38, 1743–1753.
[10] Pintar A. 2003. Catalytic processes for the purification of drinking water and industrial effluents, Catal. Today, 77, 451–465.
[11] Mantzavinos D., Hellenbrand R., Livingston A.G. and Metcalfe I.S. 1996. Catalytic oxidation of p-coumaric acid: partial oxidation intermediates, reaction pathways and catalyst leaching, Appl. Catal. B, 7, 379–396.
[12] Fortuny A., Font J. and Fabregat A. 1998. Wet air oxidation of phenol using active carbon as catalyst, Appl. Catal. B, 19, 165–173.
[13] Pintar A. and Levec J. 1992. Catalytic oxidation of organics in aqueous solutions. I. Kinetics of phenol oxidation, J. Catal., 135, 345–357.
[14] Imamura S., Fukuda I. and Ushida S. 1988. Wet oxidation catalyzed by ruthenium supported on cerium oxides, Ind. Eng. Chem. Res., 27, 718–721.
[15] Gallezot P., Chaumet S., Perrard A. and Isnard P. 1997. Catalytic wet air oxidation of acetic acid on carbon-supported ruthenium catalysts, J. Catal., 168, 104–109.
[16] Barbier Jr J., Delanoë F., Jabouille F., Blanchard G. and Duprez D. 1998. Total oxidation of acetic acid in aqueous solutions over noble metal catalysts, J. Catal., 177, 378–385.
[17] Béziat J.-C., Besson M., Gallezot P. and Durecu S. 1999. Catalytic wet air oxidation of carboxylic acids on TiO_2 supported ruthenium catalysts, J. Catal., 182, 129–135.
[18] Pintar A., Besson M. and Gallezot P. 2001. Catalytic wet air oxidation of Kraft bleaching plant effluents in the presence of titania and zirconia supported ruthenium, Appl. Catal. B, 30, 123–139.
[19] Cybulski A. and Trawczynski J. 2004. Catalytic wet air oxidation of phenol over platinum and ruthenium catalyst, Appl. Catal. B, 47, 1–13.
[20] Pintar A. and Levec J. 1992. Catalytic liquid-phase oxidation of refractory organics in waste water, Chem. Eng. Sci., 47, 2395–2400.
[21] Pintar A. and Levec J. 1994. Catalytic liquid-phase oxidation of phenol aqueous solutions. A kinetic investigation, Ind. Eng. Chem. Res., 33, 3070–3077.
[22] Eftaxias A., Font J., Fortuny A., Giralt J., Fabregat A. and Stüber F. 2001. Kinetic modeling of catalytic wet air oxidation of phenol by simulated anneling, Appl. Catal. B, 33, 175–190.
[23] Hamoudi S., Belkacemi K. and Larachi F. 1999. Catalytic oxidation of aqueous phenolic solutions catalyst deactivation and kinetics, Chem. Eng. Sci., 54, 3569–3576.
[24] Li L., Chen P. and Gloyna E.F. 1991. Generalized kinetic model for wet oxidation of organic compounds, AIChE J., 37, 1687–1697.

[25] Pintar A., Berčič G., Besson M. and Gallezot P. 2004. Catalytic wet-air oxidation of industrial effluents: total mineralization of organics and lumped kinetic modelling, Appl. Catal. B, 47, 143–152.
[26] Donlagic J. and Levec J. 1999. Wet oxidation of an azo dye: lumped kinetics in batch and mixed flow reactors, AIChE J., 45, 2571–2579.
[27] Belkacemi K., Larachi F., Hamoudi S., Tucotte G. and Sayari A. 1999. Inhibition and deactivation effects in catalytic wet oxidation of high-strength alcohol – distillery liquors, Ind. Eng. Chem. Res., 38, 2268–2274.
[28] Katzer J.R., Ficke H.H. and Sadana A. 1976. An evaluation of aqueous phase catalytic oxidation, J. Water Poll. Control Fed., 48, 920–933.
[29] Levec J. and Pintar A. 1997. Process for treating industrial waste waters with low concentrations of toxic pollutants. EP 0 664 771 B1, European Patent Office, Muenchen, 29 January 1997.
[30] Polaert I., Wilhelm A.M. and Delmas H. 1997. Phenol wastewater treatment by a two-step adsorption-oxidation process on activated carbon, Chem. Eng. Sci., 57, 1585–1590.
[31] Kapoor A. and Viraraghavan T. 1997. Nitrate removal from drinking water – Review, J. Environ. Eng., 123, 371–380.
[32] Vorlop K.-D. and Tacke T. 1989. First steps towards noble-metal catalyzed removal of nitrate and nitrite from drinking water, Chem. Ing. Tech., 64, 836–837.
[33] Hörold S., Vorlop K.-D., Tacke T. and Sell M. 1993. Development of catalysts for a selective nitrate and nitrite removal from drinking water, Catal. Today, 17, 21–30.
[34] Wärnå J., Turunen I., Salmi T. and Maunula T. 1994. Kinetics of nitrate reduction in monolith reactor, Chem. Eng. Sci., 49, 5763–5773.
[35] Pintar A. and Kajiuchi T. 1995. Catalytic liquid-phase hydrogenation of aqueous nitrate solutions, Acta Chim. Slovenica, 42, 431–449.
[36] Deganello F., Liotta L.F., Macaluso A., Venezia A.M. and Deganello G. 2000. Catalytic reduction of nitrates and nitrites in water solution on pumice-supported Pd–Cu catalysts, Appl. Catal. B, 24, 265–273.
[37] Batista J., Pintar A. and Čeh M. 1997. Characterization of supported Pd–Cu catalysts by SEM, EDXS, AES and catalytic selectivity measurements, Catal. Lett., 43, 79–84.
[38] Pintar A., Batista J., Arčon I. and Kodre A. 1998. Characterization of gamma-Al_2O_3 supported Pd–Cu bimetallic catalysts by EXAFS, AES and kinetic measurements, Stud. Sufr. Sci. Catal., 118, 127–136.
[39] Strukul G., Pinna F., Marella M., Meregalli L. and Tomaselli M. 1996. Sol–gel palladium catalysts for nitrate and nitrite removal from drinking water, Catal. Today, 27, 209–214.
[40] Prüsse U., Hähnlein M., Daum J. and Vorlop K.-D. 2000. Improving the catalytic nitrate reduction, Catal. Today, 55, 79–90.
[41] Gavagnin R., Biasetto L., Pinna F. and Strukul G. 2002. Nitrate removal in drinking waters: the effect of tin oxides in the catalytic hydrogenation of nitrate by Pd/SnO_2 catalysts, Appl. Catal. B, 38, 91–99.
[42] Roveda A., Benedetti A., Pinna F. and Strukul G. 2003. Palladium-tin catalysts on acrylic resins for the selective hydrogenation of nitrate, Inorg. Chim. Acta, 349, 203–208.
[43] Gao W., Guan N., Chen J., Guan X., Jin R., Zeng H., Liu Z. and Zhang F. 2003. Titania supported Pd–Cu bimetallic catalyst for the reduction of nitrate in drinking water, Appl. Catal. B, 46, 341–351.
[44] Epron F., Gauthard F. and Barbier J. 2002. Influence of oxidizing and reducing treatments on the metal–metal interactions and on the activity for nitrate reduction of a Pt–Cu bimetallic catalyst, Appl. Catal. A, 237, 253–261.
[45] Palomares A.P., Prato J.G., Márquez F. and Corma A. 2003. Denitrification of natural water on supported Pd/Cu catalysts, Appl. Catal. B, 41, 3–13.
[46] Matatov-Meytal Y., Barelko V., Yuranov I. and Sheintuch M. 2000. Cloth catalysts in water denitrification: I. Pd on glass fibers, Appl. Catal. B, 27, 127–135.

[47] Tacke T. and Vorlop K.-D. 1993. Kinetic characterization of catalysts for selective removal of nitrate and nitrite from water, Chem. Ing. Tech., 65, 1500–1502.
[48] Pintar A., Batista J., Levec J. and Kajiuchi T. 1996. Kinetics of the catalytic liquid-phase hydrogenation of aqueous nitrate solutions, Appl. Catal. B, 11, 81–98.
[49] Pintar A., Šetinc M. and Levec J. 1998. Hardness and salt effects on catalytic hydrogenation of aqueous nitrate solutions, J. Catal., 174, 72–87.
[50] Sell M., Bischoff M. and Bonse D. 1992. Catalytic nitrate reduction in drinking water. Results and experiences from pilot plant trials, Vom Wasser, 79, 129–144.
[51] Pintar A. and Batista J. 1999. Catalytic hydrogenation of aqueous nitrate solutions in fixed-bed reactors, Catal. Today, 53, 35–50.
[52] Lüdtke K., Peinemann K.-V., Kasche V. and Behling R.-D. 1998. Nitrate removal of drinking water by means of catalytically active membranes, J. Membr. Sci., 151, 3–11.
[53] Ilinich O.M., Gribov E.N. and Simonov P.A. 2003. Water denitrification over catalytic membranes: hydrogen spillover and catalytic activity of macroporous membranes loaded with Pd and Cu, Catal. Today, 82, 49–56.
[54] Daub K., Emig G., Chollier M.-J., Callant M. and Dittmeyer R. 1999. Studies on the use of catalytic membranes for reduction of nitrate in drinking water. Chem. Eng. Sci., 54, 1577–1582.
[55] Centi G., Dittmeyer R., Perathoner S. and Reif M. 1999. Tubular inorganic catalytic membrane reactors: advantages and performance in multiphase hydrogenation reactions, Catal. Today, 79, 139–149.
[56] Hähnlein M., Prüsse U., Daum J., Morawsky V., Kröger M., Schröder M., Schnabel M. and Vorlop K.-D. 1998. Preparation of microscopic catalysts and colloids for catalytic nitrate and nitrite reduction and their use in hallow fibre dialiser loop reactor, Stud. Surf. Sci. Catal., 118, 99–107.
[57] Pintar A., Batista J. and Levec J. 2001. Catalytic denitrification: direct and indirect removal of nitrates from potable water, Catal. Today, 66, 503–510.
[58] Pintar A., Batista J. and Levec J. 2001. Integrated ion exchange/catalytic process for efficient removal of nitrates from drinking water, Chem. Eng. Sci., 56, 1551–1559.
[59] Ghenciu A.F. 2002. Review of fuel processing catalysts for hydrogen production in PEM fuel cell systems. Curr. Opin. Solid State Mat. Sci., 6, 389–399.
[60] Shore L. and Farrauto R.J. 2003. PROX catalysts. In, Handbook of Fuel Cells – Fundamentals, Technology and Applications, Vielstich W., Lamm A. and Gasteiger A. (Eds.), Vol. 3, Wiley, Chichester, pp. 211–218.
[61] Igarashi H., Uchida H., Suzuki M., Sasaki Y. and Watanabe M. 1997. Removal of carbon monoxide from hydrogen-rich fuels by selective oxidation over platinum catalyst supported on zeolite, Appl. Catal. A, 159, 159–169.
[62] Haruta M., Tsubota S., Kobayashi T., Kageyama H., Genet M.J. and Delmon B. 1993. Low-temperature oxidation of CO over gold supported on TiO_2, α-Fe_2O_3 and Co_3O_4, J. Catal., 144, 175–192.
[63] Avgouropoulos G., Ioannides T., Matralis H.K., Batista J. and Hočevar S. 2001. CuO–CeO_2 mixed oxide catalysts for the selective oxidation of carbon monoxide in excess hydrogen, Catal. Lett., 73, 33–40.
[64] Sedmak G., Hočevar S. and Levec J. 2003. Kinetics of selective CO oxidation in excess of H_2 over the nanostructured $Cu_{0.1}Ce_{0.9}O_{2-y}$ catalyst, J. Catal., 213, 135–150.
[65] Choudhary T.V. and Goodman D.W. 2002. CO-free fuel processing for fuel cell application, Catal. Today, 77, 65–78.
[66] Oh S.H. and Sinkevitch R.M. 1993. Carbon monoxide removal from hydrogen-rich fuel cell feedstreams by selective catalytic oxidation, J. Catal., 142, 254–262.
[67] Sanchez R.M.T., Ueda A., Tanaka K. and Haruta M. 1997. Selective oxidation of CO in hydrogen over gold supported on manganese oxides, J. Catal., 168, 125–127.

[68] Choudhary T.V., Sivadinarayana C., Chusuei C.C., Datye A.K., Fackler Jr J.P. and Goodman D.W. 2002. CO oxidation on supported nano-Au catalysts synthesized from a $[Au_6(PPh_3)](BF_4)_2$ complex, J. Catal., 207, 247–255.
[69] Schubert M.M., Kahlich M.J., Feldmeyer G., Huttner M., Hackenberg S., Gasteiger H.A. and Behm R.J. 2001. Bimetallic PtSn catalyst for selective CO oxidation in H_2-rich gases at low temperatures, Phys. Chem. Chem. Phys., 3, 1123–1131.
[70] Avgouropoulos G., Ioannides T., Papadopoulou Ch., Batista J., Hočevar S. and Matralis H.K. 2002. Comparative study of $Pt/\gamma\text{–}Al_2O_3$, $Au/\alpha\text{–}Fe_2O_3$ and $CuO\text{–}Ce\text{–}O_2$ catalysts for the selective oxidation of carbon monoxide in excess hydrogen, Catal. Today, 75, 157–167.
[71] Kandoi S., Gokhale A.A., Grabow L.C., Dumesic J.A. and Mavrikakis M. 2004. Why Au and Cu are more selective than Pt for preferential oxidation of CO at low temperature, Catal. Lett., 93, 93–100.
[72] Engel T. and Ertl G. 1979. Elementary steps in the catalytic oxidation of carbon monoxide on platinum metals, Adv. Catal., 28, 1–78.
[73] Fuchs S., Hahn T. and Lintz H.-G. 1994. The oxidation of carbon monoxide by oxygen over platinum, palladium and rhodium catalysts from 10^{-10} to 1 bar, Chem. Eng. Proc., 33, 363–369.
[74] Kahlich M.J., Gasteiger H.A. and Behm R.J. 1997. Kinetics of the selective CO oxidation in H_2-rich gas on Pt/Al_2O_3, J. Catal., 171, 93–105.
[75] Han Y.-F., Kinne M. and Behm R.J. 2004. Selective oxidation of CO on $Ru/\gamma\text{–}Al_2O_3$ in methanol reformate at low temperature, Appl. Catal. B, 52, 123–134.
[76] Liu W. and Flytzani-Stephanopoulos M. 1995. Total oxidation of carbon monoxide and methane over transition metal-fluorite oxide composite catalysts, J. Catal., 153, 304–316.
[77] Sedmak G., Hočevar S. and Levec J. 2004. Transient kinetic model of CO oxidation over a nanostructured $Cu_{0.1}Ce_{0.9}O_{2-y}$ catalyst, J. Catal., 222, 87–99.
[78] Sedmak G., Hočevar S. and Levec J. 2004. CO oxidation over a nanostructured $Cu_{0.1}Ce_{0.9}O_{2-y}$ catalyst: a CO/O_2 cycling study, Top. Catal., 30–31, 445–449.
[79] Echigo M., Shinke N., Takami S. and Tabata T. 2004. Performance of natural gas processor for residential PEFC system using a novel CO preferential oxidation catalyst, J. Power Sour., 132, 29–35.
[80] Goerke O., Pfeifer P. and Shubert K. 2004. Water gas shift reaction and selective oxidation of CO in microreactors, Appl. Catal.A, 263, 11–18.

6

Design and Analysis of Homogeneous and Heterogeneous Photoreactors

Alberto E. Cassano and Orlando M. Alfano

6.1 Scope and Limitations

It is an impossible task for a short chapter to carryout a complete discussion of photoreactoreactor analysis and design unless we assume that the readers of this work have a previous background in the subject and chemical engineering fundamentals. Moreover, we can increase the chapter's feasibility if coverage is restricted to only a fraction, albeit a significant fraction, of the homogeneous and heterogeneous photoreactions. On this basis, it is possible to concentrate on those aspects that are distinctive of homogeneous photochemical and heterogeneous photocatalytic processes. For more details, the reader is referred to the original publications listed in the references.

The distinctive aspect of these reactions is the unavoidable existence of a radiation field inside the reactor, which only in very special and unusual cases can be considered uniform in space and frequently is not even constant in time. This is because inside the reaction space, besides geometrical effects produced by the characteristics of the reactor geometry, there must be absorption of radiation to produce the reaction activation. This absorption means attenuation of the incoming intensities; i.e. without attenuation, there is no photochemical reaction. In some heterogeneous systems, scattering is another source of variation in the incoming rays. Hence, spatial variations are unavoidable. These intrinsic non-uniformities, unfortunately often neglected or not properly accounted for, are responsible for the majority of the difficulties associated with photoreactor analysis and design.

Many different shapes and configurations are possible for either single-phase or multiphase reactors (Braun *et al.*, 1993; Cassano *et al.*, 1995; Puma and Yue, 1998; Ray, 1998;

Chemical Engineering Edited by Miguel A. Galán and Eva Martin del Valle

Cassano and Alfano, 2000; Alfano *et al.*, 2000). Again, we will only describe details for a few of them.

A systematic approach to the design of a reactor should start by discussing the field of velocity distributions. Much progress has been achieved in this area and the hydrodynamic characterization of a great variety of reactors is already known. For the sake of brevity, we will concentrate on two types of systems: a perfectly mixed reaction space and a fully developed unidirectional flow in a tubular reactor. In practical terms this is not a serious limitation; in computational fluid mechanics, commercially available calculating codes can be used to solve almost any other form of reactor configuration.

6.2 Mass and Energy Balances

6.2.1 Mass Conservation Equations

The general mass conservation equation is (extracted from Bird *et al.*, 2002, p. 584):

$$\underbrace{\frac{\partial C_i}{\partial t}}_{\substack{\text{Unsteady}\\ \text{state}}} + \underbrace{\nabla \cdot N_i}_{\substack{\text{All molar fluxes}\\ \text{(convection}\\ \text{and diffusion)}}} = \underbrace{R_{\text{Hom},i}}_{\substack{\text{Homogeneous}\\ \text{reactions}}} \tag{6.1}$$

Since the differential equation is valid for a single phase, in equation 6.1 only homogeneous reactions are included. Heterogeneous reactions ($R_{\text{Het},i}$), for example, in superficial, catalytic processes can be incorporated into the analysis if one considers that they are boundary conditions for equation 6.1. However, if a parallel homogeneous reaction is present, for example a photocatalytic system when direct photolysis also occurs, both $R_{\text{Hom},i}$ and $R_{\text{Het},i}$ must be included, but the second one is not part of the right-hand side of equation 6.1; it can be extracted from the second term of its left-hand side. This may be the reason why only seldom do people consider that photocatalytic reactions could be, even in slurry reactors, very often controlled by mass transport.

Equations will be derived for three representative cases: the tubular reactor of annular cross-section; the isothermal, well-mixed batch reactor; and the isothermal batch reactor inside a recirculating system. The chosen exemplifications will deal with these types of reactors that are, without doubt, the most widely used.

The tubular reactor. Consider firstly a tubular, cylindrical reactor formed by an annular space surrounding a tubular lamp (Figure 6.1). This is the simplest and more practical continuous photoreactor, particularly for artificial light illumination of a single-phase system. Under the following assumptions and operating conditions – (i) steady state; (ii) unidirectional, incompressible, continuous flow of a Newtonian fluid under fully developed laminar regime; (iii) only ordinary (concentration) diffusion is significant; (iv) azimuthal symmetry; (v) axial diffusion neglected as compared to the convective flow; (vi) constant physical and transport properties; (vii) non-permeable reactor walls; and (viii) for the moment, monochromatic operation – the following equation in cylindrical coordinates holds (extracted from Bird *et al.*, 2002, p. 850):

$$\underbrace{v_z(r)\frac{\partial C_i(z,r)}{\partial z}}_{\substack{\text{Convective flow in}\\ \text{the axial direction}}} - \underbrace{D_{\text{im}}\left[\frac{1}{r}\frac{\partial}{\partial r}\left(r\frac{\partial C_i(z,r)}{\partial r}\right)\right]}_{\substack{\text{Diffusive flux in}\\ \text{the radial direction}}} = \underbrace{R_{\text{Hom},i}(z,r)}_{\substack{\text{Homogeneous}\\ \text{reaction rate}}} \tag{6.2}$$

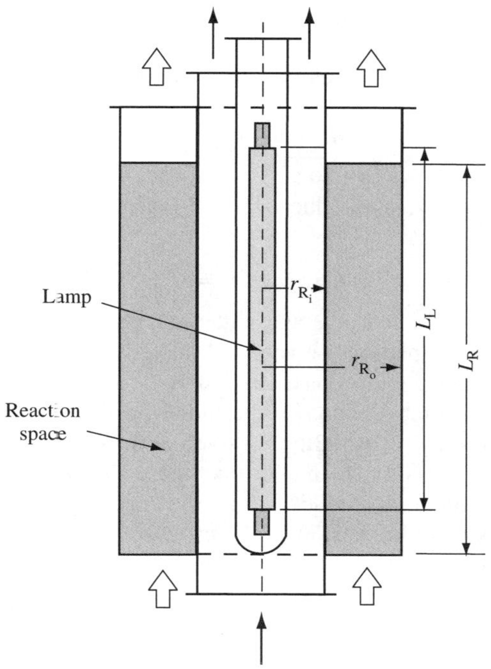

Figure 6.1 *Geometry of the continuous flow, annular photoreactor. Adapted from Cassano* et al. *(1995)*

Where v_z is the axial velocity in laminar flow, a function of the radial position and represented by the classical non-symmetric parabolic profile characteristic of annular spaces (Bird *et al.*, 2002, p. 55). The use of a pseudo-binary diffusivity is only an approximation; if more accuracy is needed the Maxwell–Stefan relationships should be used. The initial condition is

$$C_i(0, r) = C_{i,0} \tag{6.3}$$

For the stable species, the boundary conditions are

$$\frac{\partial C_i}{\partial r}(z, r_{R_i}) = 0 \tag{6.4}$$

$$\frac{\partial C_i}{\partial r}(z, r_{R_0}) = 0 \tag{6.5}$$

meaning that at the non-permeable reactor walls the mass fluxes are 0. These boundary conditions must be changed for reactive walls. For all but 0 and simple first-order reactions this equation must be solved numerically. In photochemical reactions the nature of the activation reaction eliminates the possibility of analytical solution unless one is willing to accept very crude approximations.

The plug flow reactor: Under fully developed turbulent flow regime the following approximations can be used: (i) the velocity profile is flat and equal to the average velocity, and (ii) there is perfect mixing for all stable species in the radial direction. When this is the case and the reactor walls are not permeable, concentration gradients of stable species in the radial direction can be neglected and equations 6.2, 6.3, 6.4, and 6.5 reduce to

$$\langle v_z \rangle \frac{\mathrm{d}C_i(z)}{\mathrm{d}z} = \langle R_i(z, r) \rangle_{A_{\mathrm{R,C}}} \tag{6.6}$$

Where $A_{\mathrm{R,C}}$ indicates an average value over the reactor cross-section area. The initial condition is

$$C_i(z = 0, r) = C_{i,0} \tag{6.7}$$

It should be specially noted that average values of reaction rates are needed because under no circumstances can photons be very well mixed; consequently, since in the radial direction in particular the photon concentration is normally non-uniform, usually, the reaction rate will be a strong function of the radial position. This consideration is also important if the reaction involves highly reactive intermediates, because very often their reaction characteristic time is much smaller than the hydrodynamic mixing time and the well-mixed condition cannot be extended to all the species participating in the reaction. Hence, even in approximate models, as is the case for the plug flow simplification, the radiation field non-uniformities cannot be ignored.

The isothermal, constant volume, well-stirred batch reactor. Equation 6.1 can be simplified if, due to good mixing conditions, temperature and concentrations are uniform. According to Figure 6.2, integrating in the liquid volume we get

$$\frac{\mathrm{d}C_i(t)}{\mathrm{d}t} = \langle R_{\mathrm{Hom},i}(\mathbf{x}, t) \rangle_{V_{\mathrm{R}}} \tag{6.8}$$

Note that the reaction rate is still a function of position because the radiation field, always included in the reaction rate, is usually not uniform. In equation 6.8:

$$\langle R_{\mathrm{Hom},i}(\mathbf{x}, t) \rangle_{V_{\mathrm{R}}} = \frac{1}{V_{\mathrm{R}}} \int_{V_{\mathrm{R}}} R_{\mathrm{Hom},i}(\mathbf{x}, t)\, \mathrm{d}V \tag{6.9}$$

Even in well-mixed photochemical reactors, the volume average of the reaction rate must always be calculated because usual experimental measurements never represent local values. Note that the volume of connecting lines in Figure 6.2 has been considered negligible. This stirring mechanism is suggested for laboratory reactors to avoid the effects produced by the presence of a stirrer inside the reactor thus producing a distortion of the radiation field inside the reaction space.

The isothermal, batch reactor with recycle. These systems are normally used when the reaction rate is rather slow and single-pass operation is not effective (Figure 6.3). Under the following assumptions – (i) differential operation per pass in V_{R} (slow reaction and/or very high recirculating flow rate), (ii) $V_{\mathrm{R}} \ll V_{\mathrm{T}}$, and (iii) very good mixing conditions in V_{R} and V_{Tk} – we can treat the whole system ($V = V_{\mathrm{R}} + V_{\mathrm{Tk}}$) as a well-mixed batch reactor. Then, since $\nabla \cdot N_i = 0$ (no concentration gradients, no inlet or outlet streams), integrating equation 6.1 in the total volume we get

$$(V_{\mathrm{R}} + V_{\mathrm{Tk}}) \overbrace{\frac{\langle \mathrm{d}C_i(\mathbf{x}, t) \rangle_{V_{\mathrm{T}}}}{\mathrm{d}t}}^{\text{Unknown}} = \langle R_{\mathrm{Hom},i}(\mathbf{x}, t) \rangle_{V_{\mathrm{R}}} V_{\mathrm{R}} + \underbrace{\langle R_{\mathrm{Hom},i}(\mathbf{x}, t) \rangle_{V_{\mathrm{Tk}}} V_{\mathrm{Tk}}}_{\text{Zero: No reaction in the tank}} \tag{6.10}$$

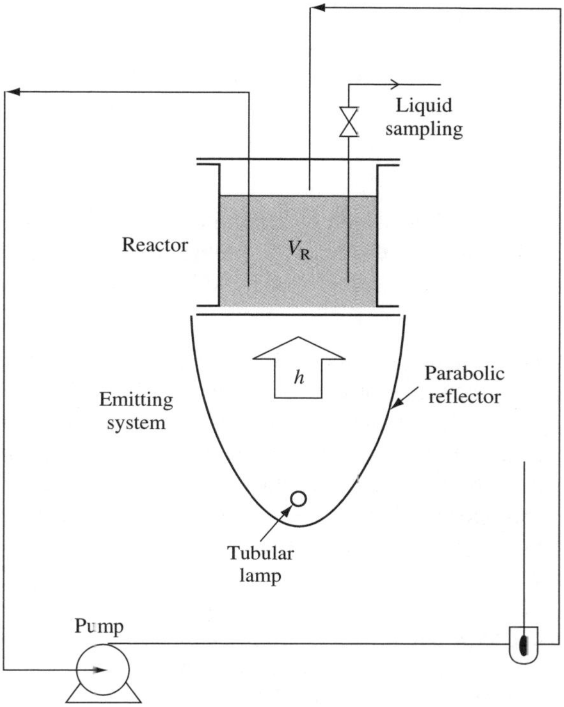

Figure 6.2 *Schematic diagram of the isothermal, constant volume, well-stirred batch reactor*

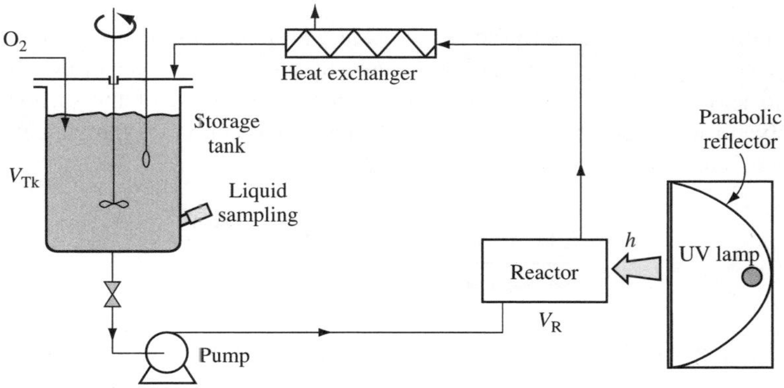

Figure 6.3 *Schematic diagram of the isothermal reactor with recirculation*

The average concentration can be divided into two parts:

$$\frac{V_R}{V_T}\underbrace{\frac{d\langle C_i(\mathbf{x},t)\rangle_{V_R}}{dt}}_{\text{In the reactor}}+\frac{V_{Tk}}{V_T}\underbrace{\frac{dC_i}{dt}}_{\text{In the tank}}=\frac{V_R}{V_T}\underbrace{\langle R_{\text{Hom},i}(\mathbf{x},t)\rangle_{V_R}}_{\text{Only in the reactor}} \tag{6.11}$$

Since $V_R/V_T \ll 1$ and the conversion per pass in the reactor is very small, the first term in equation 6.11 is negligible. Then, changes in concentration with time described by equation 6.10 can be measured directly in the tank. The final equation can be written as

$$\left.\frac{\mathrm{d}C_i(t)}{\mathrm{d}t}\right]_{\mathrm{Tk}} = \frac{V_{\mathrm{R}}}{V_{\mathrm{Tk}}}\left\langle R_{\mathrm{Hom},i}(\mathbf{x},t)\right\rangle_{V_{\mathrm{R}}} \tag{6.12}$$

A slightly different result was presented by Brandi *et al.* (2003). In this case the reaction rate is multiplied by the ratio of the reactor volume over the total volume. Under the assumption of small reactor volume employed in both derivations, both results are almost equivalent.

Heterogeneous reactions. Components of water or air pollution are usually in the fluid phase. Hence we may write equations such as equations 6.2, 6.6, 6.8, and 6.12 for the fluid. The fluid may have non-permeable boundaries (the reactor walls) and permeable boundaries (entrances and exits of the system as well as catalytic surfaces where mass fluxes must be equal to the superficial reaction rates). Usually, these reaction rates are modeled as pseudo-homogeneous and, moreover, concentration measurements are almost always made in the fluid phase. Heterogeneous reactions are the result of a process that occurs at phase interfaces. This means that for the differential equation written for the fluid phase, heterogeneous reactions (surface reactions, for example) are just boundary conditions. The problem is very simple to formulate: at steady state and at the boundary of an active surface, the normal mass or molar fluxes must be made equal to the heterogeneous, superficial reaction rate. Then,

$$\text{At } \mathbf{x} \text{ on the surface} \rightarrow \underbrace{\mathbf{N}_i\cdot\mathbf{n}}_{\text{Mass fluxes}} = \underbrace{R_{\mathrm{Het},i}\left(C_i^{\mathrm{surface}}, T, \mathrm{etc.}\right)}_{\text{Surface reaction}} = R_{\mathrm{Het},i}(\mathbf{x},t) \tag{6.13}$$

Typical examples are solid catalyzed reactions or wall reactions occurring in free radical chemistry. Usually reacting surfaces are covered by a boundary layer of the fluid. Then, it is of no surprise that the fluxes can be expressed in terms of the diffusive fluxes exclusively. In any mass balance, we usually have mass fluxes expressed in terms of $\nabla\cdot\mathbf{N}_i$. From standard definitions (Bird *et al.*, 2002, p. 537):

$$\underbrace{\mathbf{J}_i(\mathbf{x},t)\cdot\mathbf{n}}_{\substack{\text{Normal component of}\\ \text{the diffusive flux}}} = R_{\mathrm{Het},i}(\mathbf{x},t)\,[=]\,\frac{\mathrm{mol}}{\mathrm{cm}^2\,\mathrm{s}} \tag{6.14}$$

Since we are interested in pseudo-homogeneous reaction rates:

$$R_{\mathrm{Hom},i}^{\mathrm{Pseudo}} = a_{\mathrm{v}}\quad R_{\mathrm{Het},i} = C_{\mathrm{mp}}S_{\mathrm{g}}\quad R_{\mathrm{Het},i}\,[=]\,\frac{\mathrm{mol}}{\mathrm{cm}^3\,\mathrm{s}} \tag{6.15}$$

6.2.2 Thermal Energy Conservation Equation

Assuming the following heating effects due to radiative transfer and neglecting – (i) energy fluxes caused by interdiffusion of the different chemical species; (ii) heat effects produced by viscous dissipation; (iii) heat effects resulting from pressure gradients; (iv) heat conduction in the axial direction compared with the convective flow in the same direction; and assuming (v) constant physical and transport properties and (vi) steady state conditions – the balance of thermal energy for multicomponent systems in cylindrical coordinates is (extracted from Bird *et al.*, 2002, pp. 589, 848)

$$\underbrace{\rho_{\text{mix}}\hat{C}_{\text{P}}\left(v_z\frac{\partial T}{\partial z}\right)}_{\substack{\text{Thermal flow in}\\ \text{the axial direction}}} = \underbrace{k_{\text{c}}\left[\frac{1}{r}\frac{\partial}{\partial r}\left(r\frac{\partial T}{\partial r}\right)\right]}_{\substack{\text{Heat conduction in the}\\ \text{radial direction}}} + \underbrace{\rho Q^{\text{Ext}}}_{\substack{\text{Radiation}\\ \text{heat sources}}} - \underbrace{\sum_j \overline{H}_j(R_i)}_{\substack{\text{Enthalpy changes}\\ \text{due to chemical reactions}}} \tag{6.16}$$

Where Q^{Ext} is a scalar that includes all forms of heating effects produced by energy transmission without contact, i.e. from external bodies (typically, radiation). In the vast majority of photochemical reactions (employing visible and UV light), heating effects produced by radiation should not be important. However, with lamps emitting significant energy in the infrared region, if the IR radiation is not filtered (i.e. absorbed by cooling devices before entering the reactor), the Q^{Ext} term must be taken into account.

At this point an important consideration concerning photochemical reactions must be stressed. The first step of the reaction – the activation – is made by radiation absorption. The absorbed photons are usually of high energy, producing a change in the electronic state of the molecule. Thus, the alterations produced in the chemical species are not of a thermal nature (vibrations, rotations, and translations); i.e. heating effects are almost negligible. For this reason, for all practical purposes, the radiative transfer equation and the thermal energy equation can be uncoupled. In equation 6.16, $\overline{H}_j$ is the partial molar enthalpies of reactants and products. Neglecting heating effects due to radiation, this equation can be re-written in the more familiar form:

$$\rho_{\text{mix}}\hat{C}_P\left(v_z\frac{\partial T}{\partial z}\right) = k_{\text{c}}\left[\frac{1}{r}\frac{\partial}{\partial r}\left(r\frac{\partial T}{\partial r}\right)\right] - \underbrace{\sum_j \Delta H_j R_j(\mathbf{x}, t)}_{\text{Heat of reaction}} \tag{6.17}$$

In equation 6.17, the index j stands for the j different chemical reactions occurring in the system. The inlet condition is

$$T(0, r) = T_0 \tag{6.18}$$

The boundary conditions that take into account heat transfer from the reactor walls into the cooling (heating) liquid or vice versa are

$$k_{\text{c}}\frac{\partial T(z, r_{R_i})}{\partial r} = h_{\text{f}}(T - T_{\text{c}}) \tag{6.19}$$

$$k_{\text{c}}\frac{\partial T(z, r_{\text{R}_0})}{\partial r} = -h_{\text{f}}(T - T_{\text{c}}) \tag{6.20}$$

If the reactor operates under almost isothermal conditions, equations 6.17–6.20 are not needed. Under plug flow conditions we can integrate equation 6.17 over the cross-sectional area of the tubular reactor and include the boundary conditions into the differential equation. We finally have

$$\rho_{\text{mix}}\hat{C}_P\langle v_z\rangle\frac{\text{d}T(z)}{\text{d}z} + a_{\text{v}}\underbrace{h_{\text{f}}(T - T_{\text{c}})}_{\text{Heat removal}} = \underbrace{\left\langle\sum_j(-\Delta H_j)R_j(\mathbf{x}, t)\right\rangle_{A_{\text{R,C}}}}_{\text{Heat produced}} \tag{6.21}$$

6.3 Radiation Transport

When writing the rate of a photochemical reaction it is necessary to make the distinction between dark and radiation-activated (lighted) steps. To treat the dark reactions one uses the same methodology as for conventional reactors; the main difference appears when evaluating the rate of the radiation-activated step. The existence of this very particular step constitutes the main distinctive aspect (and the most important one) between thermal (or thermal catalytic) and radiation-activated reactions. The rate of the radiation-activated step is directly proportional to the absorbed, useful energy through a property that has been defined as the local volumetric rate of photon absorption (LVRPA). The LVRPA, (e^a_λ), represents the amount of photons that are absorbed by the reactant per unit time and unit reaction volume. The LVRPA depends on the radiation field (photon distribution) existing in the reaction space; hence, we must know the radiation field within the photoreactor. The value of the LVRPA is defined for monochromatic radiation but it can be extended to polychromatic fields by performing an integration over all useful wavelengths. The useful wavelength range is defined by the overlapping ranges of lamp emission, region of reactor wall good transmission properties, reactant or catalyst absorption, and, eventually, reflector reflectance (Clariá *et al.*, 1988).

The general structure for calculating the rate of the activation step may be illustrated schematically as in Figure 6.4. As was shown before, the mass balances require expressions formulating the reaction rates; be it a molecular or a free radical reaction mechanism, always some of the steps (generally one) are initiated by radiation absorption. The radiation-activated step kinetics is always written in terms of e^a_λ. The evaluation of the LVRPA is performed stating first the general radiation transport equation that requires the appropriate constitutive equations for absorption, emission, and scattering. The resulting radiative transfer equation is then successively applied to the reaction space

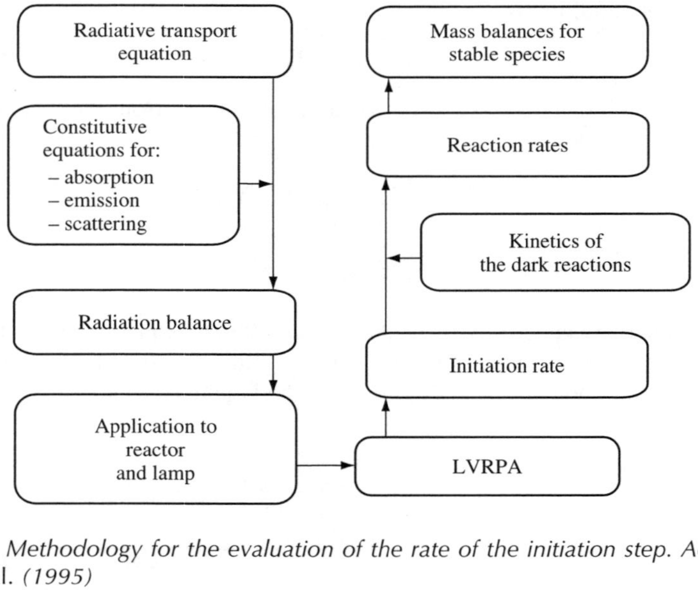

Figure 6.4 *Methodology for the evaluation of the rate of the initiation step. Adapted from Cassano* et al. *(1995)*

where there is only absorption (in homogeneous media) or absorption and scattering (in heterogeneous media), and to the lamp where emission is the prevailing phenomenon. Combining both results one can obtain, in a straightforward manner, the local value of the radiation absorption rate. With this information the rate equation is developed and incorporated into the mass balance.

6.3.1 Spectral Specific Intensity

Under usual conditions, propagation of photons may be represented by bundles of rays with a given energy. These rays may be specified by the spectral specific intensity that is the fundamental property for characterizing radiation fields (Figure 6.5). Let dE_λ be the total amount of radiative energy passing through the area dA inside the truncated cone $d\omega$ in the time dt and with an energy in the wavelength range between λ and $\lambda + d\lambda$. The spectral specific intensity (also called radiance) is defined as

$$I_{\lambda,\Omega}(\mathbf{x}, t) = \lim_{dA, d\Omega, dt, d\lambda \to 0} \left(\frac{dE_\lambda}{dA \cos\theta \, d\Omega \, dt \, d\lambda} \right) \qquad (6.22)$$

Quantum theory introduces the proportionality between frequency (or wavelength) and energy. The energy of a quantum is $e = h\nu = hc/\lambda$. To some extent, a quantum is a unit of energy, but its magnitude is not fixed because it varies with the wavelength (or the frequency). The best definition of a quantum is that it is the radiant energy equal to $h\nu$. However it is defined, when one molecule or an atom absorbs one quantum, a change in that molecule or atom from one level of energy to another will be produced; i.e. its energy will have been increased by an amount equal to one quantum. Similarly, if one mole must reach the same level of activation, the energy absorbed is Nhc/λ, where

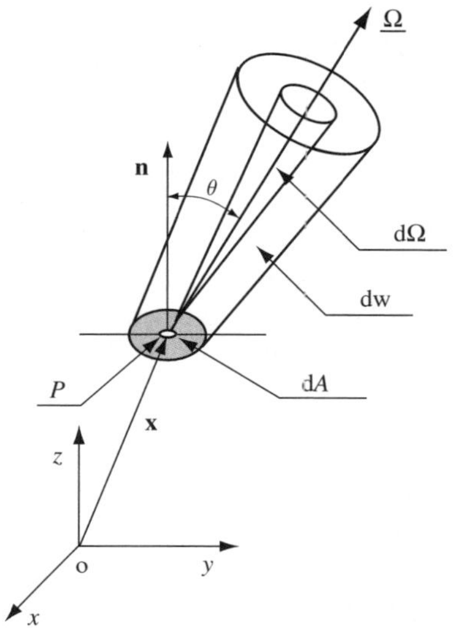

Figure 6.5 Characterization of the spectral specific intensity. Adapted from Cassano et al. (1995)

N is Avogadro's number. Hence the energy of a gram mole of a given material will be increased by Nhc/λ. The quantity of radiant energy equal to Nhc/λ is called one einstein. All units in joule (or watt) can be converted into einstein (or einstein s^{-1}) with the proper transformation. This new unit is very convenient in photochemistry because the photochemical activation is the result of the interaction of one molecule with one photon having one quantum of energy or, in other terms, one mole with one mole of photons that have an energy equal to one einstein. The transformation can be obtained as follows: 1 einstein $= [0.11964/\lambda\,(\mathrm{m})]\,\mathrm{W\,s}$.

6.3.2 Homogeneous Media

From the radiation viewpoint a homogeneous medium means that scattering does not need to be considered. This is a great simplification for modeling and design. In this case, the intensity of a monochromatic beam of radiation in any arbitrary direction Ω will be changed only by emission or absorption. Emission can be usually neglected particularly for low-temperature processes. Then, at any point in space $\mathbf{x}$ and any time t, we are left with the 3D form of the Bouguer–Lambert 'law' for monochromatic radiation absorption in homogeneous media:

$$\frac{\mathrm{d}I_{\lambda,\Omega}(\mathbf{x},t)}{\mathrm{d}s} + \kappa_\lambda(\mathbf{x},t)I_{\lambda,\Omega}(\mathbf{x},t) = 0 \tag{6.23}$$

In equation 6.23, s is measured along a chosen direction for photon transport in space (Ω). The spectral specific intensity must not be confused with radiation density fluxes. They are equal only for unidirectional irradiation, a case very distant from the general one. Radiation may be arriving at one point inside a photochemical reactor from all directions in space. For a photochemical reaction to occur, this radiation must be absorbed by an elementary reacting volume (a material point in space); thus, pencils of radiation coming from all directions must cross the whole elementary surface that bounds such an element of volume. Consequently, the important photochemical property is the spectral incident radiation (or spectral spherical irradiance) given by

$$G_\lambda(\mathbf{x},t) = \int_\Omega I_{\Omega,\lambda}(\mathbf{x},t)\mathrm{d}\Omega \tag{6.24}$$

In equation 6.24, integration for all possible directions Ω over the entire spherical space has been performed. For polychromatic radiation, integration over the wavelength range of interest must be carried out. In the elementary volume of radiation absorption, for single photon absorption, energy is absorbed according to

$$e^{\mathrm{a}}_\lambda(\mathbf{x},t) = \kappa_\lambda(\mathbf{x},t)G_\lambda(\mathbf{x},t) \tag{6.25}$$

Where e^{a}_λ is the spectral (monochromatic) LVRPA or the spectral rate of photon energy absorption per unit reaction volume. Note that since G_λ is a function of position, so is e^{a}_λ. G_λ may be a function of time for lamps operating under unsteady state conditions. The absorption coefficient may be a function of position for reactors operating under strong concentration gradients and a function of time for systems where absorption changes with the reaction progress (the reactant absorbs radiation, some reaction products absorb radiation, etc.) or when using a photocatalyst the solid semiconductor does not have stable optical properties (fouling or change in particle size). For polychromatic radiation and substituting the differential of the solid angle

$$e^{a}_{\Sigma\,\lambda}(\mathbf{x},t) = \int_{\lambda_1}^{\lambda_2}\int_{\theta_1}^{\theta_2}\int_{\phi_1}^{\phi_2} \kappa_\lambda I_{\theta,\phi,\lambda}(\mathbf{x},t)\,\sin\theta\,\mathrm{d}\phi\,\mathrm{d}\theta\,\mathrm{d}\lambda \tag{6.26}$$

Where $[\theta_1, \theta_2]$ and $[\phi_1, \phi_2]$ are the integration limits that define the space from which radiation arrives at the point of incidence. For each point of incidence, in practice, these limits are defined by the extension of the lamp (its diameter and its length). Thus, to evaluate the LVRPA we must know the spectral specific intensity at each point inside the reactor. Its value can be obtained from the photon transport equation (equation 6.23).

6.3.3 Heterogeneous Media

In more general terms, the radiative transfer equation may be rationalized considering a balance of monochromatic photons along a given direction of radiation propagation. Extending a proposal formulated by Whitaker (1977):

$$\left|\begin{array}{c}\text{Time rate of}\\ \text{change of } \Omega,\lambda\\ \text{photons in the}\\ \text{volume } V\end{array}\right| + \left|\begin{array}{c}\text{Net flux of } \Omega,\lambda\\ \text{photons leaving the}\\ \text{volume } V \text{ across}\\ \text{its bounding}\\ \text{surface } A\end{array}\right| = \left|\begin{array}{c}\text{Net gain (loss) of } \Omega,\lambda\\ \text{photons owing to absorption,}\\ \text{emission, and in- and out-}\\ \text{scattering in the volume } V\end{array}\right|$$

$$\frac{\partial}{\partial t}\int_V N_{\Omega,\lambda}\,\mathrm{d}V + \int_A N_{\Omega,\lambda}\,(c\boldsymbol{\Omega})\cdot\mathbf{n}\,\mathrm{d}A$$
$$= -\int_V N^{a}_{\Omega,\lambda}\,\mathrm{d}V + \int_V N^{e}_{\Omega,\lambda}\,\mathrm{d}V + \int_V N^{\text{s-in}}_{\Omega,\lambda}\,\mathrm{d}V - \int_V N^{\text{s-out}}_{\Omega,\lambda}\,\mathrm{d}V \tag{6.27}$$

Transforming the area integral, into a volume integral, all the terms will have the same integration limits. Then, multiplying by $(h\nu)$ and considering that $I_\lambda = c(h\nu)(N_{\lambda,\Omega})$ we can extract the differential equation in terms of specific intensities. In symbolic form (Ozisik, 1973, p. 251)

$$\frac{1}{c}\frac{\partial I_{\Omega,\lambda}}{\partial t} + \nabla\cdot\left(I_{\Omega,\lambda}\boldsymbol{\Omega}\right) = -W^{a}_{\Omega,\lambda} + W^{e}_{\Omega,\lambda} + W^{\text{s-in}}_{\Omega,\lambda} - W^{\text{s-out}}_{\Omega,\lambda} \tag{6.28}$$

Usually the first term can be neglected; i.e. at a given time the radiation field reaches its steady state almost instantaneously. However $I_{\Omega,\lambda}$ will change with time if the boundary condition associated with equation 6.28 is time-dependent (typically, a solar reactor) or if the state variables which appear in the constitutive equations for any one of the different processes $W^{a}_{\Omega,\lambda}$, $W^{e}_{\Omega,\lambda}$, $W^{\text{s-in}}_{\Omega,\lambda}$, and $W^{\text{s-out}}_{\Omega,\lambda}$ change with time. Absorption and out-scattering are modeled in the same way that absorption is accounted for in homogeneous systems. Emission should be modeled according to the particular involved process. However, in most photochemical reactions it can be neglected because the reaction temperature is usually low and more often than not there is no induced emission (fluorescence and/or phosphorescence). In-scattering is responsible for most of the complications that arise when scattering of radiation is an important phenomenon. It results from the almost unavoidable existence of multiple scattering. When scattering is not single, a photon scattered out from one direction may interact with other particles. Then, part of the radiation that is scattered in space in all directions may be incorporated into the stream of Ω, λ photons according to the scattering distribution function (the phase function). For elastic or coherent scattering there is no change in energy; then (Ozisik, 1973, p. 27)

$$W^{\text{s-in}}_{\Omega,\lambda} = \frac{1}{4\pi} \int_{\Omega'=4\pi} \sigma_\lambda(\mathbf{x}, t)\, p(\Omega' \to \Omega)\, I_{\Omega',\lambda}(\mathbf{x}, t)\, \mathrm{d}\Omega' \tag{6.29}$$

where p is the phase function. The normalizing condition for the phase function p is

$$\frac{1}{4\pi} \int_{\Omega'=4\pi} p(\Omega' \to \Omega)\, \mathrm{d}\Omega' = 1 \tag{6.30}$$

Scattering is isotropic when $p = 1$. Isotropic scattering requires, among other requirements, that at least the scattering material be homogeneous and isotropic, and that the surrounding medium be also isotropic. More details on scattering and phase functions can be found in the classical references of Van de Hulst (1957), Ozisik (1973), and Siegel and Howell (1992).

Very often the sum of the absorption coefficient and the scattering coefficient is called the extinction coefficient:

$$\beta_\lambda(\mathbf{x}, t) = \kappa_\lambda(\mathbf{x}, t) + \sigma_\lambda(\mathbf{x}, t) \tag{6.31}$$

Working photon transport equation. Going back to equation 6.28 one can neglect the transient term and substitute the different constitutive relationships. After defining a directional coordinate s along the ray path, from elementary calculus it can be written

$$\begin{aligned} \frac{\mathrm{d}I_{\Omega,\lambda}(s, t)}{\mathrm{d}s} &+ \underbrace{\kappa_\lambda(s, t)\, I_{\Omega,\lambda}(s, t)}_{\text{Absorption}} + \underbrace{\sigma_\lambda(s, t)\, I_{\Omega,\lambda}(s, t)}_{\text{Out-scattering}} = \underbrace{j^{\mathrm{e}}_\lambda(s, t)}_{\text{Emission}} \\ &+ \underbrace{\frac{1}{4\pi}\sigma_\lambda(s, t) \int_{\Omega'=4\pi} p(\mathbf{\Omega}' \to \mathbf{\Omega})\, I_{\Omega',\lambda}(s, t)\, \mathrm{d}\Omega'}_{\text{In-scattering}} \end{aligned} \tag{6.32}$$

There is an important assumption implicit in the derivation of this expression; it may be applied only to a medium that may be considered as pseudo-homogeneous. It should have a valid application when the existing heterogeneities are of small size and they are present in small concentrations. This consideration leads us to conclude that under the validity conditions already established for equation 6.32, most likely, conditions for independent scattering will prevail as well.

Perhaps one of the most important conclusions that can be drawn from this equation is that in heterogeneous reacting systems classical forms of analyzing the light distribution inside the photochemical cell (i.e. the Lambert–Beer equation) are incorrect and, very likely, useless. To integrate the radiative transfer equation (RTE) we need a boundary condition: the incoming radiation to the reaction space. It is provided by an emission model for the lamp.

6.4 Emission by Tubular Lamps[1]

6.4.1 The 3D Emission Models

Two main types of models for tubular lamps (the most widely used) will be described. There are lamps that produce an arc that emits radiation and, consequently, photons come out directly from such an arc. Emission is made by the whole lamp volume. We call this process *Voluminal Emission*. There are other types of lamps in which the discharged arc between electrodes induces an emission produced by some particular substance that has been coated on the lamp surface. We call this process *Superficial Emission*. Voluminal emission may be safely modeled as an isotropic emission; in this case the specific intensity associated with each bundle of radiation originated in some element of volume of the lamp is independent of direction, and the associated emitted energy (per unit time and unit area) is also isotropic (Figure 6.6). On the other hand, it seems that superficial emission can be better modeled by a diffuse type of emission that is also known as one that follows the Lambert's 'cosine law' of emission; in this case the emitted intensity is independent of direction but the emitted energy depends on the surface orientation and follows the 'cosine law' equation (Figure 6.7). The following assumptions are made (Irazoqui *et al.*, 1973):

1. The emitters of the radiation source are uniformly distributed over the region of emission (a volume or a surface).
2. In terms of Specific Intensities each elementary extension of emission has an isotropic emission but the outgoing radiation energy is: (i) isotropic when the emitting element is a volume or (ii) diffuse (affected by the surface orientation) when the element is a surface.

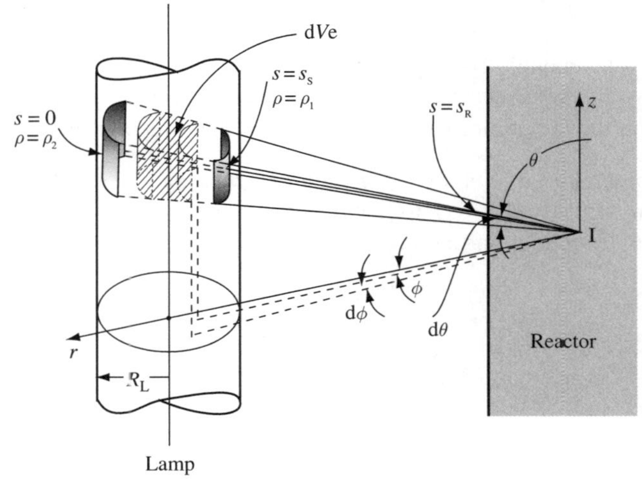

Figure 6.6 The extended source with the voluminal emission model for the lamp. Adapted from Cassano et al. *(1995)*

[1] Reprinted with permission from Cassano *et al.*, 1995, Copyright 1995 American Chemical Society.

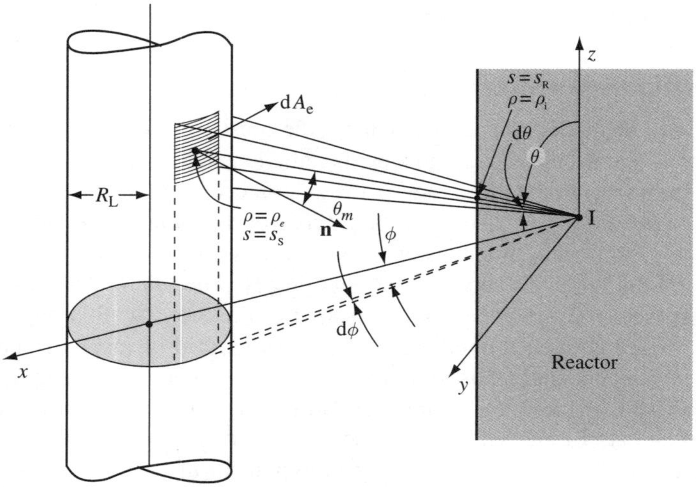

Figure 6.7 *The extended source with the superficial emission model for the lamp. Adapted from Cassano* et al. *(1995)*

3. Any emission element of the lamp emits per unit time, and for a given wavelength interval, an amount of energy proportional to its extension and independent of its position inside the lamp volume or on the lamp surface.
4. When emission is voluminal, each of the differential volumes of emission is transparent to the emission of its surroundings (a possible questionable approximation).
5. The lamp is a perfect cylinder bounded by mathematical surfaces with zero thickness. Hence, any bundle of radiation coming from inside does not change its intensity or direction when it crosses this boundary (again, an approximation).
6. The lamp is long enough; consequently, neglecting end effects, the emission produced by the lamp along its central axis is uniform. This assumption does not impose uniformity on the radiation field generated along the direction of the central axis.

Three-dimensional source with superficial diffuse emission. The E-SDE source model. From the lamp surface ($s = s_S$) to the reactor wall ($s = s_R$) there is no emission (Figure 6.7), no scattering, and no absorption (the medium is assumed to be diactinic); therefore,

$$\frac{\mathrm{d}I_\lambda(s, \Omega)}{\mathrm{d}s} = 0 \tag{6.33}$$

It follows that

$$I_\lambda^0(\theta, \phi) = I_\lambda(\mathbf{x}, \theta, \phi)|_{\rho=\rho_i} = I_\lambda(\mathbf{x}, \theta, \phi)|_{\rho=\rho_e} \Upsilon_{R,\lambda}(\Omega) \tag{6.34}$$

Since emission is uniform in space and isotropic in directions

$$I_\lambda(\mathbf{x}, \theta, \phi)|_{\rho=\rho_e} = I_\lambda^e \tag{6.35}$$

Now we must relate the value of the specific intensity of emission to the emission power of the lamp. From the definition of the spectral specific intensity

$$dP_{\lambda,S} = I_\lambda^e \, d\Omega dA_e \cos\theta_n \tag{6.36}$$

from which

$$I_\lambda^e = \frac{P_{\lambda,S}}{\int_{\Omega_S} \cos\theta_n \, d\Omega \int_{A_S} dA_e} = \frac{P_{\lambda,S}}{2\pi^2 R_L L_L} \tag{6.37}$$

According to equations 6.34, 6.35, and 6.37, the boundary condition when a lamp with superficial emission is used is given by

$$I_\lambda^0(\theta,\phi) = \frac{\Upsilon_{R,\lambda}(\Omega) P_{\lambda,S}}{2\pi^2 R_L L_L} \tag{6.38}$$

Three-dimensional source with voluminal isotropic emission. The E-VIE source model. Since emission is produced by a volume, the radiative transfer equation can be applied inside the lamp. There is no absorption (assumption 4) and no scattering. For isotropic and uniform emission

$$\frac{dI_\lambda(s,\Omega,t)}{ds} = j_\lambda^e(s,\Omega,t) = j_\lambda^e \tag{6.39}$$

Along the direction Ω, at $s = 0$ (Figure 6.6), there is no entrance of radiation; this situation provides the required boundary condition for equation 6.39:

$$s = 0 \quad I_\lambda(0,\Omega) = 0 \tag{6.40}$$

Integrating from $s = 0$ to $s = s_S$ and changing coordinates (recall Figure 6.6):

$$s = 0 \qquad \rho = \rho_2 \tag{6.41}$$

$$s = s_S \qquad \rho = \rho_1 \tag{6.42}$$

and one gets:

$$I_\lambda(s_s,\Omega) = j_\lambda^e[\Delta\rho_s(\mathbf{x},\theta,\phi)] = \rho_2(\mathbf{x},\theta,\phi) - \rho_1(\mathbf{x},\theta,\phi) \tag{6.43}$$

It must be remarked that, as should have been expected, $\Delta\rho_s$ is a function of the position $\mathbf{x}$ inside the reactor and the direction of the incoming radiation given by the spherical coordinates (θ,ϕ). Once more, from $s = s_S$ to $s = s_R$ there is no emission, no scattering, and no absorption; from assumption 5 there is no refraction or reflection at the lamp boundaries, therefore

$$\frac{dI_\lambda(s,\Omega,)}{ds} = 0 \tag{6.44}$$

Consequently, the boundary condition is

$$I_\lambda^0(\theta,\phi) = j_\lambda^e[\Delta\rho_s(\mathbf{x},\theta,\phi)]\Upsilon_{R,\lambda}(\Omega) \tag{6.45}$$

Once more, the value of j_λ^e must be related to the lamp output power $P_{\lambda,s}$. By definition, j_λ^e is the energy emitted per unit volume, unit solid angle of emission, and unit time; therefore

$$P_{\lambda,s} = \int_{\Omega_S} \int_{V_S} j_\lambda^e \, dV_e \, d\Omega \tag{6.46}$$

Since j_λ^e corresponds to an isotropic and uniform emission

$$j_\lambda^e = \frac{P_{\lambda,s}}{4\pi^2 R_L^2 L_L} \tag{6.47}$$

From equation 6.43 we must know the values of $\rho_2(\mathbf{x}, \theta, \phi)$ and $\rho_1(\mathbf{x}, \theta, \phi)$. In order to know these values one must obtain explicit expressions for the independent variable ρ, at the positions indicated by equations 6.41 and 6.42. To illustrate the procedure, the case of an annular reactor will be analyzed (Figure 6.8). Let us consider a point located in an arbitrary position I, having coordinates $\mathbf{x}(r, \beta, z)$ and look at an arbitrary direction (θ, ϕ). The equation of the boundary surface of the radiation source (a cylinder) in spherical coordinates is written as follows:

$$\rho^2 \sin^2\theta - 2\rho(\sin\theta\cos\phi)r + (r^2 - R_L^2) = 0 \tag{6.48}$$

The two solutions of this quadratic equation are precisely the values of ρ; i.e. they are the intersections of the ρ coordinate with the front and rear parts of the lamp at any value of θ and ϕ:

$$\rho_{1,2} = \frac{r\cos\phi \pm (r^2\cos^2\phi - r^2 + R_L^2)^{1/2}}{\sin\theta} \tag{6.49}$$

Finally, the following value for $\Delta\rho_S$ is obtained:

$$\Delta\rho_S = \frac{2\left[r^2(\cos^2\phi - 1) + R_L^2\right]^{1/2}}{\sin\theta} \tag{6.50}$$

The boundary conditions when a lamp with voluminal emission is employed results in:

$$I_\lambda^0(\mathbf{x}, \theta, \phi) = \frac{P_{\lambda,S}\Upsilon_{R,\lambda}(\Omega)}{2\pi^2 R_L^2 L_L} \frac{\left[r^2(\cos^2\phi - 1) + R_L^2\right]^{1/2}}{\sin\theta} \tag{6.51}$$

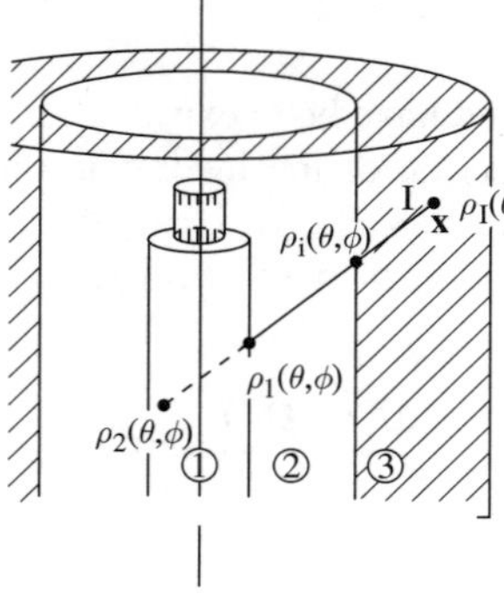

Figure 6.8 *The annular reactor. Adapted from Cassano* et al. *(1995)*

Limits of integration for the 3D emission models. When a lamp with superficial emission is used, according to equation 6.38, a constant value must be incorporated as a boundary condition. Conversely, when lamps with voluminal emission are used, according to equation 6.51, the boundary condition introduces a function of **x**, θ, and ϕ. The limits of integration for the annular reactor with the tubular lamp were derived by Irazoqui *et al.* (1973) and systematically described by Cassano *et al.* (1995). They are

$$\theta_1(\phi) = \tan^{-1}\left\{\frac{r\cos\phi - \left[r^2(\cos^2\phi - 1) + R_L^2\right]^{1/2}}{(L_L - z)}\right\} \tag{6.52}$$

$$\theta_2(\phi) = \tan^{-1}\left\{\frac{r\cos\phi - \left[r^2(\cos^2\phi - 1) + R_L^2\right]^{1/2}}{-z}\right\} \tag{6.53}$$

$$-\phi_1 = \phi_2 = \cos^{-1}\left[\frac{(r^2 - R_L^2)^{1/2}}{r}\right] \tag{6.54}$$

The limits of integration for the case in which reflecting surfaces are present in the system (for example elliptical or parabolic reflectors) require a more elaborated procedure (Cassano *et al.*, 1995).

6.5 Homogeneous Systems. Reaction Kinetics of a Pollutant Photolysis and its Application to the Prediction of the Performance of an Annular Reactor[2]

6.5.1 Reaction Kinetics

A very simple case will be used to illustrate the procedure. For many processes employing radiation having a wavelength below 300 nm, direct photolysis is often present. Hence, even if in practice an oxidant will always be used (for example, hydrogen peroxide) the parallel photolysis must be also modeled. 2,4-dichlorophenoxyacetic acid (2,4-D) is a widespread herbicide that is known to have a high level of toxicity. The reaction using UV alone with 2,4-D shows most of the features that must be taken into account to model a homogeneous reactor for AOTs. Although it is a rather slow reaction that needs to be complemented with a stronger oxidation, it can be used to illustrate some of the concepts previously developed. As reported by Cabrera *et al.* (1997a), kinetic studies were performed in a well-stirred, batch, cylindrical photoreactor irradiated from the bottom (Figure 6.2). Monochromatic light ($\lambda = 254$ nm) was used. Analyses of the results were performed as is described in the following sections.

Radiation field. Alfano *et al.* (1985) studied the above described reactor geometry using a 3D (r, z, β) model. It was found that for the used geometry and dimensions, radial and angular variations were not very significant. With this background, a 1D model (x-coordinate) can be adopted. Then, incident radiation (equation 6.24) can be described by

$$G_\lambda(x) = G_{W,\lambda}\exp\left(-\kappa_{T,\lambda}x\right) \tag{6.55}$$

[2] Reproduced with permission from Cabrera *et al.*, 1997a and Martín *et al.*, 1997; Copyright 1997 IWA Publishing.

In equation 6.55, $G_{W,\lambda}$ is the incident radiation at the wall of the reactor bottom ($x = 0$) and $\kappa_{T,\lambda}$ is the total absorption coefficient of reactant and products. It must be noted that the optical properties of the reacting medium change with the reaction evolution. Consequently, even in a first approximation, the system must be characterized by a minimum of two absorption coefficients: (i) one corresponding to the reactant (2,4-D), and (ii) a different one corresponding to the rest of the reacting mixture, both being a function of time. These values as well as $G_{W,\lambda}$ can be experimentally measured. It should be also noticed that, if desired, the incident radiation at $x = 0$ can be theoretically predicted with great accuracy with the Alfano *et al.* (1985) radiation model. The derivation of equation 6.55 needs some careful analysis. First, note that in any 1D model the intensity has the special characteristic that only one component of the 3D representation of the radiation field is different from 0. In general, with the Dirac delta function

$$I_{\lambda,\Omega}(x,t) = \bar{I}(x,t)\,\delta(\mathbf{\Omega} - \mathbf{i}) \tag{6.56}$$

Note also that the units for this 'special, one-directional, one-dimensional intensity' are $\bar{I}_\lambda(\mathbf{x},t) = \bar{I}_\lambda(x,t)\,[=]\,\text{einstein m}^{-2}\,\text{s}^{-1}$, with $\delta(\mathbf{\Omega} - \mathbf{i})\,[=]sr^{-1}$.

In this case, the incident radiation results:

$$\begin{aligned} G_\lambda(x,t) &= \int_\Omega I_{\lambda,\Omega}(x,t)\mathrm{d}\Omega = \int_\Omega \bar{I}_\lambda(x,t)\,\delta(\mathbf{\Omega} - \mathbf{i})\,\mathrm{d}\Omega \\ &= \bar{I}_\lambda(x,t)\int_\Omega \delta(\mathbf{\Omega} - \mathbf{i})\mathrm{d}\Omega = \bar{I}_\lambda(x,t) \end{aligned} \tag{6.57}$$

The LVRPA for the photolytic reaction is obtained from

$$e_\lambda^{a,D}(x,t) = \kappa_{D,\lambda} G_\lambda(x,t) = \kappa_{D,\lambda}(t) G_{W,\lambda} \exp\left[-\kappa_{T,\lambda}(t)x\right] \tag{6.58}$$

In equation 6.58, $\kappa_{D,\lambda}$ is the absorption coefficient of 2,4-D exclusively. The boundary condition for equation 6.58 is that the incident radiation at $x = 0$. It can be precisely evaluated with actinometer measurements. Potassium ferrioxalate was used according to the operating conditions reported by Murov *et al.* (1993). According to equation 6.8 a mass balance for the well-stirred, isothermal, batch reactor applied to the actinometer reaction gives for the reaction product (Fe^{2+})

$$\frac{\mathrm{d}C_{Fe^{2+}}(t)}{\mathrm{d}t} = \left\langle R_{Hom,i}(x,t)\right\rangle_{V_R} = \Phi_\lambda^{Ac}\left\langle e_\lambda^{a,Ac}(x,t)\right\rangle_{V_R} \tag{6.59}$$

$$\left\langle e_\lambda^{a,Ac}(x,t)\right\rangle_{V_R} = \frac{1}{L_R}\int_0^{L_R = V_R/A_R} \kappa_{Ac,\lambda}(t) G_{W,\lambda} \exp\left[-\kappa_{T,\lambda}(t)\,x\right]\mathrm{d}x \tag{6.60}$$

Where V_R/A_R is the radiation path. Note that both absorption coefficients are a function of t because always $\kappa_i = \kappa_i^* C_i$ and C_i for both Fe^{3+} and Fe^{2+} changes with the reaction evolution. Integrating equation 6.60 and substituting the results into equation 6.59:

$$\frac{\mathrm{d}C_{Fe^{2+}}}{\mathrm{d}t} = \Phi_\lambda^{Ac}\frac{G_{W,\lambda}}{L_R}\frac{\kappa_{Ac,\lambda}(t)}{\kappa_{T,\lambda}(t)}\left\{1 - \exp\left[-\kappa_{T,\lambda}(t)\,L_R\right]\right\} \tag{6.61}$$

In equation 6.61, Ac is the reactant (Fe^{+3}) and the reaction product is Fe^{+2}. In the batch reactor, for not too high reactant conversions, the plot of Fe^{+2} concentration versus time gives a straight line. Then, at $t \rightarrow 0$ and taking into account that at 254 nm $\kappa_{Fe^{3+}}$ is large:

$$G_{W,\lambda} = \left(\frac{V_R}{\Phi_\lambda^{Ac} A_R}\right) \lim_{t\to 0} \left(\frac{C_{Fe^{2+}} - C^0_{Fe^{2+}}}{t-0}\right) \tag{6.62}$$

2,4-D mass balance. The reactor operates under the following conditions: (i) perfect mixing, and (ii) isothermal performance. The mass balance (equation 6.8) gives

$$\frac{dC_D(t)}{dt} = \langle R_{D,\lambda}(y,t)\rangle_{L_R} \qquad C_D(t=0) = C_D^0 \tag{6.63}$$

An expression for the local reaction rate is still unknown. The following simple relationship was proposed:

$$R_{D,\lambda}(y,t) = -\Phi_{D,\lambda}\left[e_\lambda^{a,D}(y,t)\right]^n \left[C_D(t)\right]^m \tag{6.64}$$

Calculating the average reaction rate and substituting into the mass balance, after integration:

$$\frac{dC_D(t)}{dt} = -\frac{1}{L_R}\Phi_{D,\lambda}\left(G_{W,\lambda}\right)^n \frac{\left(\kappa_{D,\lambda}(t)\right)^n}{\kappa_{T,\lambda}(t)n}\left[C_D(t)\right]^m \left\{1-\exp\left[-\kappa_{T,\lambda}(t)L_R n\right]\right\} \tag{6.65}$$

This ordinary differential equation must be solved with the initial condition indicated in equation 6.63. Note that due to the required averaging procedure, the reaction order with respect to the LVRPA (n) has a rather complex relationship with respect to the time rate of change of concentrations.

Absorption coefficients. Equation 6.65 needs two optical parameters that must be obtained from independent measurements. The 2,4-D absorption coefficient can be obtained from standard measurements. To obtain the total absorption coefficient (a mixture of reactant and reaction products), it was proposed:

$$\kappa_{T,\lambda}(t) = \kappa^*_{D,\lambda} C_D(t) + \kappa^*_{Pr,\lambda} C_{Pr}(t) \tag{6.66}$$

The 'unknown-products' hypothetical concentration can be expressed in terms of the 2,4-D instantaneous concentration:

$$\kappa_{T,\lambda}(t) = \kappa^*_{D,\lambda} C_D(t) + \kappa^*_{Pr,\lambda}\left(C_D^0 - C_D\right) = \left(\kappa^*_{D,\lambda} - \kappa^*_{Pr,\lambda}\right) C_D(t) + \kappa^*_{Pr,\lambda} C_D^0 \tag{6.67}$$

With experimental information the following empirical correlation was obtained:

$$\kappa_{T,\lambda}(t) = 0.0197 C_D^0 - 0.01785 C_D(t) \tag{6.68}$$

Parameter evaluation. It is now possible to obtain the kinetic parameters from the experimental data and the proposed kinetic model in equation 6.64. We have three unknowns: the quantum yield, and the exponents 'm' and 'n'. The whole model (equation 6.65) was fed to a multiparameter, non-linear regression algorithm that is coupled with an optimization program according to the Marquardt method (Marquardt, 1963). The regression program gave the following values for the exponents: $n \cong 1$ and $m \cong 0$. With these estimations, at 25° C and 253.7 nm, the following kinetic equation was obtained:

$$R_{D,\lambda}(y,t) = -0.0262 e_{\lambda}^{a,D}(y,t) \tag{6.69}$$

This result should not be interpreted as zero-order dependence with respect to 2,4-D concentration because it participates in two parts of the variable $e_{\lambda}^{a,D}$ (see equation 6.58).

6.5.2 Reactor Analysis

A pilot plant scale, tubular (annular configuration) photoreactor for the direct photolysis of 2,4-D was modeled (Martín *et al.*, 1997). A tubular germicidal lamp was placed at the reactor centerline. This reactor can be used to test, with a very different reactor geometry, the kinetic expression previously developed in the cylindrical, batch laboratory reactor irradiated from its bottom and to validate the annular reactor modeling for the 2,4-D photolysis. Note that the radiation distribution and consequently the field of reaction rates in one and the other system are very different.

Proposed reactor (pilot plant scale). Figures 6.1 and 6.3 (the reactor substituted by the annular tube of Figure 6.1) provide a schematic representation of the employed reacting system. More details can be found in Table 6.1.

Reactor model. The reactor model was constructed according to the following sequence: (i) the annular reactor, *radiation distribution model* of Romero *et al.* (1983) was adapted for this particular set-up; (ii) the tubular lamp with voluminal and isotropic *radiation emission model* was applied to this system; (iii) a *mass balance* for an actinometric reaction carried out in a tubular reactor inside the loop of a recycling system was adapted from Martín *et al.* (1996); and (iv) the verification of the radiation model, actinometer experiments were performed in the reactor to compare theoretical predictions

Table 6.1 *Reacting system and lamp characteristics. Reproduced with permission from Martín* et al. *(1997) copyright 1997, IWA Publishing*

	Parameter	Value	Units
REACTOR	Irradiated length	48	cm
Pyrex®	Outside diameter	6.03	cm
Suprasil®	Inside diameter	4.45	cm
	Irradiated volume	624	cm^3
LAMP	Input power	30	W
Philips TUV	Output power	9	W
$\lambda = 253.7$nm	Nominal length	89.5	cm
	Diameter	2.6	cm
RESERVOIR	Volume	6000	cm^3

with actual results. This procedure permitted the verification of the quality of the radiation emission model for the lamp and the radiation distribution model for the annular reactor. Afterwards, for the photolytic reactor employing 2,4-D, the following sequence was followed: (1) a *species mass balance* for a tubular reactor inside the recycling system was written according to equation 6.12; (2) the *kinetic expression* given by equation 6.69 was incorporated into this mass balance; (3) the *radiation model* previously validated was used to predict the LVRPA in the kinetic expression (equation 6.69); (4) *radiation absorption by reactants and products* was incorporated into the radiation model according to the empirical expression represented by equation 6.68; (5) time evolution concentrations of the 2,4-D in the recycling system were predicted using steps (1), (2), (3), and (4); and (6) experimental 2,4-D concentrations in the pilot plant reactor were compared with theoretical predictions.

Radiation field. For a homogeneous medium the radiation distribution is obtained by solving equation 6.23 with the following boundary condition:

$$I_{\lambda,\Omega}(s=0) = I^{0}_{\lambda,\Omega} \tag{6.70}$$

The boundary condition was obtained from the 3D with voluminal and isotropic emission model (equation 6.51). The solution of equation 6.23 provides values of the radiation intensity as a function of position (r, z) and direction (θ, ϕ). Once I_λ is known, the incident radiation and the LVRPA can be obtained from equations 6.24 and 6.25. Since monochromatic radiation is employed, no integration over wavelength is needed. The final equation for calculating the LVRPA is

$$\begin{aligned} e^{a}_{\lambda}(\mathbf{x},t) = {} & \frac{\kappa_{i,\lambda}(\mathbf{x},t)\Upsilon_{R,\lambda}(\Omega)P_\lambda}{2\pi^2 R_L^2 L_L}\int_{\phi_1}^{\phi_2} d\phi \int_{\theta_1}^{\theta_2} d\theta (R_L^2 - r^2\sin^2\phi)^{1/2} \\ & \times \exp\left[-\kappa_{T,\lambda}\left(\frac{r\cos\phi - (r^2\cos^2\phi - (r^2 - r^2_{R,i}))^{1/2}}{\sin\theta}\right)\right] \end{aligned} \tag{6.71}$$

The integration limits for θ and ϕ for the case of the annular reactor are described by equations 6.52, 6.53, and 6.54. It must be noticed that the exponential term (attenuation) uses the reacting medium *total absorption coefficient* while only the *reactant absorption coefficient* intervenes, with a linear effect, in the value of the LVRPA. Hence i stands for the reactant.

The actinometer reaction in the annular reactor: The classic uranyl oxalate reaction was used (Murov *et al.*, 1993). According to Brandi *et al.* (2003), changes in concentration inside the recycling system were obtained from

$$\left.\frac{dC_i(t)}{dt}\right]_{Tk} = \frac{V_R}{V_T}\langle R_{Hom,Ac}(r,z,t)\rangle_{V_R} \tag{6.72}$$

$$C_i(0) = C_i^0$$

$$R_{Hom,Ac}(r,z,t) = -\Phi_{Ac,\lambda}e^{a,Ac}_{\lambda}(r,z,t) \tag{6.73}$$

Calculating the volume average of the LVRPA we get

$$\langle e_\lambda^{a,Ac}(\mathbf{x})\rangle_{V_R} = \frac{\kappa_{Ac,\lambda}\Upsilon_{R,\lambda}P_\lambda}{\pi R_L^2 L_L V_R}\int_{r_{R,i}}^{r_{R,o}} r\,dr\int_0^{L_R} dz\left\{\int_{\phi_1}^{\phi_2} d\phi\int_{\theta_1}^{\theta_2} d\theta(R_L^2 - r^2\sin^2\phi)^{1/2}\right.$$
$$\left.\times\exp\left[-\kappa_{Ac,\lambda}\left(\frac{r\cos\phi-\left(r^2\cos^2\phi-(r^2-r_{R,i}^2)\right)^{1/2}}{\sin\theta}\right)\right]\right\} \tag{6.74}$$

In this reaction, the concentration of the absorbing species remains constant for conversions below 20% (sensitized reaction). To validate the radiation model, results obtained from equation 6.74 must be compared with experiments. From equations 6.72 and 6.73 and after integration, the experimental value of the LVRPA is

$$\left[e_\lambda^{a,Ac}\right]_{Exp.} = \frac{[C_{Ac}^0 - C_{Ac}(t)]}{(t-0)}\frac{V_T}{V_R}\frac{1}{\Phi_{Ac,\lambda}} \tag{6.75}$$

Experiments were carried out at three different uranyl sulfate concentrations: 0.005, 0.001, and 0.0005 M. Oxalic acid concentrations were always five times larger. The largest error between model and experiments was smaller than 8%. Since agreement is very good, one may conclude that the radiation field of the annular reactor has been precisely represented. Note that no adjustable parameters have been used and the boundary condition was obtained with a theoretical model.

The reactor model for the 2,4-D photolysis. The simplified kinetic expression represented by equation 6.69 has the same form as equation 6.73. However, during the 2,4-D photolysis the radiation absorption characteristics of the reacting medium change. This is a very distinct phenomenon because (i) the uranyl oxalate reaction is a photosensitized reaction and the radiation absorbing species is not consumed, and (ii) conversely, not only the 2,4-D absorption coefficient changes, but absorption by reaction products increases the total absorption coefficient above the initial value. This phenomenon produces an unavoidable coupling between the steady state radiation balance and the unsteady state mass balance. The total absorption coefficient can be obtained from equation 6.68. Then;

$$\frac{dC_D(t)}{dt} = \frac{V_R}{V_T}\langle R_{Hom,D}(r,z,t)\rangle_{V_R} \tag{6.76}$$

with the I.C. $C_D(t=0) = C_D^0$ (6.77)

The reaction rate is

$$\langle R_{Hom,D}(r,z,t)\rangle_{V_R} = -\Phi_{D,\lambda}\frac{1}{V_R}\int_{V_R} e_\lambda^{a,D}(r,z,t)\,dV \tag{6.78}$$

Inserting the LVRPA into equation 6.78 and substituting the result into equation 6.76 we obtain

$$\frac{\mathrm{d}C_{\mathrm{D}}}{\mathrm{d}t} = -\Phi_{\mathrm{D},\lambda}\frac{1}{V_{\mathrm{T}}}\left[\frac{P_{\lambda}\Upsilon_{\mathrm{R},\lambda}\kappa^{*}_{\mathrm{D},\lambda}C_{\mathrm{D}}(t)}{\pi R_{\mathrm{L}}^{2}L_{\mathrm{L}}}\right]\int_{r_{\mathrm{R},i}}^{r_{\mathrm{R},0}} r\,\mathrm{d}r\int_{0}^{L_{\mathrm{R}}}\mathrm{d}z\left\{\int_{\phi_1}^{\phi_2}\mathrm{d}\phi\int_{\theta_1}^{\theta_2}\mathrm{d}\theta(R_{\mathrm{L}}^{2}-r^{2}\sin^{2}\phi)^{1/2}\right.$$
$$\left.\exp\left[-\kappa_{\mathrm{T},\lambda}\left(\frac{r\cos\phi-\left(r^{2}\cos^{2}\phi-(r^{2}-r_{\mathrm{R},i}^{2})\right)}{\sin\theta}\right)^{1/2}\right]\right\}$$
$$C_{\mathrm{D}}(t=0)=C_{\mathrm{D}}^{0} \tag{6.79}$$

Integration of this equation provides the time evolution of the 2,4-D concentration. Notice that all the lamp characteristics are incorporated in the design equation. The mass balance and the volume average procedure indicated in the equations above are greatly simplified by the differential operation in the photochemical section of the reactor. Figure 6.9 shows the results for two initial concentrations. Solid lines correspond to predictions of the 2,4-D concentrations obtained from the solution of equation 6.79. Symbols correspond to experimental values. It can be seen that agreement is fairly good. The observed discrepancies, which in some cases produce an error as large as 15%, are mainly due to the fact that the reaction kinetics of this very complex reaction has been modeled in terms of just one single variable (the 2,4-D concentration).

The ideas described in this section can be easily extended to more complex reacting systems either from the chemistry point of view – for example to include the parallel oxidation reaction with hydrogen peroxide or ozone – or to deal with other lamp-reactor configurations. A comprehensive, tutorial review for homogeneous photochemical reactors has been published (Cassano *et al.*, 1995) that provides most of the required methods.

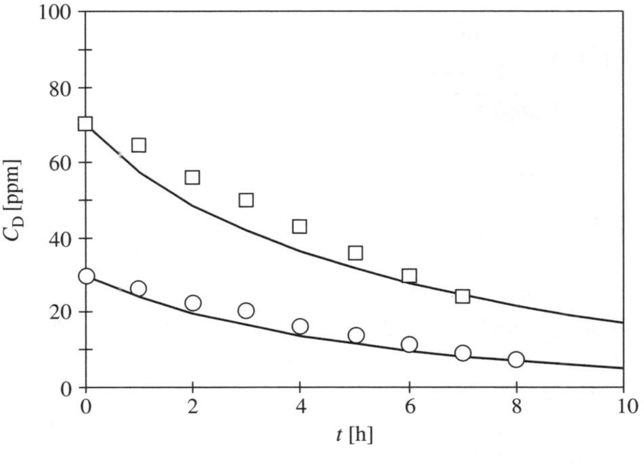

Figure 6.9 *Results of the scale-up procedure for the 2,4-D degradation process. Model predictions (–); experimental data (□, O). Reproduced with permission from Martín* et al. *(1997) copyright 1997, IWA Publishing*

6.6 Heterogeneous Systems

For the case of photocatalytic reactors employing solid semiconductors and trying to reach a compromise between length and clarity in a detailed application, it seems appropriate to concentrate effort on describing all the reactor analysis concepts that must be developed to measure (i) true quantum yields in heterogeneous slurry photoreactors, and (ii) specific procedures to obtain intrinsic kinetic models in laboratory reactors. These concepts can be immediately extended to the design of the reactor because (1) the models for the lamp emission are the same as the ones described in homogeneous systems and (2) the modeling of the reactor can, at most, include some additional complications when scattering has to be described in cylindrical geometries (Romero *et al.*, 1997, 2003). The problem of true quantum yield determination allows us to show in a rather short extension most of the main features of heterogeneous photocatalytic reactor modeling (Brandi *et al.*, 2003). How to incorporate a complex reaction scheme or mechanism into the corresponding mass balance, as has been shown by Alfano *et al.* (1997) and Cabrera *et al.* (1997b), will be also presented. Once the distribution of radiation inside reactors of different geometries is known (see e.g. Romero *et al.*, 2003, for annular reactors and Brandi *et al.*, 1999, for flat plate reactors), basic concepts already available in the chemical reactor engineering literature can be used to model other reactions and reactor types.

6.6.1 True Quantum Yield Evaluation in Slurry Reactors

The general methodology for modeling slurry photoreactors has been reviewed by Cassano and Alfano (2000). We will apply these concepts to the evaluation of absolute, true values of quantum yields.

Definition of the problem. The monochromatic, overall, true initial quantum yield is defined as

$$[\Phi_\lambda^0]_{\mathrm{TRUE}} = \frac{\left\{\left[\text{Rate of}\left\{\begin{array}{c}\text{disapearance}\\ \text{appearence}\end{array}\right\}\text{of compound `}i\text{'}(\mathbf{x},t)\right]\begin{array}{c}t\to 0\\ \lambda\end{array}\right\}_{V_{\mathrm{R}}-\mathrm{AVER}}}{\{[\text{Rate of photon absorption by the catalyst}(\mathbf{x},t)]_\lambda\}_{V_{\mathrm{R}}-\mathrm{AVER}}} = \frac{[\langle R_i\rangle_{V_{\mathrm{R}}}^0]_{\mathrm{EXPER}}}{\left[\langle e_\lambda^{\mathrm{a}}\rangle_{V_{\mathrm{R}}}\right]_{\mathrm{CALC}}} \tag{6.80}$$

The volume average of the LVRPA is very difficult to measure. However, employing rigorous mathematical modeling of photocatalytic slurry reactors it can be precisely calculated (Brandi *et al.*, 2000a,b). Consequently, in equation 6.80, (i) the numerator is the reactor volume-averaged of the reaction rate measured at initial conditions (the result of an experimental determination), (ii) the denominator is the reactor volume-averaged of the calculated spatial distribution of the LVRPA, and (iii) the LVRPA is calculated solving the radiative transfer equation (RTE) employing catalyst optical properties and light intensities arriving at the reactor window for radiation entrance, both independently measured.

Quantum yields are not unique values unless several operating conditions are precisely defined, for example:

1. Concerning the employed radiation, wavelength must be monochromatic and the employed range of radiation intensities must be defined because of the existence of different reaction order dependencies at different irradiation rates.
2. Concerning the reaction environment, several conditions must be fixed: temperature, pH, substrate initial concentration, and quality of the 'reactants' that are employed because impurities affect the photocatalytic rates.
3. Concerning the oxidative path, operating conditions must ensure excess oxygen concentration over the stoichiometric demand during the full course of the reaction.
4. Concerning the catalyst, we must define catalyst variety and catalyst concentration.
5. If there is simultaneous homogeneous photolysis, it must be treated as a parallel reaction to exclude its effects from the photocatalytic ones.

Additionally, to facilitate comparisons, quantum yields should be measured at substrate and catalyst concentrations where the reaction rate shows zero-order dependence with respect to both variables.

Methodology. To develop a rigorous model of a photocatalytic slurry reactor several steps were necessary. They are briefly described below.

1. To study scattering effects by solid particles in a fluid and adapt previous existing methods in generalized transport theory (the discrete ordinate method or DOM) (Duderstadt and Martin, 1979) to solve the RTE (Alfano *et al.*, 1995).
2. To develop a laboratory reactor that permits the easiest solution of the RTE employing the DOM (Cabrera *et al.*, 1994). It consists of a flat plate configuration (a cylinder irradiated from one of its circular surfaces).
3. To develop special methods to measure monochromatic specific (per unit catalyst mass) absorption (κ_λ^*) and scattering (σ_λ^*) coefficients of titanium dioxide slurries and obtain values for different catalysts and for the wavelength range of interest (Cabrera *et al.*, 1996).
4. To develop precise methods for obtaining intrinsic kinetic data in a batch reactor with recycle (Alfano *et al.*, 1997; Cabrera *et al.*, 1997b).
5. To include the effects of reactor wall properties into the incident radiation intensities corresponding to the boundary condition for radiation entrance. The model includes internal absorption and interfacial reflectivities (Brandi *et al.*, 1999).
6. To characterize and model the problem of reactor window fouling by titanium dioxide (Brandi *et al.*, 1999).
7. To select the best phase function for radiation scattering by titanium dioxide (Brandi *et al.*, 1999).
8. To obtain direct and precise experimental verification of the quality of the results obtained with the numerical solution of the RTE with the DOM (Brandi *et al.*, 2000a,b). For catalyst loadings above 0.25 g/L, errors were never larger than 5%.

Selection of a reactor. At this point we should decide on the experimental reactor to be used. Cabrera *et al.* (1994) proposed a new experimental reactor that – with a few

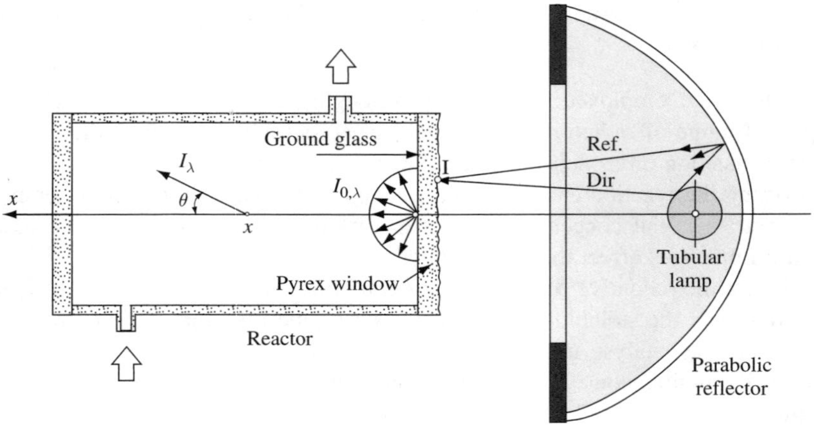

Figure 6.10 *Schematic description of the uni-dimensional photocatalytic reactor. Adapted from Cassano and Alfano, 2000*

changes – has been successfully used for detailed kinetic studies. Figure 6.10 gives a schematic description of the device. It consists of the following parts:

1. A cylindrical reactor with two flat windows made of good quality Pyrex glass (alternatively, one of them may be made of Suprasil quality quartz). The window for radiation entrance – either glass or quartz – must be modified; its external side upon abrasion with HF has the texture of ground glass. The reactor has an optical path (L_R) sufficiently large to ensure that no radiation is arriving at the flat plate facing the window of radiation entrance. Illuminating the reactor through the modified window produces diffuse irradiation inside (the irradiation boundary condition) which greatly simplifies the radiation model (Figure 6.10).
2. A tubular UV lamp of well-known characteristics: output power, radiation spectral distribution of its output energy, and geometrical dimensions. This lamp has significant peaks of emission at 313 and 365 nm.
3. A cylindrical reflector of parabolic cross-section with well- known reflecting properties (Alzac®Aluminum from Alcoa) and well-defined geometrical dimensions.
4. Monochromatic light was obtained by interposing in the radiation bundle trajectories narrow band interference filters (peaks at 313 and 365 nm).
5. A shutter placed in front of the reactor window allowed us to decide on the exact starting time of the reaction once steady state conditions had been reached.

The reacting system was operated inside the loop of a batch recycling arrangement (Figure 6.3) with provisions for (1) a storage tank, (2) an all-glass and Teflon recirculating pump with high flow rate capacity, (3) a temperature control system, and (4) a continuous feed for oxygen. A laboratory reactor must be constructed in such a way that an exact analysis of the experimental results should be simplified as much as possible. This experimental device has four important features for its modeling: (1) the tank volume is significantly larger than the reactor volume; (2) the pump has a high flow rate, thus in the reactor, conversion per pass will be very small; (3) irradiation at the inside face of the reactor window is diffuse which means that azimuthal symmetry for the direction of radiation propagation inside the reactor will be achieved; and (4) no radiation arrives at the opposite face of the reactor plate and consequently there is no radiation reflection on this face.

Calculating procedure for the reaction rate[3]. On this occasion we will use the concepts already derived in section 6.2. Consider the case represented in Figure 6.3. From equation 6.1, a local mass balance for the i component in the liquid where there is no chemical reaction (no parallel photolysis) is

$$\frac{\partial C_i(\mathbf{x}, t)}{\partial t} + \nabla \cdot \mathbf{N}_i = 0 \tag{6.81}$$

This equation can be integrated over the whole liquid volume of the system ($V_{\mathrm{L,T}}$) that, in principle, is different from the total volume of the suspension (liquid + solid) ($V_{\mathrm{Tot}} = V_{\mathrm{Tk}} + V_{\mathrm{R}} = V_{\mathrm{L,T}} + V_{\mathrm{S,T}}$). Defining

$$\langle C_i(\mathbf{x}, t)\rangle_{V_{\mathrm{L,R}}} = \frac{1}{V_{\mathrm{L,R}}} \int_{V_{\mathrm{L,R}}} C_i(\mathbf{x}, t)\, \mathrm{d}V \tag{6.82}$$

we get

$$\int_{V_{\mathrm{L,T}}} \frac{\partial C_i(\mathbf{x}, t)}{\partial t} \mathrm{d}V = V_{\mathrm{L,R}} \frac{\mathrm{d}}{\mathrm{d}t} \langle C_i(\mathbf{x}, t)\rangle_{V_{\mathrm{L,R}}} + V_{\mathrm{L,Tk}} \left. \frac{\mathrm{d}C_i(t)}{\mathrm{d}t} \right|_{\mathrm{Tk}} \tag{6.83}$$

$$\int_{V_{\mathrm{L,T}}} \nabla \cdot \mathbf{N}_i \, \mathrm{d}V = \int_{A_{\mathrm{S,T}}} \left[\underbrace{\mathbf{J}_i(\mathbf{x}, t)}_{\text{Diffusion}} + \underbrace{C_i(\mathbf{x}, t)\boldsymbol{v}}_{\text{Convection}} \right] \cdot \mathbf{n}_L \, \mathrm{d}A \tag{6.84}$$

We have considered that $A_{\mathrm{L,T}} = A_{\mathrm{S,T}}$; i.e. the total interfacial area of the liquid is equal to the total interfacial area of the solid. Noting that fluxes are different from 0 only at permeable solid surfaces, in a closed system the only permeable surfaces are those corresponding to the catalyst. Therefore;

$$\begin{aligned} V_{\mathrm{L,R}} \frac{\mathrm{d}}{\mathrm{d}t} \langle C_i(\mathbf{x}, t)\rangle_{V_{\mathrm{L,R}}} + V_{\mathrm{L,Tk}} \frac{\mathrm{d}C_i(t)}{\mathrm{d}t} &= - A_{\mathrm{S,T}} \langle [\mathbf{J}_i(\mathbf{x}, t)] \cdot \mathbf{n}_{\mathrm{L}} \rangle_{A_{\mathrm{S,T}}} \\ &\quad - \int_{A_{\mathrm{S,T}}} C_i(\mathbf{x}, t)\boldsymbol{v} \cdot \mathbf{n}_{\mathrm{L}} \, \mathrm{d}A \end{aligned} \tag{6.85}$$

In equation 6.85, $\langle [\mathbf{J}_i(\mathbf{x}, t)] \cdot \mathbf{n}_{\mathrm{L}} \rangle_{A_{\mathrm{S,T}}}$ is 'the total liquid–solid particles interface, averaged, molar diffusive flux of component i'. Note that $A_{\mathrm{S,T}} = A_{\mathrm{S,R}} + A_{\mathrm{S,Tk}}$. The second term of the right-hand side of equation 6.85 is 0 because at the catalyst interface convective fluxes are 0. On the other hand, from equation 6.13 the molar diffusive flux through the boundary layer at the liquid–solid interface must be equal to the reaction rate at the liquid–solid interface:

$$A_{\mathrm{S,T}} \langle \mathbf{J}_i(\mathbf{x}, t) \cdot \mathbf{n}_{\mathrm{L}} \rangle_{A_{\mathrm{S,T}}} = A_{\mathrm{S,R}} \langle \mathbf{J}_i(\mathbf{x}, t) \cdot \mathbf{n}_L \rangle_{A_{\mathrm{S,R}}} = -A_{\mathrm{S,R}} \langle R_{\mathrm{Het},i}(\mathbf{x}, t) \rangle_{A_{\mathrm{S,R}}} \tag{6.86}$$

---

[3] Extracted and reprinted with permission from Cabrera *et al.*, 1997b, Copyright 1997, Elsevier.

Substituting equation 6.86 into equation 6.85:

$$\varepsilon_{\mathrm{L}}\frac{V_{\mathrm{R}}}{V_{\mathrm{Tk}}}\frac{\mathrm{d}\left\langle C_i\left(\mathbf{x},t\right)\right\rangle_{V_{\mathrm{L,R}}}}{\mathrm{d}t}+\varepsilon_{\mathrm{L}}\frac{\mathrm{d}C_i(t)}{\mathrm{d}t}=\frac{V_{\mathrm{R}}}{V_{\mathrm{Tk}}}a_{\mathrm{V}}\left\langle R_{\mathrm{Het},i}(\mathbf{x},t)\right\rangle_{A_{\mathrm{S,R}}} \tag{6.87}$$

where ε_{L} is the liquid hold-up in the system that is uniform throughout and a_{V} is the catalytic surface area per unit suspension volume. Since $V_{\mathrm{R}}/V_{\mathrm{Tk}} < 1$ and the conversion per path in the reactor $\left[\frac{\mathrm{d}\langle C_i(\mathbf{x},t)\rangle_{V_{\mathrm{L,R}}}}{\mathrm{d}t}\right]$ is very small (both being design conditions);

$$\varepsilon_{\mathrm{L}}\left.\frac{\mathrm{d}C_i(t)}{\mathrm{d}t}\right|_{\mathrm{Tk}}=\frac{V_{\mathrm{R}}}{V_{\mathrm{Tk}}}a_{\mathrm{V}}\left\langle R_{\mathrm{Het},i}(\mathbf{x},t)\right\rangle_{A_{\mathrm{S,R}}} \tag{6.88}$$

$$\left\langle R_{\mathrm{Het},i}(\mathbf{x},t)\right\rangle_{A_{\mathrm{S,R}}}=\varepsilon_{\mathrm{L}}\frac{1}{S_{\mathrm{g}}C_{\mathrm{mp}}}\frac{V_{\mathrm{Tk}}}{V_{\mathrm{R}}}\left.\frac{\mathrm{d}C_i(t)}{\mathrm{d}t}\right|_{\mathrm{Tk}}[=]\frac{\mathrm{mol}}{\mathrm{cm}^2\,\mathrm{s}} \tag{6.89}$$

When the same value per unit suspension volume is needed (for equation 6.80) we must consider that (i) the cross-sectional area corresponding to the irradiated flat plate reactor is constant (a design condition), and (ii) the catalyst concentration is uniform (well-mixed system in the whole reactor volume; an established operating condition):

$$\left\langle R_{\mathrm{Hom},i}^{\mathrm{Pseudo}}(\mathbf{x},t)\right\rangle_{V_{\mathrm{R}}}=\left\langle R_{\mathrm{Hom},i}^{\mathrm{Pseudo}}(\mathbf{x},t)\right\rangle_{L_{\mathrm{R}}}=\varepsilon_{\mathrm{L}}\frac{V_{\mathrm{Tk}}}{V_{\mathrm{R}}}\left.\frac{\mathrm{d}C_i(t)}{\mathrm{d}t}\right|_{\mathrm{Tk}}[=]\frac{\mathrm{mol}}{\mathrm{cm}^3\,\mathrm{s}} \tag{6.90}$$

Equation 6.90 provides the value of the numerator of equation 6.80.

Photon absorption rate by a material particle of the suspension[4]. At this point we would like to know the LVRPA by the solid and to be able to isolate this value even if the liquid would also absorb radiation. To do this we need to model absorption by a material particle of the suspension. In the continuum mechanics sense, a material point in space is a volume for which every property can be well defined by a single value. For a catalytic suspension, it will be made of the liquid and the solid phases. Let us consider a small volume V of the suspension space representing this material particle. This volume is located at a point in space $\mathbf{x}$ (Figure 6.11). Any point inside V can be defined in terms of a local reference frame ζ.

We must now relate the LVRPA per particle, $E_{\lambda,\mathrm{Sn}}^{\mathrm{a}}(\mathbf{x}+\zeta,t)$, to the LVRPA by the suspension volume (liquid + solid), $e_{\lambda}^{\mathrm{a}}(\mathbf{x},t)$. The absorbed energy per unit wavelength interval, unit time, and unit volume of the suspension (solid plus liquid) is by definition of an average value over the total volume:

$$e_{\lambda}^{\mathrm{a}}(\mathbf{x},t)=\frac{1}{V}\int_{V}\mathrm{d}V e_{\lambda}^{\mathrm{a}}(\mathbf{x}+\zeta,t) \tag{6.91}$$

where V is the small suspension volume of the heterogeneous system (solid plus liquid) located at point $\mathbf{x}$. The right-hand side of equation 6.91 can be divided into two parts: (i) the radiation energy absorbed by the liquid, and (ii) that part of the absorbed radiation

[4] Extracted and reprinted with permission from Alfano *et al.*, 1997, Copyright 1997, Elsevier.

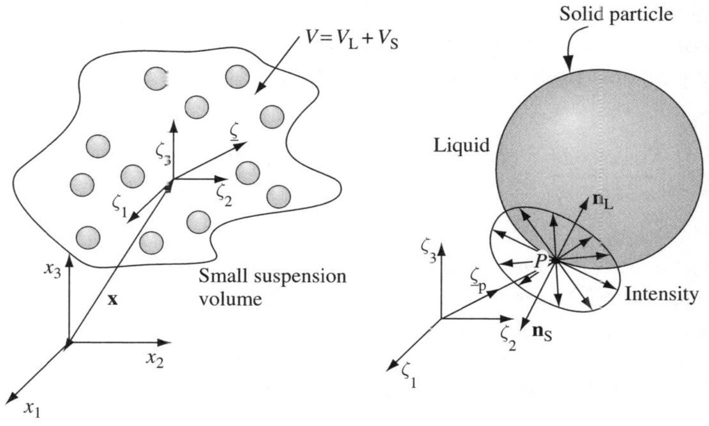

Figure 6.11 *Modeling of photon absorption by a material particle of the suspension. Adapted from Alfano* et al. *(1997)*

corresponding to the solid particles. Additionally, suppose that in the small volume V we have N solid photocatalytic particles:

$$e^{a}_{\lambda}(\mathbf{x}, t) = \frac{1}{V}\int_{V_L} dV e^{a}_{\lambda,L}(\mathbf{x}+\zeta, t) + \frac{1}{V}\left[\sum_{n=1}^{N} \underbrace{\int_{V_{Sn}} dV e^{a}_{\lambda,S}(\mathbf{x}+\zeta, t)}_{\text{Absorption by one particle}}\right] \tag{6.92}$$

Assuming that on average all particles are equal:

$$\begin{aligned} e^{a}_{\lambda}(\mathbf{x}, t) &= \varepsilon_L \underbrace{\langle e^{a}_{\lambda,L}(\mathbf{x}+\zeta, t)\rangle_{V_L}}_{\substack{\text{Average value of absorption}\\ \text{by the liquid phase}}} \\ &+ \underbrace{N_V}_{\substack{\text{Number of particles}\\ \text{per unit volume}}} \underbrace{\int_{V_{Sn}} dV e^{a}_{\lambda,S}(\mathbf{x}+\zeta, t)}_{\text{Absorption by one particle}} \end{aligned} \tag{6.93}$$

In this equation, ε_L is the liquid hold-up (V_L/V) and $N_V = N/V$ is the number of particles per unit suspension volume. Finally, from equation 6.93 we get

$$\underbrace{E^{a}_{\lambda,Sn}(\mathbf{x}, t)\, N_V}_{\text{Absorption by the solid}} = \left[\underbrace{e^{a}_{\lambda}(\mathbf{x}, t)}_{\text{Total absorption}} - \underbrace{\varepsilon_L \langle e^{a}_{\lambda,L}(\mathbf{x}+\zeta, t)\rangle_{V_L}}_{\text{Absorption by the liquid}}\right] \tag{6.94}$$

If the liquid does not absorb radiation in the wavelength range under consideration, the second term of the right-hand side is 0. When the liquid is transparent, equation 6.94

indicates that the solution provided by the RTE in terms of the absorption and scattering coefficients of the suspension can provide, directly, the value of the photon absorption rate by solid particles. Consequently, if the liquid is transparent,

$$\underbrace{e^{a}_{\lambda,\mathrm{Sol}}(\mathbf{x},t)}_{\substack{\text{Absorption by}\\ \text{the solid particles}}} = N_{\mathrm{V}} \quad E^{a}_{\lambda,\mathrm{Sn}}(\mathbf{x},t) = \underbrace{e^{a}_{\lambda}(\mathbf{x},t)}_{\text{Solution of the RTE}} \quad \left[\frac{\text{einstein}}{\text{cm}^3\,\text{s}}\right] \tag{6.95}$$

Calculating procedure for the LVRPA. In order to apply equation 6.95 we need to solve the RTE (equation 6.32) for this particular reactor set-up. As shown by Alfano *et al.* (1995) and Cabrera *et al.* (1994) the radiation field of this reactor can be modeled with a 1D, one-directional radiation model and rather simple boundary conditions (Figure 6.10). Hence, with azimuthal symmetry derived from the diffuse emission at $x = 0$:

$$\mu \frac{\partial I_\lambda(x,\mu)}{\partial x} + (\kappa_\lambda + \sigma_\lambda)\, I_\lambda(x,\mu) = \frac{\sigma_\lambda}{2} \int_{\mu'=-1}^{\mu'=1} I_\lambda(x,\mu') p(\mu,\mu')\,\mathrm{d}\mu' \tag{6.96}$$

and the following boundary conditions:

$$I_\lambda(0,\mu) = I^0_\lambda \qquad \mu > 0 \tag{6.97}$$

$$I_\lambda(L_{\mathrm{R}},\mu) = 0 \qquad \mu < 0 \tag{6.98}$$

At this point we need the value of I^0_λ. It can be obtained by two different approaches:

1. With an emission model for the tubular lamp and the parabolic reflector (Alfano *et al.*, 1985, 1986a,b). It takes into account both direct and reflected radiation. These intensities can then be transformed into fluxes and both contributions added at the external reactor wall. They were averaged over the surface of the window, affected by the experimentally measured wall transmission coefficient and transformed into direction-independent intensities according to

$$\langle q_{\mathrm{T},\lambda}(r,\beta)\rangle_{A_{\mathrm{R}}} = \frac{4}{\pi r_{\mathrm{R}}^2} \int_0^{r_{\mathrm{R}}} r\,\mathrm{d}r \int_0^{\pi/2} \mathrm{d}\beta \left[\underbrace{q_{\mathrm{D},\lambda}(r,\beta)}_{\substack{\text{Direct radiation}\\ \text{fluxes}}} + \underbrace{q_{\mathrm{Rf},\lambda}(r,\beta)}_{\substack{\text{Reflected radiation}\\ \text{fluxes}}} \right] \tag{6.99}$$

$$I^0_\lambda = \frac{1}{\pi} \Upsilon_{\mathrm{R},\lambda} \langle q_{\mathrm{T},\lambda}(r,\beta)\rangle_{A_{\mathrm{R}}} \tag{6.100}$$

2. The boundary condition may also be directly measured with homogeneous actinometry inside the reactor as was shown in the previous section of this chapter.

Equation 6.96 with boundary conditions 6.97 and 6.98 can be solved with the DOM (Duderstadt and Martin, 1979). Solution in terms of intensities can be immediately used to calculate local values of the LVRPA. Optical properties can be assumed constant (stable catalyst) and, consequently, for a transparent organic compound the κ_λ and σ_λ values are only a function of position at most. The numerical result gives monochromatic intensities as a function of position and direction. Then the following operations can be performed:

$$G_\lambda(x) = 2\pi \int_{\mu=-1}^{\mu=1} I_\lambda(x,\mu)\,\mathrm{d}\mu \quad [=] \quad \frac{\text{einstein}}{\text{cm}^2\,\text{s}} \tag{6.101}$$

The LVRPA or the photon absorption rate as a function of position results in:

$$e_\lambda^{\mathrm{a}}(x) = \kappa_\lambda\, G_\lambda(x) \quad [=] \quad \frac{\text{einstein}}{\text{cm}^3\,\text{s}} \tag{6.102}$$

The absorption coefficient has been assumed independent of position (uniform catalyst concentration) and time (stable catalyst). The reactor volume average of the LVRPA for the 1D model in the Cartesian coordinate x is

$$\langle e_\lambda^{\mathrm{a}} \rangle_{V_\mathrm{R}} = \frac{1}{L_\mathrm{R}} \int_0^{L_\mathrm{R}} e_\lambda^{\mathrm{a}}(x)\,\mathrm{d}x \tag{6.103}$$

Equation 6.103 provides the denominator of equation 6.80.

Results. Absolute values of the true monochromatic quantum yields have been recently measured for 1,4-dioxane and phenol (Brandi *et al.*, 2003). The adopted standard conditions are described in Table 6.2. Employing the method described in the previous sections, the reported values are presented in Table 6.3.

Table 6.2 *Experimental operating conditions*

Variable	Value	Units
Wavelength	313 and 365	nm
Radiation intensity for 313 nm (at clean reactor window)	2.187×10^{-10}	einstein cm^{-2} s sr
Radiation intensity for 365 nm (at clean reactor window)	4.997×10^{-10}	einstein cm^{-2} s sr
Temperature	298	K
Initial pH	3.0	–
Initial substrate concentration	20	ppm
Catalyst	Degussa P 25	
Catalyst concentration	2.0	gL^{-1}
Use pure substrates and pure water		
Work under zero-order reaction rate regimes with respect to substrate and catalyst concentrations		

Table 6.3 *Absolute quantum yield results*

Compound	Quantum yield (%) mol einstein^{-1}	
	$\lambda = 313$ nm	$\lambda = 365$ nm
Phenol	9.5	8.5
1,4-Dioxane	5.3	4.0

6.6.2 Determination of the Detailed Reaction Kinetics of an Organic Pollutant Degradation[5]

One of the most interesting processes based on titanium dioxide photocatalytic reactions in water systems is the mineralization reaction by which halogenated organic compounds can be converted into carbon dioxide, water, and halide ions. Removal of halogenated organic compounds is an important contribution to the solution of an environmental threat since most of these chemical substances exhibit rather high degrees of toxicity.

Several kinetic models for photocatalytic processes have been proposed. One of the possible reaction paths for the initiation of the degradation reaction of several hydrocarbon compounds appears to be an oxidative carbon–hydrogen bond cleavage by hydroxyl radicals attack. Hydroxyl radicals are generated as a consequence of electron transfer to the semiconductor holes by adsorbed hydroxyl ions or water molecules (Turchi and Ollis, 1990).

We are using the reaction sequence described in Table 6.4. Our approach differs from that of Turchi and Ollis in the following aspects: (i) the required assumptions to obtain a simple kinetic expression have been reduced to a minimum; and (ii) as should be expected in a practical system, due to the strong radiation extinction by the catalyst,

Table 6.4 *Reaction scheme. Reproduced with permission from Alfano* et al. *(1997) copyright 1997, Elsevier*

Activation				
	TiO_2	$\xrightarrow{h\nu}$	$e^- + h^+$	(1)
Adsorption				
	$O_L^{2-} + Ti^{IV} + H_2O$	$\longleftrightarrow$	$O_LH^- + Ti^{IV} - OH^-$	(2a)
	$Ti^{IV} + H_2O$	$\longleftrightarrow$	$Ti^{IV} - H_2O$	(2b)
	$\text{site} + R_i$	$\longleftrightarrow$	$R_{i,ads}$	(3)
	$OH\bullet + Ti^{IV}$	$\longleftrightarrow$	$Ti^{IV} \vert OH\bullet$	(4)
Recombination				
	$e^- + h^+$	$\longrightarrow$	heat	(5)
Hole Trapping				
	$Ti^{IV} - OH^- + h^+$	$\longleftrightarrow$	$Ti^{IV} \vert OH\bullet$	(6a)
	$Ti^{IV} - H_2O + h^+$	$\longleftrightarrow$	$Ti^{IV} \vert OH\bullet + H^+$	(6b)
	$R_{i,ads} + h^+$	$\longleftrightarrow$	$R_{i,ads}^+$	(7)
Electron Trapping				
	$Ti^{IV} + e^-$	$\longleftrightarrow$	Ti^{III}	(8a)
	$Ti^{III} + O_2$	$\longleftrightarrow$	$Ti^{IV} - O_2^-$	(8b)
Hydroxyl Attack				
Case I	$Ti^{IV} \vert OH\bullet + R_{i,ads}$	$\longrightarrow$	$Ti^{IV} + R_{j,ads}$	(9)
Case II	$OH\bullet + R_{i,ads}$	$\longrightarrow$	$R_{j,ads}$	(10)
Case III	$Ti^{IV} \vert OH\bullet + R_i$	$\longrightarrow$	$Ti^{IV} + R_j$	(11)
Case IV	$OH\bullet + R_i$	$\longrightarrow$	R_j	(12)

[5] Extracted and reproduced with permission from Alfano *et al.*, 1997 and Cabrera *et al.*, 1997b. Copyright 1997, Elsevier.

both the usually reported limiting cases for irradiation effects – low, and medium to high levels of irradiation – may normally coexist in different regions of the same photoreactor. Therefore a kinetic expression was obtained that will be applicable to the whole reactor regardless of the irradiation level.

Three different but connected problems must be studied: (i) the reaction kinetics model; (ii) the development of the rate of electron–hole generation in a material particle of the solid suspension; and (iii) the model for characterizing the radiation field to evaluate the local volumetric rate of photon absorption (LVRPA). Point (iii) has been already described in section 6.6.1 for quantum yield determinations. In the first part of this section, we will concentrate on problems (i) and (ii).

The kinetic model. The effect of oxygen concentration has not been included in this proposal; hence, the kinetic model should be valid if and only if oxygen concentration is not in defect in the fluid phase close to the catalytic particle (Table 6.4). A second condition of applicability is the requirement of no mass transfer limitations; this is a trivial statement that, notwithstanding, has been often neglected in many kinetic reports.

The following assumptions will be considered: (i) step 7 is not an important hole trapping reaction; (ii) whenever an adsorption process is surely involved in the sequence (reactions 3 and 4), it may be considered at equilibrium conditions; (iii) the micro steady state approximation may be applied for unstable reaction intermediates such as hydroxyl radicals and semiconductor holes; (iv) in dilute solutions of contaminants in water the concentrations of hydroxyl ions and water on the catalytic surface will be almost constant; (v) the catalytic surface (optical properties, specific surface area, and number of active sites) remains unchanged during the reaction; (vi) a rate of generation of electrons and holes can be precisely defined; (vii) the recombination reaction of electrons and holes occurs mainly in the bulk of the catalytic particle; (viii) as a good approximation, the concentration of electrons and holes can be assumed equal; and (ix) the liquid phase does not absorb radiation in the near-UV range ($300\,\text{nm} \leq \lambda \leq 400\,\text{nm}$) and, consequently, there is no parallel homogeneous reaction.

The reaction rate per particle. For a particle having a surface area equal to a_S and a volume equal to v_P the rate of disappearance of reactant R_i is given by (Table 6.4)

$$\begin{aligned} r_P = {} & k_9[\text{Ti}^{\text{IV}}|\text{OH}\bullet][R_{i,\text{ads}}]a_S^2 + k_{10}[\text{OH}\bullet][R_{i,\text{ads}}]a_S \\ & + k_{11}[\text{Ti}^{\text{IV}}|\text{OH}\bullet][R_i]a_S + k_{12}[\text{OH}\bullet][R_i] \end{aligned} \tag{6.104}$$

The adsorbed organic substance concentration, $R_{i,\text{ads}}$, and the concentration of the 'association,' $\text{Ti}^{\text{IV}}|\text{OH}\bullet$, can be obtained assuming equilibrium for reactions 3 and 4. If $[\text{Ti}^{\text{IV}}\text{OH}]$ represents the superficial concentration of bound OH^- or water that reacts with the holes, we will accept that due to the high water concentration $k'_{6+} = k_{6+}[\text{Ti}^{\text{IV}}(\text{OH})]$. Applying the micro steady state approximation (MSSA) to the $[\text{OH}\bullet]$ radicals and calling

$$\begin{aligned} \alpha_{3,i} &= \frac{k_9\, K_{3,i}[\text{site}]a_S}{k_{6-}} + \frac{k_{10}\, K_{3,i}[\text{site}]}{k_{6-}\, K_4[\text{Ti}^{\text{IV}}]} + \frac{k_{11}}{k_{6-}} + \frac{k_{12}}{k_{6-}\, K_4[\text{Ti}^{\text{IV}}]a_S} \\ \alpha_{3,j} &= \frac{k_{\text{R}j,\text{I}}K_{3,j}[\text{site}]a_S}{k_{6-}} + \frac{k_{\text{R}j,\text{II}}K_{3,j}[\text{site}]}{k_{6-}\, K_4[\text{Ti}^{\text{IV}}]} + \frac{k_{\text{R}j,\text{III}}}{k_{6-}} + \frac{k_{\text{R}j,\text{IV}}}{k_{6-}\, K_4[\text{Ti}^{\text{IV}}]a_S} \end{aligned} \tag{6.105}$$

one gets

$$[\mathrm{OH}\bullet] = \frac{k'_{6+}}{k_{6-}K_4[\mathrm{Ti^{IV}}]} \times \frac{[h^+]}{1+\alpha_{3,i}[R_i] + \sum\limits_{\substack{j=1\\(j\neq i)}}^{n} \alpha_{3,j}[R_j]} \tag{6.106}$$

Once more, the micro steady state approximation may be used to obtain the hole concentration:

$$r_{h^+} = r_g - k'_{6+}[h^+]a_S + k_{6-}[\mathrm{Ti^{IV}}\,|\mathrm{OH}\bullet]a_S - k_5[h^+][e^-]v_P \cong 0 \tag{6.107}$$

Here, r_g is the rate of electron–hole generation per particle ($r_g[=]\mathrm{mol\,s^{-1}\ particle^{-1}}$). From equation 6.106 we can extract $[h^+]$ and substitute into equation 6.107. From the resulting quadratic equation in [OH•] one obtains

$$[\mathrm{OH}\bullet] = \frac{(k'_{6+})^2 a_S}{2\,k_5\,v_P\,k_{6-}\,K_4[\mathrm{Ti^{IV}}]} \left\{ -\left(\frac{E(R_i,R_j)}{1+E(R_i,R_j)}\right) + \sqrt{\left(\frac{E(R_i,R_j)}{1+E(R_i,R_j)}\right)^2 + \frac{4\,k_5\,v_P}{(k'_{6+}\,a_S)^2} r_g} \right\} \times \frac{1}{(1+E(R_i,R_j))} \tag{6.108}$$

The final equation results in:

$$r_P = \alpha'_1 \left\{ -\left(\frac{E(R_i,R_j)}{1+E(R_i,R_j)}\right) + \sqrt{\left(\frac{E(R_i,R_j)}{1+E(R_i,R_j)}\right)^2 + \frac{2}{\alpha'_1} r_g} \right\} \frac{\alpha_{3,i}[R_i]}{(1+E(R_i,R_j))} \tag{6.109}$$

where

$$\alpha'_1 = \frac{(k'_{6+}a_S)^2}{2k_5\,v_P} \quad \text{and} \quad E(R_i,R_j) = \alpha_{3,i}[R_i] + \sum\limits_{\substack{j=1\\(j\neq i)}}^{n} \alpha_{3,j}[R_j] \tag{6.110}$$

Notice that (i) excluding the evaluation of the rate of electron–hole generation (r_g), equation 6.110 has only two constants: α'_1 and $\alpha_{3,i}$; and (ii) $\alpha_{3,i}$ is not an adsorption equilibrium constant.

The rate of electron–hole generation ($\mathrm{r_g}$). We look for an expression of r_g valid for any material point in the reacting space; i.e. a local value that is applicable to the suspension of the solid in the liquid (r_g [=] $\mathrm{mol\,s^{-1}\ particle^{-1}}$). Upon radiation absorption by the catalytic particle, electrons and holes are generated with a primary quantum yield equal to ϕ_λ.

The particles of many of the known varieties of titanium dioxide have a non-porous structure; therefore, absorption of radiation is produced in the particle volume through its bounding surface that is characterized by a unit normal vector $\mathbf{n}_S$, pointing outwards (Figure 6.11). The net radiation flux is defined as follows:

$$\mathbf{q}_\lambda (\mathbf{x}+\zeta_P, t) = \int_{\Omega=4\pi} I_{\lambda,\Omega} (\mathbf{x}+\zeta_P, t)\Omega \mathrm{d}\Omega \tag{6.111}$$

Part of this radiation may be reflected on the surface (scattered) and part may be absorbed. The rate of electron–hole generation at point $P(\mathbf{x}+\zeta_P)$ on the differential surface $\mathrm{d}A$ of particle n at a time t and for a wavelength λ (actually between λ and $\lambda+\mathrm{d}\lambda$) is

$$\begin{aligned} \mathrm{d}r_{g,n,\lambda} (\mathbf{x}+\zeta_P, t) &= \phi_\lambda [\mathbf{q}_\lambda (\mathbf{x}+\zeta_P, t) \cdot \mathbf{n}_L] \mathrm{d}A \\ &= \phi_\lambda \left[\int_{\Omega=4\pi} I_{\lambda,\Omega} (\mathbf{x}+\zeta_P, t)\Omega \cdot \mathbf{n}_L \, \mathrm{d}\Omega \right] \end{aligned} \tag{6.112}$$

for all positive values of the dot product (fluxes that are not reflected). Considering the total external area of the solid particle A_{Sn}

$$\begin{aligned} r_{g,n,\lambda} (\mathbf{x}, t) &= \phi_\lambda \int_{A_{Sn}} \mathrm{d}A \left[\int_{\Omega=4\pi} I_{\lambda,\Omega} (\mathbf{x}+\zeta_P, t)\, \Omega \cdot \mathbf{n}_L \right] \\ &= \phi_\lambda \int_{A_{Sn}} \mathrm{d}A [\mathbf{q}_\lambda (\mathbf{x}+\zeta_P, t) \cdot \mathbf{n}_L] \end{aligned} \tag{6.113}$$

According to Figure 6.11, $\mathbf{n}_L$ (the outwardly directed normal to the liquid phase) $= -\mathbf{n}_S$ (the outwardly directed normal to the solid). Applying the divergence theorem to equation 6.113 and thus transforming the surface integral into a volume one, we get

$$\begin{aligned} r_{n,g,\lambda} (\mathbf{x}, t) &= -\phi_\lambda \int_{A_{Sn}} \mathrm{d}A [\mathbf{q}_\lambda (\mathbf{x}+\zeta_P, t) \cdot \mathbf{n}_S] \\ &= -\phi_\lambda \int_{V_{Sn}} \mathrm{d}V [\boldsymbol{\nabla} \cdot \mathbf{q}_\lambda (\mathbf{x}+\zeta_P, t)] \end{aligned} \tag{6.114}$$

Following Ozisik (1973, p. 251):

$$\int_{\Omega=4\pi} \boldsymbol{\nabla} \cdot (\Omega I_{\lambda,\Omega}) \mathrm{d}\Omega = \underbrace{\underline{\nabla} \cdot \int_{\Omega=4\pi} I_{\lambda,\Omega} \Omega \mathrm{d}\Omega}_{\text{Radiation flux}} = \boldsymbol{\nabla} \cdot \mathbf{q}_\lambda = \underbrace{-\kappa_\lambda \int_{\Omega=4\pi} I_{\lambda,\Omega} \mathrm{d}\Omega}_{\text{Radiation absorption}} \tag{6.115}$$

The last term in equation 6.114 can be substituted according to

$$\boldsymbol{\nabla} \cdot \mathbf{q}_\lambda(\mathbf{x}, t) = -e^a_\lambda(\mathbf{x}, t) \tag{6.116}$$

where e^a_λ is the LVRPA. Applying equation 6.116 to the case described by equation 6.114, the monochromatic rate of electron–hole generation for particle n results in:

$$r_{g,n,\lambda}(\mathbf{x}, t) = \phi_\lambda \int_{V_{Sn}} \mathrm{d}V e^{\varepsilon}_{\lambda,S}(\mathbf{x}+\zeta, t) \tag{6.117}$$

In equation 6.117 V_{Sn} is the volume of the 'n' solid particle.

Finally, from equations 6.95 and 6.117 one gets

$$r_g(\mathbf{x}, t) = \frac{1}{N_V}\int_{\lambda_1}^{\lambda_2} d\lambda \phi_\lambda e_\lambda^a(\mathbf{x}, t) \tag{6.118}$$

In equation 6.118, the monochromatic LVRPA $[e_\lambda^a(\mathbf{x}, t)]$ is the result of the average absorption rate calculated over all the catalytic surface (area A_S corresponding to V_S) existing in the volume V. The LVRPA must be obtained from the solution of RTE applied to each particular reactor.

Evaluation of the LVRPA. The mathematical model has been presented in the section on calculating procedure for the LVRPA and the LVRPA is calculated according to equations 6.96, 6.97, and 6.98.

The final kinetic equation. Inserting a monochromatic form of equation 6.118 into equation 6.109 we get an expression of the reaction rate per particle and monochromatic radiation as follows:

$$r_P = \alpha_1' \left\{ -\left(\frac{E(R_i, R_j)}{1 + E(R_i, R_j)}\right) + \sqrt{\left(\frac{E(R_i, R_j)}{1 + E(R_i, R_j)}\right)^2 + \frac{2}{\alpha_1' N_V}\phi_\lambda e_\lambda^a(\mathbf{x}, t)} \right\} \left(\frac{\alpha_{3,i}[R_i]}{1 + E(R_i, R_j)}\right) \tag{6.119}$$

Using polychromatic radiation, if one defines

$$\overline{\phi} = \frac{\int_\lambda \phi_\lambda e_\lambda^a \, d\lambda}{\int_\lambda e_\lambda^a \, d\lambda} \tag{6.120}$$

the reaction rate results in:

$$r_P = \alpha_1' \left\{ -\left(\frac{E(R_i, R_j)}{1 + E(R_i, R_j)}\right) + \sqrt{\left(\frac{E(R_i, R_j)}{1 + E(R_i, R_j)}\right)^2 + \frac{\alpha_2'}{C_{mp}}\int_\lambda e_\lambda^a(\mathbf{x}, t)\, d\lambda} \right\} \left(\frac{\alpha_{3,i}[R_i]}{1 + E(R_i, R_j)}\right) \tag{6.121}$$

In equation 6.121 we have defined

$$\alpha_2' = \frac{2 v_P \rho_P \overline{\phi}}{\alpha_1'} \quad \text{and} \quad N_V = \frac{C_{mp}}{v_P \rho_P} \tag{6.122}$$

Equation 6.122 has three constants: α_1', α_2', and $\alpha_{3,i}$.

The mass balance. A perfectly stirred continuous flow reactor inside the loop of a recirculating system, as described in section 6.6.1, s was used. As is shown in the section on calculating procedure for the reaction rate, a rigorous mass balance for this system gives

$$\varepsilon_L \frac{dC_i(\mathbf{x}, t)}{dt} = \frac{V_R}{V_{Tk}} a_V \langle r_{S,i}(\mathbf{x}, t)\rangle_{A_{S,R}} = \frac{V_R}{V_{Tk}} C_{mp} S_g \nu_i \langle r_S(\mathbf{x}, t)\rangle_{A_{S,R}} \tag{6.123}$$

Equation 6.123 can be integrated with an initial condition:

$$C_i\,(t=0) = C_{i,0} \tag{6.124}$$

The reaction rate per particle, equation 6.121, must be transformed into an expression valid for a heterogeneous catalytic reactor, i.e. a reaction per unit surface area ($r_{\mathrm{S}}(\mathbf{x}, t)[=]\mathrm{mol\,s^{-1}\,m^{-2}}$). Clearly,

$$r_{\mathrm{S}}(\mathbf{x}, t) = \frac{r_{\mathrm{P}}(\mathbf{x}, t)}{a_{\mathrm{S}}} \tag{6.125}$$

One can also define

$$\alpha_1 = \frac{\alpha_1'}{a_{\mathrm{S}}} \quad \text{and} \quad \alpha_2 = \alpha_2' \tag{6.126}$$

Substituting equation 6.121 into equation 6.125 and the result into equation 6.123, and considering that TCE is the only reacting species, finally we get

$$\varepsilon_{\mathrm{L}} \frac{\mathrm{d}C_i(t)}{\mathrm{d}t} = -\frac{V_{\mathrm{R}}}{V_{\mathrm{Tk}}} C_{\mathrm{mp}} S_{\mathrm{g}} \alpha_1 \left(\frac{\alpha_{3,i} C_i(t)}{1+\alpha_{3,i} C_i(t)} \right) \times \left\{ \left\langle -\left(\frac{\alpha_{3,i} C_i(t)}{1+\alpha_{3,i} C_i(t)} \right) + \sqrt{\left(\frac{\alpha_{3,i} C_i(t)}{1+\alpha_{3,i} C_i(t)} \right)^2 + \frac{\alpha_2}{C_{\mathrm{mp}}} \int_\lambda e_\lambda^{\mathrm{a}}(\mathbf{x}, C_{\mathrm{mp}})\,\mathrm{d}\lambda} \right\rangle_{A_{\mathrm{S,R}}} \right\} \tag{6.127}$$

Calculating the reactor volume average that in this case reduces to an average over the reactor length, equation 6.127 renders

$$\varepsilon_{\mathrm{L}} \frac{\mathrm{d}C_i(t)}{\mathrm{d}t} = -\frac{V_{\mathrm{R}}}{V_{\mathrm{Tk}}} C_{\mathrm{mp}} S_{\mathrm{g}} \alpha_1 \left(\frac{\alpha_{3,i} C_i(t)}{1+\alpha_{3,i} C_i(t)} \right) \times \left\{ \left\langle -\left(\frac{\alpha_{3,i} C_i(t)}{1+\alpha_{3,i} C_i(t)} \right) + \sqrt{\left(\frac{\alpha_{3,i} C_i(t)}{1+\alpha_{3,i} C_i(t)} \right)^2 + \frac{\alpha_2}{C_{\mathrm{mp}}} \int_\lambda e_\lambda^{\mathrm{a}}(\mathbf{x}, C_{\mathrm{mp}})\,\mathrm{d}\lambda} \right\rangle_{L_{\mathrm{R}}} \right\} \tag{6.128}$$

In equation 6.128, $C_i(t)$ and t are experimental data, C_{mp} is a reaction parameter, and $\int_\lambda e_\lambda^{\mathrm{a}}(\mathbf{x}, C_{\mathrm{mp}})\mathrm{d}\lambda$ can be computed from available information on lamp, reactor, reflector, and catalyst (Table 6.5). Then, with a non-linear regression procedure the three kinetic parameters $(\alpha_1, \alpha_2, \text{and } \alpha_{3,i})$ can be obtained.

Experiments. The equipment has already been described in the section concerning true quantum yields. The only difference is that in this work the polychromatic light was used without filters (Table 6.6).

Kinetic Parameter Evaluation. In the previous sections we have described the following:

1. A photocatalytic reactor and reactor model that permits the precise evaluation of radiation absorption and scattering. The radiation field is analyzed in terms of a 1D one-directional radiative transfer model. With this approach, the LVRPA at every point inside the reactor can be known. The same reactor has been used here.

Table 6.5 Reactor, lamp, and reflector characteristics. Reproduced with permission from Cabrera et al. (1997b), copyright 1997, Elsevier

	Parameter	Value	Units
REACTOR	Inside diameter	5.2	cm
	Length	10.0	cm
	Plate tickness	0.38	cm
	Plate radius	2.7	cm
	Reactor volume	212	cm^3
	Total volume	1950	cm^3
LAMP 360 UA-3	Nominal power	360	W
	Diameter	1.9	cm
	Arc length	15.2	cm
REFLECTOR	Parabola characteristic constant	2.5	cm
	Distance from vertex of parabolic reflector to reactor plate	14.8	cm
	Length	22.0	cm

Table 6.6 Experimental operating conditions. Reproduced with permission from Cabrera et al. (1997b), copyright 1997, Elsevier

CATALYST	Aldrich titanium dioxide (>99% anatase) sp. surf. area: $9.6 m^2 g^{-1}$ nominal diameter: 150–200 nm concn.: (0.1, 0.2 and 1.0) $\times 10^{-3}$ g cm^{-3}
REACTANT SOLUTION	Trichloroethylene Carlo Erba RSE MW = 131.36 g mol^{-1} concn.: (0.084 to 0.55) $\times 10^{-6}$ mol cm^{-3} initial pH: 6
OXIDANT	Oxygen, satured at 293 K
TEMPERATURE	293 K
LAMP	GE UA-3 Uviarc 360W Hg medium pressure $P(295-405\,nm) = 6 \times 10^{-5}$ einstein s^{-1}
RADIATION FLUX AT REACTOR WINDOW (295-405 nm)	1.14×10^{-7} einstein $cm^{-2}\,s^{-1}$

2. An extension of the Turchi and Ollis (1990) kinetic scheme for photocatalytic reactions involving hydroxyl radical attack. In this work the precise evaluation of the photon–solid catalyst interaction and the proper knowledge of the LVRPA $\left[e^{a}_{\lambda}(\mathbf{x}, C_{mp})\right]$ were incorporated into the kinetic expression. The final result can be applied to any photocatalytic reactor regardless of its irradiation level.

Experiments were carried out for TCE concentrations between 0.15×10^{-6} and 0.75×10^{-6} mol cm^{-3}, and three different levels of catalyst concentration were used: 0.1×10^{-3}, 0.2×10^{-3}, and 1.0×10^{-3} g cm^{-3}. Similarly, part of the data were obtained at three different irradiating conditions. To achieve this effect, a neutral density filter was

Table 6.7 *Results for TCE model*

Constant	Value	Units
α_1	2.46×10^{-14}	$\mathrm{mol\,cm^{-2}\,s^{-1}}$
α_2	1.57×10^{11}	$\mathrm{g\,s\,einstein^{-1}}$
α_3	6.42×10^{6}	$\mathrm{cm^3\,mol^{-1}}$

interposed between the lamp and the reactor. In this way the irradiation level was varied in the following sequence: 100%, 30% and 10%.

The experimental data so produced can be evaluated in terms of equation 6.128, where the parameters $(\alpha_1, \alpha_2, \text{and } \alpha_{3,i})$ can be obtained using the Marquardt (1963) algorithm. The numerical algorithm receives the following information. (i) TCE concentration–time relationships resulting from the integration of equation 6.128. (ii) Results of the integration of the RTE to incorporate values of the LVRPA into equation 6.128; the RTE is integrated using the DOM. For polychromatic radiation the local values of e^{a}_{λ} must be integrated over the wavelength range of interest and, afterwards, integrated once more over the reactor length to obtain the volume-averaged rate of reaction. (iii) Experimental TCE-time information. The parameter evaluator searches, with an optimization technique, the minimum differences between experimental values and theoretical predictions.

Table 6.7 gives the values of the three kinetic parameters α_1, α_2, and $\alpha_{3,i}$. In Figure 6.12 (a), (b), and (c) we represent a sample of the obtained results for changes in the boundary condition: the solid line corresponds to the model and the squares to the experimental data when the radiation flux at the wall of radiation entrance was varied as explained before. A good agreement can be observed between model predictions and experimental points.

(a)

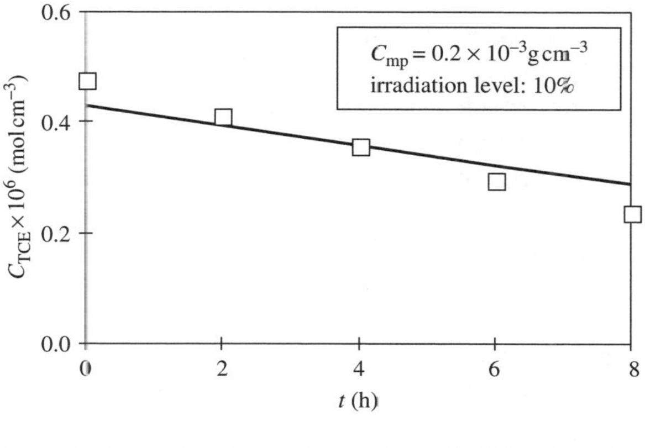

Figure 6.12 *Kinetic results (model and experiments) for the TCE degradation. Reproduced with permission from Cabrera* et al. *(1997b), copyright 1997, Elsevier*

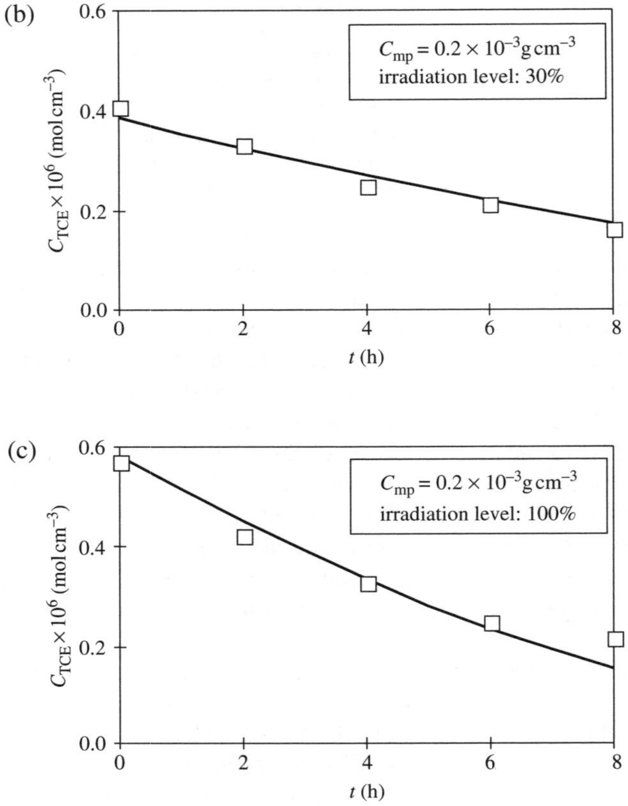

Figure 6.12 *(Continued)*

6.7 Summary

This chapter presents a condensed description of the most important technical tools that are needed to design homogeneous and heterogeneous photoreactors using computer simulation of a rigorous mathematical description of the reactor performance both in the laboratory and on a commercial scale. Employing intrinsic reaction kinetic models and parameters derived from properly analyzed laboratory information, it is shown that it is possible to scale up reactors with no additional information, avoiding costly pilot plant experiments and particularly without resorting to empirically adjusted correcting factors. The method is illustrated with examples concerning the degradation of organic pollutants as typical applications of the newly developed advanced oxidation technologies (AOT). One particular aspect of two of these examples – heterogeneous photoreactors – gives to these reactions a unique characteristic: in many cases, absorption of light by the employed solid semiconductor cannot be separated from scattering or reflection by the catalyst, making more difficult the analysis of kinetic information and the design of practical reactors. Starting always from fundamental principles and using mathematical modeling as the main tool, we show some of the methods for tackling many of the problems derived from the most difficult type of system particularities.

Acknowledgments

The authors are grateful to Universidad Nacional del Litoral, Consejo Nacional de Investigaciones Científicas y Técnicas and Agencia Nacional de Promoción Científica y Tecnológica for their support in producing this work. We acknowledge the technical assistance received from Eng. Claudia M. Romani. Thanks are also given to Academia Nacional de Ciencias Exactas, Físicas y Naturales of Buenos Aires for allowing us to use part of the material published in 'Photoreactor Analysis Through Two Examples in Advanced Oxidation Technologies', *Anales Acad. Nac. de Cs. Ex., Fís. y Nat.* 53, 84-120 (2001) (without Copyright).

Notation

a_S	particle surface area (cm^2 $particle^{-1}$)
a_V	solid–liquid interfacial area per unit reactor volume ($cm^2\,cm^{-3}$)
C_i	molar concentration of component i ($mol\,cm^{-3}$)
C_{mp}	mass catalyst concentration ($g\,cm^{-3}$)
C_p	specific heat at constant pressure ($J\ g^{-1}\,K^{-1}$)
D_{im}	diffusion coefficient of component i in the mixture ($cm^2\,s^{-1}$)
e^a	volumetric rate of photon absorption (einstein $s^{-1}\,cm^{-3}$)
G	incident radiation (also known as spherical irradiance) (einstein $s^{-1}\,cm^{-2}$)
h	film heat transfer coefficient ($J\ cm^{-2}\,s^{-1}\,K^{-1}$); also Planck constant ($J\,s$)
H_i	enthalpy of component i ($J\ mol^{-1}$)
ΔH	heat of reaction at constant pressure ($J\ mol^{-1}$)
I	specific (radiation) intensity (also known as radiance) (einstein $s^{-1}\,cm^{-2}\,sr^{-1}$)
j^e	radiation emission (einstein $s^{-1}\,cm^{-3}\,sr^{-1}$)
$\mathbf{J}_i$	molar diffusive density flux vector of component i ($mol\,s^{-1}\,cm^{-2}$)
k	kinetic constant (for different reaction steps) (units vary with type of step)
k_c	thermal conductivity ($J\ cm^{-1}\,s^{-1}\,K^{-1}$)
L	length (cm)
L_L	lamp length (cm)
LVRPA	local volumetric rate of photon absorption (einstein $s^{-1}\,cm^{-3}$)
$N_{\lambda,\Omega}$	λ, Ω photon density number (photons $cm^{-3}\,sr^{-1}$ and per unit wavelength interval)
$\mathbf{n}$	unit normal vector to a given surface
N	number of particles
N_V	number of particles per unit volume (particle cm^{-3})
N_i	molar flux of component i ($mol\,cm^{-2}\,s^{-1}$)
p	phase function (dimensionless)

P	output power from the lamp (einstein s^{-1})
q	radiation density flux for a given direction on surface orientation (also known as superficial irradiance) (einstein s^{-1} cm^{-2})
$\mathbf{q}$	radiation density flux vector (einstein s^{-1} cm^{-2})
Q^{Ext}	heat transferred from external fields (J g^{-1} s^{-1})
r	radius (cm) or radial coordinate (cm)
R_L	lamp radius (cm)
$R_{Hom,i}$	homogeneous, molar reaction rate of component i (mol cm^{-3} s^{-1})
$R_{Het,i}$	heterogeneous, molar reaction rate of component i (mol cm^{-2} s^{-1})
s	variable representing distances in a 3D space (cm)
S_g	catalyst specific surface area (cm^2 g^{-1})
t	time (s)
T	temperature (K)
v	velocity (cm s^{-1})
V	volume (cm^3)
x	cartesian coordinate (cm)
$\mathbf{x}$	vector representing position in a 3D space (cm)
z	cartesian or cylindrical coordinate (cm)

Greek letters

β	cylindrical coordinate (rad); also extinction coefficient $= \kappa + \sigma (cm^{-1})$
ε	liquid hold-up (dimensionless)
ζ	3D position vector inside a material particle (cm)
θ	spherical coordinate (rad)
κ	absorption coefficient (cm^{-1})
κ^*	specific (per unit mass) absorption coefficient (cm^2 g^{-1})
λ	wavelength (nm $= 10^{-7}$ cm)
μ	direction cosine $\mu = \cos\theta$
ν	frequency (s^{-1})
ρ_{mix}	density of mixture (g cm^{-3})
σ	scattering coefficient (cm^{-1})
σ^*	specific (per unit mass) scattering coefficient (cm^2 g^{-1})
Υ	transmission or compounded transmission coefficient (dimensionless)
ϕ	spherical coordinate (rad); also primary quantum yield (mol $einstein^{-1}$)
Φ	overall quantum yield (mol $einstein^{-1}$)
Ω	solid angle (sr, ster radian)
$\mathbf{\Omega}$	unit vector in the direction of radiation propagation

Superscripts

a	denotes absorbed energy
Dir	denotes direct radiation from the lamp
Pseudo	denotes a heterogeneous reaction expressed per unit reactor volume
0	denotes initial or inlet conditions
Ref	denotes reflected radiation from the reflector
*	denotes specific (per unit mass) properties

Subscripts

A	denotes area
Ac	denotes actinometer
Hom	denotes a homogeneous reaction
Het	denotes a heterogeneous reaction
i	denotes internal or component i
L	denotes liquid phase
o	denotes outside or external
0	denotes initial value or inlet condition
R	denotes reactor
r	denotes radius or radial direction
S	denotes solid phase
Sol	denotes solid surface
Susp	denotes suspension
T	denotes total
Tk	denotes tank
V	denotes volume
z	denotes axial direction
λ	denotes wavelength
Σ	denotes polychromatic radiation
Ω	denotes direction of radiation propagation

Special symbols

$\_$	denotes vector value
$\bar{}$	denotes average value over wavelengths
$\langle\rangle$	denotes average value over a defined special dimension

References

Alfano O.M., Romero R.L. and Cassano A.E. 1985. A cylindrical photoreactor irradiated from the bottom. I. Radiation flux density generated by a tubular source and a parabolic reflector, Chem. Eng. Sci., 40, 2119–2127.

Alfano O.M., Romero R.L. and Cassano A.E. 1986a. A cylindrical photoreactor irradiated from the bottom. II. Models for the local volumetric rate of energy absorption with polychromatic radiation and their evaluation, Chem. Eng. Sci., 41, 1155–1161.

Alfano O.M., Romero R.L., Negro C.A. and Cassano A.E. 1986b. A cylindrical photoreactor irradiated from the bottom. III. Measurement of absolute values of the local volumetric rate of energy absorption. Experiments with polychromatic radiation, Chem. Eng. Sci., 41, 1163–1169.

Alfano O.M., Negro A.C., Cabrera M.I. and Cassano A.E. 1995. Scattering effects produced by inert particles in photochemical reactors. 1. Model and experimental verification, Ind. Eng. Chem. Res., 34(2), 488–499.

Alfano O.M., Cabrera M.I and Cassano A.E. 1997. Photocatalytic reactions involving hydroxyl radical attack. I: Reaction kinetics formulation with explicit photon absorption effects, J. Catal., 172(2), 370–379.

Alfano O.M., Bahnemann D., Cassano A.E., Dillert R. and Goslich R. 2000. Photocatalysis in water environments using artificial and solar light, Catal. Today, 58, 199–230.

Bird R.B., Stewart W.E. and Lightfoot E.N. 2002. Transport Phenomena, 2nd edition. John Wiley & Sons, New York.

Brandi R.J., Alfano O.M. and Cassano A.E. 1999. Rigorous model and experimental verification of the radiation field in a flat plate solar collector simulator employed for photocatalytic reactions, Chem. Eng. Sci., 54(13–14), 2817–2827.

Brandi R.J., Alfano O.M. and Cassano A.E. 2000a. Evaluation of radiation absorption in slurry photocatalytic reactors. 1. Assessment of methods in use and new proposal, Environ. Sci. Technol., 34(12), 2623–2630.

Brandi R.J., Alfano O.M. and Cassano A.E. 2000b. Evaluation of radiation absorption in slurry photocatalytic reactors. 2. Experimental verification of the proposed method, Environ. Sci. Technol., 34(12), 2631–2639.

Brandi R.J., Citroni M.A., Alfano O.M. and Cassano A.E. 2003. Absolute quantum yields in photocatalytic slurry reactors, Chem. Eng. Sci., 58, 979–985.

Braun A.M., Jakob L., Oliveros E. and Oller do Nascimento C.A. 1993. Up-scaling photochemical reactions, Adv. Photochem., 18, 235–313.

Cabrera M.I., Alfano O.M. and Cassano A.E. 1994. Novel reactor for photocatalytic kinetic studies, Ind. Eng. Chem. Res., 33(12), 3031–3042.

Cabrera M.I., Alfano O.M. and Cassano A.E. 1996. Absorption and scattering coefficients of titanium dioxide particulate suspensions in water, J. Phys. Chem., 100(51), 20043–20050.

Cabrera M.I., Martín C.A., Alfano O.M. and Cassano A.E. 1997a. Photochemical decomposition of 2,4-dichlorophenoxyacetic acid (2,4-D) in aqueous solution. I. Kinetic study, Water Sci. Technol., 35(4), 31–39.

Cabrera M.I., Negro A.C., Alfano O.M. and Cassano A.E. 1997b. Photocatalytic reactions involving hydroxyl radical attack. II: Kinetics of the decomposition of trichloroethylene using titanium dioxide, J. Catal., 172(2), 380–390.

Cassano A.E. and Alfano O.M. 2000. Reaction engineering of suspended solid heterogeneous photocatalytic reactors, Catal. Today, 58(2–3), 167–197.

Cassano A.E., Martín C.A., Brandi R.J. and Alfano O.M. 1995. Photoreactor analysis and design: fundamentals and applications, Ind. Eng. Chem. Res., 34(7), 2155–2201.

Clariá M.A., Irazoqui H.A. and Cassano A.E. 1988. A priori design of a photoreactor for the chlorination of ethane, AIChE J., 34(3), 366–382.

Duderstadt J.J. and Martin W.R. 1979. Transport Theory. John Wiley, New York.

Irazoqui H.A., Cerdá J. and Cassano A.E. 1973. Radiation profiles in an empty annular photoreactor with a source of finite spatial dimensions, AIChE J., 19, 460–467.

Marquardt D.W. 1963. An algorithm for least-squares estimation of non linear parameters, SIAM J. Appl. Math., 11, 431–441.

Martín C.A., Baltanás M.A. and Cassano A.E. 1996. Photocatalytic reactors II. Quantum efficiencies allowing for scattering effects. An experimental approximation, J. Photochem. Photobiol. A: Chem., 94, 173–189.

Martín C.A., Cabrera M.I., Alfano O.M. and Cassano A.E. 1997. Photochemical decomposition of 2,4-dichlorophenoxyacetic acid (2,4-D) in aqueous solution. II. Reactor modeling and verification, Water Sci. Technol., 35(4), 197–205.

Murov S.L., Carmichael I. and Hug G.L. 1993. Handbook of Photochemistry, 2nd edition. Marcel Dekker, New York.

Ozisik M.N. 1973. Radiative Transfer and Interactions with Conduction and Convection. Wiley, New York.

Puma G.L. and Yue P.L. 1998. A laminar falling film slurry photocatalytic reactor. Part I – Model development, Chem. Eng. Sci., 53, 2993–3006.

Ray, A K. 1998. A new photocatalytic reactor for destruction of toxic water pollutants by advanced oxidation process, Catal. Today, 44, 357–368.

Romero R.L., Alfano O.M., Marchetti J.L. and Cassano A.E. 1983. Modelling and parametric sensitivity of an annular photoreactor with complex kinetics, Chem. Eng. Sci., 38, 1593–1605.

Romero R.L., Alfano O.M. and Cassano A.E. 1997. Cylindrical photocatalytic reactors. Radiation absorption and scattering effects produced by suspended fine particles in an annular space, Ind. Eng. Chem. Res., 36(8), 3094–3109.

Romero R.L., Alfano O.M. and Cassano A.E. 2003. Radiation field in an annular, slurry photocatalytic reactor 2. Ind. Eng. Chem. Res., 42, 2479–2488.

Siegel R. and Howell J.R. 1992. Thermal Radiation Heat Transfer, 3rd edition. Hemisphere, Washington, DC.

Turchi C.S. and Ollis D.F. 1990. Photocatalytic degradation of organic water contaminants: Mechanisms involving hydroxyl radical attack, J. Catal., 122, 178–192.

Van de Hulst H.C. 1957. Light Scattering by Small Particles. Wiley, New York.

Whitaker S. 1977. Fundamental Principles of Heat Transfer. Pergamon Press, New York, p. 381.

7

Development of Nano-Structured Micro-Porous Materials and their Application in Bioprocess–Chemical Process Intensification and Tissue Engineering

G. Akay, M.A. Bokhari, V.J. Byron and M. Dogru

7.1 Introduction

Flow through porous media has always been an important subject especially in civil, geological, and chemical engineering. However, in most cases, the structure of the porous media is large (pore size in tens to hundreds of micrometers) and the fluids involved are homogenous without a dominant microstructure. As a result, the interactions between the fluid microstructure, the flow field, and the flow field boundary can be ignored. Such interactions and their consequences are well known in many fields such as polymer processing, lyotropic or thermotropic liquid crystals, concentrated suspensions/emulsions, and hematology[1–3]. The macroscopic manifestations of the fluid microstructure/flow field interactions in micro-scale can be very important and a large number of phenomena exist, some of which have been used in process intensification (PI).

In recent years, a new processing technique, called process intensification and miniaturization (PIM), has been evolving in biological, chemical, environmental, and energy conversion processes[4,5]. The basic tenets of the process are to reduce the size of the processing volume and to increase the processing rate. Ultimately, it is possible to device stagewise processes in which each stage is essentially a well-defined pore which is connected to the others in series and/or in parallel. Such reactors are already in operation

Chemical Engineering Edited by Miguel A. Galán and Eva Martin del Valle

in nature (i.e. human body). With the currently available technology, it is now possible to obtain micro-reactors based on micro-porous structures in which the pore size can range from hundreds of micrometers to sub-micrometer, and ultimately to nanometers.

Flow of structured fluids (such as surfactant and polymer solutions) through micro-porous media has been investigated in connection with membrane separation processes. When macromolecules or surfactant molecules (in water) enter into membrane pores, their configuration can be substantially different compared with their configuration in the 'unconfined/unrestricted' environment. We have shown that such phenomena do in fact exist. The thermodynamic state of surfactants in membrane pores (size ca. 10 μm) is such that they form highly viscous stable phases which can exist only at high concentrations in the 'unrestricted' state[6,7]. Consequently, this phenomenon has been utilized recently in the intensive demulsification (separation/breakdown) of oil–water emulsions[8,9].

Alongside with sustainable/green energy technology, life sciences and biotechnology are seen as the most important emerging chemical engineering activity. Within life sciences and biotechnology, tissue engineering and bioprocess intensification (BI) are priority areas. In certain cases, sustainable and green energy technologies are also achieved through BI. Although bioconversions can be accelerated and made selective (both are the objectives of BI) through genetic engineering, additional intensification can be achieved through the understanding of interactions between the microorganisms/enzymes, kinematics of flow, and micro-environment of the bioreactors.

In order to achieve chemical processing and bioprocessing at micrometer scale, micro-porous materials with controlled pore and interconnecting hole (i.e. interconnect) sizes as well as chemical structure should be manufactured in the first place. However, when these micro-porous materials are used in tissue engineering or bioreactors as support (scaffold) for animal cells or microorganisms, they hinder mass transfer for nutrients and metabolites as well as cell proliferation, differentiation, and cell penetration into the scaffold[10–12]. These scaffolds serve as a synthetic extracellular matrix to organize cells into a 3D architecture and to provide stimuli to direct the growth and formation of a desired tissue[13]. Furthermore, it is known that the self-assembling peptide hydrogels, which consist of alternating hydrophilic and hydrophobic amino acids, form nano-patterns to promote cell adhesion, proliferation, and differentiation[13–15]. The presence of nano-scale interconnecting pores within the walls of the micro-pores (which act as the scaffold) therefore allows the diffusion of metabolic products, while the larger micro-interconnects allow cell migration and nutrient transport.

Nano-structured micro-porous materials are also useful in the intensification of catalytic conversions in chemical reactions. Once again, large micro-interconnects provide bulk transport, while the nano-pores provide an extensive surface area for catalysis [5,16].

In this study, after a brief introduction to PI we provide the bases of a technique for the preparation of polymeric micro-porous materials, known as polyHIPE polymers (PHPs) which are now used extensively in PIM, and micro-reactor technology. These polymers are prepared through the high internal phase emulsion (HIPE) polymerization route. In order to control the pore size, the flow-induced phase inversion phenomenon is applied to the emulsification technique. The metalization of these polymers and formation of nano-structured micro-porous metals for intensified catalysis are also discussed. Finally, we illustrate the applications of these materials in chemical- and bioprocess intensifications and tissue engineering while examining the existence of several size-dependent phenomena.

7.2 Process Intensification

PI represents a novel design strategy aimed at the reduction in the processing volume by at least an order of magnitude, compared with the existing technology, without any reduction in process output. This restricted view of PI is relativistic and is a design objective driven primarily by cost savings. On the other hand, process miniaturization (PM) in chemical industry has existed in the form of analytical equipment and sensors. Therefore, for a given production objective (or processing rate) PI and PM represent, respectively, top-down and bottom-up approaches in process design. These two design approaches can be integrated with the sole aim of plant size reduction to provide major savings on capital and operating costs.

However, the integration of PIM also creates synergy in the development of intensified processes, novel product forms, and size-dependent phenomena which in turn provides novel intensified processes. PIM is seen as an important element of sustainable development since PIM can deliver (i) at least a 10-fold decrease in process equipment volume; (ii) elimination of parasitic steps and unwanted by-products, thus eliminating some downstream processing operations; (iii) inherent safety due to reduced reactor volume; (iv) novel product forms; (v) energy, capital, and operating cost reduction and an environmentally friendly process; (vi) plant mobility, responsiveness, and security; and (vii) a platform for other technologies.

7.2.1 Types of Process Intensification: Physical and Phenomenon-Based Intensifications

In order to achieve PI, a driving force is necessary. This process driving force [4,5,17,18] can be achieved by operating at very high/ultra-high pressures, deformation rates, temperatures or through diffusion/conduction path length reduction in heat/mass transfer processes. PI based on these physical driving forces is termed physical process intensification.

In miniaturized systems where the transport processes occur across a length scale of 100–0.1 μm (or less), not only is diffusion/conduction path length reduced, but high process selectivity can be achieved which when repeated several thousand times in microscopic volumes, process selectivity and transport length reduction at each stage results in PI. Therefore, miniaturization or processing in microscopic scale is a prerequisite to PI and there is an underlying phenomenon associated with enhanced selectivity/activity. Such PI techniques are termed phenomenon-based PI. The interaction between process driving force, processing volume, and type of intensification is shown in Figure 7.1.

7.3 Flow-Induced Phase Inversion (FIPI) Phenomenon

Flow-induced phase inversion (FIPI) phenomenon was observed by Akay[3,19–21] and used extensively in phenomenon-based PI, especially in particle[19–26] and emulsion technologies [20,21,27–32]. FIPI is most readily observed in multi-phase systems and most unambiguously in emulsions where the effects of deformation rate and type of deformation on the phase inversion characteristics of the emulsions can be quantified [3,19,33].

Physical intensification
Phenomenon-based intensification (inherently intensive process)

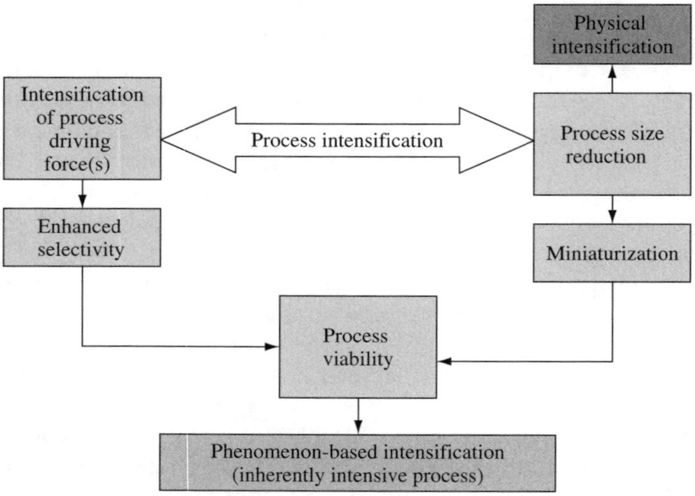

Figure 7.1 *Relationship between process intensification fields, physical process intensification, process miniaturization, selectivity, process viability, and phenomenon-based intensification*

This phenomenon was applied to the intensive structuring of materials such as agglomeration, microencapsulation, detergent processing, emulsification, and latex production from polymer melt emulsification[19–33]. Diagrammatic illustration of FIPI is shown in Figure 7.2. When a material A is mixed with material B, in the absence of any significant deformation, the type of dispersion obtained ([A-in-B] or [B-in-A]) is dictated by the thermodynamic state variables (TSVs) (concentration, viscosity of components, surface activity, temperature, pressure). If the prevailing TSVs favor the formation of [A-in-B] dispersion, phase inversion to [B-in-A] dispersion can be achieved by changing the TSVs (thermodynamically driven process).

Alternatively, the dispersion can be subjected to a well-prescribed deformation, characterized by its rate and type (deformation state variables, DSVs) in order to invert the dispersion under constant thermodynamic conditions; this phenomenon is known as FIPI. It is found that FIPI is not catastrophic and the dispersion goes through an unstable co-continuous state denoted by [AB], followed by a relatively stable multi-dispersion state denoted as [A-in-B]-in-A, before complete phase inversion to [B-in-A]. Therefore, the interchange ability of TSVs with DSVs forms the basis of FIPI processes.

The characteristics of the microstructure formed (such as emulsion droplet size) are dependent on the type of microstructure, type of deformation (shear, extension, or combined), and deformation rate as well as the TSVs. In order to maximize the fluid microstructure/flow field interactions, the flow field must be uniform which requires the generation of the flow field over a small processing volume. There are several types of equipment such as multiple expansion contraction static mixer (MECSM) or its dynamic

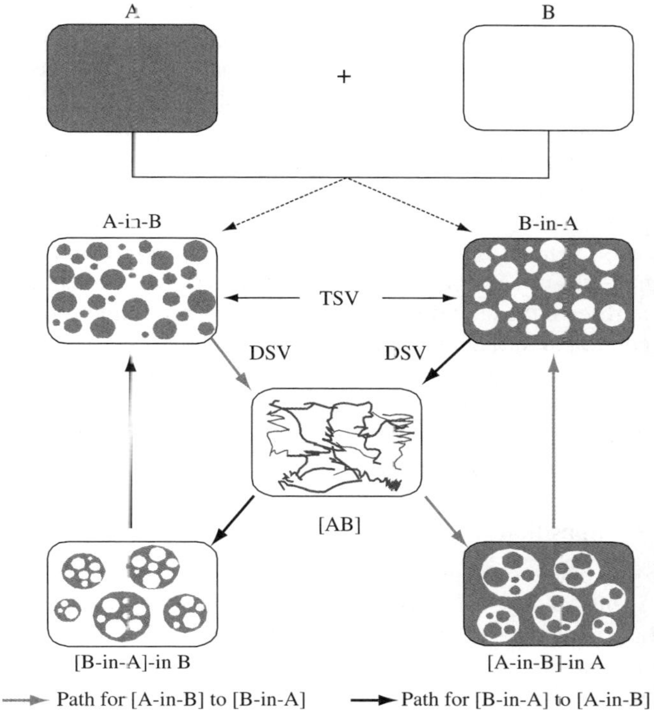

Figure 7.2 *Isothermal flow-induced phase inversion (FIPI) paths for the inversion of [A-in-B] or [B-in-A] emulsions through a co-continuous unstable emulsion phase [AB]. TSV = thermodynamic state variable; DSV = deformation state variable*

version called controlled deformation dynamic mixer (CDDM) which are most suitable for PI in the preparation of emulsions or microstructured materials[3,34]. FIPI-based PI techniques can be further facilitated by using non-isothermal FIPI[22–27].

The importance of FIPI in PI is twofold. It can be used to promote phase inversion without changing the thermodynamics of the system to obtain a higher entropy state or it is possible to delay phase inversion while reducing the system entropy[33]. These attributes of FIPI were utilized in devising intensive processes in material structuring such as agglomeration, microencapsulation, detergent processing, emulsification, and latex production from polymer melt emulsification[19–32]. FIPI was also used in the preparation of HIPEs which were subsequently polymerized to produce micro-porous polymers with controlled pore size[10,32] and used in PI and micro-reactor technology[10,16].

7.4 High Internal Phase Emulsion (HIPE) Preparation

PHPs are prepared through a HIPE polymerization route. The continuous phase of the emulsion contains monomer(s), cross-linking agent, surfactant, and in certain cases, oil-phase soluble polymerization initiator as well as 'additive(s)/filler(s)'. The dispersed phase can contain initiator as well as additives/fillers. Additive(s)/filler(s) in both phases are subsequently utilized after polymerization to functionalize PHP. In most cases, the continuous phase is the oil phase containing the monomer(s) and the dispersed phase is the aqueous phase which may also contain electrolyte. In most applications, the dispersed phase volume is in the range of 80–95% and therefore the inclusion of the additive(s)/filler(s) within the dispersed phase is more practical.

The most important characteristics of PHP are the average pore size (D), and average interconnecting hole size (interconnect) (d) which can be evaluated by examining the fracture surface of PHP using scanning electron microscopy (SEM). PHP constructs containing animal cells or bacteria can also be examined using SEM although they need to be pre-treated to reserve the integrity of the microorganisms. Both of these characteristics (D and d) are controlled through the composition of the phases, processing as well as the polymerization conditions. In this study, we examine the structure formation in PHP as a result of these parameters. In this process, the composition of the phases is fixed, except when the effects of additives/fillers are considered and the emulsification is carried out using the same batch mixing equipment.

7.4.1 Phase Composition

Typically, the oil phase contained 78% monomer/co-monomer, 8% divinyl benzene (cross-linking agent), and 14% non-ionic surfactant Span 80 (Sorbitan monooleate), while the aqueous phase contained 1% potassium persulfate as the initiator. In most cases studied here, monomer is styrene and when elasticity of the polymer is required, 2-ethylhexyl acrylate (2EHA) was used (styrene/2EHA ratio is 1:4). Whenever additives/fillers are placed in the aqueous phase their amounts are stated as weight percent while the phase volume of the aqueous phase remains constant. In some cases, the aqueous phase contains 0.5% hydroxyapatite and 15% phosphoric acid which is used to dissolve the hydroxyapatite, or alternatively, the aqueous phase may contain varying amounts of water-soluble polymer, such as polyethylene glycol or polyethylene oxide. If the styrene-based PHP is to be sulfonated to obtain ionic-hydrophilic foam, the pre-dispersion of sulfuric acid within the pores is useful, if not essential, and in that case, acids (typically 10%) can be used as the internal phase[26,32].

7.4.2 Equipment

Emulsification was carried out at various temperatures (up to 80° C), depending on the desired pore size, using a stirred stainless steel vessel (12-cm diameter) with heating jacket. The oil phase was held in the mixing vessel and the aqueous phase separately heated to a specific temperature and then delivered by two peristaltic pumps to four feed points at a constant rate during dosing period. The mixing was carried out using two flat impellers (diameter 8 cm) at 90° to each other so that the final level of the emulsion is about 1 cm above the top impeller. The lowest impeller on the stirrer shaft is as close to the bottom surface of the vessel as possible. In each experiment the amount of internal phase was typically 225 mL.

7.4.3 Characterization of HIPE Processing

The processing of HIPE can be divided into two stages. In the first stage, the dispersed (aqueous) phase is continuously dosed into the continuous (oil) phase which is already placed in the mixing vessel. The addition of the aqueous phase also creates mixing and therefore care is taken to minimize jet mixing of the phases. Owing to the rotation of the impellers during dosing, there is a reduction in the aqueous phase droplet size. The second stage of processing starts after the completion of dosing when further mixing can be carried out in order to reduce aqueous phase droplet size (i.e. size of the pores after polymerization) and to obtain HIPE with narrow droplet size distribution. If the dosing rate is very low, there will be no need for the additional mixing (homogenization) stage.

The relative dosing rate which has the dimension of deformation rate is used to characterize the dosing rate of the aqueous phase:

$$R_{\mathrm{D}} = \frac{V_{\mathrm{A}}}{t_{\mathrm{D}} V_{\mathrm{O}}}$$

where V_{A} is the volume of aqueous phase added over a period of time, t_{D}, and V_{O} is the volume of the oil phase placed in the batch mixer.

The total mixing time t is defined as

$$t = t_{\mathrm{D}} + t_{\mathrm{H}}$$

where t_{H} is the homogenization time.

The mixing rate is defined as

$$R_{\mathrm{M}} = \frac{D_{\mathrm{I}}}{D_{\mathrm{O}}} \Omega$$

where D_{O} is the diameter of the batch mixer, D_{I} is the diameter of the impellers, and Ω is the rotational speed of the impellers.

If the relative dosing rate is very large and the mixing rate is small, HIPE does not form, but a dilute (low) internal oil-in-water emulsion is obtained. When HIPE is stable, it can be polymerized without phase separation. Polymerization is carried out at 60° C for 8 h and the resulting polymers are washed in alcohol and double distilled water and finally dried before being used. These polymers are also used in the evaluation of pore and interconnect size using SEM. Figure 7.3 illustrates the mapping of the HIPE formation for various HIPEs which do not contain any additives/fillers. The phase volume ranged from 80 to 95%.

Operations under the phase inversion line result in stable emulsion formation which can be polymerized. Figure 7.3 clearly defines the role of mixing in HIPE formation and illustrates the FIPI phenomenon. This stability diagram is useful in obtaining large-pore-size polymers since subsequent homogenization results in reduced water droplet size as illustrated in Figure 7.4 where the average pore size initially decays rapidly with increasing mixing time and it reaches a plateau as the mixing time becomes long.

7.4.4 Effect of Emulsification Temperature on Pore Size

The phase inversion diagram (Figure 7.3) is useful in obtaining large-pore PHPs which are often necessary in tissue engineering and biotechnology. However, since the emulsion

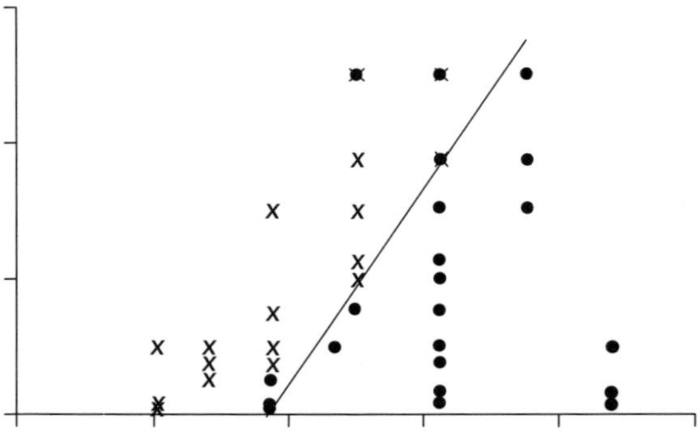

Figure 7.3 High internal phase emulsion (HIPE) stability diagram. (x) indicates phase inversion to oil-in-water emulsion, while (•) indicates stable water-in-oil HIPE formation. Dispersed phase volume is 80% and the phases do not contain any additives or fillers

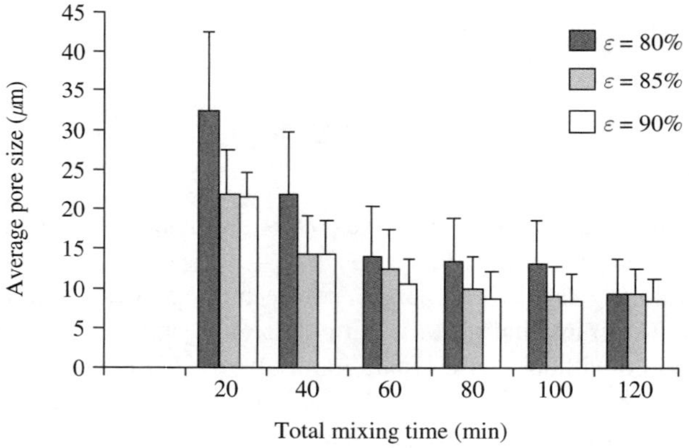

Figure 7.4 Variation of average pore size (D) with total mixing time (t) as a function of dispersed phase volume fraction (ε). Dosing time is 10 min, impeller speed $\Omega = 300$ rpm, emulsification temperature $T = 25°$ C. Pore size is evaluated from the scanning electron micrographs of the polymers and the raw data are corrected to compensate for the random space distribution of the pores

produced by the above technique does not go through a homogenization stage, the pore size distribution is wide in the resulting PHP. This can be overcome by operating at high temperatures where the resulting PHP has large pores as shown in Figure 7.5. Therefore, Figures 7.4 and 7.5 are useful in controlling the pore size especially when the required pore size is large. Several techniques are available to control the interconnect size (d) [10].

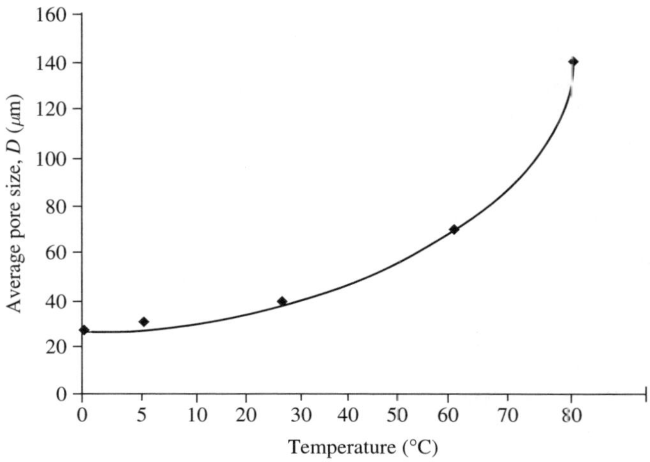

Figure 7.5 *Variation of average pore size with emulsification temperature when dosing time = 40 s, total mixing time = 100 s, impeller speed = 300 rpm, phase volume = 90%*

7.4.5 Effect of Additives/Fillers and Formation of Coalesce Pores

The inclusion of additives in the aqueous phase has several objectives. These additives can be used to control the pore and interconnect sizes but, most importantly, they can be used to chemically modify the polymer after polymerization. Surface as well as bulk modification of PHP cannot be achieved through post-polymerization impregnation especially when the pore/interconnect sizes are small and when the sample is thick. Since the additives are uniformly distributed in the aqueous phase droplets, at the post-polymerization modification stage, the additives are uniformly distributed within the pores. As the aqueous phase is the major emulsion phase, large quantities of desired substances can be incorporated within PHP at levels comparable to that of the polymer phase.

The inclusion of oil-soluble fillers in the continuous phase aims to produce co-polymers or to bulk and/or surface modify the polymer. When the additive does not take part in the polymerization reaction, it can be deemed as 'filler' which can then be leached out to provide nano-porosity to the walls of PHP. Such nano-pores are important in tissue engineering and BI since such nano-pores allow the transport of small molecules (nutrients/metabolites) to and from the microorganisms.

When the additives/fillers are incorporated in small quantities, stable emulsions are formed which do not separate or coalesce during polymerization. However, as the concentration of the additives/fillers are increased, HIPE becomes unstable during polymerization, which results in water droplet coalescence leading to larger pores and eventual phase separation if the competing polymerization and cross-linking reactions do not arrest emulsion separation in time. Therefore, the resulting PHP can have very large macroscopic pores approaching several millimeters. Figure 7.6 illustrates the types of pore structures encountered in PHP. The primary pores in Figure 7.6 (a) and (b) represent

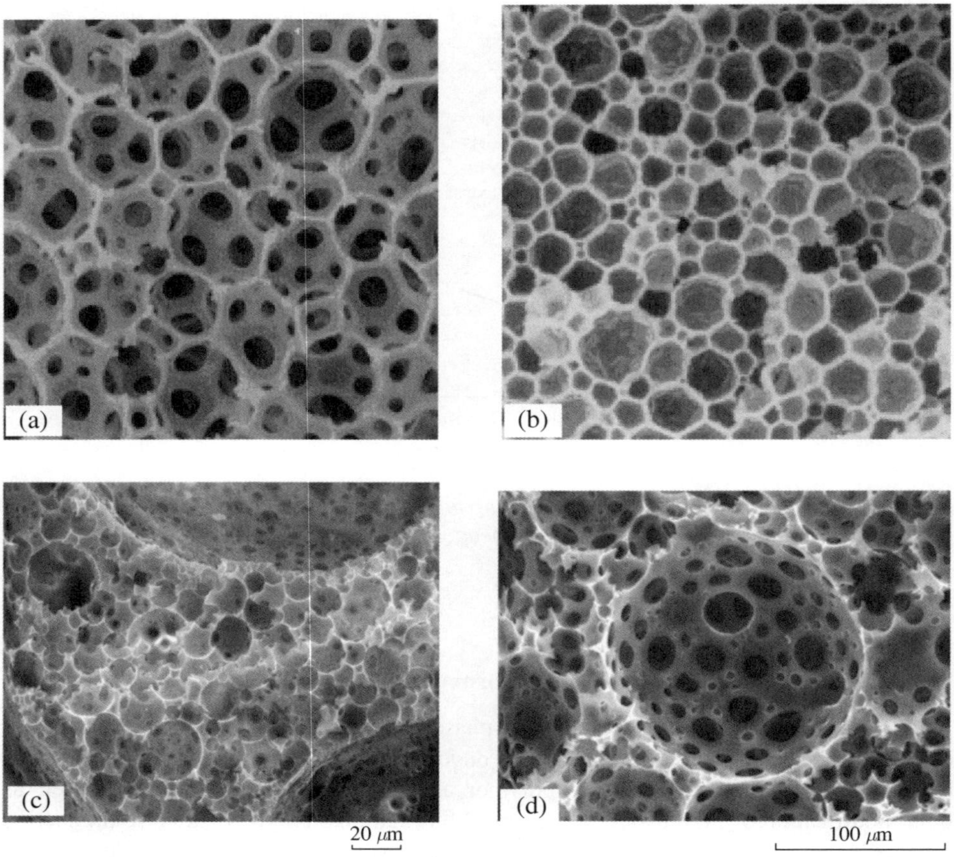

Figure 7.6 *Basic polyHIPE polymer structures: (a) primary pores with large interconnecting holes; (b) primary pores with nano-sized interconnecting holes; (c) large coalescence pores (three such pores are partially shown) dispersed into the primary pores in the process of coalescence; and (d) detail of the coalescence pores. Note that these pore structures can be prepared over a wide size range*

open and close pores, respectively. In the apparently closed-pore PHP, the interconnect size is very small to be identified at this magnification. By using fillers in the oil phase, it is possible to obtain open-pore PHP with nano-porous pore walls. Figure 7.6 (c) and (d) illustrates the coalescence pores at low and high magnification. The coalescence pores are dispersed into the primary pores which also show signs of coalescence in progress.

The effect of oil-phase additives on average pore size is illustrated in Table 7.1. Here the additives are water-soluble polymers with varying relative molecular mass. The size of the coalescence pores increases with increasing molecular mass as well as concentration. Such polymers are useful in modifying the surface of the hydrophobic polymer for tissue engineering.

Table 7.1 *The variation of coalescence pore size with water-soluble polymer concentration as a function of molecular weight. The phase volume is 85%, impeller speed of 300 rpm, and dosing and homogenization times of 600 s each, temperature of emulsification is 25° C*

Relative molecular mass	Water-soluble polymer concentration wt.% of aqueous phase	Pore size (μm)
Control	0	18
Sodium carboxylmethyl cellulose (CMC)		
90 000	0.5	22
90 000	1.0	260
250 000	1.0	1200
Polyethylene oxide (PEO)		
200 000	1.0	420
400 000	1.0	4300

As seen in Figure 7.6 (a–d), the interconnections in coalescence pores are different compared with those of the primary pores. In order to have primary pore structure in the presence of additives/fillers, the concentration of the additives must be low. The example below illustrates one such case. In this example, the aqueous phase contains 0.5 wt.% hydroxyapatite dissolved in 15% phosphoric acid solution. After emulsification and polymerization, PHP is soaked in 1 M NaOH to precipitate hydroxyapatite and subsequently washed in water to obtain pH = 7. These materials are then washed in isopropanol to remove residual surfactant, toxic monomer residues, and electrolytes. Polymer samples were finally dried in a vacuum oven and then sterilized in an autoclave before use as support in micro-bioreactors or tissue culture studies.

7.5 Bio-process Intensification Using PHP – Micro-Bioreactors

The application of PI to biotechnology is one of the major challenges of bioreaction engineering. Processes can be intensified either through the application of high/ultra-high processing field (physical intensification) or through the enhancement of selectivity (phenomenon-based PI). In both cases, the processing volume is reduced. In phenomenon-based intensification (also known as inherently intensive processes), selectivity is often achieved through the combination of intensified processing field as well as drastically reduced processing volume, including processing at microscopic scale [5,12]. PI in biotechnology has certain inherent restrictions in terms of 'intensification fields', or PI-driving forces [4,5,12], such as temperature, pressure, concentration of reactants/products, mechanical stresses and deformation rates, and electric field. The above-mentioned PI fields/driving forces are commonly used in chemical PI, often in combination. The chemical PI increases with increasing field strength and therefore the PI is only limited by the limits of the reactor engineering. However, in most cases, the above-referred intensification fields cannot be used in BI directly. Owing to these restrictions on the type of PI-driving forces, BI can therefore be achieved, in the first instance, through the reduction of diffusion path for the reactants and products, and through the creation of the most suitable environment for the biocatalysts and microorganisms which can enhance selectivity thus resulting in phenomenon-based intensification. It is likely that the optimization of the strength and type of the intensification field will be required in BI.

The most suitable driving force in BI is the reduction in diffusion path which already operates in transport processes across biological bilayers. Consequently, biocatalyst membranes and specially designed bioreactors, such as jet loop and membrane reactors, are available to intensify biochemical reactions [12,35–46]. Supported biocatalysts are often employed in order to enhance catalytic activity and stability, to protect enzymes/microorganisms from mechanical degradation and deactivation[12,37,38]. Immobilization of the cells and enzymes is one of the techniques to improve the productivity for bioreactors.

Immobilized cells are defined as cells that are entrapped within or associated with an insoluble matrix. Various methods are used for immobilization, including covalent coupling, adsorption, entrapment in a 3D scaffold, confinement in a liquid–liquid emulsion, and entrapment within a semi-permeable membrane. Bioreactors with immobilized cells have several advantages over bioreactors operating with free cells or immobilized enzymes. Immobilized cell systems permit the operation of bioreactors at flow rates that are independent of the growth rate of the microorganisms employed. Catalytic stability can be greater for immobilized cells than for free cells. Some immobilized microorganisms tolerate higher concentrations of toxic compounds compared with free cells, when the cell support media act as a temporary sink for the excess toxin. However, in the current biocatalyst support technology, the presence of the support itself introduces mass transfer restrictions for the substrate/product/nutrient diffusion to and from the biocatalyst. These disadvantages are also valid when supports are used to grow animal cells *in vitro*[10–16].

Furthermore, in most cases, the 3D architecture of the support and cell distribution is also important in cell viability and cell function. The cell support system developed by Akay *et al.*[10] is designed to address these issues. In this study, we use the technique of Akay and co-workers[8,10] to obtain micro-cellular polymers of controlled micro-architecture to immobilize bacteria employing a novel seeding technique. Operated as a flow-through system in the degradation of phenol in water, this micro-bioreactor is shown to represent a clear BI.

We have chosen phenol for which the metabolic pathways are well described [46–49] as a model substrate to conduct the continuous micro-bioreactor experiments. There are various degradation studies using phenol and other aromatic compounds by free or immobilized microorganisms[46,49–53]. Phenol has a highly toxic effect on all bacteria and there are various data reported on the inhibition effect of phenol for the growth of *Pseudomonas* sp. ranging between 50 and 350 mg/L[54]. Furthermore, the degradation of phenol is aerobic and therefore it presents further challenge to the intensification of bioconversion. Recently we have also studied the degradation of phenol using immobilized enzymes[41,42] and immobilized bacteria[12,46] using a micro-porous cell support system originally developed for animal cell support for tissue engineering applications [5,10,11,15]. Phenol degradation studies using immobilized *Pseudomonas* sp.[46] also provide us with a direct base line to compare the BI achieved in the present study.

BI was demonstrated by using a well-known bioconversion, namely the degradation of phenol using *Pseudomonas syringae*. In the control experiments bacteria was immobilized on PHP beads and degradation was carried out in a packed bed[46]. It was found that after a prolonged operation, biofilm formation on the surface of the support PHP particles prevented bacterial penetration into the pores of the support. In the case of BI studies, a small PHP disk housed in a sealed reactor was used as the monolithic support. The experimental set-up is shown in Figure 7.7 (a), while the detail of the micro-reactor is illustrated in Figure 7.7 (b). Initially, bacteria was force-seeded within the pores and subsequently allowed to proliferate followed by acclimatization and phenol degradation at various initial substrate concentrations and flow rates. Two types of micro-porous polymer were used as the monolithic support. These polymers differ with respect to their pore and interconnect sizes, macroscopic surface area for bacterial support, and phase volume. The polymer with a nominal pore size of 100 μm and with phase volume of 90% (with highly open pore structure) yielded reduced bacterial proliferation, while the polymer with a nominal pore size of 25 μm and with phase volume of 85% (with small interconnect

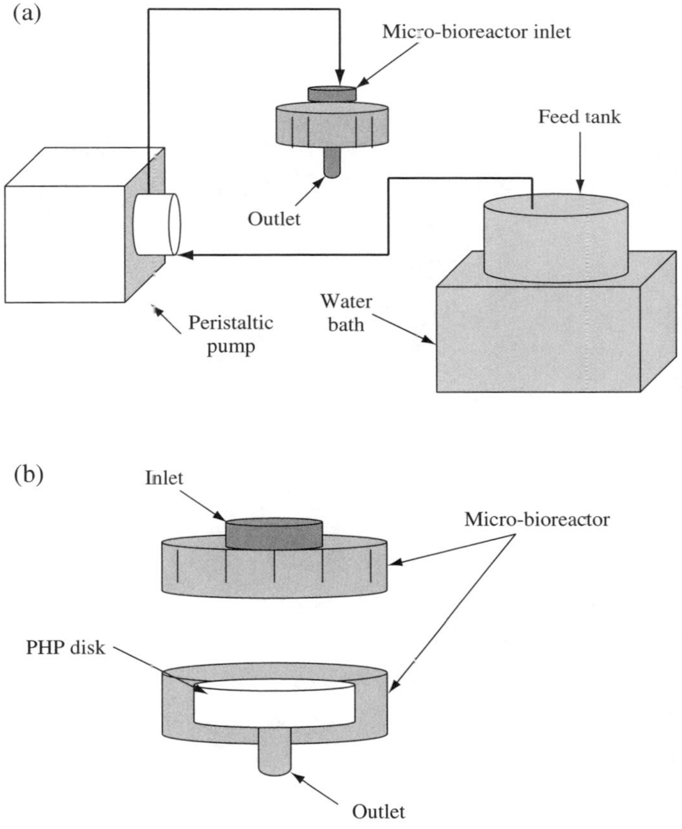

Figure 7.7 *Micro-bioreactor: (a) flow diagram, and (b) housing detail for the PHP support for bacteria or animal cells*

size and large pore area for bacterial adhesion) yielded monolayer bacterial proliferation. Bacteria within the 25-μm polymer support remained monolayered without any apparent production of extracellular matrix during the 30-day continuous experimental period. The monolayer bacterial growth and the lack of extracellular matrix are shown in Figure 7.8. The micro-bioreactor performance was characterized in terms of volumetric utilization rate and compared with the published data, including the case in which the same bacteria was immobilized on the surface of micro-porous polymer beads and used in a packed bed during continuous degradation of phenol. The results are summarized in Table 7.2. It is shown that at similar initial substrate concentration, the volumetric utilization in the micro-reactor is at least 20-fold more efficient than the packed bed depending on the flow rate of the substrate solution. The concentration of the bacteria within the pores of the micro-reactor decreases from 2.25 cells per μm^2 on the top surface to about 0.4 cells per μm^2 within 3-mm reactor depth. The variation of the bacteria concentration with distance is shown in Figure 7.9. If the bacteria-depleted part of the micro-reactor is disregarded, the volumetric utilization increases by a factor of 30-fold compared with the packed bed. This efficiency increase is attributed to the reduction of diffusion path for the substrate and nutrients and enhanced availability of the bacteria for bioconversion in the absence of biofilm formation as well as the presence of flow over the surface of the monolayer bacteria.

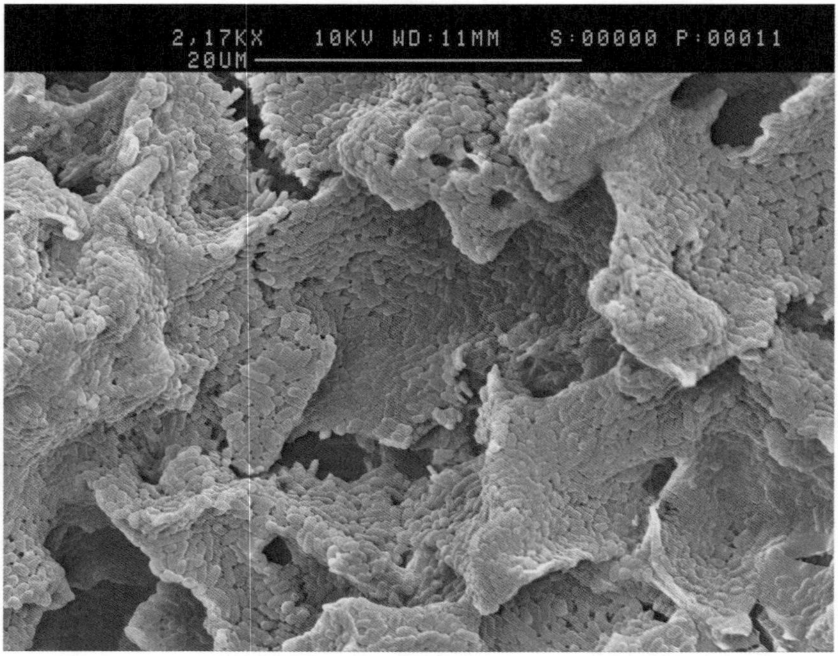

Figure 7.8 *SEM micrograph of the micro-bioreactor at the end of the 30-day continuous cultivation: cross-section at a distance of 2000 μm from the inlet surface showing monolayer bacterial coverage and absence of any extracellular matrix*

Table 7.2 *Comparison of micro-bioreactor performance and its comparison with other systems*

Reference	Q (L/h) Flow rate	C_o (mg/L) Initial substrate concentration	U(g/m^2 min) Utilization rate	R (g/L h) Volumetric utilization rate	Reactor type	F(L/m^2 min) Flux rate	A (cm^2) Reactor cross-sectional area	H (cm) Reactor height
[52]	Batch (benzene)	135	–	0.01	Fibrous bed	Batch	11	45
[52]	Batch (toluene)	201		0.02	Fibrous bed	Batch	11	45
[42]	–	500	0.78	–	Membrane immobilized enzyme	2.76 (cross-flow)	176	0.3
[46]	0.27	720	8.43	1.33	PHP-immobilized bacteria	15	3.01	38
[46]	0.6	200	5.54	1.28	PHP-immobilized bacteria	33	3.01	38
[46]	1.1	200	1.03	0.24	PHP-immobilized bacteria	60	3.01	38
[41]	–	467	0.092	–	Membrane-immobilized enzyme	0.56 (dead-end flux)	45.3	–
[53]	13.32	39	0.93	0.24	Three-phase fluidized bed reactor	23.80	93.26	30
[12]	0.12	650	1.73	31.50	Micro-bioreactor	5.78	3.46	0.5
[12]	0.12	380	2.19	39.9	Micro-bioreactor	5.78	3.46	0.5
[12]	0.18	380	2.16	39.45	Micro-bioreactor	8.67	3.46	0.5

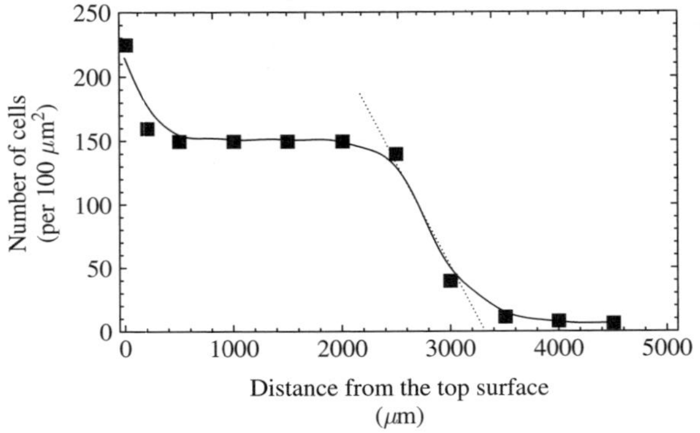

Figure 7.9 *Variation of bacteria coverage density (number of cells per 100 square μm) as a function of distance from the micro-reactor inlet. The intercept of the dashed line indicates the optimum thickness of the micro-reactor monolithic support*

7.6 PHP Scaffolds for Tissue Engineering

Tissue engineering has the potential to repair/replace damaged/diseased tissue [10,11,13–15,55]. In creating a replacement tissue such as bone or cartilage, it is possible to grow the tissue *in vitro* initially and transplant it afterwards. In this approach, a porous scaffold that seeds the cells and serves as a template for tissue regeneration is necessary. The structure of the biomaterial should meet the criteria of tissue engineering. It is important to be able to control the scaffold's spatial arrangement, which in turn contributes substantially to its mechanical characteristic limits. The architectural design of the scaffold is also important to determine the quantity and shape of the substratum surfaces available for colonization by *in vitro* cell seeding. It is also important that the cells can actively migrate into the scaffold, and that cell attachment occurs by the production of extracellular matrix proteins. The cells should also retain their phenotype within the scaffold.

The tissue engineering approach to bone regeneration is based on the hypothesis that healthy progenitor cells, either recruited or delivered to an injured site and then under tight regulation, can ultimately regenerate lost or damaged tissue. Recently many cell–polymer constructs are being developed and evaluated for tissue engineering. Polymers have the advantage of being biocompatible, non-toxic, and can be made biodegradable and can also contain micro-channels for the transport of the nutrients and metabolites as well as for growing nerves for neural excitation [10].

7.7 Cartilage Tissue Engineering

Articular cartilage is composed of cells, chondrocytes, embedded in a resilient matrix of collagen, proteoglycan, and water. There are three main groups of cartilage: hyaline, elastic, and fibrocartilage. Elastic cartilage is found at sites such as the external ear and epiglottis. Fibrocartilage has a matrix with type I collagen fibers and is located at the pubic symphysis and many tendon insertions into bone. Hyaline cartilage contains type II collagen fibers in its matrix and is found in the nasal and respiratory tract and at most synovial joints (articular cartilage). It is articular or hyaline cartilage which we consider here. An important feature of hyaline is the production of collagen II as opposed to collagen I. The chondrocytes that were used in this study were primary cells obtained from a bovine metacarpalphalangeal joint.

Chondrocytes were seeded onto the two types of polymer. The first polymer was a plain styrene/2EHA PHP which had not been chemically modified. The second PHP (HA-PHP) had been chemically modified to include a coating of hydroxyapatite. Hydroxyapatite is a mineral found in large quantities in bone and other tissues. Details of the technique are found in Ref.[56]. Primary cells were seeded onto the PHP support and they were cultured for various time spans. These polymer/cell constructs were then examined through histology or SEM to determine the penetration of the cells and their morphology as a function of time. The penetration of the chondrocytes into the polymer was determined by histology. The production of glycosaminoglycans (GAG) which are found in the extracellular matrix was determined by a colorimetric assay and the type of collagen produced in the matrix was determined with immunocytochemistry. This matrix (GAG) gives cartilage its load-bearing capacity and is crucial to the development of chondrocytes into cartilage tissue. As seen from the SEM micrograph in Figure 7.10, rounded morphology of the chondrocytes are preserved within the pores of the hydroxyapatite-coated PHP. The DNA concentration obtained from the chondrocytes/polymer constructs is used to quantify the cell growth rate. It was found that the cell growth rate could be threefold higher in hydroxyapatite-coated PHP supports compared with non-coated PHP after 10 days in culture.

In order to test the effect of pore size on the cell penetration and collogen II production rates, five types of styrene/2EHA PHP were produced with average pore sizes (D) of 8, 17, 24, 31, 45, and 89 μm. The corresponding average interconnect sizes (d) in these

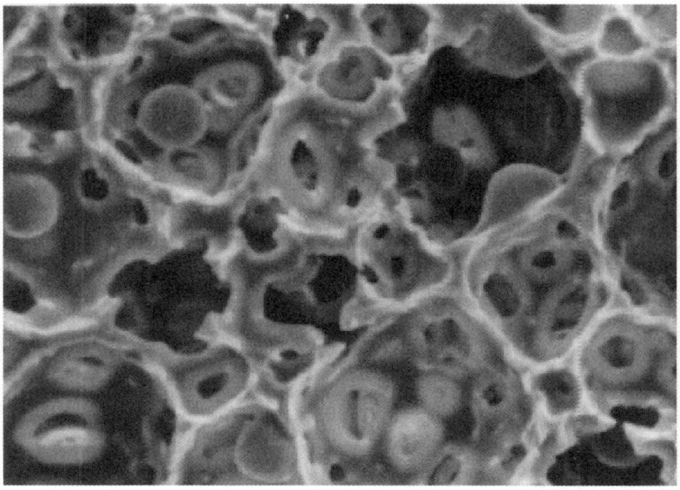

Figure 7.10 SEM micrograph showing the rounded morphology of chondrocytes on the surface pores of HA-PHP

polymers were 2.5, 5, 6, 6, 7, and 7 μm, respectively, so as to eliminate the effect of interconnect size in the experiments. These samples were then seeded with chondrocytes and the effects of pore size on cell penetration and GAG production rate were determined after 21 days in culture. The results are shown in Figures 7.11 and 7.12. As seen in Figure 7.11, maximum cell penetration occurs when $D = 24\,\mu$m and $d = 6\,\mu$m, and contrary to the expectations, larger pore and interconnect size PHPs do not facilitate cell penetration. Similarly, GAG production is maximum when the 24-μm polymer is

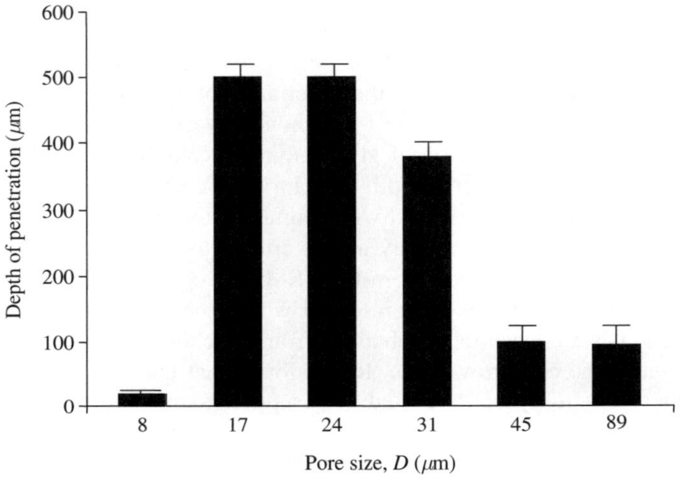

Figure 7.11 *Effect of pore size on the depth of penetration of chondrocytes, measured from histological sections of chondrocytes on various pore-sized hydroxylapatite-coated 2EHA/styrene (copolymer) polyHIPE polymer after 21 days in culture. Each bar represents the mean ± SE for eight samples with six sections being taken from each sample*

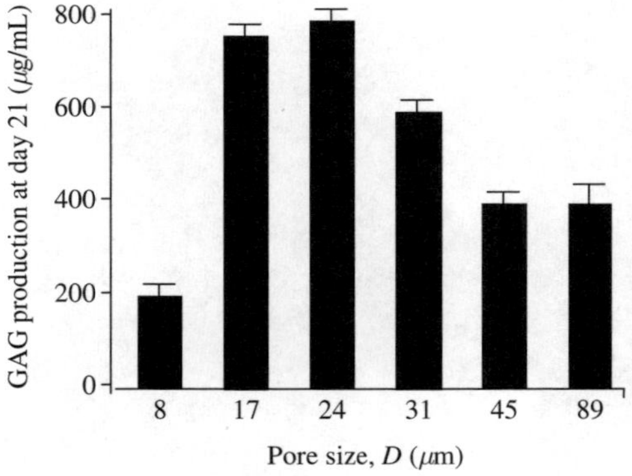

Figure 7.12 *Effect of pore size on the production of total glycosaminoglycan (GAG) protein (cell associated + released into the medium) by chondrocytes cultured on various pore-sized hydroxylapatite-coated 2EHA/styrene (copolymer) polyHIPE polymer after 21 days in culture*

used. These experiments were also conducted using tissue culture plastic (TCP) which is non-porous but coated with bioactive proteins to enhance cell growth. TCP-grown chondrocytes yielded GAG production levels similar to those observed for the 89-μm polymer. However, at longer time spans, chondrocytes on TCP become flattened. These results indicate that there is an optimum pore size to maintain the chondrocyte morphology and to optimize the production of collagen II. At lower pore sizes, the cells cannot penetrate the polymer and at higher pore sizes the cells' morphology changes from rounded to flat and fibroblastic in appearance. These fibroblastic cells proliferate rapidly and form a layer on the surface rather than penetrating the polymer.

7.8 Bone Tissue Engineering

The experimental technique in the study of bone tissue engineering is similar to that of the cartilage tissue engineering described earlier. However, in order to accelerate the cell penetration into the PHP, we employed a forced seeding technique which was also employed in the seeding of the bacteria into the micro-bioreactor as shown in Figure 7.7 (b). In these studies, we used styrene only in the support system in order to enhance and match the mechanical properties of the support with that of the bone. Pore volume fraction was 95%. Three different pore sizes were used: 40, 60, and 100 μm. However, in this case the corresponding interconnect sizes were 15, 20, and 30 μm, respectively. Once again, some of the samples were coated with hydroxapatite with an estimated coating thickness ranging from 25 nm for the 40-μm pore material to 40 nm for the 100-μm pore size PHP [11,15,55,56].

Statically seeded disks were cultured for periods between 7 and 35 days on PHP and subsequently prepared for analysis. SEM was used to examine cell morphology on the polymer surface and inside of the polymers. SEM analysis (Figure 7.13) shows surface and transverse sections seeded with primary rat osteoblasts and demonstrates that the cells reached confluence on the surface after 14 days (Figure 7.13 (a)) and formed a continuous, thick layer that at later time points became multilayered sheets, with fibrous matrix present (Figure 7.13 (b)). Osteoblasts were also observed within the polymers (Figure 7.13 (c)), but these were fewer in number and were present as either individual cells or in isolated colonies.

The effect of including hydroxyapatite during the preparation of PHP was further investigated by comparing its performance with the unmodified polymers. In these experiments rat osteoblasts were seeded onto the surface of 40-, 60-, and 100-μm pore size polymers that were either unmodified or modified with hydroxyapatite and cell growth and migration assessed by histological analysis.

The amount of bone nodule formation was quantified from the surfaces of unmodified and modified polymers at 28 and 35 days in culture under osteogenic conditions, for each of the three different pore-sized polymers. Histomorphometric analysis shows that the surfaces of modified polymers had significantly increased areas of Von Kossa staining (Figure 7.14) compared to unmodified polymers. Increased amounts of nodule formation can be observed for each of the different pore sizes between days 28 and 35 but this is only significant for 100-μm pore size. There was no evidence that nodule area was significantly influenced by the three pore sizes evaluated in this study.

Image analysis reveals that when polymers are modified with hydroxyapatite there is a significant increase in the penetration of cells into the polymer compared to unmodified

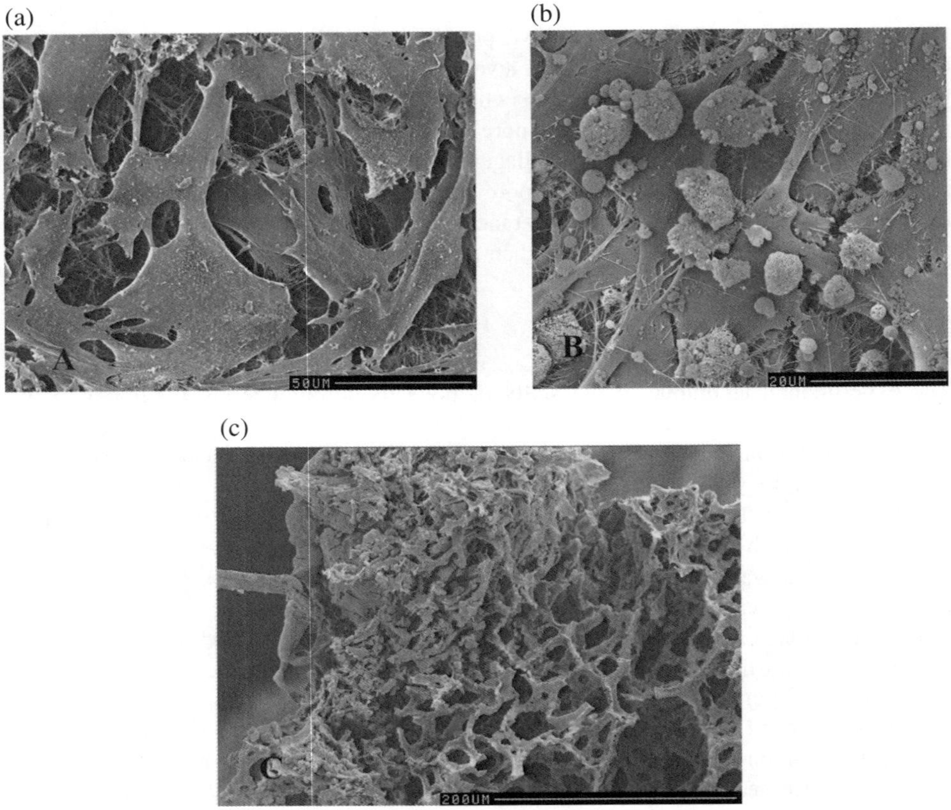

Figure 7.13 *Scanning electron micrographs of primary rat osteoblasts cultured in hydroxylapatite-coated PHP. (a) Surface appearance after 14 days; (b) surface appearance after 35 days; and (c) transverse section after 35 days illustrating the penetration of bone formation*

controls that were independent of pore size. Relatively few cells penetrated deeper than 1 mm in any of the PHPs tested, regardless of pore size. Between the different pore sizes evaluated, significantly more cells could be found beneath the surface of 40-μm modified PHP compared to both 60- and 100-μm modified PHPs.

There was little overall effect of pore size on the maximal depth of cell penetration into the polymer at 35 days (approximately 1.4 mm). Cell movement into all the PHPs tested progressed with time, but the rate was notably quicker with the 100-μm pore size polymer in comparison to the 40- and 60-μm pore size polymers.

The effect of pore size on cell concentration in the polyHIPE support at different times was determined from DNA analysis of the cell/polymer constructs. The results are shown in Figure 7.15 which indicates that DNA content per construct increases with increasing pore size. Since the pore size span in the experiments is not large (2.5-fold) the differences between the DNA contents are not as marked as the case for chondrocytes where the pore size span was 10-fold. Nevertheless, the present results also indicate that for a given cell type, an optimum pore size is present for cell proliferation as indicated by DNA content in

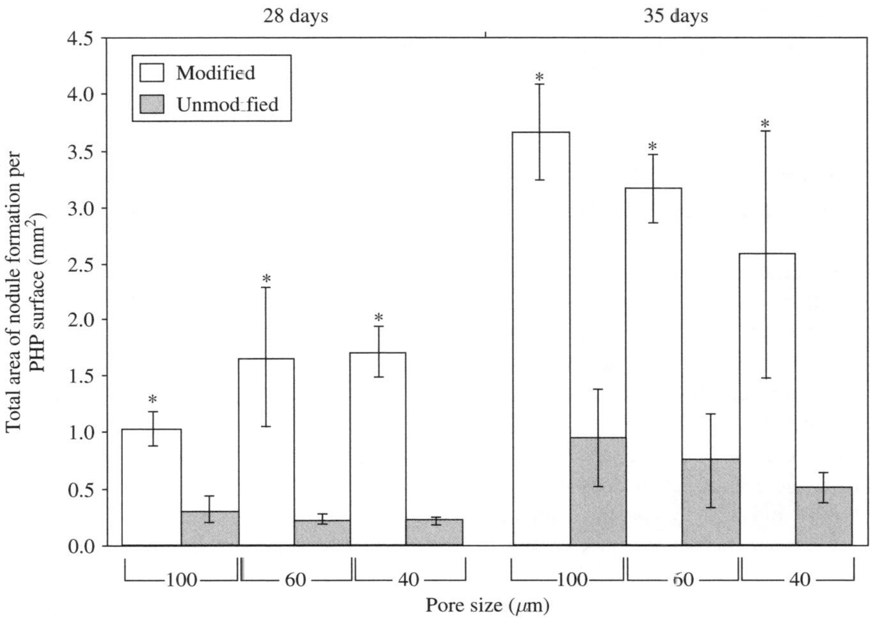

Figure 7.14 Effect of support pore size (D = 100, 60, 40 μm) on mineralized nodule formation of rat osteoblast cells cultured in vitro after 28 or 35 days using hydroxylapatite-coated (modified) or uncoated (unmodified) styrene PolyHIPE Polymer cell supports

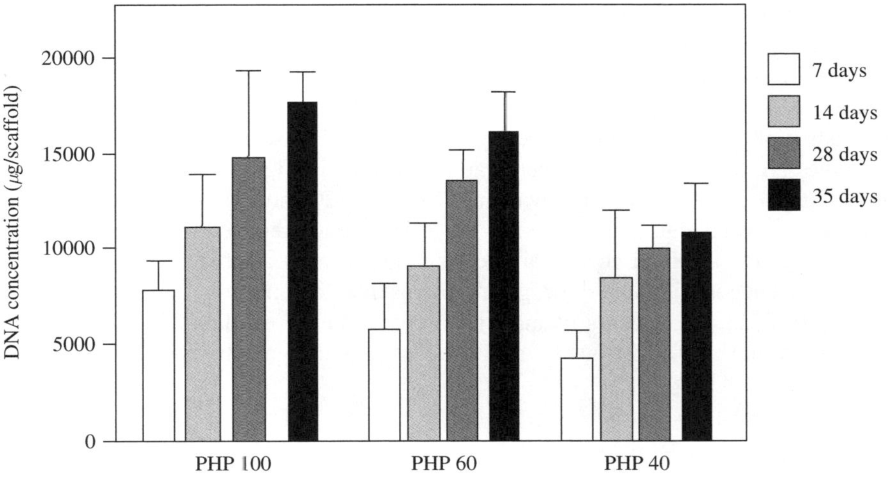

Figure 7.15 Effect of pore size on DNA concentration as a function of time in culture of rat osteoblast cells in hydroxyapatite-coated polyHIPE polymer support, illustrating the dependence of cell proliferation rate on support pore size

the supports. The results also indicate that the coating of the polymers with hydroxyapatite increases cell penetration, and proliferation and further enhancement is possible when the hydroxyapatite-coated PHPs are coated with self-assembling peptide hydrogel[15,57].

7.9 Nano-Structured Micro-Porous Metals for Intensified Catalysis

Surface area and its accessibility are important both in catalysis and gas cleanup. Nano-structured micro-porous catalysts or catalyst supports offer intensified catalysis since they provide an enhanced surface area which is accessible to the reactants and products through a network of arterial channels feeding into the regions of catalytic activity. In non-structured catalysts, although the surface area might be large, as determined by gas adsorption, they are often not accessible as a result of surface fouling and the diffusion resistance can slow down the rates of reactions. Catalysts are either deposited as a thin film on a support or they are used as pellets. These two techniques have certain drawbacks: in coated systems, catalyst adhesion can be non-uniform and weak while the accessibility of the active sites within the interior of the catalyst is hindered due to low porosity.

In this study we summarize the recent developments in catalyst development in which nano-porous catalytic sites are accessible through a network of arterial micro-pores. These catalysts are obtained through a solution deposition of metals on a micro-porous polymeric template which is subsequently heat-treated to obtain porous metallic structures where the size of the pores ranged from tens of micrometers to tens of nanometers thus eliminating the problems of accessibility and rapid pore fouling and closure. The technique differs fundamentally from the compression-based systems where the porosity is reduced as a result of compaction. It also differs from the well-known wash-coating or chemical vapor deposition techniques. Furthermore, the mechanisms of metal deposition within micro-pores and nano-structure formation are novel. The importance and current fabrication techniques of porous metallic systems can be found in Refs.[5,16].

Figure 7.16 illustrates the micro-structure of these novel materials. The overall skeleton (Figure 7.16a) is formed by fused metallic (nickel alloy) grains of ca. 8 μm forming micro-pores of ca. 30-μm arterial passages (channels). As seen in Figure 7.16 (a), grains themselves are porous which can be clearly identified in Figure 7.16 (b) from where the size of the surface pores can be evaluated to be ca. 200 nm. However, further detailed examination reveals that these materials have finer structures well below 100 nm as seen in Figure 7.16 (c). All of these parameters – grain, arterial channel – and grain pore sizes – can be controlled. As the grains themselves are porous, they provide a large available surface area. Details of material preparation are available[16]. These materials are based on nickel which is useful as a catalyst in many gas phase reactions.

7.10 Concluding Remarks

This paper summarizes the recent developments in micro-fabrication and its applications in PI and tissue engineering which is complementary to phenomenon-based BI. In such intensified processes, miniaturization is essential and therefore micro-reactors

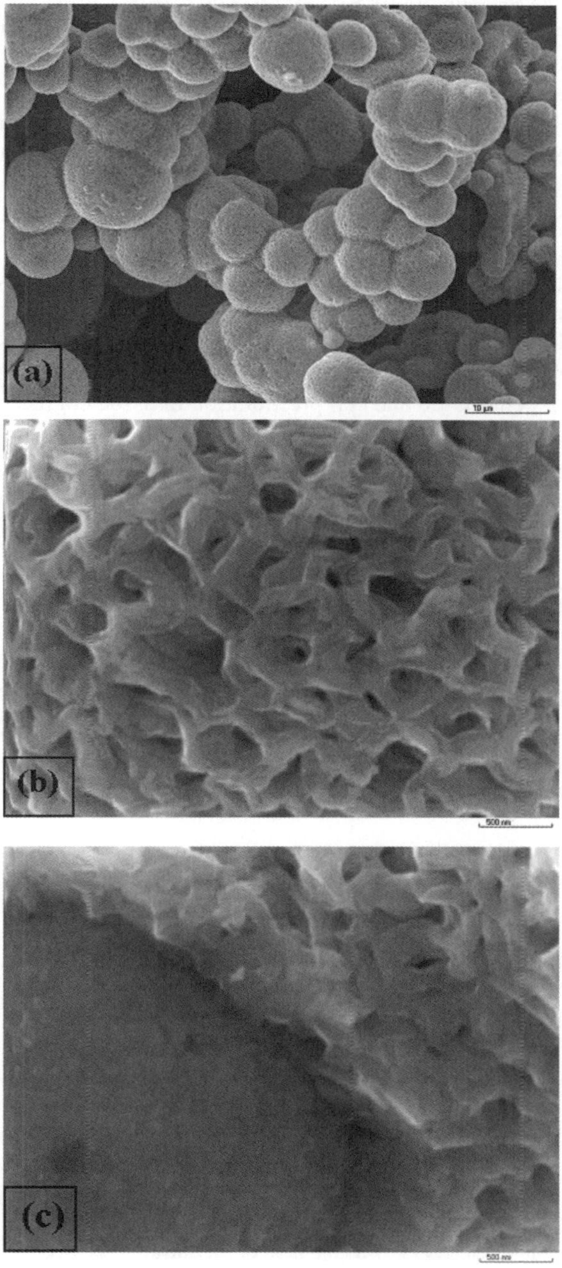

Figure 7.16 *Micro-structure of nickel-based intensified catalyst/catalyst support showing the hierarchy of the pore sizes. (a) arterial micro-pores; (b) nano-structure of the surface pores of the fused metallic grains; and (c) nano-structure of the inside of the fused metal grains showing even smaller pore structure than the surface pores*

with stagewise processes in micro-scale represent a novel processing approach. Such micro-reactors/unit-operations are already available in nature, in the form of organs where micro-channels and neural network between the collection of cells provide the necessary facilities for the organ function. In these systems, diffusion takes place across nano-scale lipid bilayers. Such miniaturized systems can be achieved *in vitro* by using micro-porous materials with a network of capillaries having nano-porous walls separating different domains. Monolithic PHPs with a network of capillaries have been proposed[10] to mimic the micro-architecture of organs. Metallic versions of such micro-reactors are also described in this study.

We have also provided evidence that the behavior of microorganisms in confined micro-environment is substantially different and that their desired metabolic activities can be maximized through the modification of the surface characteristics as well as the size of the pores. These characteristics can therefore be utilized in BI as well as in the enhancement of cell penetration and cell proliferation in tissue engineering and when such polymers are grafted. Micro-fabrication technique has also been used in the development of highly porous catalysts with arterial channels feeding nano-pores which provide an extended surface area. Such materials can be used as micro-reactors as well as catalysts.

Acknowledgements

We are grateful to the UK Engineering and Physical Sciences Research Council (EPSRC), UK Department of Trade and Industry, Avecia/Cytec, BLC Research, BP Amoco, Exxon, Intensified Technologies Incorporated (ITI), Morecroft Engineers Ltd., Safety-Kleen Europe, Triton Chemical Systems, and Willacy Oil Services Ltd. for their support. We also thank Burak Calkan, Omer Calkan, and Zainora Noor for their help in the experiments.

References

[1] Akay G. 1986. Rheological properties of thermoplastics. In, Encyclopedia of Fluid Mechanics, Cheremisinoff N.P.(Ed.), Vol. 1, Chapter 35, Gulf Publishing, Houston, USA pp. 1155–1204.

[2] Agarwal U.S., Dutta A. and Mashelkar R.A. 1994. Migration of macromolecules under flow: The physical origin and engineering implications, Chem. Eng. Sci., 49, 1693–1717.

[3] Akay G. 1998. Flow-induced phase inversion in the intensive processing of concentrated emulsions, Chem. Eng. Sci., 53, 203–223.

[4] Akay G., Mackley M.R., Ramshaw C. 1997. Process intensification: Opportunities for process and product innovation. In, IChemE Research Event, Chameleon Press, London, pp. 597–606.

[5] Akay G. 2005. Bioprocess and Chemical Process Intensification. In, Encyclopaedia of Chemical Processing, Lee S. (Ed.), Marcel Dekker, NY.

[6] Akay G., Wakeman R.J. 1994. Mechanism of permeate flux decay, solute rejection and concentration polarisation in crossflow filtration of a double chain ionic surfactant dispersion, J. Membr. Sci., 88, 177–195.

[7] Akay G., Odirile P.T., Keskinler B., Wakeman R.J. 2000. Crossflow microfiltration characteristics of surfactants: The effects of membrane physical chemistry and surfactant phase behavior on gel polarization and rejection. In, Surfactant Based Separations: Science and Technology, Scamehorn J.F. and Harwell J.H. (Eds.), ACS Symposium Series, 740, 175–200.

[8] Akay G., Vickers J. 2003. Method for separating oil in water emulsions, European Patent, EP 1 307 402.
[9] Akay G., Noor Z.Z., Dogru M. 2005. Process intensification in water-in-crude oil emulsion separation by simultaneous application of electric field and novel demulsifier adsorbers. In, Microreactor Technology and Process Intensification, Wang Y. and Halladay J. (Eds.), Chapter 23, ACS Symposium Series, Oxford University Press, Oxford.
[10] Akay G., Dawnes S., Price V.J. 2002. Microcellular polymers as cell growth media and novel polymers, European Patent, EP 1 183 328.
[11] Akay G., Birch M.A., Bokhari M.A. 2004. Microcellular Polyhipe polymer (PHP) supports osteoblastic growth and bone formation *in vitro*, Biomaterials, 25, 3991–4000.
[12] Akay G., Erhan E. and Keskinler B. 2005. Bioprocess intensification in flow through micro-reactors with immobilized bacteria, Biotechnol. Bioeng., 90, 180–190 (in press). Published on line, 1 March 2005.
[13] Yang S., Leong K-F., Du Z. and Chua C-K. 2001. The design of scaffolds in tissue engineering. Part 1. Traditional factors, Tissue Eng., 7, 679–689.
[14] Zhang S. 2003. Fabrication of novel biomaterials through molecular self assembly, Nat. Biotechnol., 21, 1171–1178.
[15] Bokhari M.A., Akay G., Birch M.A., Zhang S. 2005. The enhancement of osteoblast growth and differentiation *in vitro* on a peptide hydrogel – PolyHIPE Polymer hybrid support material. Biomaterials, 26, 5198–5208 (in press).
[16] Akay G., Dogru M., Calkan B. and Calkan O.F. 2005. Flow induced phase inversion phenomenon in process intensification and micro-reactor technology. In, Microreactor Technology and Process Intensification, Wang Y. and Halladay J.(Eds.), Chapter 18, ACS Symposium Series, Oxford University Press, Oxford.
[17] Stankiewicz A., Moulijn J.A. 2002. Process intensification, Ind. Eng. Chem. Res., 41, 1920–1924.
[18] Stankiewicz A., Drinkenburg A.A.H. 2004. Process intensification: History, philosophy, principles. In, Chemical Industries, Marcel Dekker, NY pp. 1–32.
[19] Akay G. 1991. Agglomerated abrasive material compositions comprising same, and process for its manufacture, US Patent, US 4 988 369.
[20] Akay G. 1995. Flow-induced phase inversion in powder structuring by polymers. In, Polymer Powder Technology, Narkis M. and Rosenzweig N.(Eds.), Chapter 20, Wiley pp. 542–587.
[21] Akay G. 2001. Stable oil in water emulsions and a process for preparing same, European Patent, EP 649 867.
[22] Akay G., Tong L., Addleman R. 2002. Process intensification in particle technology: Intensive granulation of powders by thermo-mechanically induced melt fracture, Ind. Eng. Chem. Res., 41, 5436–5446.
[23] Akay G., Tong L. 2003. Process intensification in particle technology: Intensive agglomeration and microencapsulation of powders by non-isothermal flow induced phase inversion process, Int. J. Transp. Phenomena, 5, 227–245.
[24] Akay G., Tong L. 2003. Process intensification in polymer particle technology: Granulation mechanism and granule characteristics J. Mater. Sci., 38, 3169–3181.
[25] Akay G. 2004. Upping the ante in the process stakes, Chem. Eng., 752, 37–39.
[26] Akay G. 2004. International Patent Application, Method and apparatus for processing flowable materials and microporous polymers, WO 2004/004880.
[27] Akay G., Tong L. 2001. Preparation of low-density polyethylene latexes by flow-induced phase inversion emulsification of polymer melt in water, J. Colloid Interface Sci., 239, 342–357.
[28] Akay G., Tong L., Hounslow M.J., Burbidge A. 2001. Intensive agglomeration and microencapsulation of powders, Colloid Polym. Sci., 279, 1118–1125.
[29] Akay G., Irving G.N., Kowalski A.J., Machin D. 2001. Process for the production of liquid compositions, European Patent, EP 799303.

[30] Tong L., Akay G. 2002. Process intensification in particle technology: Flow induced phase inversion in the intensive emulsification of polymer melts in water, J. Mater. Sci., 37, 4985–4992.

[31] Akay G., Tong L., Bakr H., Choudhery R.A., Murray K., Watkins J. 2002. Preparation of ethylene vinyl acetate copolymer latex by flow induced phase inversion emulsification, J. Mater. Sci., 37, 4811–4818.

[32] Akay G. 2004. International Patent Application, Microporous polymers, WO 2004/005355.

[33] Akay G. 2000. Flow induced phase inversion. In, Recent Advances in Transport Phenomena, Dincer I. and Yardim F. (Eds.), Elsevier, Paris pp. 11–17.

[34] Akay G., Irving G.N., Kowalski A.J., Machin D. 2002. Dynamic mixing apparatus for the production of liquid compositions, US Patent, US 63445907.

[35] Chisti Y., Moo-Young M. 1996. Bioprocess intensification through bioreactor engineering, Trans. IChemE, 74, 575–581.

[36] Giorno L., Drioli E. 2000. Biocatalyst membrane reactors: Applications and perspectives, TIBTECH, 18, 339–349.

[37] De Bartolo L., Morelli A., Bader D., Drioli E. 2001. The influence of polymeric membrane surface free energy on cell metabolic functions, J. Mater. Sci., Materials in Medicine, 12, 959–963.

[38] Bayhan Y.K., Keskinler B., Cakici A., Levent M., Akay G. 2001. Removal of divalent heavy metal mixtures from water by *Saccharomyces cerevisiae* using crossflow microfiltration, Water Res., 35, 2191–2200.

[39] Nuhoglu A., Pekdemir T., Yildiz E., Keskinler B., Akay G. 2002. Drinking water denitrification by a membrane bioreactor, Water Res., 36, 1155–1166.

[40] Erhan E., Keskinler B., Akay G., Algur O.F. 2002. Removal of phenol from wastewater by using membrane immobilized enzymes: Part 1. Dead end filtration, J. Memb. Sci., 206, 361–373.

[41] Akay G., Erhan E., Keskinler B., Algur O.F. 2002. Removal of phenol from wastewater by using membrane-immobilized enzymes: Part 2. Crossflow filtration, J. Memb. Sci., 206, 61–68.

[42] Pekdemir T., Keskinler B., Yildiz E., Akay G. 2003. Process intensification in wastewater treatment: ferrous iron removal by a sustainable membrane bioreactor system, J. Chem. Technol. Biotechnol., 78, 773–780.

[43] Yildiz E., Keskinler B., Pekdemir T., Akay G., Nihoglu A. 2005. High strength wastewater treatment in a jet loop membrane bioreactor: Kinetics and performance evaluation, Chem. Eng. Sci., 60, 1103–1116.

[44] Keskinler B., Yildiz E., Erhan E., Dogru M., Akay G. 2004. Crossflow microfiltration of low concentration- non-living yeast suspensions, J. Memb. Sci., 233, 59–69.

[45] Keskinler B., Akay G., Pekdemir T., Yildiz E., Nuhoglu A. 2004. Process intensification in wastewater treatment: Oxygen transfer characteristics of a jet loop reactor for aerobic biological wastewater treatment. Int. J. Environ. Technol. Manag., 4, 220–235.

[46] Erhan E., Yer E., Akay G., Keskinler B., Keskinler D. 2004. Phenol degradation in a fixed bed bioreactor using micro-cellular polymer-immobilized *Pseudomonas syringae*, J. Chem. Technol. Biotechnol., 79, 196–206.

[47] Dagley S. 1971. Catabolism of aromatic compounds by microorganisms. Adv. Microbial. Physiol., 6, 1–46.

[48] Buswell J.A. 1975. Metabolism of phenol and cresols by *Bacillus stearothermophilus* J. Bacteriol., 124, 1077–1083.

[49] Kline J., Schara P. 1981. Entrapment of living microbial cells in covalent polymeric networks. II. A quantitative study on the kinetics of oxidative phenol degradation by entrapped andida tropicalis cells, Appl. Biochem. Biotechnol., 6, 91–117.

[50] Bettmann H., Rehm H.J. 1984. Degradation of phenol by polymer entrapped microorganisms, Appl. Microbiol. Biotechnol., 20, 285–290.

[51] Schröder M., Muller C., Posten C., Deckwer W.D., Hecht V. 1996. Inhibition kinetics of phenol degradation from unstable steady-state data, Biotechnol. Bioeng., 54, 567–576.

[52] Shim H., Yang S.T. 1999. Biodegradation of benzene, toluene. Ethylbenzene, and o-xylene by a coculture of Pseudomonas putida and *Pseudomonas fluorescens* immobilized in a fibrous-bed bioreactor, J. Biotechnol., 67, 99–112

[53] Hecht V., Langer O., Deckwer W.D. 2000. Degradation of phenol and benzoic acid in a three-phase fluidized bed-reactor, Biotechnol. Bioeng., 70, 391–399.

[54] Hill G.A., Robinson C.W. 1975. Substrate inhibition kinetics: Phenol degradation by *Pseudomonas putida*, Biotechnol. Bioeng., 17, 1599–1615.

[55] Bokhari M., Birch M., Akay G. 2003. Polyhipe polymer: A novel scaffold for in vitro bone tissue engineering, Adv. Exp. Med. Biol. (Tissue Engineering, Stem Cells, and Gene Therapies), 534, 247–254.

[56] Byron V.J. 2000. The development of microcellular polymers as support for tissue engineering, PhD Thesis, University of Newcastle, Newcastle upon Tyne, UK.

[57] Bokhari M.A. 2003. Bone tissue engineering using novel microcellular polymers, PhD Thesis, University of Newcastle, Newcastle upon Tyne, UK.

8

The Encapsulation Art: Scale-up and Applications

M.A. Galán, C.A. Ruiz and E.M. Del Valle

8.1 Control Release Technology and Microencapsulation

Controlled release technologies are invaluable scientific tools for improving the performance and safety of chemicals. They involve materials such as barriers surrounding active materials to deliver the latter at the optimum time and rate needed[1,2].

The technical objective of this science is to find and use judiciously barriers, usually specially designed polymers (but may include adsorption inorganics or complexing with certain chemicals). Such formulations have also provided drug manufacturers, in particular, with a method for the measured, slow release of drugs as well as a business tactic for extending patent life and usefully differentiating their product. Methods include the following[1,2]:

- Designing the barrier surrounding the active chemical so as to change its permeability for the extraction and thus provide a tortuous path. Means include designing the barrier material to swell or slowly dissolve in the extracting fluid.
- Selection of an inorganic material which will adsorb the active material within its layered or porous structure, thus again providing a torturous path for the extracting fluid.
- Designing a chemical which will complex the active material and release it at a controlled rate under the right environmental conditions.
- Controlling the chemistry of the active material itself to release only under certain environmental conditions.

The application may require release in different ways: (a) constant release over time; (b) release rate diminishing with time; and (c) 'burst release', where all of the active

Chemical Engineering Edited by Miguel A. Galán and Eva Martin del Valle

material is released suddenly at a particular time, such as after the drug has passed through the stomach into the intestines. Designing the barrier makes use of the dissolving rates or swelling rates of the barrier polymer. In turn, the dissolving or swelling rates and permeability may depend on the pH, moisture, and temperature of the environment, chemical properties of the encapsulating polymer, its size, shape, and thickness.

Microencapsulation is an important sub-category of controlled release technology. Active materials are encapsulated in micrometer-sized capsules of barrier polymers designed to control the rate of release of the active materials that are encapsulated. The term 'microencapsulation' is often confused with 'controlled release' but the latter is much more inclusive, as indicated above[1,3].

8.2 The Microcapsule

In its simplest form, a microcapsule consists of a small ball surrounded by a homogeneous coating. The material enclosed in the microcapsule is called[4]:

- *Core*, or *nucleus*, internal phase, encapsulated, active substance, etc.

The coating is also called:

- *Shell*, envelope, external phase, membrane.

Although this is the most common type, there are several other kinds of microcapsules, depending on the technology used in their production. Their size usually varies from 1 to 2000 μm.

Capsules smaller than 1 μm are called nano-capsules because their size is measured in nanometers. When the core and the coating are not really separated, the microcapsule, or nanocapsule, is called microparticle, or nanoparticle.

The architecture of microcapsules is generally divided into several arbitrary and overlapping classifications. One such design is known as matrix encapsulation. Where the matrix particle resembles that of a peanut cluster. The core material is buried to varying depths inside the wall material. The most common type of microcapsule is that of a spherical or reservoir design. It is this design that resembles a hen's egg. It is also possible to design microcapsules that have multiple cores which may be an agglomerate of several different types of microcapsules[1,3,5]. If the core material is an irregular material, such as occurs with a ground particle, then the wall will somewhat follow the contour of the irregular particle and one achieves an irregular microcapsule. The last well-known design for a microcapsule is that of a multiple wall. In this case, multiple walls are placed around a core to achieve multiple purposes related to the manufacture of the capsules, their subsequent storage and controlled release.

8.2.1 Properties of the Core and the Capsule

The core, which is the substance to microencapsulate, can have different characteristics. It may belong to various categories of chemical substances. It can be liquid or solid, acid or basic, in powder or rough crystals. In addition to the requirement to induce microencapsulation of certain substances, the choice of the microencapsulation method and the coating material also depends on the characteristics of the active principle[6,7].

The choice of the coating material (capsule) often depends on the purpose, or the purposes, of microencapsulation. Not all membranes are able to confer specific properties onto the microencapsulated product. The choice of the right coating is often crucial in achieving the microencapsulation purpose. Some one hundred suitable substances have been described (and already used) to form a microcapsule film. The most commonly used substances are collected in Table 8.1.

Therefore, the first process consists of forming a wall around the core material. The second process involves keeping the core inside the wall material so that it does not release. Also, the wall material must prevent the entrance of undesirable materials that may harm the core. And finally, it is necessary to release the core material at the right time and at the right rate.

Microencapsulation is like the work of a clothing designer. He selects the pattern, cuts the cloth, and sews the garment in due consideration of the desires and age of his customer, plus the locale and climate where the garment is to be worn. By analogy, in microencapsulation, capsules are designed and prepared to meet all the requirements in due consideration of the properties of the core material, intended use of the product, and the environment of storage.

In a discussion of microencapsulation technology, particularly when one is talking about quantities and cost, it is necessary to understand that encapsulation is a volume process, independent of the density or value of the core material. Thus, microencapsulators frequently state that it is just as expensive on a volume basis to encapsulate diamond as graphite. Likewise, on a volume basis it is just as expensive to encapsulate paraffin wax as tungsten metal. Also, when experimenting with or acquiring microcapsules, it should be emphasized that it is necessary to use common, consistant terminology because of the preference for discussing microcapsules in terms of the core material, particularly when one is discussing the cost of production[7–9].

8.2.2 Microcapsules Uses

The uses of microcapsules since the initial coacervation work in the 1940s are many and varied. A good early review of these uses that also includes pharmaceuticals and

Table 8.1 *Materials used as coating material*

• *Agar*	• *Albumin*
• *Cellulose and its derivatives*	• *Gelatin*
• *Arabic gum*	• *Hydrogenated fats*
• *Glutens*	• *Glycerides*
• *Polyamides*	• *Acrylic polymers*
• *Polyesters*	• *Polyvinyl pyrrolidone*
• *Polyethylene glycols*	• *Polystyrene*
• *Starch*	• *Stearic acid*
• *Paraffins*	• *Waxes*
• *Polyvinyl, myristic, stearyl alcohols, etc.*	• *Others*

agricultural materials is provided by Gutcho[6]. The uses of microcapsules that are of interest here include the following[10]:

1. Reduce the reactivity of the core with regard to the outside environment, for example oxygen and water.
2. Decrease the evaporation or transfer rate of the core material with regard to the outside environment.
3. Promote the ease of handling of the core material:
   a. prevent lumping;
   b. position the core material more uniformly through a mix by giving it a size and outside surface matching the remainder of the materials in the mix;
   c. convert a liquid to a solid form; and
   d. promote the easy mixing of the core material.
4. Control the release of the core material so as to achieve the proper delay until the right stimulus.
5. Mask the taste of the core.
6. Dilute the core material when it is only used in very small amounts; but, achieve uniform dispersion in the host material.

8.2.3 Release Mechanisms

A variety of release mechanisms have been proposed for microcapsules; but, in fact, those that have actually been achieved and are of interest here are rather limited. These are as follows:

1. A compressive force in terms of a 2 point or a 12 point force breaks open the capsule by mechanical means.
2. The capsule is broken open in a shear mode such as that in a Waring blender or a Z-blade type mixer.
3. The wall is dissolved away from around the core such as when a liquid flavoring oil is used in a dry powdered beverage mix.
4. The wall melts away from the core releasing the core in an environment such as that occurring during baking.
5. The core diffuses through the wall at a slow rate due to the influence of an exterior fluid such as water or by an elevated temperature[1–4].

8.2.4 Release Rates

The release rates that are achievable from a single microcapsule are generally '0' order, 1/2 order, or 1st order. '0' order occurs when the core is a pure material and releases through the wall of a reservoir microcapsule as a pure material. The 1/2-order release generally occurs with matrix particles. 1st-order release occurs when the core material is actually a solution. As the solute material releases from the capsule the concentration of solute material in the solvent decreases and a 1st-order release is achieved. Please note that these types of release rates occur from a given single microcapsule. A mixture of microcapsules will include a distribution of capsules varying in size and wall thickness. The effect, therefore, is to produce a release rate different from '0', '1/2', or '1' because of the ensemble of microcapsules. It is therefore very desirable to examine carefully on an experimental basis the release rate from a collection of microcapsules and to recognize that the deviation from theory is due to the distribution in size and wall thickness[9,11].

8.2.5 Microcapsule Formation

The general technology for forming microcapsules is divided into two classifications known as physical methods and chemical methods. The physical methods are generally divided into the following[1–3,5,6]:

Spray coating

Pan coating: This is a mature, well-established technology initially patented by a pharmacist in the 19th century by the name of Upjohn. Generally, it requires large core particles and produces the coated tablets that we are familiar with.
Fluid bed coating: One version of this coating is known as Wurster coating and was developed in the 1950s and 1960s. The Wurster coater relies upon a bottom-positioned nozzle spraying the wall material up into a fluidized bed of core particles. Another version sprays the wall material down into the core particles.

Annular jet. This technology was developed by the Southwest Research Institute and has not been extensively used in the food industry. It relies upon two concentric jets. The inner jet contains the liquid core material. The outer jet contains the liquid wall material, generally molten, that solidifies upon exiting the jet. This dual fluid stream breaks into droplets much as water does upon exiting a spray nozzle.

Spinning disk. A new method was developed by Professor Robert E. Sparks at Washington University in St Louis. This method relies upon a spinning disk and the simultaneous motion of core material and wall material exiting from that disk in droplet form. The capsules and particles of wall material are collected below the disk. The capsules are separated from the wall particles (chaff) by a sizing operation.

Spray cooling. This is a method of spray cooling a molten matrix material containing minute droplets of the core materials. This method is practiced by the Sunkist Company.

Spray drying. Spray dryers can be used from small to very high productions depending on their design. They can reach evaporative capacities of up to 15 000 lb/h. Even though the cost of the equipment is expensive, the cost of maintenance is low due to the small number of moving parts and the use of resistant materials. The purity of the product will be maintained since the food particles do not have any contact with the surface of the equipment until they are dried, minimizing problems in sticking and corrosion. The simple operating system and the cleaning conditions for spray dryers contribute to the low labor cost. Another advantage of using the spray drying method is that a low bulk density of the product can be obtained.

Spray chilling. This is a process of spray chilling the wall around an atomized core. The resulting capsules move countercurrent to a flow of tempered air and are collected in a large container below the spray nozzle. It is practiced currently by the Durkee Company.

Co-extrusion processes. Liquid core and shell materials are pumped through concentric orifices, with the core material flowing in the central orifice, and the shell material flowing through the outer annulus. A compound drop forms that is composed of a droplet of core fluid encased in a layer of shell fluid.

The centrifugal nozzle technique was developed by SwRI. It requires that, initially, both the core and the wall are pumpable liquids (or thin slurries). Both fluids are fed into a special nozzle so that a coaxial stream is formed at the nozzle. This nozzle is spun rapidly, which stretches out the liquid ligaments and breaks off individual droplets. Surface tension pull the 'shell' material around the core droplet to form a complete covering. The wall material must be selected so that it will solidify before the particle is collected.

Use of supercritical fluids. Supercritical fluid (SCF) technology is now considered as a very innovative and promising way to design particles, especially for therapeutic drug formulation [12].

The advantages of SCF technology include use of mild conditions for pharmaceutical processing (which is advantageous for labile proteins and peptides), use of environmentally benign nontoxic materials (such as CO_2), minimization of organic solvent use, and production of particles with controllable morphology, narrow size distribution, and low static charge[12].

SCF technology is making in-roads in several pharmaceutical industrial operations including crystallization, particle size reduction, and preparation of drug delivery systems, coating, and product sterilization. It has also been shown to be a viable option in the formulation of particulate drug delivery systems, such as microparticles and nanoparticles, liposomes, and inclusion complexes, which control drug delivery and/or enhance the drug stability.

The number of methods for *chemical encapsulation*[1–5] is actually far less. They are necessary because they are very effective in encapsulating liquids and small core sizes. In particular, it is possible to encapsulate flavors and fragrances down to 10 μm in size.

1. Coacervation: This is a term borrowed from colloid chemistry to describe the basic process of capsule wall formation. The encapsulation process was discovered and developed by colloid chemist Barrett K. Green of the National Cash Register (NCR) Corporation in the 1940s and 1950s. Actually, coacervative encapsulation (or microencapsulation) is a three-part process: *particle or droplet formation*; *coacervative wall formation*; and *capsule isolation*. Each step involves a distinct technology in the area of physical chemistry. The first coacervative capsules were made using gelatin as a wall in an 'oil-in-water' system. Later developments produced 'water-in-oil' systems for highly polar and water-soluble cores.

   Simple coacervation involves the use of either a second more water-soluble polymer or an aqueous non-solvent for the gelatin. This produces the partial dehydration/desolvation of the gelatin molecules at a temperature above the gelling point. This results in the separation of a liquid gelatin-rich phase in association with an equilibrium liquid (gelatin-poor), which under optimum separation conditions can be almost completely devoid of gelatin.

   Complex coacervation was conceived in 1930s, B.K. Green, a young chemist just out of school, was intrigued by the dearth of information in the collaid field of liquids dispersed in solids. It was the first process used to make microcapsules for carbonless copy paper. In complex coacervation, the substance to be encapsulated is first dispersed as tiny droplets in an aqueous solution of a polymer such as gelatin. For this emulsification process to be successful, the core material must be immiscible in the aqueous phase. Miscibility is assessed using physical chemistry and thermodynamics. The emulsification is usually achieved by mechanical agitation, and the size distribution of the droplets is governed by fluid dynamics.

2. Organic phase separation: Sometimes, this technique is considered as a reversed simple coacervation; a polymer phase separates and deposits on a 'core' that is suspended in an organic solvent rather than water.
3. Solvent evaporation: A polymer is dissolved in a volatile solvent. The active material is then suspended in this fluid. The mixture is added to carrier, and the solvent is evaporated, precipitating the polymer on the active and forming microspheres.
4. Interfacial polymerization: Includes a number of processes in which a wall is formed from monomers at the interface of a core and the suspension medium.

8.2.6 Use of Supercritical Fluids for Particle Engineering

In recent years, the crystal and particle engineering of pharmaceutical materials and drug delivery systems with SCF technology has gained momentum due to the limitations of conventional methods[13–19]. This technology offers the advantage of being a one-step process, and appears to be superior to other conventional incorporation methods such as emulsion evaporation methods[17–21].

Application of SCF is now the subject of increasing interest especially in the pharmaceutical industry and there are three aims[16–18]: increasing bioavailability of poorly soluble molecules; designing sustained-release formulations; and formulation of active agents for new types of drug delivery that are less invasive than parental delivery (oral, pulmonary, transdermal). The most complex challenge is related to therapeutic delivery, as it is extremely difficult to obtain a satisfactory therapeutic delivery effect due to biomolecule instability and very short half-life *in vivo*.

We will describe the general SCF techniques used for particle engineering, examples of drug delivery systems prepared with SCF processes, and factors influencing the characteristics of SCF products, and scale-up issues associated with SCF processes[22,23].

8.3 Supercritical Fluids

At the critical temperature (T_c) and pressure (P_c), a substance's liquid and vapor phases are indistinguishable. A substance whose temperature and pressure are simultaneously higher than at the critical point is referred to as a supercritical fluid (Figure 8.1). Of particular interest for SCF application are the ranges $1 < T/T_c < 1.1$ and $1 < P/P_c < 2$[24]. In this region, the SCF exists as a single phase with several advantageous properties of both liquids and gases. The physical and thermal properties of SCFs fall between those of the pure liquid and gas. SCFs offer liquid-like densities, gas-like viscosities, gas-like compressibility properties, and higher diffusivities than liquids. The properties of SCFs, such as polarity, viscosity, and diffusivity, can be altered several-fold by varying the operating temperature and/or pressure during the process. This flexibility enables

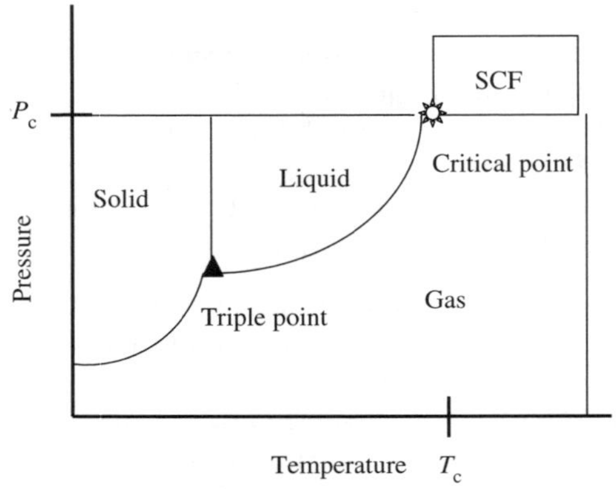

Figure 8.1 *Pressure–temperature phase diagram for a pure component*

the use of SCFs for various applications in the pharmaceutical industry, with the drug delivery system design being a more recent addition. Commonly used supercritical solvents include carbon dioxide, nitrous oxide, ethylene, propylene, propane, *n*-pentane, ethanol, ammonia, and water. Of these, CO_2 is a widely used SCF in the pharmaceutical processing due to its unique properties[21]:

- Behaves like a hydrocarbon solvent. An excellent solvent for aliphatic hydrocarbons with an estimated 20 carbons or less and for most aromatic hydrocarbons. Cosolvents, such as methanol and acetone, enhance the solubility of polar solutes in CO_2. Organic solvents, such as halocarbons, aldehydes, esters, ketones, and alcohols, are freely soluble in supercritical CO_2.
- Allows the processing of thermolabile compounds due to its low critical temperature.
- Does not react strongly (chemically) with many organic compounds.
- Can be used as solvent or antisolvent.
- Diffusion coefficients of organic solvents in supercritical CO_2 are typically one to two times higher than in conventional organic solvents.
- Easy to recycle at the end of the process.
- Nontoxic, noninflammable, and inexpensive.

8.4 Engineering Particle

Particle formation by supercritical methods is emerging as a viable platform technology for pharmaceuticals and drug delivery systems. Several requirements should be considered for an ideal particle-formation process:

- Operates with relatively small quantities of organic solvent(s).
- Molecular control of the precipitation process.
- Single-step, scalable process for solvent-free final product.
- Ability to control desired particle properties.
- Suitable for a wide range of chemical types of therapeutic agents and formulation excipients.

- Capability for preparing multi-component systems.
- Good manufacturing performance (GMP) compliant process.

SCF processing is recognized as achieving many of these objectives, particularly with recent developments in the scale of operation[25]. However, although move literature is appearing, fundamental mechanistic understanding of the SCF solvent and antisolvent processes is in its infancy[26–28]. Studies in progress that couple computational fluid dynamics with advanced laser-based 'real-time' particle imaging techniques under supercritical conditions, such as particle imaging velocimetry[29], will undoubtedly improve basic knowledge, process design, and define the boundaries and limitations of SCF particle-formation processes. Indeed, several recent reports have highlighted situations where specific particle design and crystal engineering targets have not been completely met[30,31]. Improved understanding of the complex interplay of the rapid physical, chemical, and mechanical processes taking place during particle formation by SCF techniques will help resolve such situations. Nevertheless, major benefits for SCF processing from the viewpoint of drug delivery have been demonstrated over recent years.

The different SCF particle-formation processes can be divided into six broad groups.

1. RESS: This acronym refers to 'rapid expansion of supercritical solutions'; this process consists of solvating the product in the fluid and rapidly depressurizing this solution through an adequate nozzle, causing an extremely rapid nucleation of the product into a highly dispersed material. Known for a long time, this process is attractive due to the absence of organic solvent use; unfortunately, its application is restricted to products that present a reasonable solubility in supercritical carbon dioxide (low polarity compounds).
2. Supercritical anti-solvent and related processes (GAS/SAS/ASES/SEDS): In these processes, the SCF is used as an antisolvent that causes precipitation of the substrate(s) dissolved initially in a liquid solvent. This general concept consists of decreasing the solvent power of a polar liquid solvent in which the substrate is dissolved, by saturating it with carbon dioxide in supercritical conditions, causing the substrate precipitation or recrystallization. Depending on the desired solid morphology, various methods of implementation are available:

    a) GAS or SAS, gas antisolvent or supercritical antisolvent, recrystallization: This process is used mostly for recrystallization of solids dissolved in a solvent with the aim of obtaining either small size particles or large crystals, depending on the growth rate controlled by the antisolvent pressure variation rate.
    b) ASES, aerosol solvent extraction system: This name is used when micro- or nanoparticles are expected; the process consists of pulverizing a solution of the substrate(s) in an organic solvent into a vessel swept by an SCF.
    c) SEDS, solution-enhanced dispersion by supercritical fluids: A specific implementation of ASES that consists of co-pulverizing the substrate(s) solution and a stream of supercritical carbon dioxide through appropriate nozzles.

3. PGSS: This acronym refers to 'particles from gas-saturated solutions (or suspensions)'. This process consists of dissolving an SCF into a liquid substrate, or a solution of the substrate(s) in a solvent, or a suspension of the substrate(s) in a solvent followed by a rapid depressurization of this mixture through a nozzle causing the formation of solid particles or liquid droplets according to the system.

4. DELOS: This acronym refers to 'depressurization of an expanded liquid organic solution', where an SCF is used for the straightforward production of micrometer-sized crystalline particles from organic solution. In this process, SCF acts as a co-solvent, being completely miscible, at a given pressure and temperature, with the organic solution to be crystallized. The role of the SCF is to produce, through its evaporation, a homogeneous sub-cooling of the solution with particle precipitation[32].
5. SAA, 'supercritical assisted atomization': This process is based on the solubilization of a fixed amount of supercritical carbon dioxide in the liquid solution; then the ternary mixture is sprayed through a nozzle, and as a consequence of atomization, solid particles are formed[33].
6. CAN-BD: Carbon dioxide assisted nebulization with a bubble dryer is a new-patented process that can generate a dense aerosol with small droplet and microbubble sizes that are dried to form particles less than 3-μm diameter[34].

Several reviews have considered these alternative varieties of SCF particle-formation processes[26,35–37]. However, a critical requirement to direct both process understanding and selection of required working conditions for targeted particle properties is the phase behavior of the different SCF methods. A recent review has addressed this topic[26], highlighting the importance of linking the underlying thermodynamics of phase equilibria operating under defined SCF processing conditions with changes in crystallization/precipitation mechanisms for products with desired properties.

8.4.1 Processes for Particle Design

Rapid expansion of supercritical solutions. RESS (Figure 8.2) consists of saturating an SCF with the substrate(s), then depressurizing this solution through a heated nozzle into a low-pressure chamber in order to cause an extremely rapid nucleation of the substrate(s) in the form of very small particles – or fibers, or films when the jet is directed against a surface – that are collected from the gaseous stream[36].

The pure carbon dioxide is pumped into the desired pressure and preheated to extraction temperature through a heat exchanger. The SCF is then percolated through the extraction

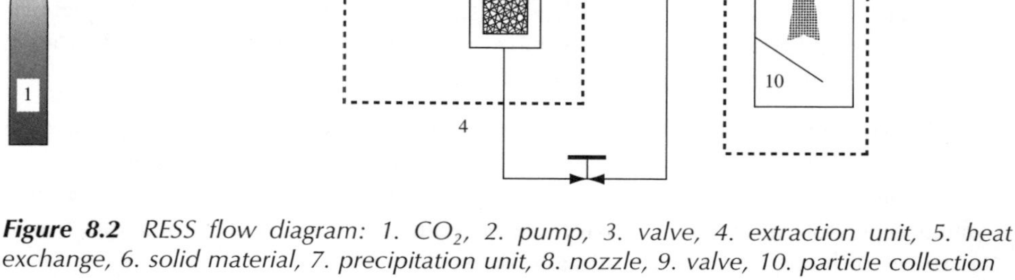

Figure 8.2 *RESS flow diagram: 1. CO_2, 2. pump, 3. valve, 4. extraction unit, 5. heat exchange, 6. solid material, 7. precipitation unit, 8. nozzle, 9. valve, 10. particle collection*

unit packed with one or more substrate(s), mixed in the same autoclave or set in different autoclaves in series. In the precipitation unit, the supercritical solution is expanded through a nozzle that must be reheated to avoid plugging by substrate(s) precipitation.

The morphology of the resulting solid material depends both on the material structure (crystalline or amorphous, composite or pure, etc.) and on the RESS parameters (temperature, pressure drop, distance of impact of the jet against the surface, dimensions of the atomization vessel, nozzle geometry, etc.)[38–51]. It is to be noticed that the initial investigations consisted of 'pure' substrate atomization in order to obtain very fine particles (typically of 0.5–20 μm diameter) with narrow diameter distribution; however, the most recent publications are related to mixture processing in order to obtain microcapsules or microspheres of an active ingredient inside a carrier.

This technology can be implemented in relatively simple equipment although particle collection from the gaseous stream is not easy. But the applications are limited as most attractive substrates are not soluble enough into the SCF to lead to profitable processes: a co-solvent may be used to improve this solubility, but it will be eliminated from the resulting powder, which is not simple and cheap[27].

Processing equipment requires a source of SCF, which passes through an extractor unit to a restricted orifice positioned in a particle collection-precipitation vessel held at a lower temperature and pressure (often ambient) than the extractor unit. There are several primary factors[52] found to influence the physical properties of particle size, shape, and surface topography of products. Those factors are

- dimensions of orifice (expansion device),
- time scale (typically 10^{-5} s),
- pressure/temperature conditions in precipitator,
- agglomeration phenomena during SCF solution expansion,
- phase process path followed during expansion.

For pharmaceutical organic materials studied for processing by RESS, SCF CO_2 is the preferred solvent. As SCF CO_2 is non-polar, those organics that are also non-polar can be expected to dissolve in SCF CO_2 and thus be suitable candidates for RESS processing. Examples include lovastatin[27], stigmasterol[51], salicylic acid, and theophylline[53]. Expansion of solutions to pressure conditions above ambient, and thereby at lower levels of supersaturation, can result in agglomeration of particles, whereas increased supersaturation during expansion leads to extremely rapid nucleation rates and micrometer- and sub-micrometer-sized particles. Several reports have considered using the RESS process for the direct formulation of drug:polymer systems by a coprecipitation strategy[54,55], with the objective of embedding drug molecules in a polymeric-core particle to provide a modified drug-diffusional flux. With evidence of phase separation for a lovastatin:poly(D, L-lactic acid) system[42,43], particles of a poly(L-lactic acid) coating on a core of naproxen have been prepared by careful control of processing conditions. Whilst most pharmaceutical compounds, produced by synthesis or natural compounds, exhibit solubilities below 0.01 wt.% under moderate processing conditions (below 60°C and 300 bar)[56], several low molecular weight hydrophobic compounds, including some steroids and biodegradable polymers, have been prepared in crystalline, micrometer-sized form with narrow distributions. However, predictive control of particle size and morphology remains a major challenge, along with processing and scale-up factors to eliminate particle aggregation and nozzle blockages caused by cooling effects on solution expansion. This process can also be used with suspensions of active substrate(s) in a polymer or other carrier substance leading to composite microspheres.

RESS is a very attractive process as it is simple and relatively easy to implement at least at a small scale when a single nozzle can be used. However, extrapolation to a significant production size or use of a porous sintered disk through which pulverization occurs is extremely difficult to carry out. The reason is that in both cases, particle size distribution is not easy to control, and may be much wider than in the case of a single nozzle. Moreover, particle harvesting is complex, as it is in any process leading to very small particles[25].

However, the most important limitation of RESS development lies in the too low solubility of compounds in SCFs, which precludes production at acceptable costs, as, in most cases, use of a co-solvent to increase solubility in the fluid is not feasible[26].

Supercritical antisolvent and related processes (GAS/SAS/ASES/SEDS). Precipitation using SCFs as non-solvents or antisolvents utilizes a similar concept to the use of antisolvents in solvent-based crystallization processes.

In those processes, the SCF is used as antisolvent that causes precipitation of the substrate(s) dissolved initially in a liquid solvent. This general concept consists of decreasing the solvent power of a polar liquid solvent in which the substrate is dissolved, by saturating it with carbon dioxide in supercritical conditions, causing substrate precipitation or recrystallization[57]. Depending on the desired solid morphology, various ways of implementation are available.

GAS or SAS recrystallization: A batch of solution is expanded several-fold by mixing with a dense gas in a vessel (Figure 8.3). Owing to the dissolution of the compressed gas, the expanded solvent has a lower solvent strength than the pure solvent. The mixture becomes supersaturated and the solute precipitates in microparticles. This process has been called gas antisolvent or supercritical antisolvent recrystallization. As shown in

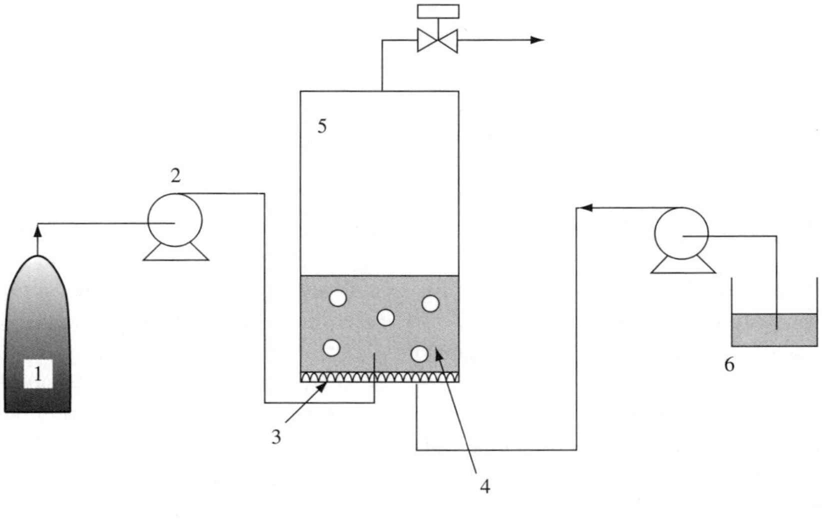

Figure 8.3 *GAS flow diagram: 1. CO_2, 2. pump, 3. particles, 4. expanded solution, 5. precipitator, 6. solution*

Figure 8.3 the precipitator is partially filled with the solution of the active substance. CO_2 is then pumped up to the desired pressure and introduced in the vessel, preferably from the bottom to achieve a better mixing of the solvent and antisolvent. After a holding time, the expanded solution is drained under isobaric conditions to wash and clean the precipitated particles.

With high solubilities of SCFs in organic solvent, a volume expansion occurs when the two fluids make contact, leading to a reduction in solvent density and parallel fall in solvent capacity. Such reductions cause increased levels of supersaturation, solute nucleation, and particle formation. This process, generally termed gas antisolvent recrystallization, thus crystallizes solutes that are insoluble in SCFs from liquid solutions, with the SCF, typically SCF CO_2, acting as an antisolvent for the solute. The GAS process was initially developed for crystallizing explosive materials[25].

Typically, the GAS process is performed as a batch process. Particles are formed in the liquid phase[37] and are then dried by passing pure SCF over product in the pressure vessel for extended periods. This situation, coupled with problems associated with heat generation during the addition of an SCF to solvent or solution[58], has resulted in modification of the process by several research groups[26] to improve both process and product control and to achieve a semi-continuous operation. In general, the developments involve spraying or aerosolizing the organic solvent drug solution into a bulk or flowing stream of SCF as the antisolvent. This is to maximize exposure of small amounts of solution to large quantities of SCF antisolvent to dissolve rapidly the solvent in the SCF, leading to dry particles and thereby reducing the drying stage in the GAS process.

ASES (aerosol solvent extraction system): This is the first modification of the gas antisolvent process and involves spraying the solution through an atomization nozzle as fine droplets into compressed carbon dioxide (Figure 8.4). The dissolution of the SCF into the liquid droplets is accompanied by a large volume expansion and, consequently,

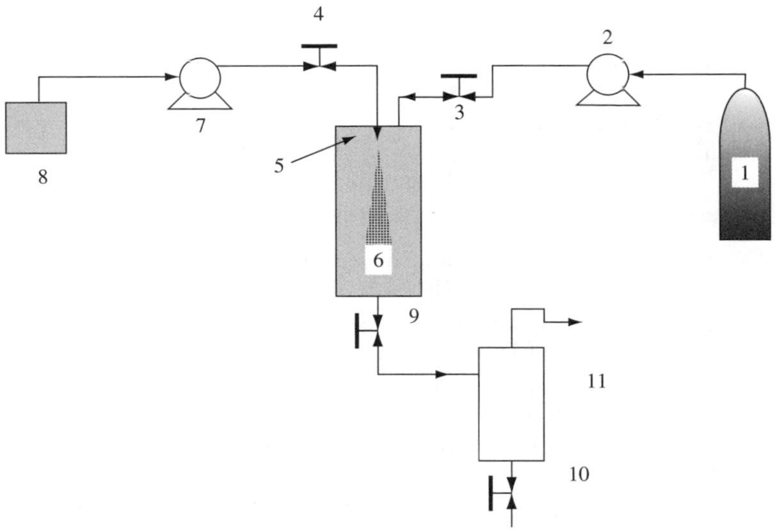

Figure 8.4 *ASES flow diagram: 1. CO_2, 2, 7. pump, 3, 4, 9, 10. valves, 5. nozzle, 6. high-pressure vessel, 8. active material+solvent, 11. low-pressure tank*

a reduction in the liquid solvent power, causing a sharp rise in the supersaturation within the liquid mixture, and the consequent formation of small and uniform particles[59]. The SCF is pumped to the top of the high-pressure vessel by a high-pressure pump. Once the system reaches steady state (temperature and pressure in precipitator), the active substance solution is introduced into the high-pressure vessel through a nozzle[60,61]. To produce small liquid droplets in the nozzle, the liquid solution is pumped at a pressure higher (typically ~20 bar) than the vessel operating pressure. Particles are collected on a filter at the bottom of the vessel. The fluid mixture (SCF plus solvent) exits the vessel and flows to a depressurization tank where the conditions (temperature and pressure) allow gas–liquid separation. After collection of a sufficient amount of particles, liquid solution pumping is stopped and pure SCF continues to flow through the vessel to remove residual solvent from the particles. This spray process has been called the aerosol solvent extraction system process[62].

SEDS (solution-enhanced dispersion by supercritical fluids): The second modification of the gas antisolvent process known as solution-enhanced dispersion by SCFs was developed by the Bradford University[61] in order to achieve smaller droplet size and intense mixing of SCF and solution for increased transfer rates. Indeed the SCF is used both for its chemical properties and as 'spray enhancer' by mechanical effect: a nozzle with two coaxial passages allows the introduction of the SCF and a solution of active substance(s) into the particle-formation vessel where pressure and temperature are controlled (Figure 8.5). The high velocity of the SCF allows breaking up the solution into very small droplets. Moreover, the conditions are set up so that the SCF can extract the solvent from the solution at the same time as it meets and disperses the solution. Similarly, a variant was recently disclosed by the University of Kansas[63], where the nozzle design leads to development of sonic waves leading to very tiny particles, around 1 μm.

Factors influencing particle properties when prepared by the SCF–GAS process[64,65] are

- solute solubility in organic solvent,
- solute insolubility in SCF,
- degree of expansion of organic solvent in SCF,
- organic solvent/SCF antisolvent ratio,
- rate of addition of SCF antisolvent,
- pressure of temperature conditions in precipitator,
- phase process path followed during particle nucleation.

In the SAS method, a solution of compound in an organic solvent is sprayed via a capillary-tube nozzle into a bulk of SCF[64,65], with pharmaceutical applications including polymers and proteins. A modification of the SAS process is the ASES, in which a drug or polymer solution is sprayed into a volume of SF for a period of time[66]. This step is followed by lengthy drying periods by flowing fresh SF over the particulate product. The precipitation with compressed antisolvent (PCA) process is similar in principle, with a liquid solution of a polymer delivered via a capillary tube into the antisolvent in a liquid (subcritical) or supercritical state[67]. Alternative polymeric particle topography and shapes have been reported, depending upon process paths followed in the phase diagram because of the polymer:SCF phase separation[68].

As with RESS, a range of pharmaceutical and polymer–drug systems have been successfully prepared, including micrometer-sized particles, albeit generally on a laboratory

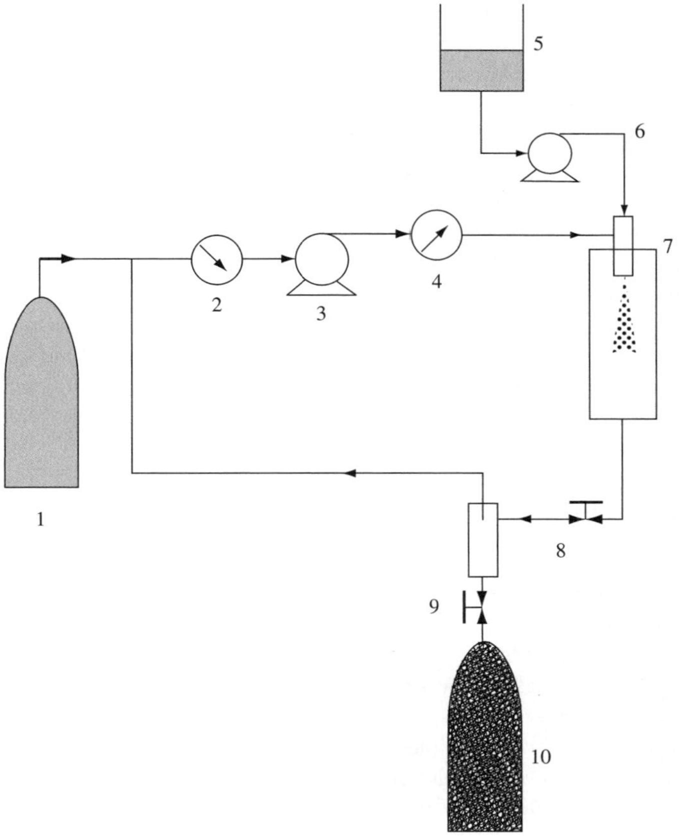

Figure 8.5 *SEDS flow diagram: 1. CO_2, 2. cooler, 3, 6. pumps, 4. heat exchange, 5. active substance solution, 7. particle formation vessel, 8, 9. valves, 10. solvent*

scale. Thus, the benefits of these SCF-antisolvent processes include a totally enclosed single-step process that requires reduced levels of organic solvent compared with conventional crystallization[63,64,69].

In the SAS and ASES techniques, the mass transfer of the SCF into the sprayed droplet determines the rate of particle formation, whereas particle agglomeration and aggregation phenomena are influenced by the rate of solvent mass transfer into the SCF from the droplet. The former mass transfer is dependent upon atomization efficiency and the latter on dispersing and mixing phenomena between the solution droplet and the SCF[37]. Thus, to minimize the particle agglomeration frequently observed and to reduce or eliminate drying times, increased mass-transfer rates are required. This has been successfully achieved in the SEDS process[67], which uses a coaxial nozzle design with a mixing chamber. This arrangement provides a means whereby the drug in the organic solvent solution interacts and mixes with the SCF antisolvent in the mixing chamber of the nozzle prior to dispersion, and flows into a particle-formation vessel via a restricted orifice. Thus, high mass-transfer rates are achieved with a high ratio of SCF to solvent, and the high velocities of the SCF facilitate break-up of the solution feed.

Over recent years, this process has been used for several drug and drug formulation applications, with evidence of successful process scale-up[25,37]. This process has been further developed to process water-soluble materials, including carbohydrates[37] and biologicals, with a three-component coaxial nozzle[28]. The metered and controlled delivery of an aqueous solution, ethanol, and SCF into the modified coaxial nozzle overcomes the problems associated with limited water solubility in $SFCO_2$, the most common SCF for pharmaceutical processing. These advances, particularly for scale-up and operation with aqueous solutions and SCF CO_2, strengthen the viability and industrial potential of SCF processing for pharmaceuticals.

There is no doubt that antisolvent processes have a bright future, especially for drug delivery systems, as they permit the monitoring of the properties and composition of the particles with great flexibility and for almost any kind of compounds. Nevertheless, scale-up is presently foreseen only for high-value specialty materials (pharmaceuticals, cosmetics, superconductors) with productions ranging from a few kilograms to a few hundred kilograms per day.

Regarding intellectual property, the situation may lead to some limitations of process applications until it will be cleared to ensure that no patent infringement is to be feared by potential users[26].

Particles from gas-saturated solutions/suspensions (PGSS). As the solubilities of compressed gases in liquids and solids like polymers are usually high, and much higher than the solubilities of such liquids and solids in the compressed gas phase, this process consists of solubilizing supercritical carbon dioxide in melted or liquid-suspended substance(s), leading to a so-called gas-saturated solution/suspension that is further expanded through a nozzle with formation of solid particles, or droplets[70,71] (Figure 8.6). Typically, this process allows the formation of particles from a great variety of substances that need not be soluble in supercritical carbon dioxide, especially with some polymers that absorb a large

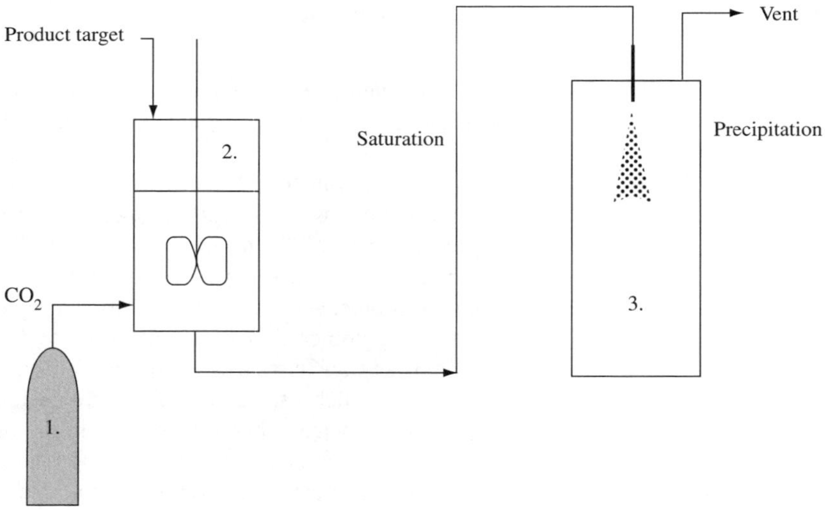

Figure 8.6 *PGSS flow diagram: 1. CO_2, 2. reactor, 3. precipitator*

concentration (10 – 40 wt.%) of CO_2 that either swells the polymer or melts it at a temperature far below (~10–50°C) its melting/glass transition temperature. A further variety of this process has been developed for controlling the porosity of polymer particles[68]. This procedure, called pressure-induced phase separation (PIPS), depends on a controlled expansion of a homogeneous solution of polymer and SCF in liquid or supercritical phase. By varying the polymer concentration and depressurizing via alternative phase transitions in the metastable or spinodal region of the phase diagram[70,71], the porosity of resulting polymeric particles can be increased with higher initial concentrations in the feed solution to the expansion orifice. As before, the process requires adequate solubility of the polymer and solute, if present, in the chosen SF for pharmaceutical vitality[72].

Particle design using the PGSS concept is already widely used at large scale, and is different from other process concepts presently under development yet. The simplicity of this concept, leading to low processing costs, and the very wide range of products that can be treated (liquid droplets or solid particles from solid material or liquid solutions or suspensions) open wide avenues for development of PGSS applications, not only for high-value materials but also perhaps for commodities, in spite of limitations related to the difficulty of monitoring particle size[72].

Recently, many patents were successively granted. Most are related to paint application (pulverization of suspensions to make coatings) and powder coating manufacture (combination of chemical reaction and pulverization of a suspension); more surprisingly, the basic PGSS process patent filed in 1995[73] for the formation of solid particles from polymer or solid substances has been successfully granted in Europe and recently in the US. Moreover, a patent for aerosol drug delivery[74] was also granted, describing several different processes and apparatus: the 'tee' process and equipment it is not clear that the pulverization it only caused by the mechanical effect of gas expansion well known for long, and a portable device for static nebulization using RESS or PGSS concepts.

Depressurization of an expanded liquid organic solution (DELOS). This new technology was developed by Ventosa *et al.*[32]. An SCF is used for the straightforward production of micrometer-sized crystalline particles from organic solution. In this process SCF acts as co-solvent being completely miscible, at a given pressure and temperature, with the organic solution to be crystallized. The role of the SCF is to produce, through its evaporation, a homogeneous sub-cooling of the solution with particle precipitation[32].

The driving force of a DELOS crystallization process is the fast, large, and extremely homogeneous temperature decrease experienced by a solution, which contains an SCF, when it is depressurized from a given working pressure to atmospheric pressure. In contrast to other already reported high-pressure crystallization techniques (RESS, GAS, PCA, PGSS), in a DELOS process the SCF behaves as co-solvent over the initial organic solution of the solute to be crystallized. Through a DELOS process it is possible to produce fine powders of a compound provided that a system 'compound/organic solvent/SCF' in a liquid one-phase state is found. In order to compare DELOS and GAS procedures, Ventosa *et al.*[32] crystallized 1,4-bis-(*n*-butylamino)-9,10-anthraquinone from 'acetone/CO_2' mixtures by both methods. The crystallization results obtained were analyzed upon the solubility behavior of 1,4-bis-(*n*-butylamino)-9,10-anthraquinone in 'acetone/CO_2' mixtures with different composition. They showed that in those ternary systems where the CO_2 behaves as co-solvent over a wide range of X_{CO_2}, like the case of 'colorant 1/acetone/CO_2'

system, this new process can be an alternative to the already reported GAS and PCA crystallization methods, where the CO_2 is used as antisolvent. In a DELOS process the extent of CO_2 vaporization at any point of the liquid solution is exactly the same; as a consequence the solution temperature decrease and the evolution of the supersaturation profile are extremely uniform over the entire system. Therefore, the design of the stirring system, which is usually a problem in many industrial processes performed in solution, is not a key point because the characteristics of the particles produced do not depend on the mixing efficiency.

Summarizing, the DELOS process is a promising new high-pressure crystallization technique, which can be a useful processing tool in the particle engineering of different compounds and materials of industrial interest.

Supercritical assisted atomization (SAA). This process is based on the solubilization of a fixed amount of supercritical carbon dioxide in the liquid solution; then the ternary mixture is sprayed through a nozzle, and as a consequence of atomization, solid particles are formed.

One of the prerequisites for a successful SAA precipitation is the complete miscibility of the liquid in the SCF CO_2, and the insolubility of the solute in it. For these reasons SAA is not applicable to the precipitation of water-soluble compounds due to the very low solubility of water in CO_2 at the operating conditions commonly used.

Reverchon and Della Porta[33] used SAA to produce tetracycline (TTC) and rifampicin (RF) microparticles with controlled particle size and particle size distributions in the range of aerosolizable drug delivery.

Water was used as a liquid solvent for TTC and methanol was used for RF; heated nitrogen was also delivered into the precipitator in order to evaporate the liquid droplets and generate the microparticles. SAA of these compounds was optimized with respect to the process parameters; then, the influence of the solute concentration in the liquid solution on particle size and particle size distribution was studied. The produced powders were characterized with respect to their morphologies and particle size: spherical particles with controlled particle size ranging between 0.5 and 3 μm were obtained for both drugs at optimized operating conditions[33].

Carbon dioxide assisted nebulization with a bubble dryer (CAN-BD®). CAN-BD®[34] can dry and micronize pharmaceuticals for drug delivery.

In this process, the drug, dissolved in water or an alcohol (or both), is mixed intimately with near-critical or supercritical CO_2 by pumping both fluids through a low volume to generate microbubbles and microdroplets, which are then decompressed into a low-temperature drying chamber, where the aerosol plume dries in seconds. CO_2 and the solution are mixed in the tee at room temperature and microbubbles and microdroplets formed are dried rapidly at lower temperatures (25–80° C) than are used in traditional spray drying processes. The residence time of the particles in the lab scale 750-mL glass drying chamber is less than 3 s. The primary advantage of this process is that there is less decomposition of thermally labile drugs. Secondly, no high-pressure vessels are needed in the CAN-BD process, except for the syringe pump, the 1/16 in. (outer diameter) stainless steel tubing, the low volume tee, and the flow restrictor, which allow fluid mixing at a moderate pressure (i.e. between 80 and 100 bar) and the expansion of the microbubbles and microdroplets to atmospheric pressure. Thirdly, these particles (hollow or solid) are generally formed in the optimum size range for pulmonary delivery to alveoli (typically 99% are less than 3 μm in diameter). They have synthesized and measured the aerodynamic diameters of dried hollow and solid particles of various drugs and model compounds. Particles can be easily prepared and collected in a CAN-BD unit. Samples as small as 1 mL in volume can be dried for formulation studies, and scale-up of the CAN-BD process is in progress.

8.5 Factors Influencing Particle Properties

The characteristics of the particles produced using SCF technology are influenced by the properties of the solute (drug, polymer, and other excipients), type of SCF used, and process parameters (such as flow rate of solute and solvent phase, temperature and pressure of the SCF, pre-expansion temperature, nozzle geometry, and the use of coaxial nozzles)[12,14]. The influence of drug and polymer properties is discussed below.

Drug properties, such as solubility and partitioning of the drug into SCF, determine the properties of the particles formed. When SCF is intended as an antisolvent, if the drug is soluble in SCF under the operating conditions, it will then be extracted into SCF and will not precipitate out[75].

Similarly, during encapsulation of a drug in a polymer matrix, the properties of the drug influence the drug loading. Poly(Lactic acid) (PLA) microparticle formation using an antisolvent process with supercritical CO_2 indicated that an increase in liophilicity decreases the loading efficiency as well as release rate, possibly because lipophilic drugs can be entrained by supercritical CO_2 during SCF precipitation. Nucleation and growth rate influence the effective encapsulation and morphology of the particles. If the initial nucleation and growth rate of the drug are rapid and the polymer precipitation rate is relatively slow, then drug needles encapsulated in polymeric coat may be formed[16].

Polymer properties, such as polymer concentration, crystallinity, glass transition temperature, and polymer composition, are important factors that determine the morphology of the particles[12]. Increase in the polymer concentration may lead to formation of less spherical and fiber-like particles[20]. In an antisolvent process, the rate of diffusion of antisolvent gas is higher in a crystalline polymer compared to an amorphous polymer. This is because the crystalline polymer has a more ordered structure than the amorphous polymer. This leads to high mass-transfer rates in crystalline polymers, producing high supersaturation ratios and small particles of narrow size distribution. SCFs act as plasticizers for polymer by lowering their glass transition temperatures (T_g). Therefore, polymers with a low T_g tend to form particles that become sticky and aggregate. A change in polymer chain length, chain number, and the use of chain composition can alter polymer crystallinity, and, hence, the particle morphology.

8.6 Drug Delivery Applications of SCFs

Microparticles and nanoparticles. Drug and polymeric microparticles have been prepared using SCFs as solvents and antisolvents. Krukonis[15] first used RESS to prepare 5- to 100-μm particles of an array of solutes including lovastatin, polyhydroxy-acids, and mevinolin. In the past decade, simultaneous coprecipitation of two solutes, a drug and an excipient, gained interest. An RESS process employing CO_2 was used to produce

PLA particles of lovastatin and naproxen[17]. In these studies, supercritical CO_2 was passed through an extraction vessel containing a mixture of drug and polymer, and the CO_2 containing the drug and the polymer was then expanded through a capillary tube. A GAS process was used to produce clonidine-PLA microparticles[56]. In this process, PLA and clonidine were dissolved in methylene chloride, and the mixture was expanded by supercritical carbon dioxide to precipitate polymeric drug particles.

SCF technology is now claimed to be useful in producing particles in the range 5–2000 nm[69]. This patent covers a process that rapidly expands a solution of the compound and phospholipid surface modifiers in a liquefied gas into an aqueous medium, which may contain the phospholipid[76]. Expanding into an aqueous medium prevents particle agglomeration and particle growth, thereby producing particles of a narrow size distribution. However, if the final product is a dry powder, this process requires an additional step to remove the aqueous phase.

Intimate mixture under pressure of the polymer material with a core material before or after SCF solvation of the polymer, followed by an abrupt release of pressure, leads to an efficient solidification of the polymeric material around the core material. This technique was used to microencapsulate infectious bursal disease virus vaccine in a polycaprolactone (PCL) or a poly(lactic-*co*-glycolic acid) (PLGA) matrix[43].

Microporous foams. Using the SCF technique, Hile *et al.*[77] prepared porous PLGA foams capable of releasing an angiogenic agent, basic fibroblast growth factor (bFGF), for tissue engineering applications. These foams sustained the release of the growth factor. In this technique, a homogenous water-in-oil emulsion consisting of an aqueous protein phase and an organic polymer solution was prepared first. This emulsion was filled in a longitudinally sectioned and easily separable stainless steel mold. The mold was then placed into a pressure cell and pressurized with CO_2 at 80 bar and 35° C. The pressure was maintained for 24 h to saturate the polymer with CO_2 for the extraction of methylene chloride. Finally, the set-up was depressurized for 10–12 s, creating a microporous foam.

Liposomes. Liposomes are useful drug carriers in delivering conventional as well as macromolecular therapeutic agents. Conventional methods suffer from scale-up issues, especially for hydrophilic compounds. In addition, conventional methods require a high amount of toxic organic solvents. These problems can be overcome by using SCF processing. Frederiksen *et al.*[78] developed a laboratory-scale method for preparation of small liposomes encapsulating a solution of FITC dextran (fluorescein isothiocyanatedextran), a water-soluble compound using supercritical carbon dioxide as a solvent for lipids[78]. In this method, phospholipid and cholesterol were dissolved in supercritical carbon dioxide in a high-pressure unit, and this phase was expanded with an aqueous solution containing FITC in a low-pressure unit. This method used 15 times less organic solvent to get the same encapsulation efficiency as conventional techniques. The length and inner diameter of the encapsulation capillary influenced the encapsulation volume, the encapsulation efficiency, and the average size of the liposomes. Using the SCF process, liposomes, designated as critical fluid liposomes (CFL), encapsulating hydrophobic drugs, such as taxoids, camptothecins, doxorubicin, vincristine, and cisplatin, were prepared. Also, stable paclitaxel liposomes with a size of 150–250 nm were obtained. Aphios Company's patent[79] (US Patent No. 5,776,486) on SuperFluidsTM CFL describes a method and apparatus useful for the nanoencapsulation of paclitaxel and campothecin in aqueous liposomal formulations called TaxosomesTM and CamposomesTM, respectively. These formulations are claimed to be more effective against tumors in animals compared to commercial formulations.

Inclusion complexes. Solubility of poorly soluble drugs, such as piroxicam, can be enhanced by forming inclusion complexes with cyclodextrins. For many non-polar drugs, previously established inclusion complex preparation methods involved the use of organic solvents that were associated with high residual solvent concentration in the inclusion complexes[80]. Cyclodextrins had previously been used for the entrapment of volatile aromatic compounds after supercritical extraction[81]. On the basis of this principle, Van Hees *et al.*[82] employed SCFs for producing piroxicam and β-cyclodextrin inclusion complexes. Inclusion complexes were obtained by exposing the physical mixture of piroxicam-β-cyclodextrin (1:2.5 mol:mol) to supercritical CO_2 and depressurizing this mixture within 15 s. Greater than 98.5% of inclusion was achieved after 6 h of contact with supercritical CO_2 at 15 MPa and 150° C.

Solid dispersions. SCF techniques can be applied to the preparation of solvent-free solid dispersion dosage forms to enhance the solubility of poorly soluble compounds. Traditional methods suffer from the use of mechanical forces and excess organic solvents. A solid dispersion of carbamazepine in polyethyleneglycol 4000 (PEG4000) increased the rate and extent of dissolution of carbamazepine[83]. In this method, a precipitation vessel was loaded with a solution of carbamazepine and PEG4000 in acetone, which was expanded with supercritical CO_2 from the bottom of the vessel to obtain solvent-free particles.

Powders of macromolecules. Processing conditions with supercritical CO_2 are benign for processing macromolecules, such as peptides, proteins, and nucleic acids. Debenedetti[35] used an antisolvent method to form microparticles of insulin and catalase. Protein solutions in hydroethanolic mixture (20:80) were allowed to enter a chamber concurrently with supercritical CO_2. The SCF expanded and entrained the liquid solvent, precipitating sub-micrometer protein particles. Because proteins and peptides are very polar in nature, techniques such as RESS cannot be used often. Also, widely used supercritical antisolvent processing methods expose proteins to potentially denaturing environments, including organic and supercritical nonaqueous solvents, high pressure, and shearing forces, which can unfold proteins, such as insulin, lysozyme, and trypsin, to various degrees[84]. This led to the development of a method wherein the use of the organic solvents is completely eliminated to obtain fully active insulin particles of dimensions 1.5–500 μm. In this development, insulin was allowed to equilibrate with supercritical CO_2 for a predetermined time, and the contents were decompressed rapidly through a nozzle to obtain insulin powder. Plasmid DNA particles can also be prepared using SCFs[85]. An aqueous buffer (pH 8) solution of 6.9-kb plasmid DNA and mannitol was dispersed in supercritical CO_2 and a polar organic solvent using a three-channel coaxial nozzle. The organic solvent acts as a precipitating agent and as a modifier, enabling non-polar CO_2 to remove the water. The high dispersion in the jet at the nozzle outlet facilitated rapid formation of dry particles of small size. Upon reconstitution in water, this plasmid DNA recovered 80% of its original supercoiled state. Such macromolecule powders can possibly be used for inhalation therapies.[85,86].

Coating. SCFs can be used to coat the drug particles with single or multiple layers of polymers or lipids[24]. A novel SCF coating process that does not use organic solvents has been developed to coat solid particles (from 20 nm to 100 μm) with coating materials, such as lipids, biodegradable polyester, or polyanhydride polymers[82]. An active substance in the form of a solid particle or an inert porous solid particle containing an active substance can be coated using this approach. The coating is performed using a solution of a coating material in SCF, which is used at temperature and pressure conditions that do not solubilize the particles being coated.

Product sterilization. In addition to drug delivery system preparation, SCF technology can also be used for other purposes, such as product sterilization. It has been suggested that high-pressure CO_2 exhibits microbicidal activity by penetrating into the microbes, thereby lowering their internal pH to a lethal level[85]. The use of supercritical CO_2 for sterilizing PLGA microspheres (1, 7, and 20 μm) is described in US Patent No. 6,149,864[87]. The authors indicated that complete sterilization can be achieved with supercritical CO_2 in 30 min at 205 bar and 34° C.

Protein and biological materials. The considerable growth in biotechnology-derived therapeutic agents, including peptides, proteins, and plasmid DNA, has generated interest in non-oral routes of drug administration to bypass the damaging gastrointestinal effects for such materials[88]. The promise of using alternative delivery routes, via nasal, respiratory, transdermal via powder delivery and parenteral routes, is frequently constrained by requirements for stable, powdered products with specific particle size requirements. The complexity and sensitivity of these biologically sourced materials necessitate careful processing to ensure stability of product and provide appropriate physical characteristics[89]. The conventionally used and complex processes of freeze drying and spray drying are far from ideal. Taken together with the problems associated with downstream sieving or milling products prepared by these drying operations to achieve target particle sizes and size distributions, particle formation by SCF methods represents an attractive option. The application of SCF antisolvent methods has shown considerable promise in this field over recent years[38].

The low solubility of water in $SFCO_2$ has forced workers to use organic solvents including dimethylformamide (DMF) and dimethylsulfoxide (DMSO) as nonaqueous media for biological materials such as proteins[65,78,90]. Such solvents have limitations because proteins have low solubility and potential loss of secondary and tertiary protein structure in solution of these agents. Nevertheless, although extensive perturbation was evidenced in DMSO solutions and was partially present in the solid protein particles, micrometer-sized particles of insulin, lysozyme, and trypsin prepared by the SAS process essentially recovered biological activity on reconstitution[78,79].

The processing of labile biological materials from aqueous solutions is clearly preferred and modifications to the nozzle arrangement in the SEDS process have achieved this objective[12]. In this modification, the aqueous solution containing the biological material is only contacted momentarily with a potentially damaging organic solvent and SF in a three-component coaxial nozzle. This approach has been successfully applied to aqueous lysozyme solutions[91], with microfine product showing a spherical morphology with free flowing powder-handling properties.

8.7 Scale-up Issue

In recent years[52], a number of particle-formation techniques have shown considerable promise, only to falter on scale-up studies and in trying to achieve strict GMP requirements for the process.

At the same time, increased vigilance is being expressed by the regulatory agencies in facilities used for the preparation of drug substances in particulate form. This is occurring against a background of ambition by the pharmaceutical industry of global harmonization of material preparation and consistency of properties of powdered materials.

In many ways, SCF processing and controlled particle formation satisfies most of these demands directly by virtue of the inherent features of the process. As a single step, enclosed operation with mass balance, high yields of very consistent products can be achieved.

Materials of construction of equipment components are high-grade pharmaceutical-grade stainless steel; there are no moving parts and organic solvent requirements can often be reduced compared with crystallization processes[92].

To achieve commercial success, any method/technique developed should be scaled to produce quantities in batches for conducting further research or to market the product. From the perspective of scale-up, SCF technology offers several advantages. The processing equipment can be a single-stage, totally enclosed process that is free of moving parts and constructed from high-grade stainless steel, allowing easy maintenance and scale-up. It offers reduced solvent requirements and particle formation occurs in a light-, oxygen-, and possibly moisture-free atmosphere, minimizing these confounding factors during scale-up.

Some advances have been made in mechanistic understanding of SCF particle-formation processes and rigorous descriptions of mass transfer and nucleation processes are being developed[27]. The advances in the understanding of the mechanism of supercritical particle formation and SCF mass transfer are forming the basis for efficient scale-up of the laboratory-scale processes.

Such knowledge will form the basis for efficient scale-up of the laboratory-scale processes generally reported to date. The majority of studies deal with milligram quantities of product prepared by a batch process. For significant commercial viability, demonstration that the processes can be scaled to produce sufficient quantities of material for clinical trials and production batches is required[93].

While many investigators in the laboratory were only able to produce milligrams of the product, Thies and Muller[85] developed a scaled process of ASES capable of producing 200 g of biodegradable PLA microparticles in the size range 6–50 μm. On the other hand, industrial units, such as Bradford Particle Design, have resources for the production of up to 1 ton per year of GMP, cGMP compliant material. Scale-up studies with SEDS SCF processing underpinned by research into the physics, physical chemistry, and engineering of the process using a pilot plant to a cGMP small manufacturing plant have been straightforward[37]. Process conditions optimized at laboratory scale have been directly transferred to the larger scale equipment[52].

Engineering pharmaceutical particles by SCF SEDS processing, and the scalability of the process, provides much theoretical understanding of the process. As ever-increasing demand is made of particles by chemists, formulators, and regulators, for example in terms of chemistry of composition, size, and shape, as well as purity and low residual solvent, the SCF approach is likely to provide wide-ranging opportunities to meet such needs. With a proven ability to process delicate biological materials into stable and active particulates, the SEDS process also provides a much needed simplified and efficient alternative to both spray- and freeze-drying operations.

From a GMP perspective, several additional attractive features can be recognized for SCF particle-formation processes. For the antisolvent-based systems, the processing equipment for the single-stage, totally enclosed process, which is free of moving parts, is constructed from high-grade stainless steel with 'clean in place' facilities available for larger-scale equipment. As well as having reduced solvent requirements compared with conventional crystallization, particle formation occurs in a light- and oxygen-free environment and, if required, moisture-free atmosphere. Although further engineering input is necessary to achieve true continuous collection and recovery of material at operating pressures, 'quasi-continuous' processing is already feasible with a switching device to parallel mounted particle-collection vessels[12].

Cost of manufacturing in pilot scale with SCF technology is comparable with (or may be better than) conventional techniques, such as single-stage spray drying, micronization, crystallization, and milling batch operations. Much has been achieved in a relatively short period since its introduction to pharmaceutical particle engineering, and the future looks attractive for SCF processing.

8.8 Conclusions

SCF technology can be used in the preparation of drug delivery systems and/or to improve the formulation properties of certain drug candidates. SCFs can be used to formulate drug carrier systems due to their unique solvent properties, which can be altered readily by slight changes in the operating temperature and pressure. In recent years, many pharmaceutical and drug delivery companies, some of which are listed in Table 8.2, have adopted SCF technology to obtain drug delivery solutions. The challenges being addressed with this technology include the formulation of poorly water-soluble compounds, obtaining particles of uniform size and shape, avoiding multistep processes, and reducing the excessive use of toxic organic solvents. SCF technology was successfully applied in the laboratory to the preparation of microparticles and nanoparticles or liposomes that encapsulate drug in a carrier, inclusion complexes, solid dispersions, microporous foams, and powders of macromolecules.

As requirements and specifications for 'smart' (those particles that deliver the drug in a controlled way) particles for drug delivery systems become more demanding, the traditional particle preparation and pretreatment procedures are often found to be unsuitable and inadequate. Key issues for emerging replacement technologies are that they provide opportunities for crystal engineering and particle design to be defined scientifically so that by manipulating the process the product can be fine-tuned, and that the process is readily scaled for manufacturing purposes according to the GMP principles and requirements. Recent research[25], development, and applications studies have shown that SF methods for pharmaceutical particle formation provide such a base technology. The SCF antisolvent principle and the SEDS process in particular provide wide scope for the diverse range of organic and biological materials used in single- and multicomponent particulate form in drug delivery systems. Products with targeted properties such as particle size or purity enhancement have been produced. In addition, several studies[40] have successfully addressed the important issue of scale-up and the inherent features of the SCF process enable GMP requirements to be readily accommodated. Indeed, many of the features recognized for an 'ideal' particle-formation process are substantially met by SCF technology.

Table 8.2 Pharmaceutical and drug delivery companies using supercritical fluid technologies

Company name	Market capital
Iomed	$9.69
Gentronics	$21.61
Flamel	$35.64
endorex	$37.70
Antares	$38.49
AP Pharma	$46.70
Elite	$57.77
Access	$58.53
AeroGen	$59.41
StemCells	$63.38
Sonus	$63.71
MexMed	$65.12
DepoMed	$65.70
MacroChem	$88.44
Boject	$99.22
Sheffield	$100.47
Generex	$103.39
Amarin	$114.19
Nastech	$120.91
Cygnus	$140.62
Aradigm	$142.36
Novavax	$215.27
Penwest	$283.86
Emisphere	$326.22
Cima	$356.47
Noven	$368.30
Atrix	$413.54
SkyePharma	$424.20
Durect	$449.79
Inhale	$724.85
Alkermes	$1684.80
Enzon	$2116.67
Andrx	$3445.73
Elan	$4772.94
Average	$503.40
Median	$108.79

However, whilst having many attractive features, further research and development of SCF processing for pharmaceuticals is required to consolidate current understanding and achieve, ultimately, predictive capability for particle design With progress being made, attention should also continue to be directed to modeling the expansion and particle nucleation events in RESS processes, and the rapid cascade of the overlapping physical, mechanical, and chemical events occurring during SCF antisolvent particle-formation methods. Progress in these areas, coupled to engineering studies on plant design for continuous operation for SCF procedures, will undoubtedly strengthen rational approaches to particle design for drug delivery systems and facilitate confident installation of manufacturing scale plant. Indeed, with continuing research in this expanding field, new possibilities are likely to be opened up, especially for biomolecules and bioreactions that are only possible by SF processing.

References

[1] Gutcho M.H. 1976. Microcapsules and Microencapsulation Techniques (Chemical Technology Review series). Noyes Data.

[2] Benita S. 1996. Microencapsulation: Methods and Industrial Applications (Drugs and the Pharmaceutical Sciences: a Series of Textbooks and Monographs). Marcel Dekker.

[3] Baxter G. 1974. In, Microencapsulation: Processes and Applications, Vandegaer J.E. (Ed.). Plenum Press, New York.

[4] Whateley T.L.. 1992. Microencapsulation of Drugs. Drug Targeting and Delivery, Vol. 1. T&F STM.

[5] Lim F. 1984. Biomedical Applications of Microencapsulation. CRC Press.

[6] Gutcho M.H. 1972. Capsule Technology and Microencapsulation. Noyes Data.

[7] Kuhtreiber W.M., Lanza R.P., Chick W.L. and Lanza R.P. 1999. Cell Encapsulation Technology and Therapeutics, 1st edition. Birkhauser, Boston.

[8] Benita S. 1997. Recent advances and industrial applications of microencapsulation. Biomedical Science and Technology, Proc. Int. Symp., 4th, Meeting Data, pp. 17–29.

[9] Brazel C.S. 1999. Microencapsulation: offering solutions for the food industry, Cereal Foods World, 44, 388–390.

[10] Heintz T., Krober H. and Teipel U. 2001. Microencapsulation of reactive materials, Schuettgut 7.

[11] Bencezdi D. and Blake A. 1999. Encapsulation and the controlled release of flavours, Leatherhead Food RA Ind. J., 2, 36–48.

[12] Kiran E., Debenedetti P.G. and Peters C.J. 2000. Supercritical Fluids Fundamentals and Applications, Chapter 7. Kluwer Academic Publishers.

[13] Kondo T. 2001. Microcapsules: their science and technology part III, industrial, medical and pharmaceutical applications, J. Oleo Sci., 50, 143–152.

[14] Kompella U.B. and Koushik K. 2001. Preparation of drug delivery systems using supercritical fluid technology, Crit. Rev. Ther. Drug Carrier Syst., 18(2), 173–199.

[15] Krukonis V.J. 1984. Supercritical fluid nucleation of difficult to comminute solids. Paper 104f presented at AIChE Meeting in San Francisco, California, November 1984.

[16] Tom J.W., Lim G.B., Debenedetti P.G. and Prod'homme R.K. 1993. Applications of supercritical fluids in controlled release of drugs. In, Supercritical Fluid Engineering Science. ACS Symposium Series 514, Brennecke J.F. and Kiran E. (Eds.). American Chemical Society, Washington, DC, p. 238.

[17] Kim J.H., Paxton T.E. and Tomasko D.L. 1996. Microencapsulation of naproxen using rapid expansion of supercritical solutions, Biotechnol. Prog., 12(5), 650–661.

[18] Bastian P., Bartkowski R., Kohler H. and Kissel T. 1998. Chemo-embolization of experimental liver metastases. Part I: distribution of biodegradable microspheres of different sizes in an animal model for the locoregional therapy, Eur. J. Pharm. Biopharm., 46(3), 243–254.

[19] Benoit J.P., Fainsanta N., Venier-Julienne M.C. and Meneibed P. 2000. Development of microspheres for neurological disorders: from basics to clinical applications, J. Control. Release, 65(1–2), 285–296.

[20] Taguchi T., Ogawa N., Bunke B. and Nilsson B. 1992. The use of degradable starch microspheres with intra arterial for the treatment of primary and secondary liver tumours – results of phase III clinical trial, Reg. Cancer Treat., 4, 161–165.

[21] Cohen S. and Bernstein H. (Eds.). 1996. Microparticulate Systems for the Delivery of Proteins and Vaccines (Knutson *et al.* chapter). Marcel Dekker, New York.
[22] Jallil R. and Nixon J.R. 1990. Microencapsulation using poly(L-lactic acid). III. Effect of polymer weight on the microcapsule properties, J. Microencapsul., 7(1), 41–52.
[23] Langer R. and Folkman J. 1976. Polymer for the sustained released of proteins and other macromolecules, Nature, 263, 797–800.
[24] Benoit J.P., Rolland H., Thies C. and Vande V.V. 2000. Method of coating particles and coated spherical particles. US Patent No. 6-087-003.
[25] York P. *et al.* 1998. In, Proc. Resp. Drug Delivery VI, Hilton Head, USA, pp. 169–175.
[26] Jung J. and Perrut M. 2001. Particle design using supercritical fluids: Literature and patent survey, J. Supercritical Fluids, 20, 179–219.
[27] Larson K.A. and King M.L. 1986. Evaluation of supercritical fluid extraction in the pharmaceutical industry, Biotechnol. Prog., 2, 73–82.
[28] Shekunov B.Yu. *et al.* 1999. Crystallization process in turbulent supercritical flows, J. Cryst. Growth, 198/199, 1345–1351.
[29] Shekunov B.Yu. *et al.* 1998. Pharm. Res., 15, S162.
[30] Ruchatz F., Kleinebudd P. and Müller B. *et al.* 1997. Residual solvents in biodegradable microparticles. Influence of process parameters on the residual solvent in microparticles produced by the aerosol solvent extraction system (ASES) process, J. Pharm. Sci., 86, 101–105.
[31] Phillips E.M. and Stella V.J. 1993. Rapid expansion from supercritical solutions: application to pharmaceutical processes, Int. J. Pharm., 94, 1–10.
[32] Ventosa N., Sala S. and Veciana J. 2003. DELOS process: a crystallization technique using compressed fluids, 1. Comparison to the GAS crystallization method, J. Supercritical Fluids, 26, 33–45.
[33] Reverchon E. and Della Porta G. 2003. Micronization of antibiotics by supercritical assisted atomization, J. Supercritical Fluids, 26, 243–252.
[34] Sievers R.E., Huang E.T.C., Villa J.A., Engling G. and Brauer P.R. 2003. Micronization of water-soluble pharmaceuticals and model compounds with a low-temperature bubble dryer, J. Supercritical Fluids, 26, 9–16.
[35] Debenedetti P.G. 1994. In, Supercritical Fluids: Fundamentals for Application, Vol. 273. NATO ASI Series E, 250–252.
[36] Reverchon E., Della Porta G., Taddeo R., Pallado P. and Stassi A. 1995. Solubility and micronization of griseofulvin in supercritical CHF_3, Ind. Eng. Chem. Res., 34, 4087–4091.
[37] McHugh M. and Krukonis V.J. 1994. Special Applications in Supercritical Fluid Extraction: Principles and Practice, 2nd edition. Butterworth-Heinemann, Boston.
[38] Weidner E. *et al.* 1996. In, High Pressure Chemical Engineering, Process Technology Proceedings, Vol. 12, Elsevier, 121–124.
[39] Jung J. and Perrut M. 2001. Particle design using supercritical fluids: literature and patent survey, J. Supercritical Fluids, 20, 179–219.
[40] Palakodaty S. and York P. 1999. Phase behavioral effects on particle formation processes using supercritical fluids, Pharm. Res., 34, 976–985.
[41] Smith R.D. and Wash R. 1986. US Patent No. 4-582-731, 15 April 1986 (Priority: 1 September 1983).
[42] Matson D.W., Fulton J.L., Petersen R.C. and Smith R.D. 1987. Rapid expansion of supercritical fluid solutions: solute formation of powders, thin films, and fibers, Ind. Eng. Chem. Res., 26, 2298–2306.
[43] Matson D.W., Petersen R.C. and Smith R.D. 1987. Production of powders and films from supercritical solutions, J. Mater. Sci., 22, 1919–1928.
[44] Petersen R.C., Matson D.W. and Smith R.D. 1987. The formation of polymer fibers from the rapid expansion of supercritical fluid solutions, Polym. Eng. Sci., 27, 1693–1697.

[45] Smith R.D. 1988. US Patent No. 4-734-451.
[46] Lele A.K. and Shine A.D. 1992. Morphology of polymers precipitated from a supercritical solvent, AIChE J., 38(5), 742–752.
[47] Phillips E.M. and Stella V.J. 1993. Rapid expansion from supercritical solutions: application to pharmaceutical processes, Int. J. Pharma., 94, 1–10.
[48] Reverchon E. and Taddeo R. 1993. Morphology of salicylic acid crystals precipitated by rapid expansion of a supercritical solution. In, I Fluidi Supercritici e Le Loro Applicazioni, Reverchon E. and Schiraldi A. (Eds.), 20–22 June 1993, Ravello, Italy, pp. 189–198.
[49] Berends E.M., Bruinsma O.S.L. and van Rosmalen G.M. 1994. Supercritical crystallization with the RESS process: experimental and theoretical results. In, Brunner G. and Perrut M. (Eds.), Proceedings of the 3rd International Symposium on Supercritical Fluids, Tome 3, 17–19 October 1994, Strasbourg, France, pp. 337–342, 4087–4091.
[50] Reverchon E. and Pallado P. 1996. Hydrodynamic modelling of the RESS process, J. Supercritical Fluids, 9, 216–221.
[51] Domingo C., Wubbolts F.E., Rodriguez-Clemente R. and van Rosmalen G.M. 1997. Rapid expansion of supercritical ternary systems: solute + cosolute $+CO_2$. The 4th International Symposium on Supercritical Fluids, 11–14 May 1997, Sendai, Japan, pp. 59–62.
[52] McHugh M. and Krukonis V. 1994. Supercritical Fluid Extraction Principles and Practice, 2nd edition, Butterworth-Heinemann.
[53] Nagahama K. and Liu G.T. 1997. Supercritical fluid crystallization of solid solution. The 4th International Symposium on Supercritical Fluids, 11–14 May 1997, Sendai, Japan, pp. 43–46.
[54] Godinas A., Henriksen B., Krukonis V., Mishra K.A., Pace G.W. and Vachon G.M. 1998. US Patent No. 0-089-852.
[55] Kim J.H. *et al.* 1996. Microencapsulation of naproxen using rapid expansion of supercritical solutions, Biotechnol. Prog., 12, 650–661.
[56] Mueller B.W. and Fisher W. 1989. Manufacture of sterile sustained release drug formulations using liquefied gases. W. Germany Patent No. 3-744-329.
[57] Brennecke J.F. and Eckert C.A. 1989. Phase equilibria for supercritical fluid process design, Am. Inst. Chem. Eng. J., 35, 1409–1427.
[58] Wubbolts F.E. *et al.* 1998. In, Proc. 5th Int. Symp. Supercrit. Fluids, Soc. Adv. Sup. Fluids, Nice, France.
[59] Mishima K., Matsuyama K., Uchiyama H. and Ide M. 1997. Microcoating of flavone and 3-hydroxyflavone with polymer using supercritical carbon dioxide. The 4th International Symposium on Supercritical Fluids, 11–14 May 1997, Sendai, Japan, pp. 267–270.
[60] McHugh M.A. and Guckes T.L. 1985. Separating polymer solutions with supercritical fluids, Macromolecules, 18, 674–681.
[61] Seckner A.J., McClellan A.K. and McHugh M.A. 1988. High pressure solution behavior of the polymer–toluene–ethane system, AIChE J., 34, 9–16.
[62] Robertson J., King M.B., Seville J.P.K., Merrifield D.R. and Buxton P.C. 1997. Recrystallisation of organic compounds using near critical carbon dioxide. The 4th International Symposium on Supercritical Fluids, 11–14 May 1997, Sendai, Japan, pp. 47–50.
[63] Liau I.S. and McHugh M.A. 1985. Supercritical Fluid Technology, Elsevier Science, Amsterdam.
[64] Tom J.E. 1993. Supercritical fluid engineering science: fundamentals and applications, ACS Symp. Ser., 514, 238–257.
[65] Yeo S.D. *et al.* 1993. Formation of microparticulate protein powders using a supercritical fluid antisolvent, Biotechnol. Bioeng., 45, 341–346.
[66] Bleich J. *et al.* 1993. Aerosol solvent extraction system – a new microparticle production technique, Int. J. Pharm., 97, 111–117.
[67] Dixon D.J., Johnston K.P. and Bodmeier R.A. 1993. Polymeric materials formed by precipitation with a compressed fluid antisolvent, AIChE J., 39, 127–139.

[68] Kiran E. and Zhuang W. 1997. Supercritical fluids: extraction and pollution prevention, ACS Symp. Ser., 670, 2–36.

[69] Hanna M. and York P. 1994. Patent WO 95/01221.

[70] Liau I.S. and McHugh M.A. 1985. Supercritical Fluid Technology. Elsevier Science, Amsterdam, p. 415.

[71] Weidner E., Steiner R. and Knez Z. 1996. In, Powder Generation from Polyethyleneglycols with Compressible Fluids. High Pressure Chemical Engineering, Rudolf von Rohr P. and Trepp C. (Eds.). Elsevier Science.

[72] Sievers R.E., Miles B.A., Sellers S.P., Milewski P.D., Kusek K.D. and Kluetz P.G. 1998. New process for manufacture of one-micron spherical drug particles by CO_2-assisted nebulization of aqueous solutions, Proceedings from Respiratory Drug Delivery IV Conference, Hilton Head, South Carolina, 3–8 May 1998, pp. 417–419. In, Supercritical Fluids, Chemistry and Materials, Poliakoff M., George M.W. and Howdle S.M. (Eds.). Nottingham, 10–13 April 1999.

[73] Sievers R.E. and Karst U. European Patent No. 0-677-332, 1995. US Patent No. 5-639-441, 1997.

[74] Weidner E., Knez Z. and Novak Z. 1995. European Patent No. EP 0-744-992, February 1995. Patent WO 95/21688, July 1995.

[75] Steckel H., Thies J. and Muller B.W. 1997. Micronizing of steroids for pulmonary delivery by supercritical carbon dioxide, Int. J. Pharm., 152(1), 99–110.

[76] Pace G.W., Vachon M.G., Mishra A.K., Henrikson I.B. and Krukoniz V. 2001. Processes to generate submicron particles of water-insoluble compounds. US Patent No. 6-177-103.

[77] Hile D.D., Amirpour M.L., Akgerman A. and Pishko M.V. 2000. Active growth factory delivery from poly(D,L-lactide-co-glycolide) foams prepared in supercritical CO_2, J. Control. Release, 66(2–3), 177–185.

[78] Frederiksen L., Anton K., Hoogevest P.V., Keller H.R. and Leuenberger H. 1997. Preparation of liposomes encapsulating water-soluble compounds using supercritical carbon dioxide, J. Pharm. Sci., 86(8), 921–928.

[79] Castor T.P. and Chu L. 1998. Methods and apparatus for making liposomes containing hydrophobic drugs, US Patent No. 5-776-486.

[80] Lin S.Y. and Kao Y.H. 1989. Solid particulates of drug-b-cyclodextrin inclusion complexes directly prepared by a spray-drying technique, Int. J. Pharm., 56, 249–259.

[81] Kamihara H., Asai T., Yamagata M., Taniguchi M. and Kobayashi T. 1990. Formation of inclusion complexes between cyclodextrins and aromatic compounds under pressurized carbon dioxide, J. Ferment. Bioeng., 69, 350–353.

[82] Van Hees T., Piel G., Evrard B., Otte X., Thunus T. and Delattre L. 1999. Application of supercritical carbon dioxide for the preparation of a piroxicam-beta-cyclodextrin inclusion compound, Pharm. Res., 16(12), 1864–1870.

[83] Debenedetti P.G., Lim G.B. and Prud'Homme R.K. 2000. Preparation of protein microparticles by precipitation. US Patent No. 6-063-910.

[84] Tservistas M., Levy M.S., Lo-Yim M.Y.A., O'Kennedy R.D., York P., Humphery G.O. and Hoare M. 2000. The formation of plasmid DNA loaded pharmaceutical powders using supercritical fluid technology, Biotech. Bioeng., 72(1), 12–18.

[85] Thies J. and Muller B.W. 1998. Size controlled production of biodegradable microparticles with supercritical gases, Eur. J. Pharm. Biopharm., 45(1), 67–74.

[86] Kumagai H., Hata C. and Nakamura K. 1997. CO_2 sorption by microbial cells and sterilization by high-pressure CO_2, Biosci. Biotech. Biochem., 61(6), 931–935.

[87] Dillow A.K., Langer R.S., Foster N. and Hrkach J.S. 2000. Supercritical sterilization method, US Patent No. 6-149-864.

[88] Castor T.P. and Hong G.T. 2000. Methods for the size reduction of proteins, US Patent No. 6-051-694.

[89] Winters M.A. *et al.* 1996. Precipitation of proteins in supercritical carbon, J. Pharm. Sci., 85, 586–594.

[90] Schmitt W.J. 1995. Finely-divided powders by carrier solution injection into a near or supercritical fluid, AIChE J., 41, 2476–2486.

[91] Forbes R.T. 1998. Supercritical fluid processing of proteins. I: Lysozyme precipitation from organic solutions, In, Proceedings of the IChE World Congress on Particle Technology 3, Brighton, UK, pp. 180–184.

[92] Kamihara H., Asai T., Yamagata M., Taniguchi M. and Kobayashi T. 1990. Formation of inclusion complexes between cyclodextrins and aromatic compounds under pressurized carbon dioxide, J. Ferment. Bioeng., 69, 350–353.

[93] Steckel H. and Müller B.W. 1998. Metered-dose inhaler formulation of fluticasone-17-propionate micronized with supercritical carbon dioxide using the alternative propellant HFA-227, Int. J. Pharm., 173, 25–33.

9

Fine–Structured Materials by Continuous Coating and Drying or Curing of Liquid Precursors

L.E. Skip Scriven

9.1 Introduction

Coatings and films produced by depositing a liquid layer and subsequently solidifying it are vital ingredients of products such as papers for printing; multilayer polymer films for packaging and specialty uses; adhesive labels, patches, and tapes; photographic and graphic art materials; photoresist preparations and thin sheets of ceramic materials for microelectronics and other applications; magnetic and optical memory media; electrical conductors, photoreceptor drums, and protective and decorative surface layers in engineering, architectural, textile, and manifold consumer materials; and all sorts of laminates and many other composites. The interior of many coatings and films requires a particular microstructure or nanostructure in order to function as intended, whether optically, photochemically, electronically, magnetically, or mechanically.

'Nanostructure' commonly means scales of 999 nm (0.999 μm) downward; 'microstructure' means scales of 0.1 μm (100 nm) upward. Coatings deposited as liquids range in *solid* thickness from around 200 nm to more than 500 μm. Requisite internal structures range down to 12 nm scale, for instance in certain permselective and catalytic coatings.

The future will demand continuous processes that deliver fine-scale, intricate, precisely controlled structures. Indicators for this include the emerging technologies of flat-panel and flexible displays based on variable state encapsulates and on organic light-emitting diodes, and incipient developments of flexible nanoelectronics based on advanced polymers and colloids – all accompanied by visions of lower cost processing on flexible

Chemical Engineering Edited by Miguel A. Galán and Eva Martin del Valle

substrates that can be rolled up and unrolled during manufacture, just like many of today's coatings, green tapes, and films. Visionaries in Silicon Valley and elsewhere have been foreseeing 'plastic electronics' and 'plastic photonics', 'reel-to-reel' or 'roll-to-roll', and largely based on coating- and printing-type processing. The future needs will drive innovation and optimization. They will also drive deepening research into the fundamentals of fine-scale structure development in coating processes. Emerging technologies, added to competitive pressures across the enormous range of current applications, will intensify the mounting demand for engineers and scientists well schooled in coating science and engineering.

The *discipline* of coating science and technology has taken shape over the past 30 years. Although its center of gravity lies in the overlapping domains of chemical engineering and mechanical engineering, it is scarcely recognized in either. Also, it intersects polymer and ceramic science and engineering, colloid and interface science, and several other areas of applied physics and chemistry. The discipline is in fact interdisciplinary. But this, along with its grounding in fluid mechanics, colloidal and interfacial phenomena, mass and heat transport, phase and chemical transformations, product and process engineering, control and optimization, makes it a microcosm of chemical engineering. What follows is a sketch of the discipline and its trends and challenges.

9.2 Terms of Reference

Figure 9.1 is a schematic of the characteristic unit operations in the coating of a continuous flexible substrate by depositing a liquid layer and solidifying it.

The solid surface to be coated is called the *substrate*. It may be only slightly deformable: a slab, panel, bar, pipe, drum, wafer, disk, ball, or more complicated shape. Or, it may be thin and flexible: a sheet, film, wire, or fiber. It may in addition be compressible, as are paper, paperboard, nonwoven and woven fabrics. When the substrate is a sheet, fabric, or film that is very long or, because of on-line splicing to the next length is

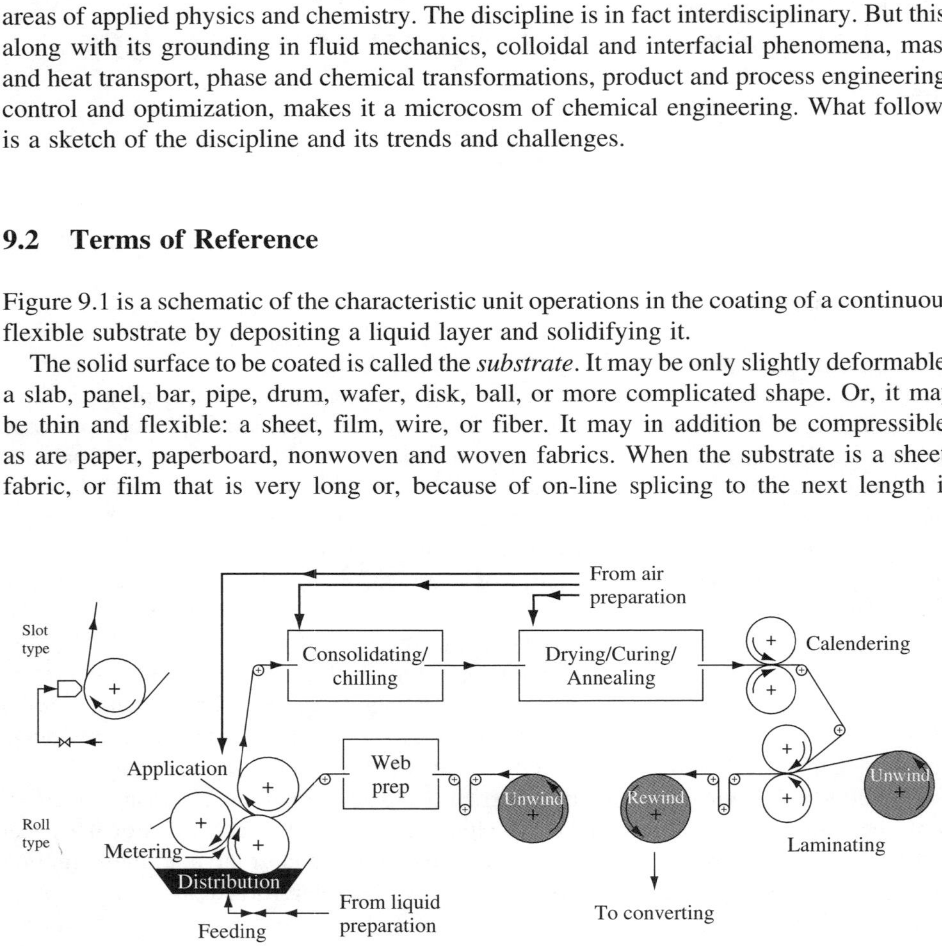

Figure 9.1 *Unit operations of generic 'roll-to-roll' coating. Roll changers with on-the-fly splicing at the unwind and rewind make the process continuous. The five basic elements are feeding, distributing, metering, applying, and solidifying (consolidating, drying, curing)*

essentially continuous, it is often called a *base*, a *web*, or a *strip*. *Web processing* means operations in which the starting material is unwound from a roll or, with splicing, a sequence of rolls, and the finished product is wound up into rolls; this is known in piece-by-piece technologies as *roll-to-roll processing*. For steel and aluminum sheet the term is *coil coating* or, in some instances, *strip coating*. The substrate may be rough or smooth, porous or impermeable. Paper, a rough and porous nonwoven fabric, is probably the substrate coated in largest amount of tonnage. Substantial amounts of paper and other fabrics are impregnated as they are coated. Steel and aluminum strip, like many other irregular or rough substrates, is often coated by first coating a rubber-covered transfer or offset roll, and then wiping that roll against the substrate to transfer a layer of liquid. But the most advanced coating technologies produce uniform layers on relatively smooth polymer films and calendered nonwovens, and on extremely smooth glass and semiconductor surfaces. In related *casting* technologies, the liquid layer is deposited on the very smooth surface of a large roll (rotating cylinder) or endless belt from which it is stripped or peeled after it has adequately solidified.

In contrast to all liquid-applied coating is continuous web, wire, or fiber coating. *Spin coating* of flat or nearly flat pieces, up to half a meter or so across, is a batch process ubiquitous in microelectronics, microphotonics, lab-on-a-chip developments, and a variety of emerging technologies. Another method of coating pieces is to dip them in liquid, then withdraw them slowly and carefully before or as they solidify; this is the batch version of *dip coating* – sometimes called *withdrawal coating*.

The liquid to be coated, or cast and stripped, may be a virtually pure compound, e.g. reactive monomer or molten block-copolymer. Usually it is a solution, a colloidal dispersion, a particulate suspension, a liquid crystal, or a melt. More and more often two or more liquids (almost always *miscible* liquids) are deposited as superposed layers simultaneously.

In formulating a liquid to be deposited by flow, a crucial factor is its *rheology*, specifically its *viscosity* and *viscoelasticity* at deposition conditions. Too low a viscosity may leave the deposited liquid layer – especially a thicker one – unacceptably susceptible to rearrangement by gravity, centrifugal force, air impingement, or drag. Too high a viscosity may unacceptably lower the deposition speed at which not enough air at the substrate surface is replaced by liquid. Or too a high viscosity may call for greater mechanical potential gradient (pressure, gravity) than can be made to work in any coating flow. Such high viscosity liquids fall in the overlapping province of *polymer extrusion*. Liquids with enough viscoelasticity fall entirely in that province: the criteria are strong enough tensile streamwise and compressive crosswise elastic stress that the relatively sharp turns and highly curved free surfaces of coating flows would cause flow instability and nonuniformities.

So the liquid to be coated may be *Newtonian* or virtually so, i.e. with viscosity independent of shear rate and extension rate (though of course sensitive to temperature and composition). Its shear viscosity may be anything from the range of water and low molecular weight organic solvents (0.01 P, or 1 mPa s) to that of concentrated suspensions and molten polymers of moderate molecular weight (upwards of 1000 P, or 100 Pa s). More often it is *shear thinning*, with high-shear-rate viscosity toward the lower part of that range. Sometimes, in cases of concentrated particulate suspensions, it is *shear thickening* and *jamming* at the highest shear rates seen in coating flows (around 10^6–10^7/s). Its *extensional viscosity* may be anything from Newtonian to moderately non-Newtonian (e.g. extension thickening), although relevant extension rates in coating flows are all too frequently higher than can be attained with contemporary rheological characterization instrumentation. The liquid may be slightly viscoelastic, even modestly so, although this, too, is still difficult to characterize. *Coating rheology* is at the frontier of research and instrumentation.

For irregular surfaces and grossly 3D shapes such as the exteriors of vehicles, appliances, and furniture, *spray coating* is often the method of choice. It is a method in which liquid solution, with or without particulate pigment, is dispensed through an atomizing nozzle and delivered stochastically as discontinuous droplets, most often with the aid of an electrostatic field. Once landed on the surface, the droplets must *coalesce* and, to some degree, level. Smoothness through *leveling* is an issue whenever a coating exists as a liquid layer before it solidifies. Ink-jet printing is a highly deterministic droplet-deposition process that has opened the way to a kind of spray coating that affords exquisite control of coating thickness and patterns down to submillimeter scale; however, coalescence and leveling remain issues.

For thicker coatings on irregular surfaces and 3D shapes the leading alternative to spray coating is *powder coating*, which avoids the former's organic or aqueous solvent. It is also employed to achieve reproducibly corrugated or 'wrinkled' coatings. Oligomer or low-molecular-weight polymer, cross-linker, and pigments are blended, heated, extruded, cooled, and ground to powder that is dispersed through an electrostatic air gun and delivered discontinuously as charged particles to the grounded surface. Arriving particles form a powder layer that is heated along with the substrate and must expel trapped air as it melts, as well as coalescing and leveling to some degree before the molten form cures to high viscosity liquid and then to solid.

The liquid phase is not involved in the sputtering, physical and chemical vapor deposition that are common in electronics, magnetics, and photonics. In this important class of processes that create ultrathin coatings – submicrometer down to atomic thicknesses – gas phase constituents are delivered continuously to the substrate surface and once there deposit, condense, or react. Typically, the structure of the coating is so thin and far from equilibrium that it cannot be attained by liquid-phase transport. But when it can be, low-density gas-phase transport and high heat-release condensation are vulnerable to replacement by liquid-phase processes such as electrodeposition from a flowing solution or even drying of a dilute solution layer delivered by a coating flow.

Coating by depositing a liquid layer is necessarily followed by some degree of *solidification*, which means the layer must acquire appreciable *elastic modulus* and *strength* against yielding and rupturing. The degree ranges from low in the case of pressure-sensitive adhesives to very high in the case of dense metal oxides. In the related process of casting a film, the stripping or peeling requires a less degree of solidification.

In continuous coating, beyond the take-away zone of the coater comes the *solidification zone*, or *zones*. Solidification may begin by *consolidation* of a suspension, for example by liquid sorption into the substrate, or by gelation of a polymer solution or colloidal dispersion, for example by chilling (as in the case of gelatin solutions) or by heating that induces particle swelling and flocculation (as in the case of plastisols). In hot melt coating and hot embossing, a kind of micromolding, cooling the molten polymer is all that is needed. More commonly the deposited liquid layer, or layers, are solidified by *solvent removal*, *phase change*, *colloidal coagulation*, *chemical reaction*, or combinations of these. Some coatings such as adhesives barely solidify: all that is needed is enough polymer entanglement or crosslinking to elevate viscosity and impart some elasticity. Chemical reactions of addition polymerization and crosslinking or of condensation polymerization, often referred to as *curing*, may be induced or hastened by activating *catalysts*. Activation can be by exposure to moisture or oxygen, by out-diffusion or breakdown of blocking agents, by heating, or by radiative bombardment with ultraviolet light or electron beam. However, the commonest means of solidification is solvent removal by hot-air *drying*, though this may be accompanied or followed by curing to arrive at the needed hardness, toughness, insolubility, or impenetrability.

If the coated liquid contains highly volatile components, the drying zone may intrude upstream into the take-away zone and even into the coating bead; this is often a leading complication of *solvent coating* as against *water-borne coating*. Likewise if the liquid is polymerizable and reaction is already initiated, the curing zone may intrude upstream; or if the liquid is actually a suspension and on the verge of colloidal instability, the flocculation zone may intrude upstream. Generally it is advantageous to separate coating and solidification in order to be able to control better each of these two parts of any coating process.

Coating and solidification are of course closely linked: typically, the more fluid the starting liquid, the easier it is to feed, distribute, meter, and deposit on the substrate; but the more drying or reaction it takes to solidify. For example, the greater the amount of dissolved polymer in a solution coating, the less solvent there is to remove by drying, but the higher the viscosity and viscoelasticity of the liquid and the harder it is to coat well. Similarly, the higher the volume fraction of particles in a suspension coating, the less solvent there is to dry, but the higher the viscosity and difficulties of coating. Similarly again, the higher the degree of oligomerization of a '100 percent solid' reactive coating, the less the polymerization, crosslinking, and stress development in the curing step, but the more viscous and difficult to coat is the precursor oligomer or polymer. On the other hand, the lower the viscosity (and the greater the thickness) of a liquid layer arriving at a drier, the weaker the convective airflow that can unlevel it, and so the more gently its drying must begin. Thus there is a basic trade-off between the coating and solidification parts of a process.

9.3 Depositing a Liquid Layer

9.3.1 Coating Flows

An astonishing variety of methods have been developed for delivering a liquid phase continuously in industrial practice. The highly simplified diagrams of Figure 9.2 include most of the methods of any importance. Those called two-layer slot, multilayer slide, and multilayer curtain are the ones by which two or more superposed layers can be deposited simultaneously. Not only do they avoid inefficiencies of coating and drying each layer of a multiple-layer structure in succession, they also make carrier layers and uncoatably thin layers possible.

Dip coating, the first two diagrams, is perhaps the simplest and probably the oldest way of depositing a precisely uniform layer of liquid less than a poise or so in viscosity. The faster the surface is withdrawn from the coating bath or the higher the liquid's viscosity, the thicker the layer that is formed, up to a limit. Consequently, dip coating

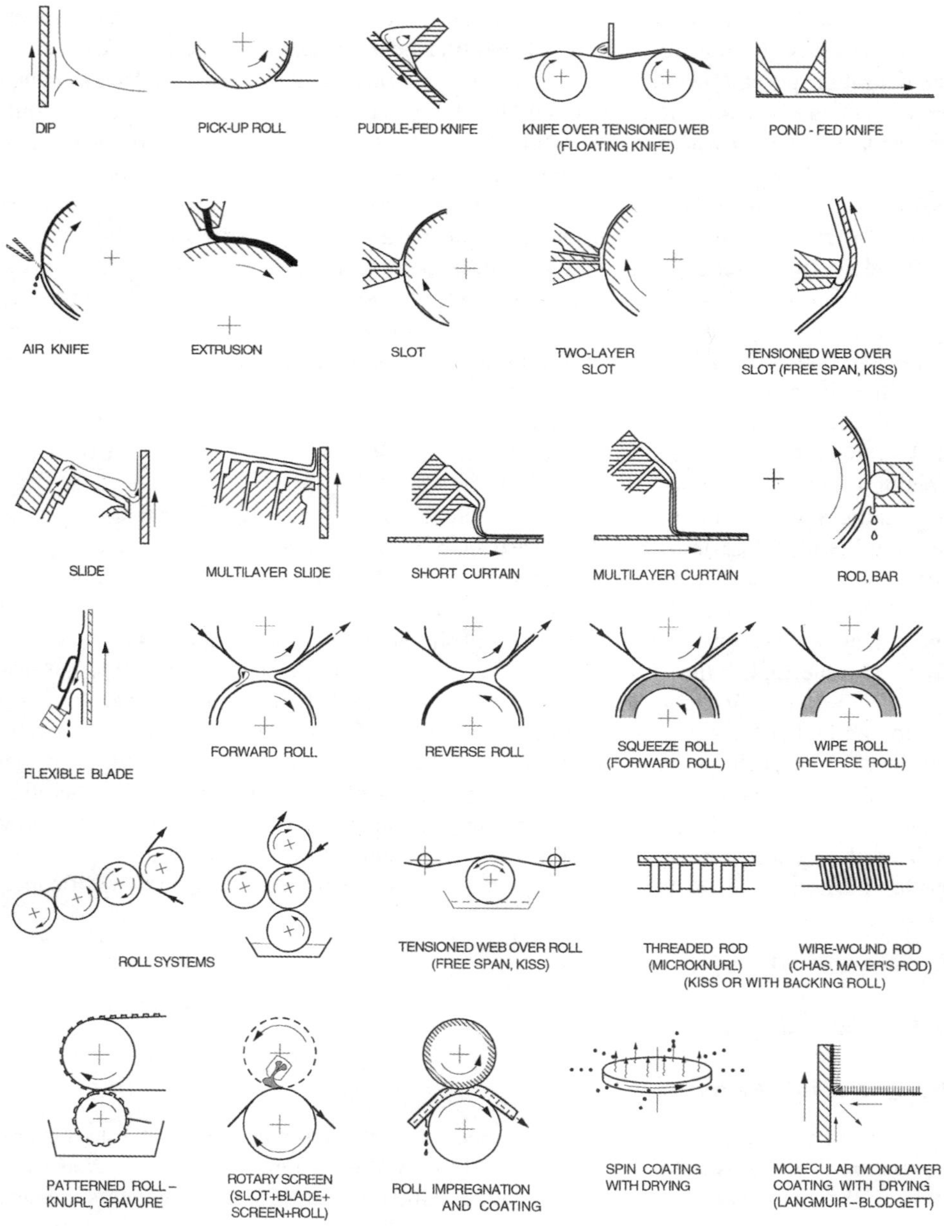

Figure 9.2 *Coating flows: simplified diagrams*

is today seldom encountered outside the laboratory, except in the second diagram where the surface being coated is a 'pick-up' roll partly immersed in a pool of liquid of modest viscosity so that as it turns it carries away a layer that can be thinned or split before it is coated or stripped.

Monolayer transfer coating, the last diagram, represents the ultimate in thin layer deposition, because anything less than a coherent molecular monolayer simply does not constitute a coating. First demonstrated by Katherine Blodgett over 60 years ago in Irving Langmuir's laboratory, monolayer transfer coating has continued to be a ticklish, slow, batch process. With the upsurge during the past decades in creating ultrathin coatings containing molecular-scale structures, studies were launched in many places that have not yet led to an industrially applicable continuous process, but may ultimately do so.

To understand the variety of methods and how they relate to one another, it is helpful to recognize that they all perform the same set of basic functions.

9.3.2 Basic Functions in Coating

All of these coating methods perform four of the five basic functions one way or another. They *feed* liquid through a pipe, or sometimes a manifold of pipes, and *distribute* it across the width of the coater, which is no less than the width of the substrate to be coated. They *meter* the flow rate or they *meter* the coated layer to the required wet thickness. They *apply* the liquid coating to, or deposit it on, the substrate, or web, displacing most of the air (or other gas) that originally contacts the surface. Indeed, to coat is to replace gas at a substrate surface by a layer of liquid. Beyond these four functions – feeding, distributing, metering, and applying – there is a fifth. That is to *solidify* the coating; it may be necessary also to *anneal* it to reduce residual stresses or develop structure. The basics of solidification are examined below.

Feeding. Whenever liquid is applied to a substrate in excess and the excess is subsequently metered off, the feed method is not critical. This situation is sometimes called post-metering, in contrast to pre-metering. Nor is the feed method critical when distribution is by a pond or pool with overflow. Whenever liquid is *not* applied in excess – in other words, when all the liquid fed is coated – the feed method does become critical. This is the situation called pre-metering. *Volumetrically controlled feeding* is most often accomplished by a positive displacement pump operated as a meter, i.e. under small enough pressure difference that the leakage flow through the pump is tolerable. The term, feeding, is also applied to arrangements that combine feeding, distribution, and in some instances metering functions as well, to supply a layer or film of coating to the entire width of the next element in a coating operation. Prime examples are slot feeding and slide feeding of curtain coating and various arrangements used in roll coating. Multilayer coating, which is the *simultaneous application* of two or more layers of coating liquid, generally requires that each layer be *separately* pre-metered.

Distribution. From the supply pipe or delivery point, the coating liquid must be spread to the width to be coated, or perhaps to somewhat greater width. Simplest is a *pool in a pan* from which liquid is withdrawn by a roll or by substrate dipping into it or by substrate passing through it. The pool's simplicity comes with potential for great difficulties, however. Open pools are prey to sloshing, bubbles and foam generated by arriving liquid, roll surface, or substrate; to waves excited by ever-present mechanical and acoustical vibrations; to contamination falling out of the air overhead; to evaporation of volatile components into the air overhead, and resulting concentration gradients in the remaining liquid; and to long mean residence time. The outcome of the last can be disastrous for coating liquid that is reacting chemically or colloidally to form, for example, extra-viscous gobs, blobs, or flocs. Consequently, there is a universal hierarchy of improvements to simple pool distribution: splashing and sloshing are reduced, the pan is covered over, the free liquid surface is made smaller, the pan is shaped and shrunk in size, and ultimately evolves into a distribution chamber. Pressure difference must be employed, and in such a way that the pressure across the width of the distribution chamber is close enough to being uniform.

Feed liquid is spread across die coaters, slot and extrusion coaters, slide and many curtain coaters by *chamber-and-slot distribution*. The liquid is introduced into a chamber or manifold as long as the coating is to be wide, which delivers the liquid in turn to a narrow slot through which it flows to an exit aperture where it may be applied to a substrate, turn and run down an apron or slide, or fall free as a curtain.

When the flow per unit width across the exit aperture of a chamber-and-slot arrangement can be made uniform enough, distribution doubles as metering. Indeed, the distribution and metering functions are inextricably combined in this manner in slot, extrusion, slide, and curtain coaters that employ chamber and slot and operate in the pre-metered feed mode. The combination is ubiquitous in multilayer coating.

Metering. The ultimate would be to generate a liquid layer of perfectly uniform thickness and microstructure on a perfectly smooth substrate. This can be approached by applying a distributed excess of coating liquid and metering off the excess, that is, by *post-metering* as in bar, rod, and blade coating. Or, it can be approached by distributing and *pre-metering* the liquid directly onto the substrate, as in extrusion coating and sometimes curtain and slide coating. Or, it can be approached via *pre-metering* onto a perfectly smooth intermediate substrate and then applying, or transferring, the layer to the final substrate, as in some kinds of roll coating.

Ideal *pre-metering* delivers a layer or layers of uniform thickness everywhere regardless of the topography of the substrate (any subsequent leveling flow might reflect that topography as it shifted liquid so that the pre-metering action was partly undone). Ideal *post-metering* delivers a single layer of thickness so modulated to the substrate topography that the surface of the liquid is uniformly smooth. Ideal *smoothing*, or *planarization* as it is known in microelectronics technology, is a type of *re-metering* that converts a liquid layer whose surface is not smooth into one whose surface is, without altering the average coat weight, i.e. the average layer thickness.

There is an intermediate category of *in situ metering* that pertains to situations where metering and application take place simultaneously, as in those kinds of roll coating that have substrate (web) in the metering gap. Then the substrate, its thickness variations, and failures to lie tight against the roll that carries it all contribute to the local thickness of the liquid layer that is deposited.

The advantage of *gravure-roll metering* can be accurate metering of a liquid layer that is on average quite thin, but at the expense of re-metering the layer by smoothing. A further cost can be that of controlling the metering action at the cell level, which may bring extensional viscous force into prominence and viscoelastic forces too if the liquid can develop them. These can give rise to deleterious filamentation and misting as coating speed rises. Electrostatic force is not infrequently deployed to assist in achieving the desired action.

Compliant-gap metering is characterized by one or both gap walls, or their mountings, being deformable by forces developed in the flowing liquid. Those forces are pressure above all, but also viscous wall shear and, in certain nonlinear and many viscoelastic liquids, extra normal forces. As a wall or its mounting deforms, its elasticity adds to any opposing force preloaded onto it, until that force sum and the hydrodynamic forces exerted by the liquid come into balance in a state that is steady – apart from unavoidable fluctuations in time, for example when rough or porous substrate is being coated. As a wall deforms, the gap through which the liquid flows changes, and so do the forces exerted by the liquid. Thus the compliance, or elastic response, of the walls and the flow of the liquid are coupled together. Such situations are the subject of elastohydrodynamics; hence compliant-gap metering can also be called *elastohydrodynamic metering*.

Application. This is the basic action of turning a more or less dry substrate or web into a wet one, and it can be carried out before or after the metering function, and even as part of the distribution function. The heart of application is the *dynamic wetting zone*, which looks like a 'line' and is often called one, although it is in fact a microscopic or submicroscopic region in which wetting forces are active and the coating liquid replaces gas at the solid surface.

Two free surfaces are present in every coating flow. One extends from a separating contact line (the 'upstream' one), a static contact line where it departs from a wall of the coating set-up, to the wetting 'line' where it meets the substrate surface. This free surface is the key player in the basic action, to coat: it becomes the interface between the liquid layer and the substrate to which it is applied. The closer the upstream separating contact line is to the dynamic wetting line, the more desirable for precision coating it is to design the coating set-up and operation so that the separating contact line is straight.

The other free surface originates at another separating contact line – the 'downstream' one – a static contact line where it departs from a wall of the coating device, and extends downstream to become the free surface of the coated liquid (where it may be subjected to smoothing or allowed to level). The shorter the distance between the downstream separating contact line and the free surface of the coated layer, the more critical for precision coating it is that the separating contact line be perfectly straight, a condition best attained by giving it a straight sharp edge to attach to and then operating so that it does. The region between the two free surfaces is, in many cases, called the 'coating bead'. There, especially, flow nonuniformity and unwanted microvortices can influence the final microstructure of multilayer coatings and coatings containing acicular, tabular, or other types of particles.

Plunging application takes place where substrate enters an open pool, pond, puddle, or fountain. Flow around the dynamic wetting line is comparatively *unconfined* so that the free surface, which is the upstream one, is fairly free to deform into a local meniscus and ultimately to entrain air. Generally, the substrate exits elsewhere and so there is no compact coating bead.

Confined flow application takes place where substrate enters the liquid through an upstream meniscus confined in a slot, or a slit sealed by a flexible blade, or a gap between walls, one or both of which are moving. Flow around the dynamic wetting line is comparatively *confined*, and the upstream free surface is less free to deform and can be manipulated by applying vacuum or electrostatic field, which may delay unacceptable air entrainment to higher substrate speeds. In slot and slide configurations, vacuum or electrostatic field can be used to draw the bead further upstream in the gap, thereby enlarging the micro-reservoir for re-metering, which may be desirable. The bead may, however, develop microvortices, which are not desirable though they seem often to be tolerated.

Impingement flow application is what is seen in curtain coating and extrusion coating, where a pre-metered layer of liquid *arrives as a free film* that attaches at the dynamic wetting line. If the curtain is not short, the falling liquid may have enough momentum to augment the pressure there and delay excessive air entrainment to higher substrate speed. Electrostatic force may be deployed around the dynamic wetting line to achieve the same end in both curtain coating ('electrostatic assist') and extrusion coating ('electrostatic pinning'). Raising the curtain height can cause the impinging flow to bulge upstream in a 'heel'; a slight bulge provides a little pre-metering, which may be desirable, but a larger one reduces impact pressure around the dynamic wetting line and ultimately develops microvortices and unwanted recirculation within the heel.

Transfer application employs a *compliant gap* to apply most of an arriving pre-metered layer to substrate passing through the gap. The compliant element can be either a soft transfer roll (often called an applicator roll), a soft backing roll, or the web tensioned over a transfer roll.

More complicated mechanisms are active in application of coating from knurled and gravure rolls used for metering and transfer. The basic action at the dynamic wetting line in both forward and reverse modes may be influenced by the pattern of grooves or cells, even favoring entrainment of microbubbles as has been reported from studies of gravure printing. The film splitting or wiping is accompanied by partial emptying, or 'pick-out' of the grooves or cells; control of the fraction transferred is critical to the metering that is combined with the application.

Because each of the four basic functions can be accomplished in multiple ways, the number of combinations into coating methods is large.

9.3.3 Physics of Coating Flows

In a coating flow the gas originally in contact with the solid surface is replaced in an action known as *dynamic wetting* (Figure 9.3). This action appears to the eye, even with the help of optical magnification, to take place at a line, or curve, a seemingly 1D 'thing'. This thing is called the *wetting line* or the *(apparent) dynamic contact line*. It is present in every coating flow and it is the biggest scientific unknown although not always the most important practical aspect. In start-up of slot and slide coating, dynamic wetting is established by liquid breaking through gas entrained by the moving substrate surface; that it does so is crucially important. The physics of dynamic wetting is still not well resolved in terms of basic principles. Whatever happens on the invisible scales of molecules and surface roughness, the arriving liquid appears to slip locally. In all other circumstances, liquid in contact with a solid surface does not appear to move relative to the surface. Thus the no-slip boundary condition that is so firmly established elsewhere in fluid mechanics (though not so firmly in polymer processing) must fail near a dynamic contact line – which is of course a 3D region, albeit a submicroscopically slender one. The visible contact angle at a wetting line or dynamic contact line is called the (apparent) *dynamic contact angle*. Generally, it differs from the (apparent) static contact angle that the same liquid and gas seem to make with the same surface when all are at rest.

At the *edges* of the coated layer its free surface ends in ordinary static contact lines. These *lateral contact lines* necessarily bend round upstream and connect with the wetting line. Thus at each edge of the layer where it is being delivered to the substrate there must be a curved segment of dynamic contact line, and the apparent slip of the liquid

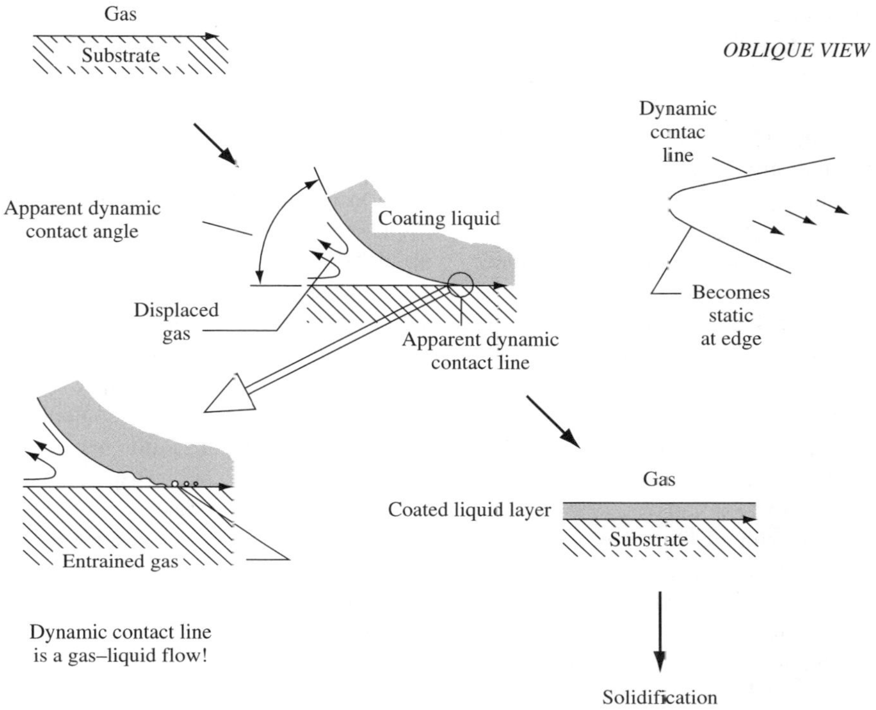

Figure 9.3 *To coat is to replace gas with liquid at a solid substrate*

must cease by the place that segment turns into a completely static, lateral contact line (Figure 9.3). A static contact line or separating line may locate other than where it is supposed to, thereby marring or destroying the uniformity of coating. Control of contact line attachment is crucial to precision coating. Solid corners of small and uniform radius (as little as 25 μm) are useful; wettability discontinuities are less so. Sloppy start-up can overwhelm a careful design.

A dynamic contact line, or wetting line, may pass visible amounts of the gas originally at the solid surface, thereby destroying the uniformity of the coating; short of this it may pass invisible amounts of gas that wreak havoc during subsequent heating and drying, or that impair product function. The issue is *air entrainment* (Figure 9.3). The onset of unacceptable air entrainment sets an ultimate limit on coating speed. At a speed less than that, the dynamic contact line may begin bending and looping, or oscillating in position, interposing another limit.

9.3.4 Experimental Analysis

Each coating method has limits within which the flow can even exist, and narrower ones within which it is close enough to 2D and steady to deliver the desired uniformity. The quality window is generally smaller, the greater the uniformity sought. Quality windows, like the operability window, are in the ranges of design parameters like lip shapes and roll runout, operating parameters like web speed and coating gap, and formulation properties like low-shear viscosity and equilibrium surface tension. Typically, the industrial approach is a 'designed experiment' to find out if a given formulation 'can be coated', i.e. whether a continuous liquid layer forms and persists, and if 'defects are excessive', i.e. whether the layer is adequately uniform. A more fruitful approach is using eyes, lights, and stroboscope, and then more advanced flow visualization techniques to see what actually happens – how the layer fails to form, how flow instabilities and other defects arise – and thereby to seek the responsible mechanisms.

Coating flows though laminar are made complex by the freedom of the liquid's surfaces and by the abrupt changes in direction and speed of the liquid where it passes from the applicator device to the solid substrate. Rigid substrates may move past at tenths of meters per second, flexible webs and fibers up to tens of meters per second. Often the flows are even more complicated by viscosity changes with the rates at which the liquid is locally sheared and extended, and by traces of viscoelastic behavior. The scale is small in two directions, along the flow and through the flow. That makes magnification necessary and illumination difficult. If the liquid or a surrogate for it is clear and transparent, marker particles or dye traces are necessary: hydrogen bubbles, aluminum flakes, and fluorescent dyes are all suitable in various circumstances. The scale is large in the third direction, across the flow, and the edges differ to some degree from the rest of the width. That makes long working length lenses and special edge plates obligatory for side views, and puts a premium on 2D sectioning by optical means.

Figure 9.4 shows Sartor and Suszynski's pioneering arrangement for slot coating where the slot is 250 μm or more across and the gap between die lips and substrate is in the same range; it drew on Schweizer's then unpublished breakthrough visualization of slide coating. Cohen and Suszynski's splendid view appears in Figure 9.5; the hydrogen bubbles and dye streaks make plain the menisci that bound the coating bead, the intense microvortex within it, and the streamlines leading to the coated layer carried away on the upward-moving substrate.

Crookedness of contact lines and crossflow nonuniformity is best assessed from plan views through transparent substrate. In tensioned-web coating methods, clear web affords

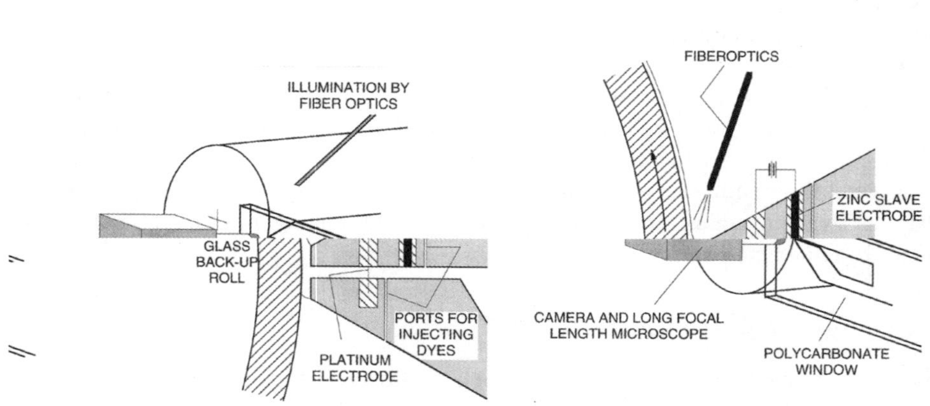

Figure 9.4 *Set-up for visualizing slot coating flow, in a cross-section to show streamlines and free surfaces*

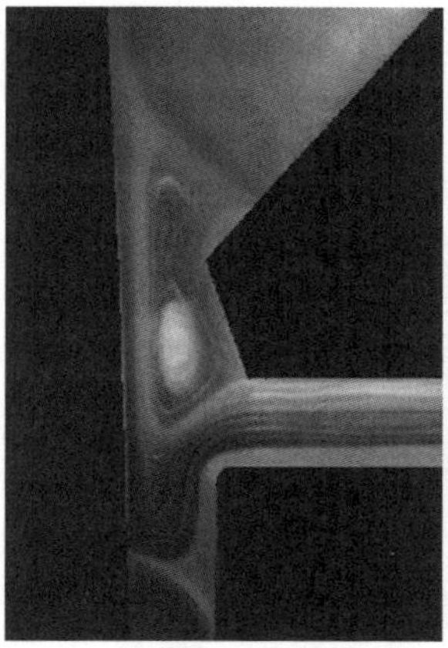

Figure 9.5 *Cross-sectional view of slot coating flow*

Figure 9.6 *Transparent glass roll finished and mounted with ±0.5 μm run-out. Visual access to plan views of flow in a coating bead is through the hollow shaft or under the cantilevered end*

the needed optical access. For other methods, the web carried on a back-up roll can be replaced by a transparent roll, as shown in Figure 9.6. The third panel of Figure 9.7 was obtained in this way (equivalent views of wire-wound rod and gravure coating with rubber-covered back-up roll were obtained by devising transparent rubber covers). Figure 9.7 summarizes the challenges of visualizing internal features of coating flows. Further information and examples are recorded by Suszynski and Scriven in *Chemical Engineering Progress*, September 1990, pp. 24–29, and by others in published theses (see Literature section) and papers.

Interpreting the results of flow visualization and measurement takes theory, and theory requires flow properties: viscosities as functions of shear and extension rates, peculiar viscoelastic parameters if the coating liquid is mildly viscoelastic solution, surface tension

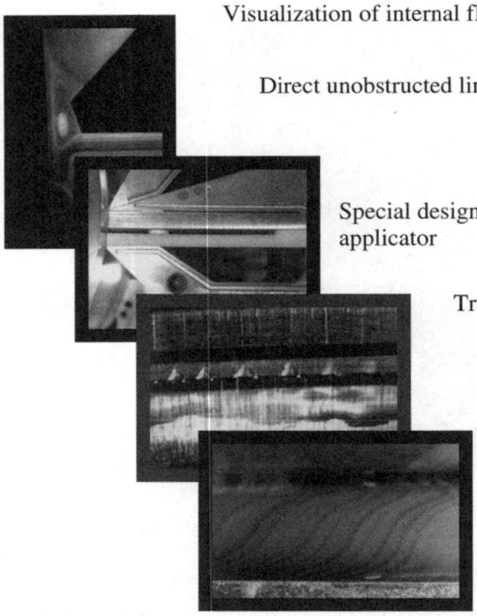

Figure 9.7 *Equipment must often be specially designed to provide optical access to the coating bead*

and its dependence on concentration if that is a factor (in which case a suite of diffusion, adsorption, and micellization properties may also be vital), appropriate contact angles at static contact lines and apparent contact angles at dynamic ones. Diffusivities and solution activities may be needed to understand growth of interlayer diffusion zones in multilayer coating. Getting the needed flow properties is often so difficult that risky guesses must suffice. Deformation rates in shear and extension in many coating flows are much higher than those attainable in laboratory rheometers. Static contact angle data tend to be crude and plagued by slow equilibration, or 'hysteresis'. Apparent dynamic contact angle is not a property, but a microscopic two-phase process, 3D and unsteady in detail; purported correlations of it can be useless for getting at limits of coating speed. Needs and opportunities in this arena are great.

9.3.5 Theoretical Analysis and Engineering

Coating flows of liquid lend themselves to practical and scientific understanding for three reasons: they are *laminar*, *steady*, and 2D to a good first approximation. They are necessarily laminar to create and maintain uniformly thick layers with desired internal structure, including stacking of multiple layers deposited simultaneously. They are necessarily steady to create uniformly thick layers in the direction of the substrate's relative motion ('downweb'); exceptions, however, are periods of start-up, shutdown, and splice passage – and the entirety of the wonderful process of spin coating. They are necessarily 2D, which here includes axisymmetric, to create layers that are uniformly thick in the direction transverse to the substrate's relative motion; exceptions are the edges of planar coatings and starts and stops in general. A second approximation is required to account for departures from steadiness and two-dimensionality caused by: out-of-flatness (or out-of-roundness), roughness, and porosity of substrate; imperfections and, in certain cases, special features of coater design and operation; and sometimes perhaps tolerable secondary flows inherent to particular circumstances. These are all challenges to analysis.

The key to understanding flow coating is the physics of the forces involved. These forces are: viscous, pressure, and capillary pressure (surface tension resultant in curved interfaces); sometimes gravity, inertia (curvilinear acceleration), surface tension gradient, and elastic responses of compliant confinement (as in 'elastohydrodynamics'); occasionally viscoelastic effects, electrostatic, and magnetic forces; always, at contact lines and small scales elsewhere, the London-van der Waals and other forces of electromagnetic origin known as 'surface forces', which give rise to disjoining pressure; and constantly in the background the very small buffeting forces of thermal fluctuations in the liquid ('Brownian forces'). Today the forces are being analyzed, predicted, and managed by design at four levels: (1) Engineering approximations are sometimes adequate – the lubricating flow and related viscocapillary flow approximations are more and more being developed and used in ranges where they have been validated by more exact Navier–Stokes or related theory. (2) Commercial computational fluid dynamics software for solving the equations of Newtonian (Navier–Stokes) flow and generalized Newtonian flow (shear-rate sensitive viscosity) is still limited but advancing and it can already be extremely useful, as in die design and certain templated 2D coating flows. (3) Advanced, tailored codes are being developed by specialists in large corporations, small entrepreneurial firms, and government laboratories, often drawing on the next level. (4) Pioneering research programs in viscous free-surface flows are being conducted in a few universities worldwide. The best validation of a solution is with whatever experimental measurements are available, whether merely the shape of a free surface, or the occurrence of a microvortex, or the streamlines and local velocities of the flow. Figure 9.8 is a steady-state example from K.S.A. Chen's research on simultaneous precision coating of multiple miscible layers assembled from successive feed slots in the inclined surface of a slide die.

For purposes of design, control, and optimization a steady-state solution – an operating state – is not enough. *Many* solutions are needed, even in the most parsimonious probing

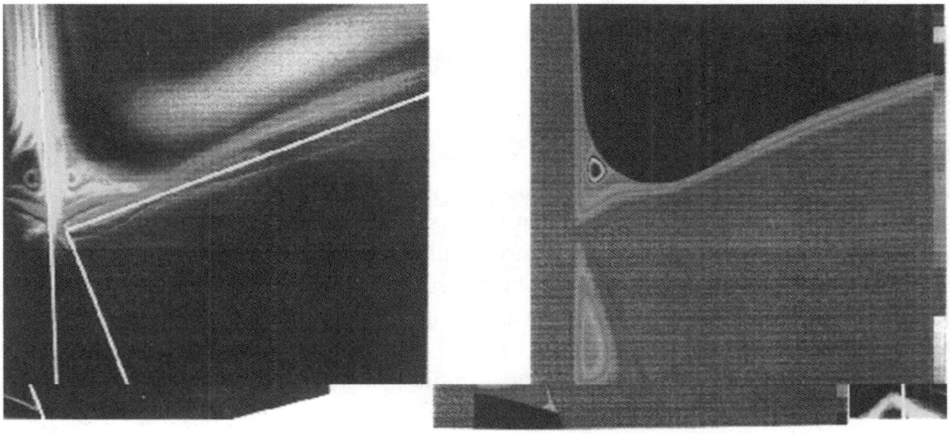

Figure 9.8 *Two-layer slide coating flows: visualization flow field versus computed flow field. Two microvortices are obvious; another may be present on the slide*

of parameter space. Many solutions are needed in order to get an idea of the ranges of design parameters and operating parameters within which steady, 2D flow states exist. Arc-length continuation also makes it possible to discover when more than one such flow state exists, a not uncommon situation, and the relative stability of the multiple states. Augmented continuation schemes make it possible to track turning points themselves through parameter space, a procedure called fold-tracking. This procedure can be used for delineating windows of feasible operation.

Within a feasibility window lie coating quality windows defined by features and sensitivities of the flow states. Similar augmented continuation schemes make it possible to track through parameter space the limiting states at which a microvortex appears or disappears, a static contact line pins to an edge or moves free, a machining flaw creates an unacceptably thick streak, and so forth – and in addition to see how the feasibility and quality windows depend on shape and dimensions of the coating applicator.

Quality windows are also delineated by set values of the damping coefficients and attenuation factors that are computed in stability and frequency analysis. These too can be traced out efficiently in parameter space by augmented continuation schemes. The same is true of the turning points and bifurcation points in parameter space, points of marginal stability. These are the guides to situations in which there is more than one stable operating state. When such situations may arise, it becomes desirable to solve repeatedly the full equation system of flow for transient behavior in order to know how different start-up procedures and upsets select among the multiple stable states.

Fold-tracking, feature-tracking, stability and frequency analysis, and transient analysis, all now demonstrated in conjunction with the computational fluid mechanics of coating flows, are potent tools for process design, control, and optimization. Edge effects and other 3D free-surface phenomena are challenges that have had scant theoretical analysis. Current theory-centered research is on aspects of coater shape optimization and active flow control. Rational engineering of coating flows, whether to advance established technology or to produce new lines of fine-structured coatings, is a field wide open to chemical engineers in industry and academia.

9.4 Solidifying the Liquid Layer

The basic function of solidification can be accomplished in a variety of ways, some of them already noted (Figure 9.9). The principal ones are: solvent removal from the surfaces of a coating; solvent movement to the surfaces from within a coating; colloidal transformations, phase separations, fusion, and binding; polymerization (both growth by chain-extension or condensation-addition reactions, and networking by crosslinking reactions) and other chemical reactions of curing; initiating and speeding the reactions by controllable catalysts and, in some cases, radiations; and energy delivery to supply the latent heat of evaporating solvent or to initiate or drive the chemical reactions. At one extreme, solidification is a simple matter of solvent diffusion to and removal from the surface of the coating; at another extreme, the coated liquid is entirely monomer or oligomer that can be polymerized in place, so that no solvent whatsoever is involved. Residual monomer or oligomer can be as difficult to remove during processing and as objectionable in storage and use as residual solvent, however.

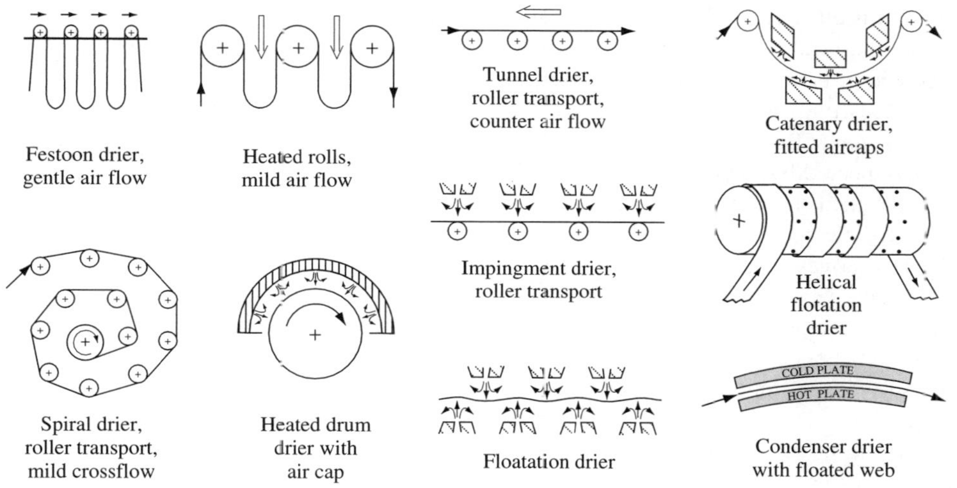

Figure 9.9 *Drying and thermal curing: simplified diagrams*

Solvent removal. Evaporation of solvent from the exposed surface of a coating into the adjacent air or other gas is the usual means of solvent removal. Because diffusion into stagnant gas is a relatively slow process, it is generally augmented by convective sweeping of the coating surface. Sometimes the sweeping is by weak and irregular *natural convection* and room *air currents*, as in the slow dip coating often used in the laboratory to coat thin films – as thin as one surfactant molecule or polymer molecule thick. A layer dipped at millimeters per minute is so thin that natural convection may be adequate to remove the tiny amount of solvent present even when the solvent is initially at high concentration. Sometimes the sweeping is by strong and reproducible *induced convection*, as in the ubiquitous spin coating commonly used in both the laboratory and the factory to coat thin films onto plates, disks, and wafers. The spinning substrate is itself a pump that draws gas around the axis of rotation toward its coated surface and then drives the gas radially outward over the coating. In dip and spin coating, the metering of the coating by flow and the solidification by evaporation can be, and frequently are, combined into a single operation. Doing this successfully can require elaborate control of the flowing gas and its solvent content in order to sequence or balance the two functions optimally.

In festoon driers and some tunnel drying operations (cf. Figure 9.9), gentle convective sweeping of the coated web by low-velocity crosswise air flow is sufficient to remove evaporated solvent. The same can be true of terminal zones of drying and curing ovens used to lower residual solvent by out-diffusion and residual stress by annealing. Where solvent is to be removed rapidly, strong sweeping by *forced convection* of air or other gas is used. The notable exception is rapid solvent removal by simple diffusion across gas flowing laminarly in a carefully controlled gap. The gap can be made so narrow that the rate of diffusion is quite high, as in condensation drying.

The mass transfer coefficient is *highest* when the gas flow is *perpendicular* to the evaporating surface because then the diffusion of vaporized solvent off the surface is most enhanced by convective action. When the drying rate is extreme, the vaporizing solvent itself may produce an appreciable flow away from the coating surface, an action sometimes known as phase-change-driven convection. But ordinarily the strong convective action is achieved by sets of *impinging sheet jets or round jets* of gas interspersed with departing streams; in other words by an array of *stagnation flows* and *reverse stagnation flows*. Both laminar and turbulent flows cause this convective action, but it can be augmented by the chaotic, fine-grained convective action of turbulent flow, provided the turbulent shear stress and pressure fluctuations are tolerable. An impinging flow is of course deflected laterally and a departing flow is recruited laterally, so that between them the gas moves more or less *parallel* to the coating surface in a flow of boundary-layer type. This type is less effective at enhancing diffusion because its direction makes close to a right angle with the solvent partial pressure gradient. Consequently, in the design of impingement driers the size, shape, spacing, and distance of nozzles or slots above the surface are important factors which could be optimized with respect to mass transfer coefficient, drier fabrication costs, and blower expense under constraints imposed by the nature of the coating. In downstream zones where internal resistance dominates, the details of the gas flow matter relatively little except in bringing the almost dried coating to the gas temperature. Presumably the various single-zone and multiple-zone *impingement drying systems* available commercially are designs that have evolved so that each is close to optimum for some class of applications.

When the gas moves in boundary layer-type flow *parallel* to the evaporating surface, the mass transfer coefficient is somewhat lower than would be provided by impinging flow at the same maximum velocity, as already noted. This shortcoming may be offset by advantages, however. The pressure gradients that can disturb a liquid coating are smaller. A more important advantage of parallel flow is that a translating web or strip can be supported on gas flowing through narrow gaps between the web and opposed rigid surfaces. Two mechanisms can be responsible: one is the *air-bearing effect* of gas carried through a converging gap; the other is the *Bernoulli effect* of gas flowing through a converging–diverging slit. These two competing mechanisms, along with the resultant web tension in a flexed web and the flexural rigidity of the web itself, are brought into balance in various ways by *floatation nozzles*, or *bars*. Several types have evolved since the invention of floatation driers and ovens around 1970.

An alternative to forced convection drying is called *condensation drying*. It involves parallel flow in another way. The heated substrate is transported past a cooled solid surface. The gap between the evaporating surface of the coating and the condensing surface of the solid is made so narrow that diffusion across the gas in the gap is rapid. Because the gas entrained in the gap by the moving web moves parallel to the web, hence perpendicular to the path of evaporating solvent vapor, the vapor transport is by diffusion alone and the mass transfer coefficient is simply the diffusion coefficient divided by the gap. In practice the gap can be made narrow enough that the mass transfer coefficient is comparable to that in impingement drying systems. The difficulty is to remove the condensing liquid rapidly enough to prevent build-up on the cooled surface and contact with the coating. A recent patent by Huelsman and Kolb discloses that the difficulty can be overcome by capillary pressure gradient-driven flow through transverse grooves cut in the cooled plate; the still proprietary realization of this is called *gap drying*. Another problem can be supersaturation and formation of solvent fog in the upper, cooler part of the gap. When it can be used, this method is comparatively simple, the amounts of gas involved are relatively small, and recovery or incineration of organic solvent is much easier. The drier zones are smaller than when forced convection is used. Moreover, the gas exerts virtually uniform shear stress and pressure on the coating's surface and so does not disturb the uniformity of its thickness.

Solvent movement within coatings. The mechanisms by which volatile solvents can move within a coating as it solidifies are few: *pressure gradient-driven flow* in a porous coating, and *diffusion* along with *diffusion-induced convection* (or 'diffusion-engendered flow') in general. Though the mechanisms are few, the complexities are many.

If a coating is, or becomes, porous, liquid can flow within it in response to differences in capillary pressure at the menisci between liquid and gas. If the liquid wets the pore walls, the flow is away from less curved menisci, which reside in larger pores, and toward more curved ones, which reside in smaller pores; if the menisci are all of about the same curvature but are at different temperatures, the flow is from the hotter ones toward the cooler ones. If liquid should happen to exceed its bubble-point and boil anywhere within the porespace, the pressure there tends to drive liquid away and out of the porespace, as in so-called impulse drying of paper. If the solid of the matrix shrinks, stress is transmitted as pressure to liquid in the porespace and the liquid flows in response to any gradient of pressure that arises. If the coating and substrate together make a porous sheet, liquid can be driven toward one side by applying high enough gas pressure on the other side. In every case the local flux of the flowing liquid is proportional to the local pressure gradient. The proportionality factor is the property called *porespace permeability*, divided by the *viscosity* of the liquid. The viscosity can be quite sensitive to liquid composition and temperature. Measurements of these properties for solidifying coatings are scarce, and estimating them may be difficult. Nevertheless, they are keys to understanding convective flow in porous coatings, as theoretical modeling of the flow processes makes clear.

Solvent in solution with other soluble components diffuses in response to differences in composition (chemical potential, actually). Within a given phase, it tends to diffuse from regions of higher concentration toward regions of lower concentration – always in binary solutions, but *not* always in multicomponent systems. In a binary solution only one mole fraction or mass fraction is independent, and so the local diffusional flux of either component is proportional to the local gradient of mole fraction, mass fraction, or concentration. The proportionality factor is the *binary diffusion coefficient*, or *diffusivity*. Generally it depends on concentration, and the dependence can be extreme when the concentration of solvent in a polymer is low – as in the late stages of drying a solvent-borne polymeric coating. Furthermore, the diffusion coefficient can lag in its response to falling concentration, giving rise to a version of what is called 'non-Fickian diffusion'. This is probably one of the causes of the phenomenon of *skinning* in which a solvent-stripped layer at the surface of a rapidly drying coating develops enough resistance to diffusion to seal off effectively solvent remaining deeper in the coating. The body of measurements of binary diffusivities relevant to coating formulations is growing and estimation procedures are advancing. The complication in multicomponent systems is that gradients of other components can contribute to the diffusion flux of a given component. This can produce puzzling distributions of solutes in drying coatings. Measurements of *ternary diffusion coefficients* are rare, and estimating them even roughly is difficult. Nevertheless, the diffusion coefficients are keys to understanding solvent transport in solidifying coatings, as theoretical modeling of drying and curing makes clear. The need for data is great.

Heat delivery. Convection and conduction from hot gas sweeping by is the leading mode of heat transfer to a drying coating to supply the latent heat of vaporization of solvent. Except when solvent evaporation is so very rapid as to produce an appreciable convective velocity away from the surface, in turbulent gas flow the mechanisms of heat transfer to and solvent transfer away from the evaporating surface are virtually identical combinations of convective action with thermal conduction on the one hand and molecular diffusion on the other. This is reflected in useful correlations, like Colburn's, of the mass transfer coefficient with the more easily measured heat transfer coefficient in turbulent flow. It is also the reason that the now fairly extensive literature on the performance and design of driers focuses on heat transfer coefficients and heat delivery rates.

However, the correlation of mass transfer with heat transfer is relevant only to the heat transferred directly to the evaporating surface. That other modes of energy delivery can be highly useful and have to be considered separately is well appreciated in the literature.

Phase separations and colloidal transformations. A liquid is coated in order to deliver material that will end up in a solid or semi-solid form on the substrate. Temporary solidification may be achieved by consolidating suspended particles, for example by liquid sorption into the substrate, or by gelling polymer solution, for example by chilling. To dry a coating is to remove enough volatile solvent that the remaining material solidifies permanently (though some coatings such as adhesives barely solidify). Whether that material is initially in the form of solutes dissolved or particulates suspended in the liquid, it tends to concentrate at the coating surface, or at menisci in a porous coating, as the volatile components evaporate and leave it behind. Diffusion in the liquid tends to drive the concentrated material away from the surface, or menisci, back into the less concentrated liquid behind. The balance between evaporation and diffusion steadily shifts in favor of the former. Sooner or later the concentration of a polymer solute reaches a level where *gelation* or *vitrification* ensues, or else the concentration exceeds the solubility limit locally – the solution becomes supersaturated – and solidification ensues, whether by *nucleation, growth, and aggregation* of precipitate or by *spinodal decomposition and consolidation* of solvent-lean material. A polymeric solution may both vitrify and precipitate, producing a partially crystalline material consisting of crystallites in an amorphous matrix. As it concentrates, a polymeric solution may also become thermodynamically unstable and either separate into *two* polymer *solution* phases, or spinodally decompose toward the two solution phase compositions. When two polymers are present, such instability is often called 'incompatibility'. If the circumstances make one of the separating phases glassy, the outcome can be a microstructured solid.

Likewise, the particulates of a colloidal suspension concentrate as the volatile components evaporate. Sooner or later their concentration exceeds the colloidal stability limit and *flocculation*, *coagulation*, and *sintering* or *fusion* ensue. A nearly monodisperse colloidal suspension may on flocculation produce regions of colloidal crystal (ordered arrays of particles) interspersed with amorphous colloidal material. In any event when the concentration of particulates falls to that of ordered or amorphous close-packing – the *critical packing condition* – they *consolidate* into a structure that is solid, at least in compression. Soft particles of colloidal polymer, or 'latex', are thereafter deformed and compacted by capillary and van der Waals forces and may fuse into a coherent coating; or else they spread over harder particles, e.g. 'pigment', to bind the latter together and to the substrate in a composite coating. Harder particles may alternatively be fixed in place by a soluble polymeric binder that phase separates as solvent evaporates. Three routes to solidification can be simultaneously active during drying of a coating that is initially a solution of solvent and polymer which also contains colloidal particles and larger particulates.

Whatever the phase behavior, the key to understanding it is the *phase diagram*, a concentration diagram on which are drawn the boundaries of relevant one-phase, two-phase, and even three-phase ranges of composition at equilibrium. *Metastable states* like *supersaturated solutions*, *subcooled* or *vitrified liquids*, and *gels* can often be placed usefully on phase diagrams. So can thermodynamically unstable ranges, which are defined by spinodal boundaries and are the provinces of *spinodal decompositions* – remarkable nonequilibrium processes in which solutions spontaneously grow local concentration variations that tend toward an interspersion of the two phases that would be present at equilibrium. Spinodal decompositions can give rise to *bicontinuous microstructures* prized for such products as photopolymer systems and permselective membranes. The idea of converting a single coated layer to a two-layer structure by segregation during spinodal decomposition is over a decade old. Recently, just such a phenomenon was discovered during basic studies of block-copolymer microstructures in layers a micrometer or less thick. The possibilities of creating ultrathin coatings in this way are intriguing. Among the possibilities are spontaneously nanopatterned coatings.

What actually happens depends on the *process path* across the phase diagram. A process path is the sequence of compositions and temperatures followed by the material as it dries at constant pressure. A process path of a drying polymer solution may not lie entirely in a one-phase region of a phase diagram; it may instead cross a binodal boundary and enter the metastable single-phase area of the two-phase region. Then droplets of the new solvent-rich phase grow if they can nucleate; if local equilibrium were obtained, the path would break into two phases of fixed compositions but one growing at the expense of the other. Eventually the remaining polymer-rich phase may vitrify, and it may still be continuous when it does. If the process path also crosses a spinodal boundary and enters the absolutely unstable single-phase area of the two-phase region, the solution may spontaneously 'undiffuse' into randomly interpenetrating solvent-richer and polymer-richer compositions. Such spinodal decomposition can give rise to valued microstructure.

In reality, the internal resistance to solvent movement sooner or later becomes appreciable. Consequently, composition (and temperature for at least a short while too) may vary appreciably with depth into the coated layer. So there can be a whole *family* of process paths, not just one (Figure 9.10). The differences between the paths depend on the diffusion rates by which volatile components reach the drying surface and less volatile ones redistribute; temperature gradients contribute to the differences during non-isothermal stages of drying. Thus mass and heat transfer rates can strongly affect process paths. 'Skinning' has to do with the path traveled rapidly by the outer portion of the coating and 'blistering' usually with the path traveled more slowly by the deepest portion.

Microstructure evolution is exemplified by the sequence shown in Figure 9.10. The surface of the coating dried so quickly that even though it followed a path that went deep in the spinodal region and then traversed the region between the spinodal and binodal, there was no time for any phase separation at all and the surface region ended up a solid skin. The near-surface region did not go as deeply into the spinodal region but spent enough time there that it decomposed spinodally and grew too viscous for

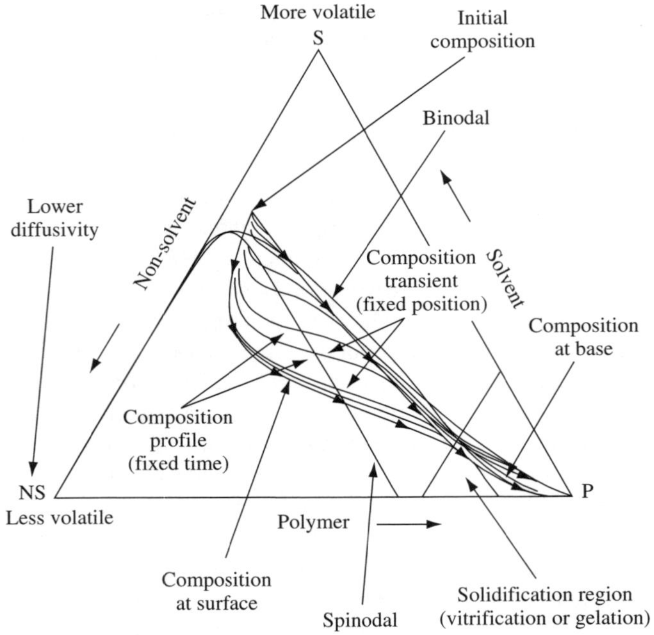

Figure 9.10 *Ternary phase diagrams showing the family of isothermal process paths ('composition transients') at different depths in a coating (from Dabral's research)*

nucleation and the growth of solvent-rich phase as it traversed the spinodal-to-binodal region; hence the near-surface region ended up a bicontinuous microstructure. The deeper regions of the coating followed paths that barely entered the spinodal region but spent lots of time in the spinodal-to-binodal region, where they developed a progressively coarser microstructure by nucleation and the growth of solvent-rich phase in the form of droplets. The phase diagram and family of process paths of Figure 9.10 could have given rise to the microstructure in Figure 9.11.

If the polymeric system *polymerizes* further or *crosslinks*, i.e. if it *cures*, its mean state again describes a path in the phase diagram: in the simplest view, the relevant thermodynamic state variable is degree of reaction, extent of crosslinking, or molecular weight. Further polymerization or crosslinking can bring on gelation or vitrification, or turn a solution thermodynamically unstable so that spinodal decomposition ensues. So, of course, can simultaneous crosslinking *and* drying. Process paths can be portrayed not only on phase diagrams, but also on diagrams of reaction (cure) temperature versus elapsed time, on which precipitation, decomposition, gelation, and vitrification transformations can be represented (TTT (time-temperature-transformation) cure diagrams). There are many, many scenarios, few of which have been examined in terms of the basic phenomena. But something of their character can be glimpsed from the examples.

Stress development. Generally, solidification is accompanied by development of in-plane tensile elastic stress. This is because departure of solvent from solution, crystallization and vitrification, consolidation of particulates, colloidal flocculation and coagulation, and the chemical reactions of curing almost always tend to produce *shrinkage* of the

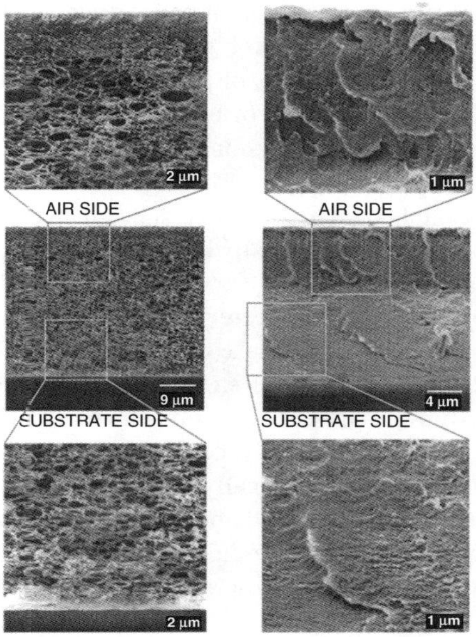

(a) Coating produced by slow drying shows a dense layer atop a porous substructure

(b) Coating produced by fast drying appears dense with no visible pores at the magnification shown

Figure 9.11 *Scanning electron micrographs of slow drying (stagnant air) versus fast drying (impinging jet of air) coatings of cellulose acetate in mixed acetone–water solvent*

stress-free, or equilibrium, state and because shrinkage in the plane of the coating is frustrated by the coating's adherence to the substrate. Hence elastic strain develops even though there is no in-plane movement: strain is the difference between the current state and the elastic stress-free state. But the *elastic* stresses that develop also tend to relax by various mechanisms, and so the state of elastic stress in a coating at any stage of solidification is the outcome of the *competition* between *elastic stress development and elastic stress relaxation*. If elastic stress grows large enough, it can produce a variety of defects, among them *stretch pattern (tensile yielding)*, *curling*, *crazing*, *cracking*, *peeling (delamination)*, and *microstructural alterations*, e.g. changes in layering, crystal state, particle location, and porosity. Swelling of the stress-free state of a coating by solvent, as when fresh solution is coated atop a (partially) dried layer, can produce elastic compressive stresses. If great enough they can drive the buckling instability that leads to wrinkling, which is usually an unwanted defect. Swelling of a surface zone, where curing is more advanced, by monomer or oligomer from deeper in the coating can also lead to surface buckling. This mechanism appears to be exploited in commercial wrinkle finishes that are prized in certain applications.

On the other hand, among contacting colloidal particles or within microporous 'gels' the local stresses that develop owing to capillary pressure when they are partially dry are extremely important to the consolidation and compaction that can result in a coherent coating. If the stresses are large enough they can *collapse* a gel, which may or may not be desired. The stronger they are, the more they may promote the *sintering*, or *fusion*, and *coalescence* into continuous coating by polymeric particles that are intended to do just that. Also *film formation*, i.e. a moderate to high degree of compaction and coalescence, generally is desired of aqueous dispersions of polymer latex particles that are to become either a coating themselves or the binder of a particulate coating. Water-borne coating formulations often incorporate such dispersions.

What is fascinating about drying polymer coatings is that once they begin developing elasticity, i.e. an elastic modulus, their stress-free state changes as drying proceeds, even if they are prevented from deforming laterally at all because they adhere to the substrate. The elastic stress-free state locally within a drying coating depends not just on the composition there. It depends too on the current configurations and entanglements of the polymer molecules in the same region. These molecules, through their thermal motions, however hindered, tend toward an elastic stress-free equilibrium state: the process is called stress relaxation. Elastic stresses take time to relax. The rate depends on the amount of solvent left and on the density or free volume, molecular configurations, and entanglements. The rate may be practically instantaneous when the composition and temperature of the polymer layer put it well above the 'glass transition range' on the one hand, or onset of plastic yielding, on the other. The rate can become quite slow below that range. Especially in the latter case, the present situation is an evolutionary result of past situations. Thus an important aspect of drying polymer coatings is that the elastic stress-free states depend on the *history* of drying, and so do the elastic stresses present during most of the time. That can be equally true of particle-laden coatings.

When elastic stresses appear the coated layer responds in several ways. Insofar as its adhesion to the substrate allows, it (and the substrate) *deforms elastically* to bring the stresses into mechanical equilibrium. This may reduce stresses in some parts and intensify them in others, especially near outside edges and internal flaws and inclusions. Insofar as time allows, the coating's molecular organization rearranges to remove stresses – *stress relaxation*. Insofar as time, adhesion to the substrate, and mobility of molecular free volume allow, the coating slowly relaxes plastically in the process called *creep*. But if at any stage a local stress grows too large, the response may be *plastic yielding and stretch pattern, crazing, cracking, peeling, delamination*, or other unwanted phenomena. The elastic stresses from continued drying which are not relieved by elastic deformation are 'frozen in' as an internal stress state, or 'residual stresses', if nothing worse happens immediately. Later, though, the residual stresses can give rise to unwanted phenomena like curl and creep, and even to patterned swelling if the dried coating has to be exposed to solvent during its use (as when photographic film being developed suffers from 'reticulation' patterns).

Stress development in drying polymer layers is not understood well enough that defects can be confidently diagnosed and corrected in the many manufacturing technologies for producing coatings. These include photographic films, optical films, paint films and protective coatings, adhesive coatings and laminates, and a host of other products. Hence, an important challenge is to understand the fundamentals of stress development well enough to identify the mechanisms responsible for defects, to predict when they can be active, and to show how to avoid them.

Solidifiability, e.g. *dryability/curability*, is, like coatability, many-sided. There is more than simple windows in the parameter space: generally, there are advantages to varying conditions along the solidification path, so that in place of a coating window there is a whole progression of windows, a solidification corridor. In fact, the situation is yet more complex because in solidification the state of the coating at any stage along the path can depend on what has happened earlier along that path (e.g. the effects of 'skinning'). Each depth in the coating should remain within the corridor. The walls that define the corridor depend on the levels of sensitivities that product quality can tolerate, given the degree to which disturbances and contamination can be reduced and controlled. This goes back to the type of drier, oven, or reactor and its specific design, beginning with zoning or compartmentalization along its length. Thus the corridor of quality is a limited set of process paths through equipment designed to control flows, temperatures, radiation, and partial pressures in multiple zones.

9.4.1 Experimental Analysis

Each method of solidifying a liquid layer is a trajectory through the zones of drying or curing and annealing. All along the trajectory the operating variables have limits within which the process is close enough to deliver the desired uniformity of microstructure, smoothness of surface, and low levels of residual solvent or monomer and stress. In the first zones, including the path from coater to oven or radiation curing unit – 'Zone Zero' – the hazard is often rearrangements by pressures and drag forces of nonuniform gas flow that results in 'mottle'. In condensation driers, cool fog may form and settle, or rain may fall from the cold surfaces, both causing local nonuniformities or even 'craters' by microscopic surface tension gradient-driven flows. Examples of hazards in the later zones are: superheating of the coating's base, which can produce local boiling and 'blistering'; supercooling of the coating's surface, which can condense moisture droplets and cause 'blushing'; stratification of mobile constituents as in 'blooming' and 'binder migration'; premature solidification with stress development, or diffusivity loss in the coating's surface, termed 'skinning'; unwanted depthwise gradients of pore dimensions and other structural features; and growth of in-plane shrinkage stress that outruns its relaxation and causes curling, cockling, or delamination. The industrial approach is typically to examine the coating after it is solidified and try to deduce, sometimes with results of 'designed experiments', why it does not meet specifications.

The most potent means of diagnosis are visualization and depthwise probing. These can be of coating as it is transported past successive fixed stations on a production or pilot line; or of samples selected after the line has been abruptly halted; or of samples arrested at successive times of a laboratory batch approximation to the solidification process. Once arrested, the samples can be spatially sectioned by fracturing, microtoming, confocal microscopy, etc., to examine depthwise and transverse variations.

This last approach to documenting and understanding solidification by 'time-sectioning' needs a coating of little more than a representative area. Establishing that it works may require lengthy, painstaking effort. The visualization method of choice for microstructure and nanostructure of coatings is scanning electron microscopy (SEM), which provides topographic contrast and sometimes compositional contrast. Any liquid or volatile components have to be deeply frozen to halt Brownian motion and drop vapor pressures. Freezing has to be rapid enough to avoid excessive crystal growth and rearrangement of the sample's structure. Figure 9.12 depicts the technique developed by Sheehan, Sutanto, Ming, Huang, Ma, and others at Minnesota to fast-freeze (by extremely subcooled nucleate boiling) thin samples of coating in liquid ethane at its freezing point so that they can be fractured, etched as appropriate, transferred to a cold stage, and

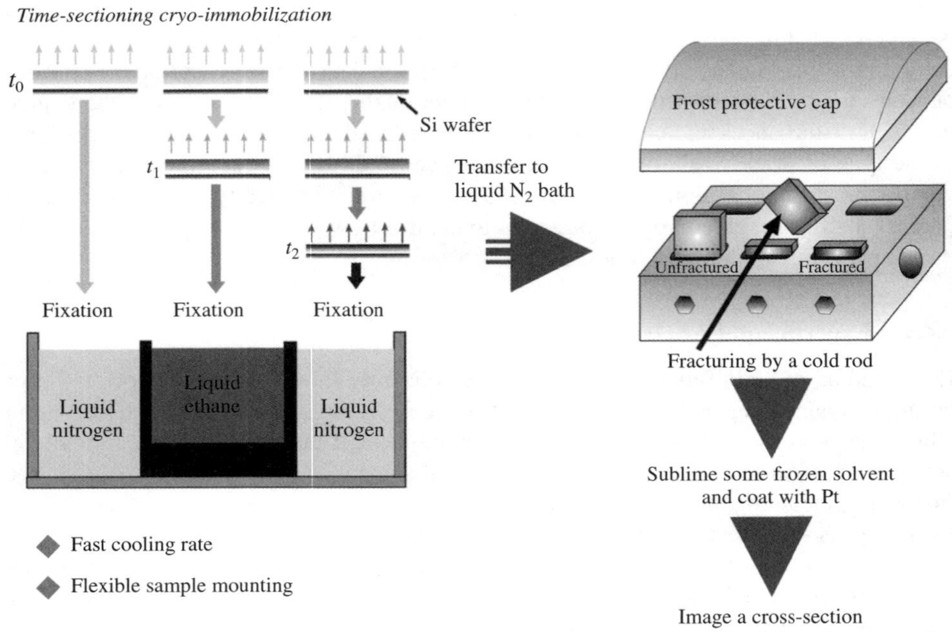

Figure 9.12 *Fast-freeze cryogenic scanning electron microscopy for visualizing structure development in drying or curing coatings*

imaged by SEM with a resolution often of 10 nm or better – all at temperatures close to that of boiling nitrogen. The first cryo-system for doing this is shown Figure 9.13.

One of its applications was by Prakash with Francis and Scriven to examine structures developing in partially dried, re-immersed phase-separating 'asymmetric membrane' (dry–wet phase inversion membrane) still adhering to the substrate on which the liquid was coated, or 'cast', before drying began: a composite image of the surface of a fracture through graded polysulfone coating 143 μm thick appears in Figure 9.14 and offers clues to how the 'macrovoid' defect arose.

Another example, by Ma with Davis and Scriven, is shown in Figure 9.15: the surface of a fracture through a coated layer of 850-nm polymer 'latex' spheres in concentrated aqueous suspension. The layer has dried and consolidated just enough that air has begun to invade from the top surface (which shows obliquely) – surely by Haines jumps, which are initiated by meniscus instability. The fracture ran through air-filled porespace, ice-filled porespace, and pendular rings of ice around sphere–sphere contacts otherwise surrounded by air. This is one of the first images that provided direct proof that drying latex coatings do indeed obey the principles of the science of porous media.

Suitable microscopic transducers of local stress state do not yet exist; the means of choice for measuring the in-plane stress in solidifying coating is the deflection of it and its substrate when they are cantilevered in a small-scale batch approximation to the process. Payne and Vaessen with Francis and McCormick devised the versatile apparatus shown in Figure 9.16 to control temperature and solvent partial pressure and to visualize sample surface besides monitoring deflection (and mass loss in a parallel procedure).

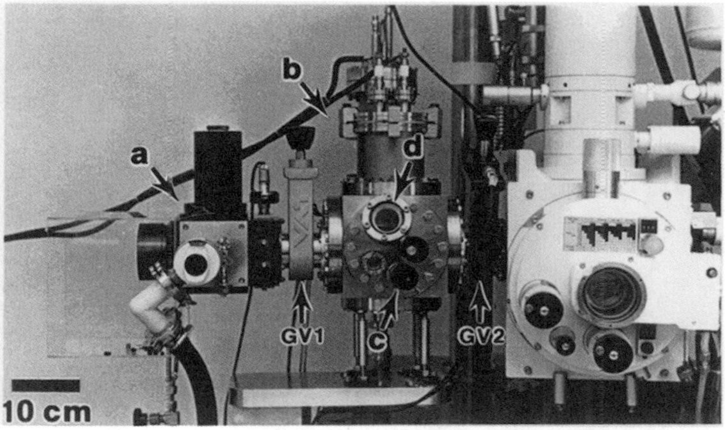

SEM CS attached to the JEOL JSM-840 SEM: (a) prepump chamber; (b) metal-coating chamber; (c) shutter manipulator; (d) viewing window; (GVI) and (GV2) gate valves; (RPV) rotary pump valve

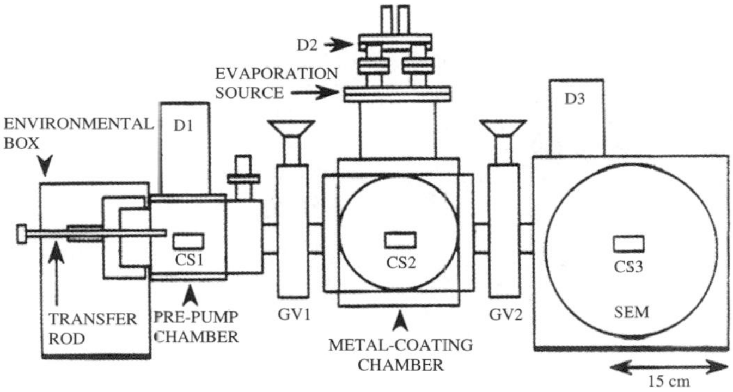

Schematic of the SEM CS: (CS1), (CS2) and (CS3) cold stages; (D1), (D2) and (D3) liquid nitrogen vessels; (GV1) and (GV2) gate valves

Figure 9.13 *Cold-stage system for fracturing frozen specimens, etching the fracture surfaces by sublimation, and coating them with a few nanometers of conductive metal*

Interpreting the results hinges on theoretical analysis that requires a wide variety of properties of solidifying coatings: vapor–liquid equilibria, solution densities and activities, multicomponent diffusivities and cross-diffusivities, colloidal stability measures, polymerization and crosslinking rate parameters in thermal and radiation-initiated curing, shrinkage behavior, elastic and yield moduli, viscoelastic parameters, adhesion and fracture strength, and so on. Data on most of these are still exceedingly scarce, theory for estimating many is undeveloped, and measuring some is more difficult than has seemed justified. For getting at mechanical properties *in situ*, scanning probe and microindentation ('nanoindentation') techniques are a promising advance; applications were made

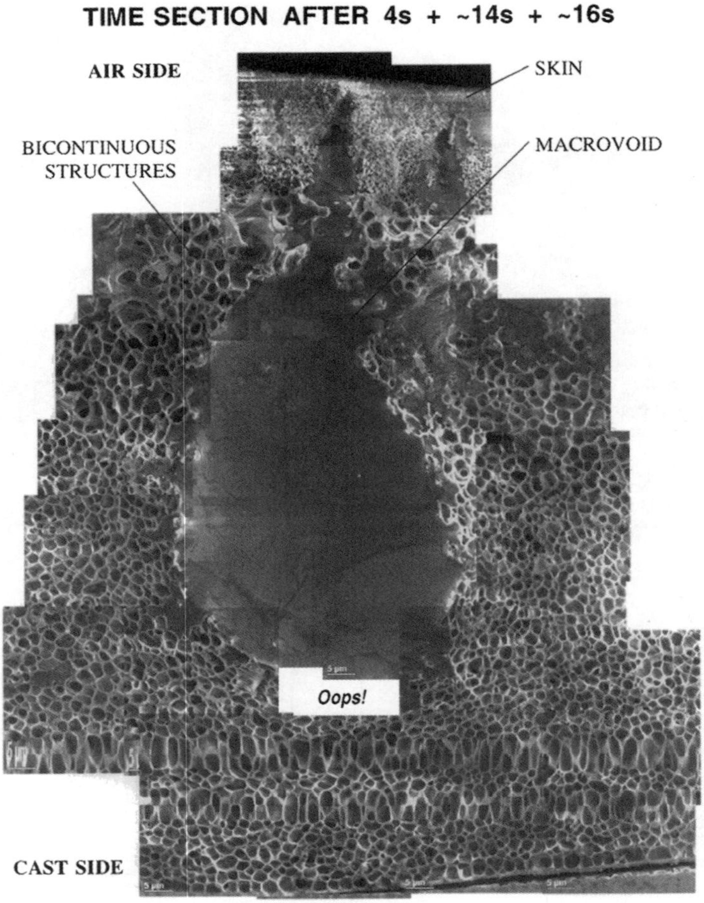

Figure 9.14 Composite image (missing a small part) of a fracture surface of a polysulfone dry–wet phase immersion coating after 4s forced drying, 14s free drying, 16s immersion, and quick withdrawal before cryo-immobilization

early at the University of Minnesota. There, the current focus is on achieving finely structured coatings from latex, latex–ceramic and latex–semiconductor composites, and particle–polymer composites of the sort employed in magnetic recording. Overall the needs and opportunities related to solidification border on the enormous.

9.4.2 Theoretical Analysis and Engineering

Two sets of keys open the way to understanding solidification. One is the basics of phase equilibria – liquid–vapor, liquid–liquid, and liquid–solid – of reaction equilibria, polymer gelation and vitrification, and colloidal transitions. The other is heat and mass transport processes; reaction, transformation, and shrinkage kinetics; and stress phenomena in polymeric systems and polymer–particulate composites. Engineering approximations

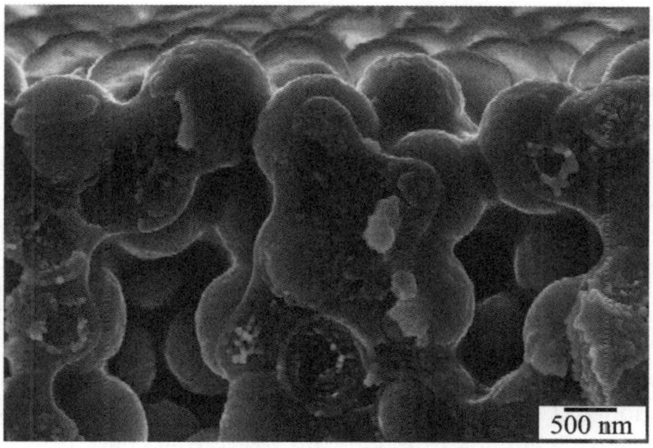

Figure 9.15 *Cryo-SEM image of the fracture surface of a coated layer of poly(styrene-co-acrylic acid, T_g 50° C) spheres 850 nm in diameter dried for 8 min in room air. The adjacent top surface shows obliquely*

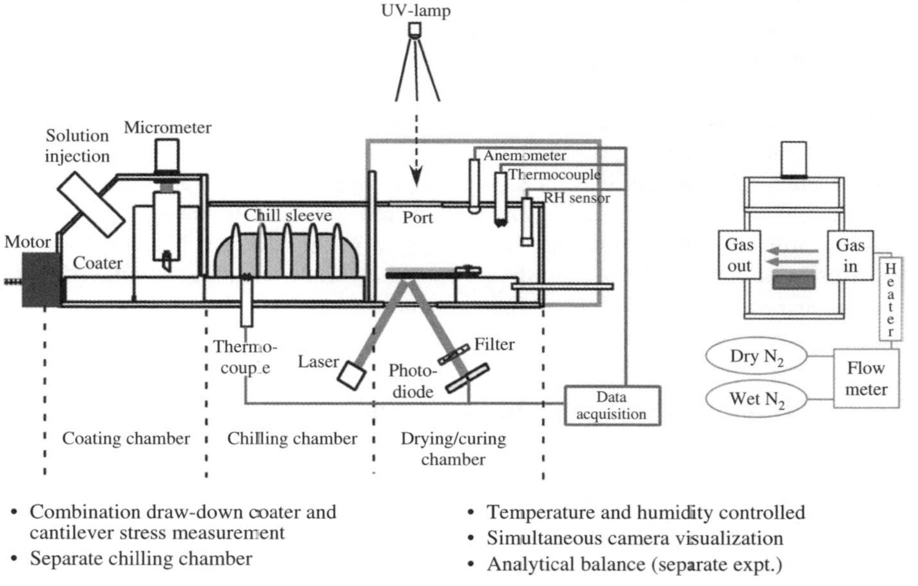

Figure 9.16 *Schematic of apparatus for measuring in-plane elastic stress in coatings drying or curing in controlled conditions*

may sometimes adequately describe these. Commercial software based on such approximations or sounder theory is still scarce and severely limited in capability. Advanced, tailored codes are being developed in only a few corporate and government laboratory settings. Research programs in academia are still narrowly focused, with a few exceptions. Rational analysis, design, control, optimization, and comparison of solidification and microstructuring processes on the basis of comprehensive 2D, much less 3D, theory is still way off.

Solidification is, strictly, appearance of appreciable elastic modulus in a coating, i.e. simultaneously an internal stress-free reference state, elastic or recoverable strain from that state, and elastic stress proportional to (or monotonic in) the elastic strain. The simplest and most often used criterion of solidification in this sense is S.G. Croll's: elastic modulus appears locally at a critical value of falling solvent content or advancing extent of curing. Rubbery behavior in a solidifying coating is most simply modeled by neo-Hookean theory, i.e. elastic stress proportional to a quadratic measure of elastic strain. Glassy behavior is most simply modeled by ordinary Hookean theory, i.e. elastic stress proportional to (small) linear strain. Stress relaxation by plastic yielding seems most reasonably modeled by von Mises' yield criterion and the associated 'flow rule' for post-yield viscous-like change in the elastic stress-free state. Relaxation by yield-free change in the stress-free state seems most reasonably modeled by one of two variants of linear viscoelastic theory. One is the standard elastoviscous solid, which behaves purely elastically at very short and, with lower modulus, at very long times and so cannot reach the elastic stress-free state by relaxation alone. The other is the standard viscoelastic liquid, which ultimately can reach an elastic stress-free state. The initiation of fracture, i.e. stress reduction by surface cracking or edge delamination, is most easily modeled by Griffith's criterion that release of the coating's elastic free energy by crack opening be greater than the elastic surface free energy of the newly formed surface of the crack.

A current focus is theoretical analysis of how particle ordering, coating porosity, and tensile strength arise during the drying-driven consolidation, compaction, and binding or fusion of submicrometer particles in a coated layer of suspension. Hard particles require polymeric binder to make a coating, or 'form a film': particulate magnetic coatings are examples. Soft ones of polymer can flatten against each other and bind themselves by interdiffusion or welding: latex coatings are examples. Understanding the roles of particle sizes and shapes lies in the future of fine-structured coatings.

9.5 Trends and Opportunities in the Discipline

Higher speed coating, greater web width, and lower solvent content (hence less drying but thinner wet coating, higher liquid viscosity, and riskier rheology) impact productivity directly. So does defect reduction, which drives progress toward eliminating heterogeneity and contamination – of arriving coating liquid, of web or other substrate, and of ambient air as well as of the coating machine and its enclosure. The thinner a liquid layer, the more susceptible it is to deleterious local flows driven by transitory surface tension gradients caused by trace surface-active substances arriving with airborne or substrate-borne particles and droplets. Every coating process must feed and distribute the liquid across the web width, meter it to the desired thickness, apply it to the web, and solidify it. It is a natural progression from feeding and gravity-driven distribution with an open pool or pond, as in most dip, knife, bar, and roll coating, to using a liquid enclosure, pressure-driven distribution, and premetering, as with a die in slot, slide, and curtain coating. Premetered application, whether direct or via a transfer roll, lends itself to more accurate coating thickness control, more latitude in coating width, and to simultaneous multilayer coating – three strong trends. Slot coating accommodates stripe and patch coating as well as the widest range of viscosities. Slide and slide-fed curtain are more restricted but can deliver stacks of many liquid layers of appropriate surface tension and rheology and not too great a viscosity. Pressures on solidification are to raise speed and lower cost, and, particularly for new applications, to create more sophisticated fine-scale structuring of coatings.

The main routes to microstructuring are: controlled phase separation, an extreme case that is modulated by the substrate surface; multiple-layered coatings; shear- and extension-mediated particle orientation; gravure and other patterning methods from printing (which is fine-grained discontinuous coating!); suspension coating with drying-controlled consolidation, compaction, and fusion or coalescence; and microembossing and microreplication. An attractive way to replicate a patterned surface is to coat it with conforming reactive liquid polymer, cure it until it has adequate elastic modulus, and then peel it from the surface. Three main trends avoid volatile organic solvents. One is to switch to a water-borne colloidal polymer that adheres and coalesces well enough upon drying and subsequent treatment. Another is to employ a reactive liquid monomer or oligomer that can be cured, i.e. polymerized and crosslinked by heat, ultraviolet or electron-beam radiation. The third is to pulverize a reactive polymer of low enough molecular weight that after it is coated as dry powder and heated, it melts, coalesces, adequately levels, and cures. A natural trend is toward more and better measurement and control along the length of a continuous solidification process – for example, temperature and composition sensors and control measures in drying. The approach is usually by data logging, statistical analysis, and perhaps designed experiments. Less common is science-based, computer-aided optimization of driers and curing units.

9.5.1 State of Understanding

Beyond descriptive books like Harazaki's (1987) and those edited by Satas (1984, 1989), Cohen and Gutoff (1992), and Walter (1993), the first scientific treatise appeared in 1997: *Liquid Film Coating*, edited by Kistler and Schweizer. The largest body of information on the basics of coating are the edited and published Ph.D. theses from the Coating Process Fundamentals Program at the University of Minnesota. These are listed in the Literature section, and are readily available. They are complemented by M.S. theses at Minnesota which are not so easily obtained. Sound technical papers as well as patents that truly teach are appearing more frequently. The biannual coating symposia, first in the United States, then in Japan, and later in Europe, are unique forums of coating science and technology.

9.5.2 Challenges for the Future

The principles for developing, analyzing, designing, and comparing coating processes vis-à-vis the coated product seem clear in outline. Experimental studies and companion theoretical analyses so far have brought out some of the relative merits and drawbacks of different methods. The goal is to be able to design and control each contending method optimally and then to choose the *best* from among the best versions of each method. That requires accurate theoretical models, the physical and chemical properties they call for, and modern computer-based techniques. Ultimately the goal is to include the coating formulation itself as part of the optimization, adapting it to each contending coating process in turn. Not enough of the information needed for process engineering and product formulation of this order is available yet: even with astute, sustained, cooperative research the goal appears years away, particularly for novel fine-structured materials.

New products drive advances in the technology even more than modifications of old ones and competitive pressures to achieve higher quality and greater productivity. Ongoing trends and the need for basic understanding are clear. Besides faster coating, less or no solvent, and fewer defects, looming challenges are thinner coatings and thinner layers in multilayer coatings, finer in-plane patterning, and more intricate structure at micrometer to nanometer scale. For thin coatings, there is already an instructive variety of approaches; only a couple are recent innovations. The cutting edge of coating science and technology is currently at high-performance membranes, ceramic sheets, batteries, and proliferating displays – liquid crystal, electronic paper, and large-area organic light-emitting diodes. It is progressing toward electronic and photonic products based on polymeric and colloidal structures down to nanometers in size – for example, vast arrays of 'quantum dots' and 'quantum wires' that are now at the frontiers of science. Beyond them lie macromolecule-based 'molecular electronics'. Beyond that lies the dream of harnessing electron spin states in a yet vaguer 'spintronics' from a current fringe of science. Rather as the surging of photographic and graphic art technologies impelled advances in substrates, coating, solidification, and ancillary processing in the twentieth century, these looming developments seem destined to be the drivers in the twenty-first.

Literature

No single adequate book exists about coating by depositing liquid and solidifying it. Few books have been written, in fact, despite the enormous technological and scientific importance of coating processes. The soundest one is focused on coating flows and draws heavily on University of Minnesota researches and sequels:

Kistler S.F. and Schweizer P.M. 1997 (Eds.). Liquid Film Coating. Chapman & Hall, London. The first book to focus on scientific principles and their technological implications. Fifteen chapters by 31 authors cover physics and material interactions, including surfactants, wetting, and dewetting; experimental and mathematical methods; theory and some practical aspects of nine coating flows. An expensive but valuable source.

Seven other books that warrant attention are:

Walter Jan B. 1993 (Ed.). The Coating Processes, TAPPI Press, Atlanta. Devoted to paper coating, this comprehensive volume is technological and descriptive. It also has illuminating retrospectives of paper coating technology by George Booth and other highly experienced authors.

Cohen E.D. and Gutoff E.B. 1992 (Eds.). Modern Coating and Drying Technology. VCH Publishers, New York. Articles at a variety of levels by highly experienced authors who range from technological to scientific orientation. Based on lectures in a companion short course. A closely related volume by Gutoff and Cohen (1995) is Coating and Drying Defects. John Wiley & Sons, New York.

Harazaki Y. 1987. Progress in Coating Technology. Sogo Gijyutsu Center Co. (General Technology Center), 4-5-12 Shiba, Minato-Ku, Tokyo 108 (in Japanese). Technological and descriptive. The Principal in Harazaki Consulting, the author graduated in chemistry from Osaka University of Science and Technology, and earned his Ph.D. from

Tokyo Institute of Technology. He wrote earlier books on Coating Technology (1971), Basic Science of Coating (1977), Coating Technology, New Edition (1978), and Coating Methods (1979) (all in Japanese).

Satas D. 1984 (Ed.). Web Processing and Converting Technology and Equipment. Van Nostrand Reinhold Publishing Co., New York. Articles by highly experienced authors on various modes of coating as well as related operations. More up-to-date and more nearly comprehensive than Booth's book, but just as purely technological and descriptive.

Booth G.L. 1970. Coating Equipment and Processes. Lockwood Publishing Co., New York. Purely technological and descriptive, it covers a lot of ground but falls far short of being comprehensive. In particular, precision coating methods are not mentioned. Out of print.

Weiss H.L. 1977/1983. Coating and Laminating Machines. Converting Technology Co., Milwaukee. Diagram-laden, information-packed volume privately published by a broadly experienced consultant to the web processing and converting industry. Up-to-date for its time, e.g. interchangeable modular coating stations, and fairly comprehensive (excepting precision coating and high-speed coating), though entirely technological and descriptive.

Middleman S. 1977. Fundamentals of Polymer Processing. McGraw-Hill Book Co., New York. Chapter 6 on Extrusion, Chapter 8 on Coating, Chapter 9 on Fiber Spinning, Chapter 14 on Elastic Phenomena, and Chapter 15 on Stability of Flows, cover fragments of the subject as it was understood in the mid-1970s, and show how simple models can sometimes be useful (but not how such models can be misleading – and they sometimes are).

There is a more extensive literature about drying, but not much of it deals with drying and curing processes for continuously coated liquid layers and multilayers. The books edited by Cohen and Gutoff and by Satas, listed above, contain useful chapters on industrial practice. Also useful is Satas's own chapter on drying in the book he edited in 1989, *Handbook of Pressure Sensitive Adhesive Technology*. A recent encyclopedia article of general nature is *Drying* by P.Y. McCormick in Volume 8 of the *Kirk-Othmer Encyclopedia of Chemical Technology* (Fourth Edition, 1993). Leading general texts are

Kröll K. and Kast W. 1989 (Eds.). Trocknungstechnik, Vol. 3. Springer-Verlag, Berlin. A recent monograph on drying technology, written (in German) from fairly scientific points of view of industrial processes, for example the drying of colloids and gels such as gelatin systems.

Krischer O. and Kast W. 1978. Die wissenschaftlichen Grundlagen der Trocknungstechnik, 3rd edition, Vol. 1, Trocknungstechnik. Springer-Verlag, Berlin. A thorough account of the fundamentals of heat and mass transport in drying processes.

Kröll K. 1978. Trockner und Trocknungsverfahren, 2nd edition, Vol. 2, Trocknungstechnik. Springer-Verlag, Berlin. A thorough account of drying equipment and drying processes as of the 1970s: compare the 1989 book by Kröll and Kast.

Keey R.B. 1972. Drying Principles and Practice. Pergamon Press, Oxford. Still probably the best available textbook in English on the subject, but it gives little attention to drying of paper or any sort of coated sheet or web.

Majumdar A.S. 1987 (Ed.). Handbook of Industrial Drying. Dekker, New York. Despite its title, this is a collection of articles, not a handbook, and the articles tend to be academic. There is a descriptive chapter on drying of pulp and paper, and a few pages on drying of coated webs.

Vergnaud J.-M. 1992. Drying of Polymeric and Solid Materials. Springer-Verlag, London. Wide-ranging, reflecting the breadth of the author's researches, this book does treat modeling of plane sheets, rubbery and thermosetting coatings, but often with over-simplified models.

There seems to be no book that covers the science and technology of solidification by processes other than freezing, and surely none that does so comprehensively. A useful volume on the fundamentals of curing is:

Stepto R.F.T. 1998 (Ed.). Polymer Networks: Principles of Their Formation, Structure and Properties. Blackie Academic & Professional, London.

The Ph.D. theses (below) from the Coating Process Fundamentals Program at the University of Minnesota are published by UMI (University Microfilms International, Ann Arbor, MI). The editors are variously H.T. Davis [HTD], M.C. Flickinger [MCF], L.F. Francis [LFF], W.W. Gerberich [WWG], C.W. Macosko [CWM], M.L. Mecartney [MLM], A.V. McCormick [AVM], L.E. Scriven [LES], H.K. Stolarski [HKS], and K. Takamura [KT]:

Coating flows

Huh C. 1969. Capillary Hydrodynamics. See also Huh and Scriven (1971) J. Colloid Interf. Sci., 35, 85–101. Hydrodynamic model of steady movement of a solid/liquid/fluid contact line.

Ruschak K.J. 1974. Fluid Mechanics of Coating Flows (by Asymptotic Methods). [LES]

Orr F.M. Jr 1976. Numerical Simulation of Viscous Flow with a Free Surface (by Galerkin's Method with Finite Element Basis Functions). [LES]

Silliman W.J. 1979. Viscous Film Flow with Contact Lines: Finite Element Simulation. [LES]

Higgins B.G. 1980. Capillary Hydrodynamics and Coating Beads. [LES]

Bixler N.E. 1982. Stability of a Coating Flow. [LES]

Kistler S.F. 1983. The Fluid Mechanics of Curtain Coating and Related Viscous Free Surface Flows with Contact Lines. [LES]

Kheshgi H.S. 1983. The Motion of Viscous Liquid Films. [LES]

Teletzke G.F. 1983. Thin Liquid Films: Molecular Theory and Hydrodynamic Implications. [HTD, LES]

Coyle D.J. 1984. The Fluid Mechanics of Roll Coating: Steady Flows, Stability, and Rheology. [CWM, LES]

Saita F.A. 1984. Elastohydrodynamics and Flexible Blade Coating. [LES]

Papanstasiou A.C. 1984. Coating Flows and Processing of Viscoelastic Liquids: Fluid Mechanics, Rheology and Computer-Aided Analysis. [CWM, LES]

Secor R.B. 1988. Operability of Extensional Rheometry by Stagnation, Squeezing, and Fiber-Drawing Flows: Computer-Aided-Analysis, Viscoelastic Characterization, and Experimental Analysis. [CWM, LES]

Bornside D.B. 1988. Spin Coating. [LES, CWM]

Christodoulou K.C. 1988/1990. Physics of Slide Coating Flow. [LES]

Pranckh F.R. 1989. Elastohydrodynamics in Coating Flows. [LES]

Schunk P.R. 1989. Polymer and Surfactant Additives in Coating and Related Flows. [LES]

Sartor L. 1990. Slot Coating: Fluid Mechanics and Die Design. [LES]

Navarrete R.C. 1991. Rheology and Structure of Flocculated Suspensions. [CWM, LES]

Chen K.S.A. 1992. Studies of Multilayer Slide Coating and Related Processes. [LES]

Cai J.J. 1993. Coating Rheology: Measurements, Modeling and Applications. [CWM, LES]

Benjamin D.F. 1994. Roll Coating Flows and Multiple Roll Systems. [LES]

Carvalho M.S. 1995. Roll Coating Flows in Rigid and Deformable Gaps. [LES]

Hanumanthu R. 1996. Patterned Roll Coating. [LES]

Gates I.D. 1998. Slot Coating Flows: Feasibility, Quality. [LES]

Dontula P. 1999. Polymer Solutions in Coating Flows. [CWM, LES]

Pasquali M. 2000. Polymer Molecules in Free Surface Coating Flows. [LES]

Pekurovsky M.L. Air Entrainment in Liquid Coating: From Incipient to Apparent. [LES]

Brethour J.M. 2000. Coating with Deformable and Permeable Surfaces: Focus on Rotary Screen Coating. [LES]

Fermin R.J. 2000. Electrohydrodynamic Coating Flows. [LES]

Musson L.C. 2001. Two-Layer Slot Coating. [LES]

Apostolou K. 2004. Slot Coating Start-Up. [LES]

Owens M. 2004. Misting in Forward-Roll Coating: Structure, Property, Processing Relationships. [CWM, LES]

Drying, curing

Sanchez J. 1994. Kinetics and Models of Silicon Alkoxide Polymerization. [AVM]

Cairncross R.A. 1994. Solidification Phenomena During Drying of Sol-to-Gel Coatings. [LFF, LES]

Pan S.X. 1995. Liquid Distribution and Binder Migration in Drying Porous Coatings. [HTD, LES]

Rankin S. 1998. Kinetic, Structural, and Reaction Engineering Studies of Inorganic–Organic Sol–Gel Copolymers. [AVM]

Dabral M. 1999. Solidification of Coatings: Theory and Modeling of Drying, Curing and Microstructure Growth. [LFF, LES]

Pekurovsky L.A. Capillary Forces and Stress Development in Drying Latex Coating. [LES]

Wen M. 2000. Designing Ultraviolet-Curing of Multifunctional (Meth) Acrylate Hard Coats. [AVM, LES]

Rajamani V. 2004. Shrinkage, Viscoelasticity, and Stress Development in Curing Coatings. [AVM, LFF, LES]

Arlinghaus E. 2004. Microflows, Pore and Matrix Evolution in Latex Coatings. [LES]

Radhakrishnan H. 2004. Solidification by Drying: Effect of Non-Uniformities. [LES]

Microstructuring

Hamlen R.C. 1991. Paper Structure, Mechanics, and Permeability: Computer-Aided Modeling. [LES]

Bailey J.K. 1991. The Direct Observation and Modeling of Microstructural Development in Sol–Gel Processing of Silica. [MLM]

Sheehan J.G. 1993. Colloidal Phenomena in Paper Coatings Examined with Cryogenic Scanning Electron Microscopy. [HTD, KT, LES]

Pozarnsky G.A. 1994. NMR Investigations of Oxide Formation by Sol–Gel Processing. [MLM]

Li P. 1995. Crystallization Behavior of Sol–Gel Derived Lithium Disilicate Powders and Coatings. [LFF]

Kim Y.-J. 1995. Sol–Gel Processing and Characterization of Macroporous Titania Coatings. [LFF]

Ming Y. 1995. Microstructure Development in Polymer Latex Coatings. [HTD, KT, LES]

Wara N.M. 1996. Processing of Macroporous Ceramics Through Ceramic-Polymer Dispersion Methods. [LFF]

Cooney T. 1996. Materials Integration and Processing Development for Microelectromechanical Systems (MEMS). [LFF]

Craig B.D. 1997. Interpenetrating Phase Ceramic/Polymer Composite Coatings. [LFF]

Daniels M. 1999. Colloidal Ceramic Coatings with Silane Coupling Agents. [LFF]

Lyngberg O.K. 1999. Development of Ultra Thin Thermostabile Biocatalytic Composite Coatings Containing Latex and Metabolically Active Cells. [MCF, LES]

Prakash S. 2000. Microstructure Evolution in Phase Inversion Membranes by Time-Sectioning Cryo-SEM. [LFF, LES]

Grunlan J.C. 2001. Carbon Black-Filled Polymer Composites: Property Optimization with Segregated Microstructures. [LFF, WWG]

Huang Z. 2001. Continuous Coatings from Particulate Suspensions: Polymer Powder and Latex Coatings. [HTD, LES]

Ma, Y. 2002. High-Resolution Cryo-Scanning Electron Microscopy of Latex Film Formation. [HTD, LES]

O'Connor A.E. 2003. The Influence of the Coating Process on the Structure and Properties of Block-Copolymer-Based Pressure-Sensitive Adhesive. [CWM]

Ge H. 2004. Microstructure Development in Latex Coatings: High Resolution Cryo-Scanning Electron Microscopy. [HTD, LES]

Stress effects, mechanical properties

Tam S.-Y. 1996. Stress Effects in Drying Coatings. [HKS, LES]

Wang F., 1996. Mechanical Properties of Interfaces: Adhesion and Related Interfacial Phenomena. [LFF]

Payne J.A. 1998. Stress Evolution in Solidifying Coatings. [LFF, AVM]

Lei H. 1999. Flow, Deformation, Stress and Failure in Solidifying Coatings. [LES, LFF, WWG]

Strojny A. 1999. Mechanical Characterization of Thin Coatings Using Nanoindentation. [WWG]

Xia X. 2000. Micro-Nanoprobing Measurement of Polymer Coating/Film Mechanical Properties. [WWG, LES]

Vaessen D.M. 2002. Stress and Structure Development in Polymeric Coatings. [LFF, AVM]

The M.S. theses from the same program (listed below) are available only through the University of Minnesota. The editors are from the same set:

Coating flows

Scanlan D.J. 1990. Two-Slot Coater Analysis: Inner Layer Separation Issues in Two-layer Coating. [LES]

Lund M.A. 1990. Nonhomogenous Behavior of Barium Ferrite Dispersions. [CWM, LES]

Fukuzawa K. 1992. Reverse Roll Coater Model. [LES]

Yoneda H. 1993. Analysis of Air-Knife Coating. [LES]

Palmquist K.E. 1993. Studies of Slide Coating Start-Up Flow. [LES]

Cohen D. 1993. Two-Layer Slot Coating: Flow Visualization and Modeling. [LES]

Nagashima K. 1994. Slide Coating Flow. Splice Passage. [LES]

Anderson T.J. 1996. Reverse Deformable Roll Coating Analysis. [LES]

Dinh K.T.T. 1997. In-line Mixing and Cleaning Processes. [LES]

Kazama K. 1998. Leveling of Multilayer Coating. [LES]

Nagai R. 1999. Modeling Slot Coating Start-Up Flow. [LES]

Tada K. 2001. Curtain Coating Cylinders. [LES]

Lee H. 2002. Tensioned Web Slot Coating. [LES]

Drying, curing

Yapel R.A. 1988. A Physical Model of the Drying of Coated Films. [LES]

Zheng Y. 1996. Photoinitiated Polymerization of Multifunctional Acrylates and Methacrylates. [AVM, LES]

Microstructuring

Limbert A. 1995. Microstructure Development in Coatings by Cryo-SEM. [LFF, LES]

LeBow S.M. 1997. Sequential Study of the Coalescence of Carboxylated Latex Polymer Films. [HTD, LES]

Lei M. 2002. Hardness and Microstructure of Latex/Ceramic Coatings. [LFF, LES]

Gong X. 2004. Role of van der Waals Force in Latex Film Formation. [HTD, LES]

Stress effects

Mountricha E. 2004. Stress Development in Two-layer Polymer Coatings. [LFF, AVM]

10

Langmuir–Blodgett Films: A Window to Nanotechnology

M. Elena Diaz Martin and Ramon L. Cerro

10.1 Langmuir–Blodgett Films and Nanotechnology

The last decade has seen an increased interest in molecular monolayers, alongside the development of micro- and nano-technologies. Langmuir–Blodgett (LB) films, first reported by Blodgett (1935), are films of amphiphilic molecules, one molecule thick, that present many interesting optical and biological properties, consistent thickness, and are essentially smooth. These films can also be used to coat surfaces without harming the surface they are coating or the substances that surround the coated surface.

Fueling this interest in LB films is the commercial potential of these thin films as part of the microelectronics revolution that calls for ultrathin fabrication methods. The first international conference on LB films was held in 1979 and since then the use of this technique has been increasing worldwide. LB films have been utilized in many applications such as lubrication, wetting, adhesion, electronics, and construction of chemical, physical, and biological sensing devices (Roberts, 1990).

Electronic devices usually rely on the properties of inorganic materials. However, organic materials can show increased versatility in device design when their molecular and bulk structures are tailored to produce bulk material holding specific functional properties. Molecular microelectronics deals with the development and exploitation of novel organic materials in electronic and opto-electronic devices, i.e. organic semiconductors. Of particular interest to microelectronics technology is the ability to deposit very thin films on a wafer surface. The LB technique offers the possibility of depositing films of organic molecules a few nanometers thick with remarkably low levels of defects (Peterson, 1996). More recently, LB films have been used in the development

Chemical Engineering Edited by Miguel A. Galán and Eva Martin del Valle

of super-molecular electronic assemblies where the transport and control of electronic signals are performed at the nanometer scale (Roberts, 1990). The LB technique has been used to build 3D circuits and molecular-scale switches, which can be used in both logic and memory circuits, based on carefully arranged assembled molecules in mono-molecular films.

A biological membrane is a structure particularly suitable for study by the LB technique. The eukaryotic cell membrane is a barrier that serves as a highway and controls the transfer of important molecules in and out of the cell (Roth *et al.*, 2000). The cell membrane consists of a bilayer or a two-layer LB film (Tien *et al.*, 1998). Lipid bilayers are composed of a variety of amphiphilic molecules, mainly phospholipids and sterols which in turn consist of a hydrophobic tail, and a hydrophilic headgroup. The complexity of the biomembrane is such that frequently simpler systems are used as models for physical investigations. They are based on the spontaneous self-organization of the amphiphilic lipid molecules when brought in contact with an aqueous medium. The three most frequently used model systems are monolayers, black lipid membranes, and vesicles or liposomes.

Another interesting area of application of LB films is electron beam micro-lithography. The degree of resolution of an electron beam system depends on electron scattering characteristics of the lithographic film. To achieve higher resolution very thin resists must be used because the thinner the film, the better the resolution. These films can be built on suitable materials by means of the LB technique. The magnetic coating industry has been searching for many years for a coating with good permeability and that can act as a lubricant. In this field, LB films can lead to very interesting applications. Seto *et al.* (1985) showed dramatic decrease of frictional coefficient by coating an evaporated cobalt tape with an LB film of barium stearate. A closed LB multilayer also limits moisture penetration and can be used as a micro-encapsulating agent at the same time.

Micro- and nano-sensors as well as a vehicle for the formation of nano-particles (Elliot *et al.*, 1999) configure important and challenging applications of LB films. The chemical component of the sensor surface is chosen specifically to react with a given substance. The reaction should be detected by a change in electrical properties and other physical properties, especially weight change. Weight change can be measured by monitoring changes in frequency of a cantilevered quartz oscillator coated with an LB film. In most cases, the response time of sensors depends on the thickness of the sensing layer. Because the LB films are extremely thin, they promise to produce very fast response times. However, the high order of molecular structure in high-quality films slows down the rate of diffusion through the film and in turn the response time.

10.2 The Langmuir–Blodgett Technique

The transfer onto the surface of a solid substrate of successive monolayers of divalent soaps compressed on the surface of water in a Langmuir trough was described by Blodgett (1935). A Langmuir trough is a container with moving barriers for manipulation of the film at the air–water interface. The solid substrata, in the original experiments glass microscope slides, are moved up and down, out and into the water. The term Langmuir–Blodgett technique is currently used to denote the deposition of mono-molecular layers by transfer from the air–water interface onto a solid surface.

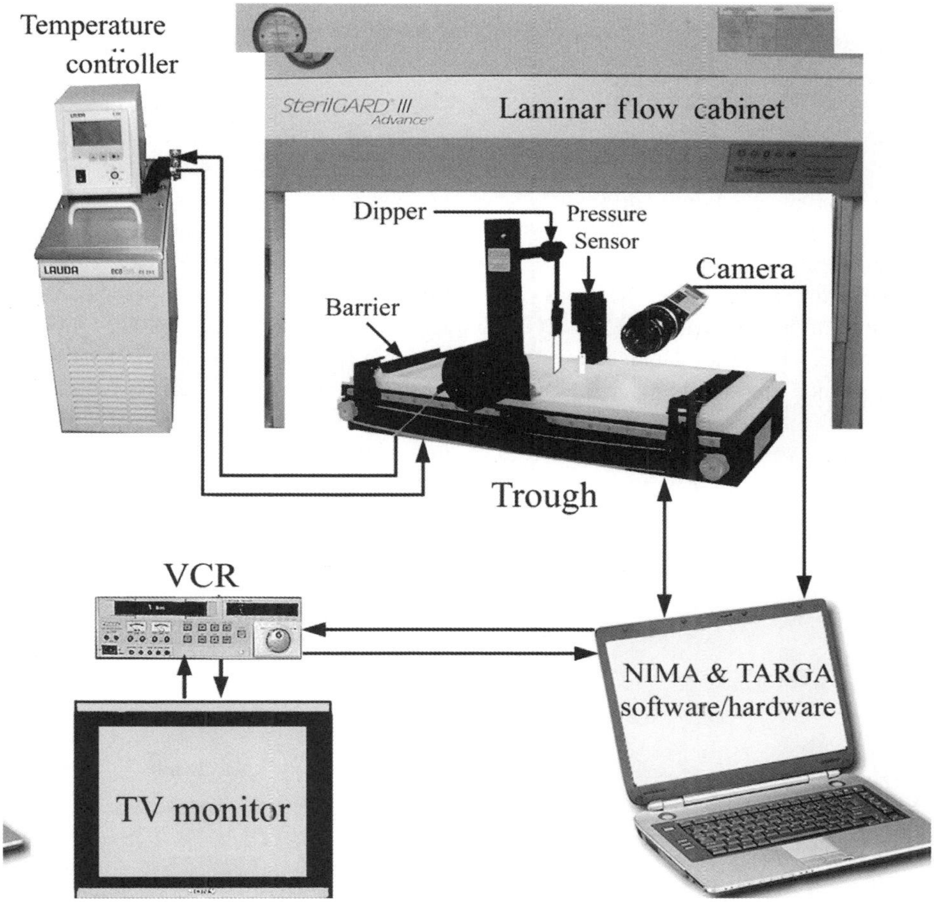

Figure 10.1 Schematic representation of the experimental setup. The LB trough is a NIMA Standard with computerized barrier control. The image analysis for contact angle detection includes a CCD camera and a professional Targa® board for data acquisition

A sketch of the experimental device for deposition of LB films is shown in Figure 10.1. A few drops of a dilute solution of an amphiphilic compound showing a hydrophobic tail and a hydrophilic headgroup are spread out on top of the water surface in a Langmuir trough. The solution of the amphiphilic compound is usually made in a volatile organic solvent, such as chloroform or ethyl ether. The solvent evaporates leaving a low concentration of the amphiphilic compound on the water surface due to the repulsive forces on their hydrophobic ends. A moving barrier is used to compress the layer of molecules on the water surface. A Wilhelmy plate attached to a force balance is used to measure the interfacial tension and the signal from the balance controls the moving barriers. By moving a plate in and out of the water, the monolayer can be transferred to the surface of the plate. Typical coating techniques are based on vertical movement of the plate, but angles other than 90° have been used successfully. A CCD camera with a zoom lens, attached to an image analysis system, is pointed at the interface near the contact line and is used to observe details of the movement of the contact line as well as to measure contact angles during deposition.

Compression of the monolayer changes the concentration of molecules of the amphiphilic compound on the air–water interface and in turn changes interfacial tension. The difference between the surface tension of pure water, σ_o, and the interfacial tension of the liquid–air interface, σ_{int}, is defined as the surface pressure, $\pi = \sigma_o - \sigma_{\text{int}}$. The surface pressure as a function of the area occupied by the amphiphilic molecules, $A(\text{Å}^2/\text{molecule})$, presents a remarkable similitude with PVT (pressure-volume-temperature) phase diagrams. An experimental surface pressure versus area per molecule for arachidic acid ($C_{19}H_{39}COOH$) over a 10^{-4} M solution of $ZnSO_4$ is shown in Figure 10.2. The region labeled *Gas*, corresponds to a sparse concentration of molecules where surface pressure is inversely proportional to area. The region labeled *Liquid*, corresponds to a high concentration but without packing. Small changes in area within the liquid region correspond to large changes in surface pressure. Finally, the region labeled *Solid*, corresponds to a tightly packed layer of molecules such that very small changes in area/molecule bring large changes in surface pressure. The surface pressure versus area/molecule, as will be explained in this chapter, is a very important element of data for understanding deposition behavior.

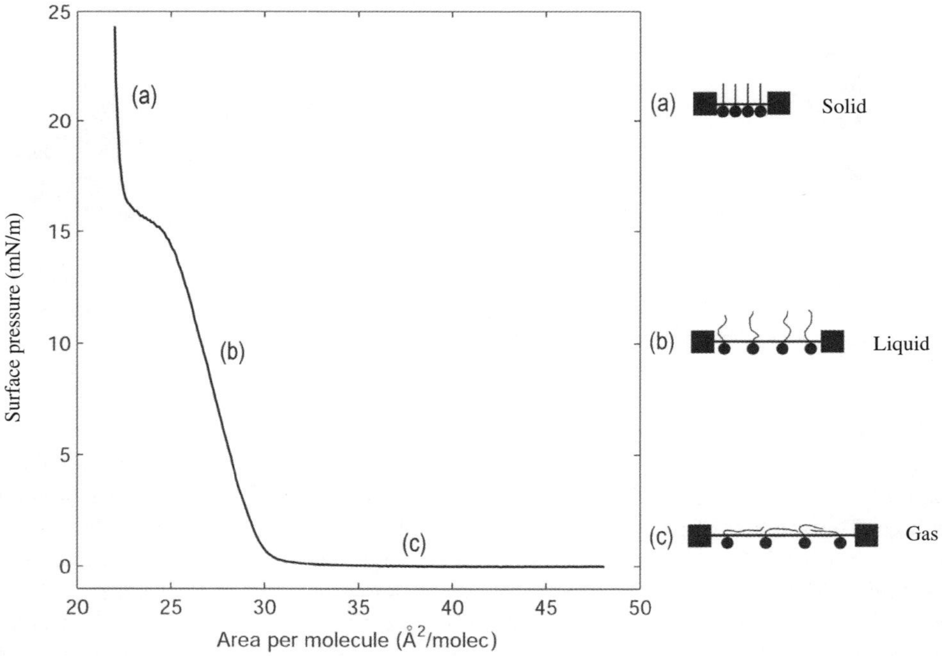

Figure 10.2 *Surface pressure versus area per molecule isotherm of arachidic acid on a subphase of 10^{-4}M $ZnSO_4$ at pH = 5.6 and T = 25°C. The drawings on the right are a schematic representation of the orientation of the amphiphilic fatty acid molecules on the water surface*

The similitude between πA diagrams and PVT diagrams was exploited in the development of equations of state for the films deposited on the air–water interface (Gaines, 1966). Using an analogy with ideal gas theory, a simple model assumes that in the gas region, molecules at the air–water interface have a mobility dependent on their thermal kinetic energy

$$\pi A = k\ T \tag{10.1}$$

where k is Boltzmann's constant and T is the temperature. This simple model ignores the effect of the liquid subphase, but its use as a limiting equation for large values of A has been established with different variations. Modeling of the expanded and condensed regions has not been as well established but progress has been made in describing and modeling shape transitions (Keller *et al.*, 1987).

Collection of monolayers deposited during immersion only are described as X-type films, multilayers deposited during removal only are described as Z-type films, and multilayers deposited sequentially during both immersion and removal are described as Y-type films. A sketch of the deposition of different types of LB film is shown in Figure 10.3. Figure 10.3 (a) shows the typical behavior during deposition of a monolayer during immersion. Notice that the contact angle, i.e. the angle formed between the liquid phase and the solid substrate, is shown to be larger than 90° since deposition during immersion

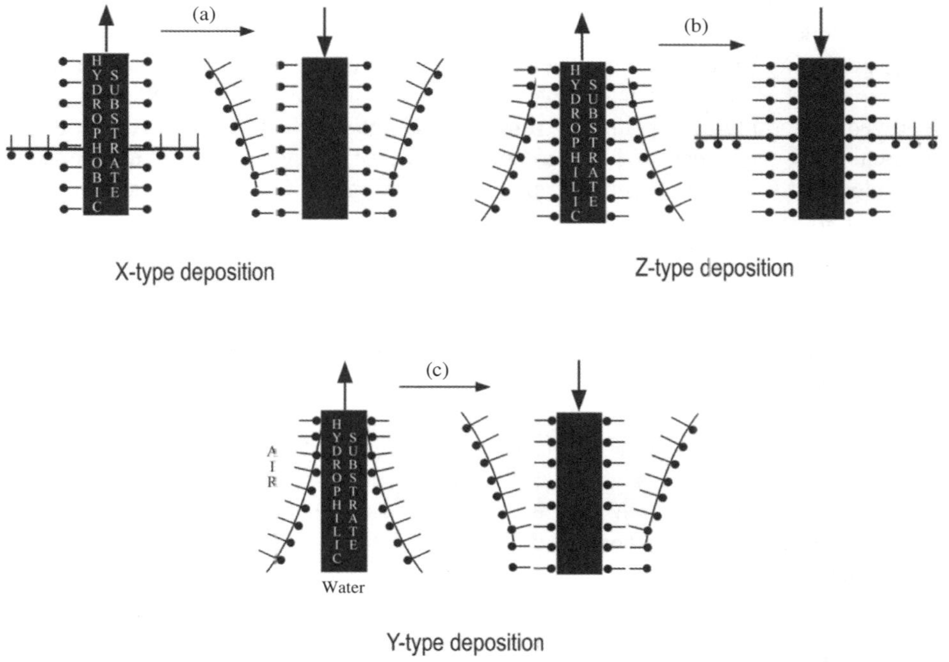

Figure 10.3 *Schematic representation of the different types of multilayer deposition of LB films. (a) X-type films deposited during immersion only. (b) Z-type films deposited during removal only. (c) Y-type films deposited during immersion and removal*

can only be accomplished on hydrophobic surfaces, i.e. for contact angles $\phi_S > 90°$ (Bikerman, 1939). A sketch of the deposition of a monolayer during removal, i.e. a Z-type deposition, is shown in Figure 10.3 (b). Notice that the contact angle is shown to be smaller than 90°, since deposition during removal can only be accomplished on hydrophilic surfaces, i.e. for $\phi_S < 90°$. A sketch of the successive deposition of monolayers during immersion and removal to create a Y-type film is shown in Figure 10.3 (c).

Naturally and ideally, the most prevalent LB films should be Y-type films. When a monolayer is deposited during immersion, the hydrophobic tails attach to the solid substrate and the hydrophilic ends of the amphiphilic molecules remain on the outside in contact with the liquid subphase creating a hydrophilic surface and the necessary condition for deposition during removal. On the other hand, when a monolayer is deposited during removal, the hydrophilic ends attach to the substrate and the hydrophobic tails are in contact with the air creating a hydrophobic surface, which will in turn determine the necessary condition for deposition of the next monolayer during immersion. In practice, under experimental conditions, forces generated by electrical double layers between the carboxylic acid end and the cations of the metal salts of the liquid subphase deteriorate Y-type films and allow the formation of X- or Z-type films.

The precise thickness of mono-molecular assemblies and the degree of control over their molecular architecture have firmly established LB films as an essential building block of micro- and nano-technologies. However, despite the superior properties of mono-molecular films deposited via the LB technique, the development of applications has been hindered by slow deposition speeds and by what is perceived as a lack of reliability or reproducibility. On the basis of a hydrodynamic model of the moving contact line we have developed a framework to analyze the physical–chemical and hydrodynamic processes governing LB depositions. The development of a hydrodynamic model, including fundamental phenomena such as molecular and double-layer force effects, has greatly improved our ability to explain experimental findings that had otherwise confused LB researchers for many years. When a completely reliable and precise deposition technique is developed, LB films will certainly be the favorite choice for the manufacture of biosensors (Barraud *et al.*, 1993), nonlinear optic devices, and photo-lithography patterns for MEMS and NEMS fabrication (Bowden and Thompson, 1979).

10.3 Contact Angles and Flow Patterns near the Moving Contact Line

Static contact angles change with surface treatment, liquid phase composition, surface tension, and pH. Dynamic contact angles depend also on substrate speed and liquid phase transport properties such as density and viscosity. There are indications in the LB literature that researchers recognized the effect of dynamic contact angles on monolayer transfer ratios and the change of dynamic contact angles with the speed of the solid substrate. Transfer ratios are defined as the ratio of the area covered by the monolayer on the solid substrate to the decrease of area occupied by the monolayer on the liquid interface. However, without information on flow patterns near the three-phase contact line, it is not possible to explain the mechanics of LB deposition to dynamic contact angles and the movement of the gas–liquid interface. In the LB literature, it is customary to report dynamic contact angles measured in the liquid phase.

The role of contact angles in LB film deposition was outlined by Bikerman (1939) who recognized the need of hydrophobic surfaces, i.e. contact angles $\phi_S \geq 90°$ for successful immersion deposition (X-type), and of hydrophilic surfaces, i.e. contact angles $\phi_S \leq 90°$ for successful removal deposition (Y- or Z-type). Bikerman (1939) used a simple geometrical argument, essentially the zipper mechanism that has been accepted as the basic deposition mechanism (Roberts, 1990).

Gaines (1977) performed the first systematic study of dynamic contact angles during the transfer of monolayers. Static contact angles were measured using the meniscus height method on treated glass substrates. Typical immersion dynamic contact angles were $\phi_S \approx 110°$ and greater. Typical removal contact angles were $20° \leq \phi_S \leq 60°$. However, with no information on the hydrodynamics of the three-phase contact line, there was not a mechanism to explain the effect of wetting on the mechanics of LB deposition.

Petrov *et al.* (1980) analyzed the causes for the entrapment of water between the solid substrate and the monolayer in Z-type depositions. This phenomenon has many common features with film thinning processes found during foam and emulsion breakdown and it is dependent on interfacial properties and on molecular interactions between the solid substrate and the monolayer. Petrov *et al.* (1980) measured the maximum speed of removal of the solid substrate before entrainment of a water layer and found it to be dependent on pH and ionic strength. There is no record in the publication of the measurement of dynamic contact angles.

An experimental demonstration that the effect of pH on transfer ratios can be explained by the effect of pH on dynamic contact angles was provided by Aveyard *et al.* (1992) for substrata moving at constant deposition speeds.

Peng *et al.* (1985) recognized the value of the dynamic contact angle as a useful characterization and diagnostic tool and used a Wilhelmy plate technique to measure dynamic contact angles during multilayer deposition of lead stearates on mica substrates. Experimental results show clear trends of transfer ratio dependence with contact angles, including the interesting feature that transfer ratios during removal decrease with increasing dynamic contact angle.

A remarkable analysis of the role of hydrodynamics in LB depositions was done by de Gennes (1986). This analysis concerns only deposition during removal of the solid substrate. de Gennes recognizes that the only flow pattern that would allow Y-deposition is a split-ejection streamline in the liquid phase. However, he uses as a reference the work of Huh and Scriven (1971), but their hydrodynamic theory predicts a rolling motion in the liquid phase.

Petrov and Petrov (1998) developed a molecular hydrodynamic theory of film deposition during removal. Their theory correctly assumes a flow pattern – which we identified as a split streamline – between the solid substrate and the monolayer in Figure 10.5 (c). This pattern is indeed the necessary pattern for successful deposition during removal, but it is not the only flow pattern for solid removal at all dynamic contact angles. Petrov and Petrov (1998) address the kinetics of water removal between the solid and the monolayer and the formation of wet or dry monolayers depending on the amount of water entrained.

Zhang and Srinivasan (2001) performed a hydrodynamic analysis of water entrainment based on the augmented version of the film evolution equation, where molecular forces and Marangoni effects can be introduced.

Contact angles and flow patterns are nothing but symptoms of the effect of highly asymmetric molecular and structural force fields in the vicinity of a contact line. The solid substrate and the liquid phase are dense phases with large molecular concentrations, while the gas phase is essentially a void. In the wedge-like shape of the moving contact line, this asymmetric geometry gives rise to unbalanced surface forces that in turn translate into shear stresses at the interface. Recently, Fuentes *et al.* (2005) have shown that discrepancies between experiments and the purely hydrodynamic contact line theory cannot be explained without introducing molecular and structural forces within the proximal region of the moving contact line.

In a stable dynamic situation, the velocity of the contact line must match the velocity of immersion, or removal, of the solid. In the classical hydrodynamic theory two parameters determine the dynamics of the moving contact line: (1) the dynamic contact angle, ϕ_D, and (2) the ratio of viscosities of the two fluids, $R = \mu_A/\mu_B$ (Huh and Scriven, 1971). The three basic flow patterns near a moving contact line are shown schematically in Figures 10.4 and 10.5. Figure 10.4 shows the flow patterns during immersion of the solid substrate, while Figure 10.5 shows the flow patterns during removal. The first flow patterns in Figure 10.5 (a)–(c) are the mirror image of the patterns occurring during immersion. An additional flow pattern is shown in Figure 10.5 that does not correspond to a moving contact line. It is the dip-coating flow pattern, Figure 10.5 (d), corresponding to very small or zero contact angles where a liquid film is continuously entrained with the substrate during removal. Additional, more complicated transition patterns have been predicted theoretically (Diaz and Cerro, 2003), but the size of the vortex is too small for experimental confirmation.

At a moving or dynamic contact line, the contact angle is modified by stresses of hydrodynamic origin developed in fluid motion. We will assume that in a region very close, i.e. within 2 mm of the contact line, the interface is a straight line and that the contact angle is the dynamic contact angle. In general, one of the fluids has a rolling flow pattern, while the other fluid shows a splitting streamline. In the first flow pattern of the immersion sequence,

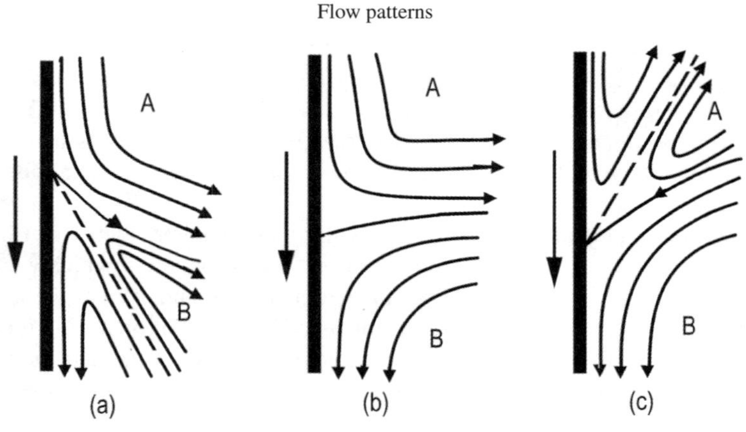

Figure 10.4 *Schematic representation of flow patterns near a moving contact line during immersion of a solid substrate into a pool of liquid. (a) Split-injection streamline in phase B and rolling pattern in phase A. (b) Transition flow pattern with motionless interface and rolling motion in phases A and B. (c) Rolling motion in phase B and split-ejection streamline in phase A*

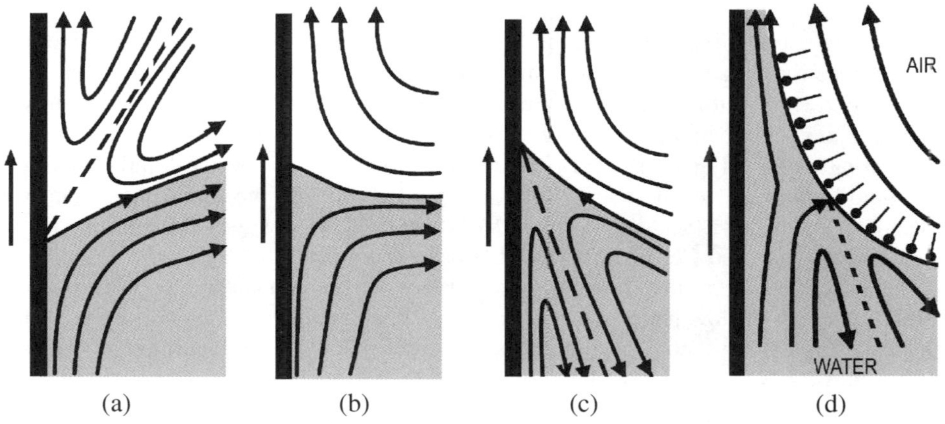

Figure 10.5 *Schematic representation of flow patterns near a moving contact line during removal of a solid substrate from a liquid pool. (a) Rolling motion in the liquid phase and split-injection streamline in the gas phase. (b) Transition flow pattern with motionless interface. (c) Split-ejection streamline in the liquid phase and rolling motion in the vapor phase. (d) Dip-coating flow pattern with stagnation point in the air–water interface and a liquid film entrained on the solid surface*

Figure 10.4 (a), there is a split-injection streamline in the liquid phase and the air–liquid interface moves away from the contact line. The contact angle measured in the liquid phase is purposely represented as a small contact angle, i.e. $\phi_D < 90°$. The third pattern, Figure 10.4 (c) shows a rolling pattern in the liquid phase and a split-ejection pattern in the gas phase such that the air–liquid interface moves toward the contact line. Notice that the contact angle is purposely represented as a large contact angle, $\phi_D > 90°$. The intermediate pattern, shown in Figure 10.4 (b), is a transition pattern and the air–liquid interface is motionless. The contact angle in this case is an intermediate value between the contact angles in Figure 10.4 (a) and (c), typically around 90°. The first flow pattern in the liquid phase, Figure 10.4 (a), is described as a split-injection streamline pattern to signify that the fluid near the contact line is being displaced by fluid coming from the bulk of the liquid phase. In the third sketch, Figure 10.4 (c), the flow pattern in the gas phase is described as a split-ejection streamline to signify that the fluid near the contact line leaves along the splitting streamline. During removal, as shown in Figure 10.5 (a)–(c) the flow patterns are reversed and for small contact angles the liquid phase shows a split-ejection pattern, while for large contact angles the liquid phase is in a rolling motion.

Experimental evidence of the occurrence of these flow patterns and quantitative dependence with contact angles was provided by Savelski *et al.* (1995). During LB depositions one of the fluids is always a liquid, predominantly water, and the other fluid is air. Viscosity ratios are small when the solid is immersed as shown in Figure 10.4, $R = \mu_{air}/\mu_{water} \approx 0.02 \ll 1$. Similarly, viscosity ratios are large when the solid is removed from the water bath as shown in Figure 10.5, $R = \mu_{water}/\mu_{air} \approx 50 \gg 1$.

When a wetting solid is removed from the liquid phase, the flow pattern in the liquid phase is the split-ejection streamline pattern shown in Figure 10.5 (c). The interface liquid–air moves toward the contact line and a Z-type LB deposition is possible. During removal, transfer ratios of the monolayer show a strong dependence with the relative velocity of the interface that is also a strong function of the contact angle. When a non-wetting solid is removed from the liquid phase, the flow pattern is a rolling motion, the interface moves away from the contact line, and LB deposition is not possible.

For a perfectly wetting system, i.e. for very small or zero contact angles, a continuous film of liquid is entrained on the solid substrate, creating a dip-coating flow pattern. The dip-coating pattern, Figure 10.5 (d), has a stagnation point on the gas–liquid interface. The liquid entrained on the solid surface comes from inside the liquid phase and the lower part of the interface moves away from the solid. Under these conditions, it has been shown (Diaz and Cerro, 2004) that LB deposition cannot take place.

The experimental results showing flow patterns versus contact angles were used to configure a map of flow regions as a function of flow parameters, shown in Figure 10.6. The x-axis is dynamic contact angles, ϕ_D, and the y-axis is the logarithm of the viscosity ratios, $R = \mu_A/\mu_B$. For the formulation of the hydrodynamic theory (Huh and Scriven, 1971), dynamic contact angles are measured on the advancing fluid and can vary from perfect wetting, $\phi_D = 0$, to entrainment, $\phi_D = \pi$. Following this convention, in Figure 10.6, contact angles are measured in the liquid phase during immersion and contact angles are measured in the air phase during removal.

Flow region I in Figure 10.6 corresponds to the flow pattern of Figure 10.4 (a) and shows a split-injection streamline in the lower fluid (i.e. liquid) and rolling motion in the upper fluid (i.e. air). The lower part of region I, below the line $R = 1$, takes place during immersion of wetting solids, that is for smaller dynamic contact angles. Notice that in region I, the interface is moving away from the contact line.

Region II shows a split streamline in the air phase and rolling motion in the liquid phase. This pattern is found when a non-wetting solid is immersed in a liquid or for larger immersion speeds. The interface is moving toward the contact line. Region III corresponding to the flow pattern of Figure 10.4 (c), presents a split-ejection streamline in the liquid

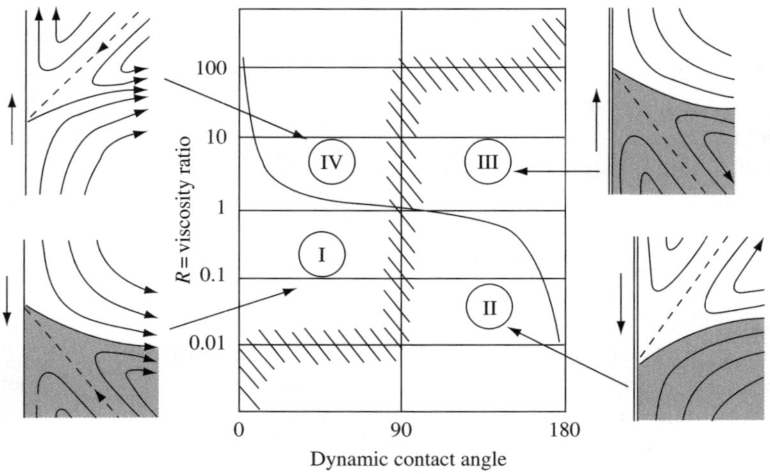

Figure 10.6 *Map of flow regions as a function of viscosity ratio, $R = \mu_A/\mu_B$, and dynamic contact angle, ϕ_D. The solid line is the locus of the motionless interface transition pattern as predicted by the hydrodynamic theory (Huh and Scriven, 1971). The dashed line is the experimental locus of transition patterns*

phase and a rolling pattern in the air phase. Region III represents the typical flow pattern during removal of wetting solids, for small removal speeds. Notice that in region III the interface moves toward the contact line. Region IV shows a split-injection streamline in the air phase and a rolling motion in the liquid phase. This pattern is typical of the removal of a non-wetting solid from a liquid. The air–liquid interface moves away from the contact line.

The fuzzy line dividing regions I and IV from regions II and III is the locust of points where a stagnant interface – flow pattern of Figure 10.4 (b) – is found. This is a transition pattern and the fuzzy line is used to represent typical experimental uncertainty for a clean two-fluid system. The locus of the motionless interface is usually found for dynamic contact angles near $\phi_D \sim 90°$. In practice, large dynamic contact angles typical of treated substrata are more sensitive to movement of the contact line than clean surfaces and exhibit considerable advancing–receding hysteresis (Blake and Ruschak, 1997).

The dependence of dynamic contact angles on moving contact line velocity has been the subject of many experimental studies since this is an important parameter in high-speed coating. For an early review of experimental data on dynamic contact angles, refer to Dussan (1979). For more recent reviews on dynamic contact angles, refer to the excellent contributions of Blake (Chapter 5) and Kistler (Chapter 6) in Berg (1993).

When a solid is immersed into a liquid, such as in X-type film deposition, dynamic contact angles increase with immersion speed. Because of contact angle hysteresis, the static contact angle is not uniquely defined. The static contact angle, ϕ_S, regardless whether it can be determined experimentally, is a concept that results in the application of a variational principle, i.e. any change in contact angle results in an increase in the entropy of the system. The contact angle defined by the variational principle complies with the definition of the thermodynamic static contact angle. The limit at vanishing r and contact line speed of the dynamic contact angle, $\phi_D\,(N_{Ca}, N_{Re})$, depends on the dynamic variables. This limit, in general, is not equal to the thermodynamic static contact angle and depends on the movement history of the system and on the geometry of the solid surface. At very low immersion speed, i.e. when $U_s \sim 0$, the contact angle measured in the liquid phase approaches the static contact angle from above, $\phi_D \rightarrow \phi_S + \Delta\phi$, i.e. during immersion, contact angles are larger than static contact angles. When the immersion velocity increases, the contact angle increases steadily until it reaches $\phi_D \sim \pi$. In Figure 10.6 this is the right-end side of the map where a small air film would be entrained between the solid and the LB film and determines the maximum speed of operation of high-speed coating. The phenomena of air entrainment have been determined experimentally in high-speed coating situations (Gutoff and Kendrick, 1982), but it is not a problem encountered in LB deposition where coating speeds are very small.

When the solid is removed from the liquid, as in Y-type deposition, the dynamic contact angle measured on the liquid phase decreases steadily with increasing speed. At very low removal speeds, i.e. when $U_s \approx 0$, the contact angle approaches the static contact angle from below, $\phi_D \rightarrow \phi_S - \Delta\phi_S$, i.e. during removal dynamic contact angles are smaller than static contact angles. At increasing removal speeds, the dynamic contact angle decreases until $\phi_D = 0$. If one measures the contact angle in the advancing phase, then the contact angle $\phi_D = \pi$. Notice that this is also the right-end side of Figure 10.6 for viscosity ratios larger than 1. At this point, a thin film of liquid would be entrained between the solid and the LB film. Water entrainment between the solid substrate and the monolayer is one of the biggest challenges in LB deposition. The causes for this phenomenon have been explored experimentally (Petrov *et al.*, 1980; Peterson *et al.*, 1983; Srinivasan *et al.*, 1988) and theoretically (de Gennes, 1986; Zhang and Srinivasan, 2001).

If we further increase the removal speed, continuous entrainment of the liquid phase develops into the dip-coating regime (Diaz and Cerro, 2004). During the dip-coating regime, shown schematically in Figure 10.5 (d), a continuous liquid film is removed, attached to the solid, and there are steady stagnation streamlines in the liquid and gas phases, slightly above the level of the liquid surface. The streamline pattern for the dip-coating regime is a deformed 2D four-vortex pattern where the gas–liquid interface moves *away* from the contact line. This transition has been demonstrated experimentally (Diaz and Cerro, 2004).

10.4 Windows of Operation for Successful LB Deposition

The experimental results described in this section, linking flow patterns to dynamic contact angles, can be used to define the *windows of operation* during LB depositions. A qualitative map of the windows of operation is shown in Figure 10.7. The ordinates of the map are immersion/removal speeds and the x-coordinates are static/dynamic contact angles. The values on the x-axis, i.e. for $U_s = 0$, are static contact angles. The lines separating the regions on the map are drawn under the assumption that the dynamic contact angles depart from the static contact angle (i.e. $\phi_D = f(N_{Ca}, \phi_S)$; Gutoff and Kendrick, 1982). At the typical speed of deposition of LB films, departure from static-advancing or

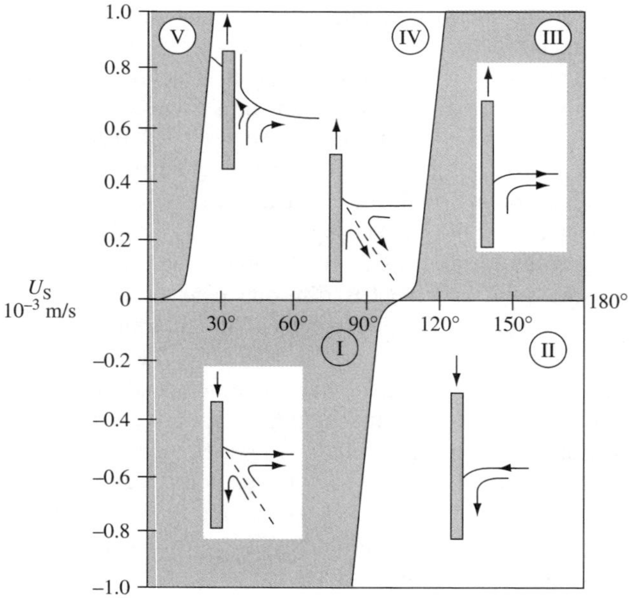

Figure 10.7 Sketch illustrating a qualitative picture of the windows of operation of the LB technique. Removal velocities are positive and immersion velocities are negative. The shadowed regions determine regions where LB deposition is not possible

static-receding contact angles is very small. Negative substrate speeds indicate immersion where $\phi_D \geq \phi_S$, and positive substrate speeds denote removal of the solid substrate where $\phi_D \leq \phi_S$.

The map lines and the values of contact angle and removal speeds must be taken only approximately. The lines dividing the flow regions in Figure 10.6 were drawn as fuzzy lines to highlight two facts: (1) It is difficult to measure with precision the dynamic contact angle, in the region very close to the moving contact line. Even with our best video images there is a $+/-5°$ uncertainty. (2) There may be small variations in dynamic contact angles due to the presence of the LB film. There is no comprehensive and systematic experimental data on the limits for good coating conditions. This has been one of the sources of great frustration to researchers in this area. By identifying the problem we hope to focus experimental efforts on this topic.

10.4.1 Conditions for X-Type Depositions

Region I in Figure 10.7 is typical of flow patterns during immersion of solids on wetting liquids and corresponds to region I in Figure 10.6. At dynamic contact angles smaller than $\phi_D \approx 95°$ the liquid phase shows a split-injection streamline and the interface moves away from the contact line. This flow situation is typical of systems, such as glass and water showing very small static contact angles. Thus, during immersion of a wetting solid surface, such as glass in water, the flow pattern near the contact line shows a split-injection streamline in the water phase, the interface moves away from the contact line, and X-type deposition is not possible. The cutoff value of $\phi_D \approx 95°$ for X-type depositions determined by Gaines (1977) agrees with the dynamic contact angles showing a transition flow pattern, determined by Savelski *et al.* (1995). One must remember that at low and even at intermediate removal speeds ($U_s < 1$ cm/s) for clean substrata, the dynamic contact angle is essentially equal to the static-advancing contact angle.

Region II in Figure 10.7 corresponds to region II of Figure 10.6. In region II, dynamic contact angles are larger than approximately 95°, the liquid phase is in a rolling motion, the interface moves toward the contact line, and deposition is possible. A glass surface can be treated to obtain static contact angles larger than 95°. Gaines (1977) showed that it is possible to deposit X-type LB films on treated, non-wetting glass surfaces.

The interface moves toward the contact line at a speed that is not necessarily the same as the velocity of the solid surface (Savelski *et al.*, 1995). For dynamic contact angles about 95°, the velocity of the interface is close to 0 because it is within the region where the transition flow pattern (Figure 10.4 (b)) is found. Under these conditions transfer ratios from the liquid interface to the solid surface can be very low. Honig (1973) recognized the effect of contact angles on transfer ratios and supplied a set of observations summarized as follows: (a) the transfer ratio is constant in successive layers but not equal to 1, (b) the transfer ratio depends on the number of layers already deposited on the solid, and (c) the transfer ratios are not equal for the upward or downward depositions. Some of these observations cannot be substantiated by later experiments of others nor our own experiments and highlight the lack of a sound framework to relate transfer, contact angles, and flow patterns. For contact angles below 95°, no deposition is possible during the downward stroke. For contact angles slightly larger than 95°, the transfer ratio will be small because the interface moves slower than the solid surface. For larger contact angles, i.e. 120°, during the downward stroke the transfer ratio will approach unity because the speed of the interface approaches the speed of the solid substrate.

X-type films deposited on the downward stroke are hydrophilic because the hydrophobic part of the fatty acid molecules is deposited on the solid surface. Similarly, Z-type films deposited on the upward stroke are hydrophobic because the hydrophilic end of the molecules is deposited on the substrate. Obviously, there is no reason for the resulting contact angles to generate equivalent transfer ratios. Finally, contact angles on multilayered Y-type films depend not only on the type of hydrophilic or hydrophobic molecule ends but also on the number of layers on the composite film (Gaines, 1977).

For low contact line speeds, i.e. $U_s \rightarrow 0$, the dynamic contact angle is close to the static-advancing contact angle. There is a theoretical possibility that X-type LB monolayers could be deposited on a solid where $\phi_S < 95°$ if the immersion speed is large enough to increase the dynamic contact angle. Further increasing the contact line speed may cause the dynamic contact angle to approach $\phi_D \rightarrow 2\pi$ and air entrainment will take place. For an air–water system, however, air entrainment will occur at speeds well above the typical operation speeds of LB deposition.

10.4.2 Conditions for Y- and Z-Type Depositions

Region III in Figure 10.7 represents the typical flow patterns found during removal of non-wetting solids, i.e. when the dynamic contact angles measured in the liquid phase are larger than 90°/100°. The liquid phase is in rolling motion and the interface moves away from the contact line making deposition infeasible. Notice that dynamic contact angles measured in the liquid phase *decrease* with increasing removal speed. Thus, it is theoretically possible to deposit LB films on solids with static contact angles larger than 90° as long as the speed of the contact line results in dynamic contact angles smaller than 90°.

Region IV is the window of operation for successful deposition of Y-type films. The flow pattern in this region is typical during removal of solids with dynamic contact angles $0 < \phi_D \leq 90°$.The split-ejection streamline is in the liquid phase and the interface moves toward the contact line. The interface, however, moves at a speed lower than the removal velocity of the solid substrate. For contact angles closer to but smaller than 90°, the flow approaches the transition profile (Figure 10.4 (b)) and the speed of the interface is nearly 0. As the contact angle decreases, the speed of the interface increases. Thus, transfer ratios during removal deposition are largely affected by contact angles. Petrov and Petrov (1998) incorrectly assumed that by maintaining a constant surface pressure a transfer ratio of unity follows. Constant surface pressure indicates the integrity of the film on top of the interface, but it does not assure that the solid and the interface move at the same speed. The effect of contact angles on transfer ratios was demonstrated experimentally by Sanassy and Evans (1993; Evans *et al.*, 1994) and by Diaz and Cerro (2004) for a different system. These authors showed a sharp decline in transfer ratios for dynamic contact angles between ~35° and 50°, corresponding to static contact angles between ~65° and 80°. Regardless of the large difference between static and dynamic contact angles, since flow patterns evolve toward the motionless interface for increasing contact angles, experimental transfer ratios decline from ~100% to ~0%.

For very small dynamic contact angles, the liquid is not completely removed by the split streamline and it is entrained between the film and the solid surface, creating what is known as a *wet LB film*. Water trapped between the solid surface and the LB monolayer prevents adhesion and is a leading cause of monolayer instability. Petrov *et al.* (1980) sketched the flow pattern near the moving contact line. The flow pattern is the one described here for region IV. The authors, however, reference Huh and Scriven (1971) ignoring that the hydrodynamic theory of the moving contact line predicts a rolling pattern in the liquid phase.

Petrov *et al.* (1980) define a limiting value for the withdrawal speed, U_{max}, such that for removal speeds larger than U_{max} a continuous film of water is entrained. The argument follows the original definition of Langmuir as *fast* and *slow* films as those emerging dry or wet for a given removal speed. The authors mention adherence of the film to the solid surface as the main cause for fast removal of the entrapped liquid layer. The variation of values of U_{max} with ionic strength reported by Petrov *et al.* (1980), however, indicates that these phenomena cannot be explained by a simple adherence argument based on double-layer forces. Using the same model flow pattern, and again ignoring the fact that Huh and Scriven's (1971) solution predicts the wrong flow pattern, de Gennes (1986) performed a balance of forces to find an expression for U_{max}. Similarly, Petrov and Petrov (1998) developed a molecular hydrodynamic description of Y-type depositions based on the incorrect assumption that the interface moves at the same speed as the solid substrate. This assumption is correct only for a relatively narrow range of dynamic contact angles, $10° \leq \phi_D \leq 35°$ (Diaz and Cerro, 2004). The split-ejection streamline is still visible in this pattern, but the stagnation point has moved from the contact line to the gas–liquid interface. This is not a Landau film as suggested by de Gennes (1986) because in a dip-coating flow pattern the interface moves away from the stagnation point, as indicated in Figure (new). The entrainment of a film of water, as shown in Figure 10.5 (d), was suggested by Miyamoto and Scriven (1982) as a way to relieve the shear stress singularity at the contact line. Air entrainment at large coating speeds is one of the main operating problems for the film industry. Water entrainment, however, occurs at relatively low speeds and determines the upper limit of coating speeds in LB deposition. When the thickness of the entrained film exceeds a small upper limit, the flow pattern of Figure 10.5 (c) evolves into a dip-coating flow pattern (Figure 10.5 (d)) region V in Figure 10.7. During the dip-coating regime the interface moves away from the contact line and LB deposition is not possible.

Peterson *et al.* (1983) addressed the problem of slow LB depositions by focusing on the maximum speed that could be attained before the film stability is compromised, in

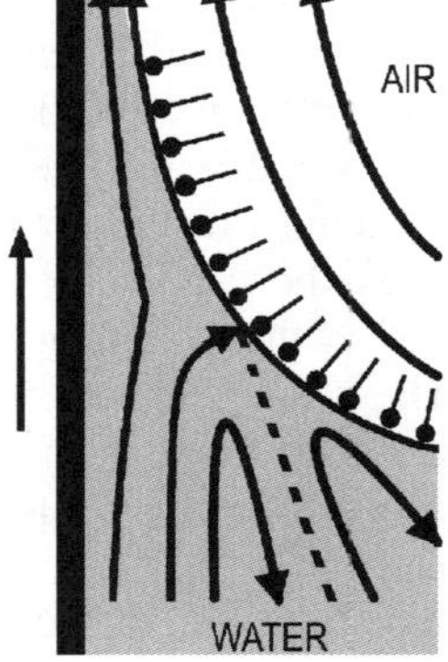

Figure (new) *Entrainment of a thin film of water. This flow pattern was suggested by Miyamoto and Scriven (1982)*

down- and upstrokes. The authors report the need for a hydrophobic surface in downstroke deposition and the absence of a maximum speed limit. These findings are consistent with contact line dynamics since larger immersion speeds increase the dynamic contact angle, the velocity of the interface approaches the velocity of the substrate, and consequently transfer ratios increase. The upper limit to this process, on the right-hand side of region II, would be entrainment of air. However, air entrainment would only occur at speeds much larger than the top speed of conventional Langmuir troughs. Interestingly, Peterson *et al.* (1983) report a *lower* limit for downstroke deposition (about 400 μm/s) which could be traced back to contact line instabilities. The effect of electrolytes (Petrov *et al.*, 1980) and pH (Peng *et al.*, 1985) on the maximum deposition speed during removal is another example of how the dynamic of moving contact lines affects LB deposition and can be traced back to the flow transition to a dip-coating regime.

Transfer ratio dependence with contact angle also determines the orientation of the films deposited on the surface, as shown in Figure 10.3. Films deposited during downstroke have the hydrophobic chain close to the solid, while films deposited during upstroke have their hydrophobic chains away from the solid surface, creating a headgroup to headgroup bilayer. The fact that some films rearrange after deposition to create the hydrophobic–hydrophobic and hydrophilic–hydrophilic arrangement has prompted some researchers to postulate the bilayer as the basic unit of an ideal LB film (Schwartz, 1997). One could further speculate that the bilayer with external hydrophobic chains is the stable structure *outside* the water subphase but the bilayer with external hydrophilic groups is the stable structure *inside* the water subphase.

10.5 Marangoni Effects Due to the Presence of Langmuir Films

Transfer ratios for LB deposition during removal are largely affected by contact angles (Aveyard *et al.*, 1992, 1995; Diaz and Cerro, 2004). The actual window for the high-efficiency deposition of LB films, i.e. transfer ratios near 100%, is relatively narrow and confined to a range of dynamic contact angles, $10° \leq \phi_D \leq 25°$. Within this region, however, the LB film plays a large role in stabilizing deposition from the gas–liquid interface to the solid substrate.

To demonstrate this effect an augmented version of the film evolution equation was developed, introducing the disjoining pressure, Π, due to molecular forces as well as the elasticity of the interface (Diaz and Cerro, 2003). The film evolution equation was developed for solving the pressure variable by integration of the component of the momentum balance, normal (Y direction) to the direction of flow (X direction). The expression for the pressure includes the capillary component, $(2H)$, through surface tension and surface curvature as well as the disjoining pressure computed on the basis of molecular and structural forces. The expression for pressure is subsequently substituted into the component of the momentum balance in the direction of flow and integrated across the film thickness. To express this equation solely as a function of film thickness, $H(X)$, a generic velocity profile function, $u(X, Y)$, must be introduced. In our version of the film evolution equation, the velocity profile includes a shear stress source term at the gas–liquid interface that is computed as a function of molecular and structural forces and is part of the *jump* momentum balance at the interface. Additionally, this velocity profile verifies the conditions of non-slip at the solid surface and the conservation of mass, that is, a constant net flow, Q. The resulting expression for the film evolution equation is as follows:

$$\frac{d(2H)}{dX}=\frac{-d\Pi}{dX}+N_{Bo}+3N_{ca}\left(\frac{-1}{2}\frac{(\partial u/\partial Y)_{Y=H(X)}}{H(X)}-\frac{1}{H(X)^2}+\frac{Q}{H(X)^3}\right) \tag{10.2}$$

where N_{ca} and N_{Bo} are the capillary and Bond number respectively that are defined as:

$$N_{ca}=\frac{\mu u_w}{\sigma_0},\quad N_{Bo}=\frac{\rho g h_\infty^2}{\sigma_0}$$

begin ρ the density of the liquid phase, u the bulk Newtonian viscosity, u_w the liquid velocity at the wall, σ_0 the surface tension of the pure liquid, of the gravity acceleration and h_∞ the thickness of the liquid film far from the liquid level at the bath.

The disjoining pressure term is computed taking into account molecular, double layer and structural forces as a function of the distance from the interface to the solid surface.

$$\tilde{\Pi}=\left[\sum_{m=4}^{6}\left(\frac{-(m-3)\,\overline{\zeta}_m(\alpha)}{H(X)^{m-2}}\right)+\frac{\overline{\zeta}_3(\alpha)\,\sin(\alpha)}{H(X)}\right]\cos(\alpha) \tag{10.3}$$

$$\overline{\zeta}_m(\alpha)=\frac{\rho^{(s)}\overline{\Gamma}_m(\alpha)}{L_c^{m-2}\sigma} \tag{10.4}$$

where angle α is the angle between the interface and solid surface and $\Gamma_m(\alpha)$ are functions of the inclination angle for different values of m, the exponent of the binary interaction potential.

$$\overline{\Gamma_m}(\alpha)=-n_L\beta_{LL}^{(m)}\overline{G}_m(\alpha)+n_S\beta_{SL}^{(m)}\left(\overline{G}_m(\alpha)+\overline{G}_m(\pi-\alpha)\right) \tag{10.5}$$

where

$$\overline{G_6}(\alpha)=\frac{\pi}{12}\left(1+\frac{3}{2}\frac{\cot(\alpha)}{\csc^3(\alpha)}+\cos^3(\alpha)\right)\quad \overline{G_5}(\alpha)=\frac{1}{3}\left(\sin(\alpha)\cos(\alpha)+(\pi-\alpha)\right)$$

$$\overline{G_6}(\alpha)-\overline{G_6}(\pi-\alpha)=\frac{\pi}{6}\qquad \overline{G_5}(\alpha)-\overline{G_5}(\pi-\alpha)=\frac{\pi}{3} \tag{10.6}$$

and

$$\overline{G_4}(\alpha)=\pi\cos^2\left(\frac{\alpha}{2}\right)\qquad \overline{G_3}(\alpha)=2(\pi-\alpha)$$

$$\overline{G_4}(\alpha)-\overline{G_4}(\pi-\alpha)=\pi\qquad \overline{G_3}(\alpha)-\overline{G_3}(\pi-\alpha)=2\pi \tag{10.7}$$

Finally at the air–water interface, the shear stress source term has two components

$$\left(\frac{\partial U}{\partial Y}\right)_{Y=H(X)}=T_s+\left(\frac{\partial\Sigma}{\partial X}\right) \tag{10.8}$$

The first component, T_s, is a resultant of molecular, double layer and structural forces on the molecules at the air–water interface. The second component, $(\partial\Sigma/\partial X)$, is due to the elasticity of the LB film at the interface and is written in dimensionless form

$$\left(\frac{\partial\Sigma}{\partial X}\right) = N_{\text{El}}\frac{1}{U_{\text{S}}(X)}\left(\frac{\partial U_{\text{S}}(X)}{\partial X}\right) \tag{10.9}$$

where Σ is the dimensionless interfacial tension ($\Sigma = \sigma/\mu u_{\text{w}}$), U_s is the dimensionless velocity at the interface and the elasticity of the interface is defined as

$$N_{\text{El}} = \left(\frac{\partial\sigma_{\text{GL}}}{\partial \ln A}\right)\left(\frac{1}{\mu U_{\text{W}}}\right) \tag{10.10}$$

The first term in equation 10.10 is the derivative of the interfacial tension with respect to the logarithm of the area occupied by the molecules, that is the negative of the slope of the surface pressure versus area per molecule given in Figure 10.2. The shear stress source term is formulated using the jump momentum balance at the interface. For films within the solid region (Figure 10.2), the elasticity number is a very large number, typically 10^5–10^6. Thus, when the interface is stretched due to fluid motion, a Marangoni-like effect creates a large force that pulls the interface up to the speed of the liquid interface. The resulting flow pattern is shown in Figure 10.8 (a). The velocity profile is a quadratic function of the coordinate normal to the movement of the solid, and an internal stagnation point develops as shown in Figure 10.8 (a). The streamline patterns computed using the film evolution equation are shown in Figure 10.8 (b). The interface velocity as a function of vertical position is shown in Figure 10.8 (c). Notice that the velocity of the interface is remarkably constant and almost identical to the velocity of the solid substrate, indicating a transfer ratio, TR $\sim$ 100%. For comparison we show also the velocity of a dip-coating flow that would occur at the same capillary number.

Even at the peak of the shear stress source term, these are of the order of 10^{-2} Pa. Thus, outside of a certain range of contact angles, hydrodynamic forces are larger than the forces generated by the stretching of the film at the interface, despite the large value of the elasticity number, and the interface velocity is smaller than the velocity of the solid surface, resulting in transfer ratios smaller than 100%. Figure 10.9 shows the values of experimental transfer ratios of zinc arachidate films for different dynamic contact angles. Subphase temperature was kept at 26° C, stroke speed was 19 mm/min, concentration of the spreading solution was 1 mg/mL of arachidic acid in chloroform, concentration of the subphase was 10^{-4} M of $ZnSO_4$, and surface pressure was 25 mN/m. Surface treatment of the solid substrate and pH were varied in order to cover a wide range of values of dynamic contact angles. Figure 10.9 shows transfer ratios for clean glass slides and glass slides coated with diluted Sigmacote® during upstrokes at pH = 5.3 as a function of dynamic contact angles. The largest transfer ratios during upstroke are obtained for dynamic contact angles of about 10°–25° using clean glass slides at pH = 5.3. In this region the transfer ratios are close to 100%. For these range of contact angles the liquid subphase flows in a split-ejection flow pattern and the interface moves toward the solid at a velocity close or equal to the velocity of withdrawal of the solid substrate. As the contact angle is increased, by using glass slides treated to become only partially wetting, the transfer ratio decreases. The transfer ratio decreases with increased contact angles because the velocity of the air–liquid interface decreases steadily until it becomes

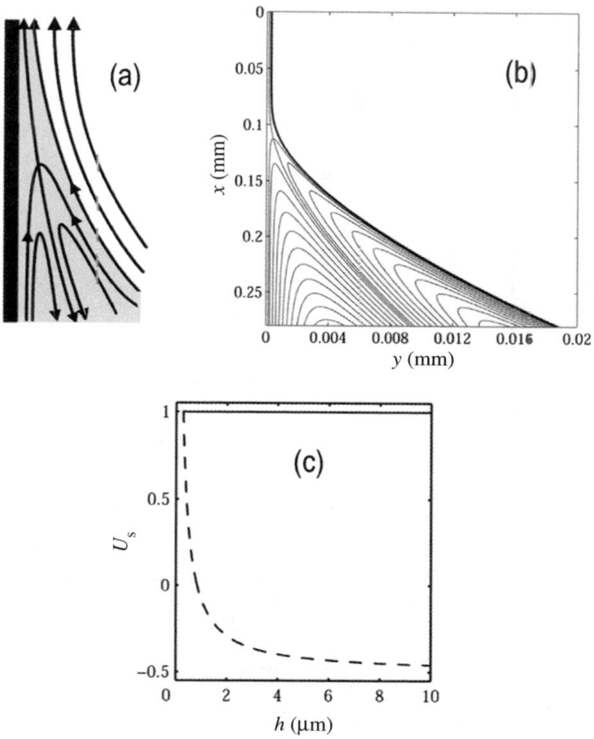

Figure 10.8 *Flow pattern during removal–deposition of an LB film.* $N_{Ca} = 1.90.10^{-5}$ *and* $N_{El} = 4.01.10^{5}$. *(a) Schematic representation; (b) computation results for the streamlines; (c) relative interface velocity (solid line) and relative dip coating velocity (dashed line) versus the film thickness*

0 near 90°. At this point the split-ejection flow pattern (Figure 10.5 (c)) evolves into the transition flow pattern (Figure 10.5 (b)). Further increase in dynamic contact angle promotes a transition to the rolling flow pattern in the liquid subphase and the transfer ratio is 0. These results agree very well with experimental data presented by Evans and coworkers (Sanassy and Evans, 1993; Evans *et al.*, 1994) for gold-coated substrates immersed in mixture thiols, and using pure water as the liquid subphase.

Using clean glass slides only, pH was increased in several steps adding a dilute solution of ammonium hydroxide to the 10^{-4} M zinc sulfate subphase. A monolayer of arachidic acid is partially ionized at pH = 5.3. Increasing the pH of the subphase to 6.6 and 7.7 causes an increase of ionization of the carboxylic headgroups. At higher pH, the monolayer is completely ionized. No area loss of monolayer is observed over the experimental period for all the values of pH. Transfer ratios at pH 6.6 and 7.7 are essentially the same as at pH = 5.3, that is, close to 100%. However, when pH reaches 8.7, the dynamic contact angle is very small or 0 and no transference of monolayer to the solid substrate is recorded while it is evident that a thin film of liquid leaves with the solid substrate. Similar results were obtained for pH 9.4, 10.1, and 10.6. These results are shown on the left-end side of Figure 10.9.

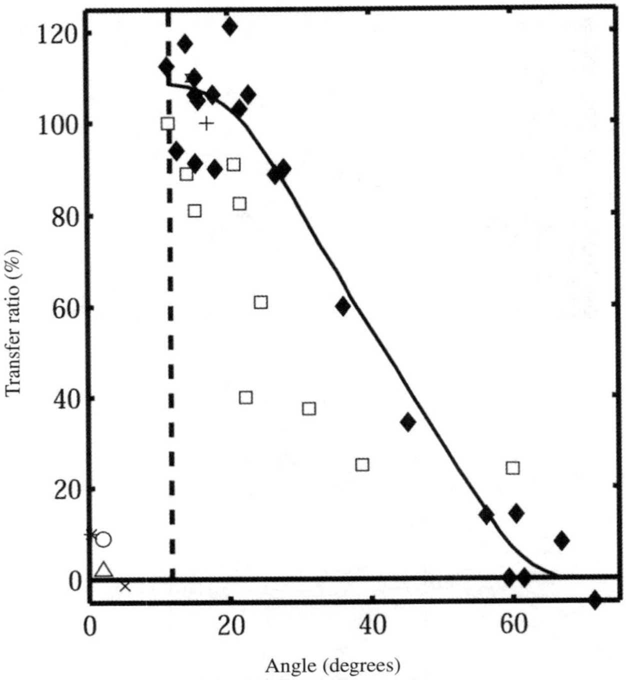

Figure 10.9 *Experimental transfer ratios (TR) versus dynamic contact angles during upstrokes for different pH values. Open squares: Sanassy and Evans (1993) results with pure water. Results from this work: deposition were made at surface pressure of 25 mN/m, T = 26° C and a deposition speed of 19 mm/min. The values of the pH were as follows: pH = 5.3 (diamonds); pH = 6.6 (star); pH = 7.7 (cross); pH = 8.7 (open circle); pH = 9.4 (asterisk); pH=10.1 (×); pH = 10.6 (open triangle). The line is just a guide to the eye*

During immersion a hydrophobic surface, i.e. a large contact angle between the liquid and the solid surface, is needed for successful LB depositions. Downstroke experiments were performed using glass slides coated with diluted Sigmacote® and with ferric stearate to create dynamic contact angles larger than 90°. All experiments were done at 26° C, 65 mm/min stroke velocity, and pH = 5.3. These experimental results are shown in Figure 10.10. At dynamic contact angles ranging from 50° to 90° transfer ratios are essentially 0 because the flow pattern in the liquid subphase shows a split-injection streamline with the interface moving away from the contact line. Under these conditions, no LB deposition is possible. For dynamic contact angles between 120° and 130° transfer ratios are essentially 100%. For large dynamic contact angles the flow pattern in the liquid subphase is a rolling motion and deposition is possible. It is not clear why during downstroke transfer ratios go from 0 to 100% with no apparent intermediate values. The argument that monolayer compression promotes the movement of the interface, as the compressing barriers move in, is unlikely because this effect is not apparent during upstroke where the transition from an interface moving with the speed of the solid substrate to a motionless interface occurs smoothly as the dynamic contact angle increases (Figure 10.9), regardless of the pressure applied on the monolayer. We were unable to

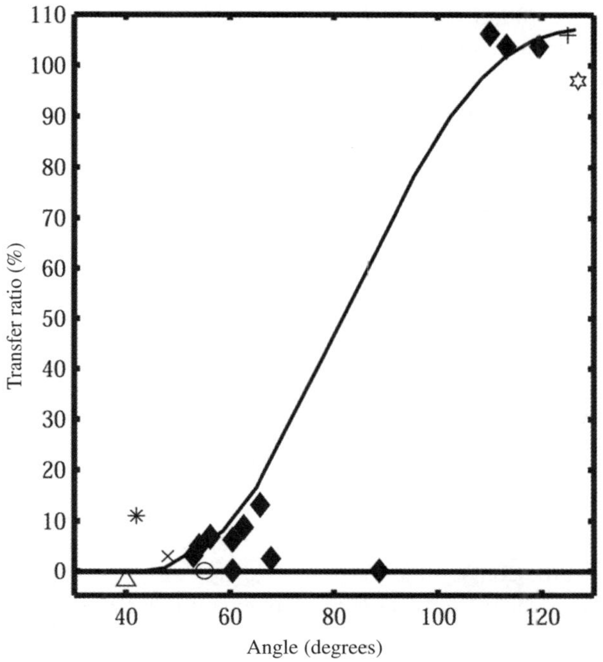

Figure 10.10 *Experimental transfer ratios (TR) versus dynamic contact angles during down-strokes for different pH values. The conditions of deposition were as follows: a surface pressure of 25 mN/m, 26° C, and a deposition speed of 19 mm/min. The values of the pH were as follows: pH = 5.3 (diamonds); pH = 6.6 (star); pH = 7.7 (cross); pH = 8.7 (open circle); pH = 9.4 (asterisk); pH = 10.1 (×); pH = 10.6 (open triangle). The line is just a guide to the eye*

create surfaces with contact angles ranging from 90° to 120°, thus it is not possible to determine if, within this range, there is a smooth variation from 0 transfer ratios to 100%.

10.6 Role of Molecular, Structural, and Electrical Double-Layer Forces

Short-range molecular forces, i.e. with ranges up to 3 nm, can account for a variety of macroscopic phenomena. Short-range molecular forces account for capillarity, the shapes of macroscopic liquid droplets on surfaces, the contact angle between coalescing soap bubbles, and the breakup of a jet of water into spherical droplets (Israelachvili, 1985). Long-range structural forces, namely hydrophobic forces, can be accounted for at distances as long as 300 nm (Colic and Miller, 2000). Electrostatic double-layer forces control the macroscopic properties of slurries and cause large differences in pressure drop during slurry filtration. The recognition that long-range structural forces are significant beyond tens and hundreds of molecular diameters allowed the explanation of macroscopic phenomena such as capillarity, the stability of colloidal and particle suspensions, and the breakup of liquid films on solid surfaces (Derjaguin *et al.*, 1987).

Independently and almost simultaneously the effect of molecular forces near a three-phase contact line was analyzed by Miller and Ruckenstein (1974) and Jameson and Garcia del Cerro (1976). While both papers point to the presence of asymmetric force fields, mainly generated by the presence of two dense phases (solid and liquid) and a gas phase, Miller and Ruckenstein (1974) developed the concept of a resulting force to explain the movement of a contact line and Jameson and Garcia del Cerro (1976) balanced the resultant force with an interfacial tension gradient.

The need to introduce molecular and structural forces to explain flow patterns near a moving contact line was recently recognized (Fuentes *et al.*, 2005) and flow patterns near moving contact lines have been used to explain the windows of operation of LB depositions (Cerro, 2003). When a contact line moves, unbalanced molecular and structural forces produce a residual shear stress at the solid–liquid and fluid–fluid interface. In the presence of surfactants or contaminants at the interface, motion can generate Marangoni-like effects due to changes in surface concentration. On clean interfaces an elasticity component results, from stretching and compressing the interface (Edwards *et al.*, 1991). Thus, it is generally accepted that, regardless of the source, there is a strong force asymmetry at the contact line.

Unlike colloidal particles, moving contact lines are inherently asymmetric. Electrical potentials of double layers in a solid–liquid interface can be different in magnitude and in sign to the potentials of the air–liquid interface on the same liquid pool. Thus, we may have attractive or repelling forces at the contact line, depending on the nature of the interfaces and on the pH of the liquid phase. The nature and the effect of these forces and their effect on contact angles have been analyzed (Churaev, 1995).

The interaction of the double layers near the moving contact line and its effect on LB depositions, has been recognized (Petrov, 1986) but the connection between double layers and flow patterns has not been explored until recently (Fuentes *et al.*, 2005). We develop here a framework for analyzing the hydrodynamics of LB depositions under the presence of electrical double layers on the solid–liquid interface and on the air–liquid interface. These interfaces are subject to movement, the motion disrupts the diffusion layer, essentially creating streaming potentials. In addition, both interfaces show a large concentration of charged molecules, behaving essentially as 2D solids.

In general, when two phases are in contact, electrons or ions will be attracted in different ways by the different phases and dipolar molecules will be oriented selectively (Hunter *et al.*, 1981). When a solid phase is in contact with a liquid subphase the solid surface may be charged and surrounded by ions of opposite sign (counterions). This is the typical arrangement of an electrical double layer as described by the Gouy–Chapman–Grahame–Overbeek theory of potential and charge distributions in electrical double layers (Grahame, 1947; Overbeek, 1952). In LB depositions, there are two interfaces that acquire charge, the substrate–subphase interface and the LB film–subphase interface. The substrate–subphase interface is a typical solid–liquid interface with the difference that it may or may not have a deposit consisting of one or many layers of the amphiphilic compound that makes up the LB film. For short, we will call this surface the solid–liquid interface. The LB film–subphase is for all purposes also a solid–liquid interface, with one layer of the amphiphilic compound compressed to the point that is essentially a 2D solid. We will denote this as the film–liquid interface. At the contact line, that is, the line where the three phases coincide, the double layers of the solid–liquid and film–liquid interfaces overlap. The sign and magnitude of the electrical forces created by the presence of the electrical double layers determine the dynamic contact angle and the flow patterns near a moving contact line.

At least two of the recognized mechanisms for the formation of electrical double layers (Hunter, *et al.* 1981; Russel *et al.*, 1989) are relevant to LB film depositions: (1) ionization of carboxylic acid group and amphoteric acid groups on solid surfaces, and (2) differences between the affinities of two phases for ions or ionizable species. The latter mechanism includes the uneven distribution of anions and cations between two immiscible phases, the differential adsorption of ions from an electrolyte solution to a solid surface, and the differential solution of one ion over the other from a crystal lattice. Since the solid–liquid and the film–liquid interfaces are flat, large surfaces and since both have a large, solid-like concentration, the analysis that follows applies to both interfaces. For an interface conformed by a thin film of an amphiphilic compound with the hydrophilic end of the molecule in contact with the water subphase, the equilibrium of charges is based on pH and subphase concentration. The effect of pH is highlighted by the definition of the pK_a of the carboxylic acid:

$$H_2O + RCOOH \rightleftharpoons \{RCOO\}^- + [H_3O]^+ \tag{10.11}$$

The equilibrium constant for this reversible ionization process is written in terms of ionic concentrations:

$$K_{eq} = \frac{[H_3O]^+ \{RCOO\}^-}{[RCOOH]\,[H_2O]} \tag{10.12}$$

By definition, pK_a is the pH when the concentration of ionized carboxylic acid is equal to the concentration of non-ionized acid groups:

$$K_a = \frac{[H_3O]^+}{[H_2O]};\ pK_a = pH \tag{10.13}$$

where pK_a is equal to the pH only when half of the carboxylic acids are ionized. Values of pK_a depend on the length of the hydrocarbon chain attached to the carboxylic group and are also mildly modified by the presence of a metal subphase counterion and its matching group. When an amphiphilic carboxylic acid and its subphase are at the pK_a, the intermolecular distances between molecules at the air–water interface reach a minimum (Kanicky *et al.*, 2000) and macroscopic properties, such as foam height and stability as well as surface viscosity, are at a maximum value. The ion–dipole interactions taking place between ionized and non-ionized carboxylic groups are somehow complemented by the presence of divalent cations in the liquid subphase (i.e. Cd^{++}, Zn^{++}, etc). There is experimental evidence that monovalent (i.e. NH_4^+) or trivalent (i.e. Fe^{+++}) cations have very different effects on monolayer stability (Gaines, 1966).

Although the theoretical tools for modelling and interpreting double-layer properties and electrokinetic behavior have been around for a long time (Kruyt, 1952), it was not until recently that ζ-potentials and electrokinetic properties could be measured accurately (Gu and Li, 2000; Usui and Healy, 2002). Regardless of the particular charge of the LB film and of the thickness of the electrical double layer, the following facts must be stressed: (1) the thickness of the electrical double layer, i.e. the width of the region where the electrical potential decays to within 2% of its maximum value, is of the same order of magnitude as the size of the *proximal* region (de Gennes *et al.*, 1990) in the description of moving contact lines; (2) the sign and magnitude of the charge at the solid–liquid and the film–liquid interfaces depend on pH and subphase salt concentration; and (3) the relative movement of liquid with respect to the solid substrate, due to the immersion/removal of the solid surface and to the movement of the mechanical barriers to compress the LB film, affects the integrity of the double layer and creates a streaming potential.

The thickness of the electrical double layer can be estimated in tens and perhaps hundreds of nanometers. When the solid–liquid and film–liquid interfaces approach the three-phase contact line, the double layers overlap and interact. This interaction leads to attraction–repulsion forces that determine contact angles (Churaev, 1995) and flow patterns (Fuentes *et al.*, 2005) near the moving contact line. Flow patterns, in turn, have been linked to the ability to deposit an LB film on a moving substrate (Cerro, 2003). Double-layer sign and magnitude of the charge depend on subphase pH and concentration of the metal salts in the subphase. During successive deposition of LB films, the same compound is deposited on the solid–liquid and film–liquid interfaces. If the liquid subphase is at the pK_a of the carboxylic acid, interaction between ionized and non–ionized acid groups keeps the film at a relatively small charge level. Experimental measurements of ζ-potential of stearic acid films (Usui and Healy, 2002) show at $pK_a = 4.8$ negative charges, $\Delta V = -80\,\text{mV}$, smaller than the maximum value of $\Delta V = -150\,\text{mV}$, at $\text{pH} \sim 8.5$. It is important to point out that these measurements were made for stearic acid layers in the presence of NH_4NO_3–NH_4OH subphases, i.e. for a monovalent ion.

Carboxylic acid films in the solid–liquid or in the film–liquid interface submerged in the same water subphase should develop a similar type of double layer, with similar sign and magnitude of charge. Thus, during immersion the hydrophobic hydrocarbon chains can account for a large contact angle, allowing deposition, but during removal the carboxylic acid ends will have same sign charges creating repulsion and a large contact angle, preventing LB deposition. Consequently, a Y to X film transition arises because deposition during removal cannot take place.

In addition, immersion and removal speeds play a definite role in LB depositions because they disturb equilibrium of charges at the electrical double layer. When the solid is immersed, the double layer begins to form and ions must diffuse toward and away from the solid surface before equilibrium is reached. When the solid is removed, the double layer is partially wiped out by the flow but metal cations are retained on the film deposited on the solid substrate. On the other hand, during film deposition, the mechanical barriers designed to keep the surface pressure constant move the film on the air–water interface and disturb the double layers under the film. At this point, we do not have a quantitative way to estimate the disturbance to the double layer caused by the movement of the solid substrate, but we can introduce the following assumptions:

(1) The movement of the substrate disturbs the double layer and generates a streaming potential.
(2) The streaming potential drives a streaming current in the reverse direction to restore charges and approach equilibrium.

(3) The magnitude of the streaming potential and the streaming currents are directly proportional to the ζ-potential of the double layer and to the velocity of the solid substrate (Gu and Li, 2000).
(4) The dynamics of double-layer formation are controlled by the electrokinetic velocity of ions and it takes a certain amount of time for the double layer to re-establish.

Taking into account what we know about electrical double layers and their effect on dynamic contact angles and flow patterns, we attempt an explanation of Y to X film transitions, on the basis of our experimental data. Figure 10.11 shows deposition of successive layers of arachidic acid at a pH equal to the $pK_a = 5.5$ of arachidic acid. Subphase is a solution of $CdCl_2$ at 2.0 10^{-4} M and glass slides treated with ferric stearate were immersed and removed at constant speed, $U_s = 60\,mm/min$, and a constant surface pressure of $\Pi = 25\,mN/m$. Contact angles were consistently low, about 30° during removal, and consistently high, about 110° during immersion, assuring successful deposition for at least 18 monolayers without apparent change in deposition effectiveness. Figure 10.12 shows similar experiments with identical substrate and subphase conditions but for a different pH = 6.1. Clearly, deposition during immersion remains high, and contact angles, not shown in the figure, remain over 100°. Contact angles during removal are small for the first three cycles but they soon reach values of the order of 60°–70°, and deposition during removal stops.

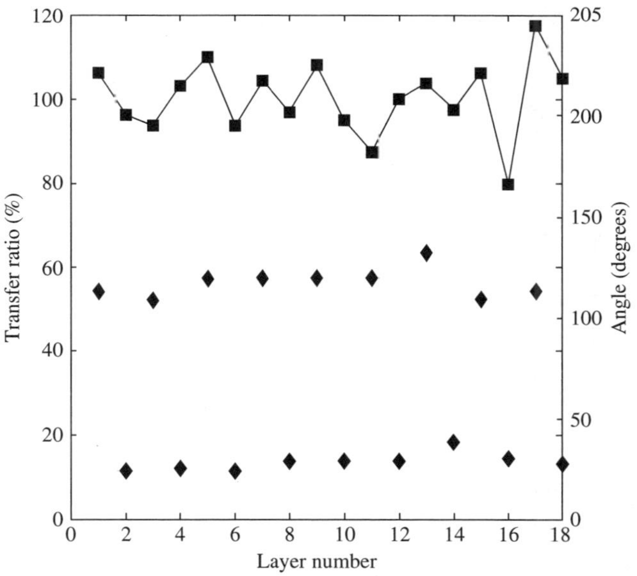

Figure 10.11 *Variation of transfer ratio and dynamic contact angle with number of layers for the deposition of arachidic acid on a subphase of 2.10^{-4} M $CdCl_2$ at pH = 5.5, $\pi = 25$ mN/m, 60 mm/min, and T = 25°C. Squares and solid lines: transfer ratios; diamonds: dynamic contact angles*

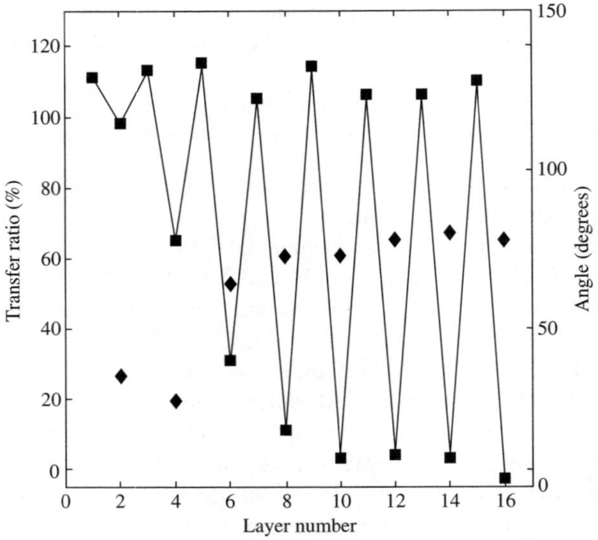

Figure 10.12 *Variation of transfer ratio and dynamic contact angle with layer number for the deposition of arachidic acid on a subphase of 2.10^{-4} M $CdCl_2$ at pH = 6.1, π = 25 mN/m, 60 mm/min, and T = 25° C. Squares and solid lines: transfer ratios; diamonds: dynamic contact angles*

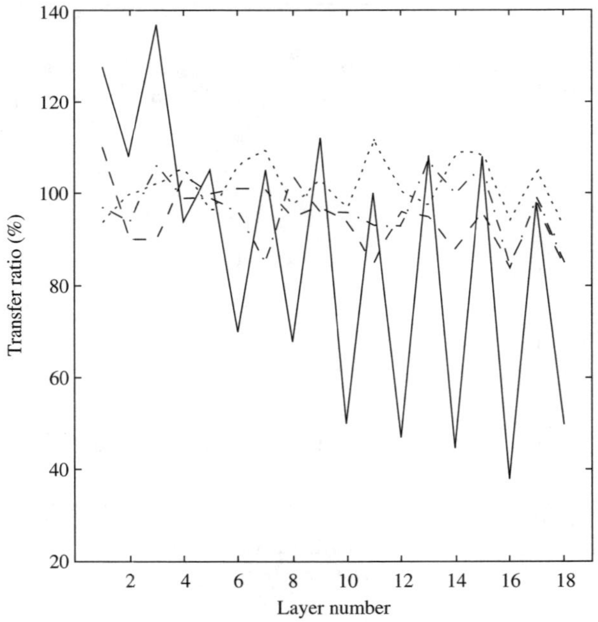

Figure 10.13 *Variation of transfer ratio with layer number for the deposition of arachidic acid on a subphase of 2.10^{-4} M $CdCl_2$ at pH = 5.7, π = 25 mN/m, and T = 25° C. Dotted line: 60 mm/min; dashed-dotted line: 15 mm/min; dashed line: 9 mm/min; solid line: 3 mm/min*

Figure 10.13 shows successive deposition of films of arachidic acid at $pH = pK_a = 5.5$ at different removal speeds. Experiments were done with identical glass slides treated with ferric arachidate, and similar subphase concentrations of $CdCl_2$ of 10^{-4} M. The removal speeds were 3, 9, 15, and 60 mm/min. Notice that successive depositions at 15 and 60 mm/min show transfer ratios close to 100%. Depositions at 9 mm/min show signs of decreasing transfer ratios after 10 layers. However, at 3 mm/min transition from Y to X films takes place after a couple of cycles. Assuming that the immersion/removal cycles take place without stopping at the higher and lower positions of the substrate, and that the length of the substrate run is 20 mm, the times elapsed for one cycle are 13.3, 4.4, 2.6, and 0.6 min for the 3, 9, 15, and 60 mm/min substrate speeds. To test the premise that electrokinetic effects determine the speed at which the double layer is being restored, slides were let to rest inside the liquid subphase for a period of 10 min after immersion. Experimental results allowing the substrate to equilibrate for 10 min inside the liquid subphase are shown in Figure 10.14. Notice that deposition of LB films during removal for films left to *age* under the liquid subphase is affected in a way similar to the lower deposition speed shown to be affected in Figure 10.13.

On the basis of experimental evidence and taking into account the role of electrical double layers on contact angles and flow patterns, one can certainly argue that the most successful conditions for multilayer Y-type LB deposition take place at a pH equal to the pK_a of the carboxylic acid, with a subphase where a minimum concentration of a

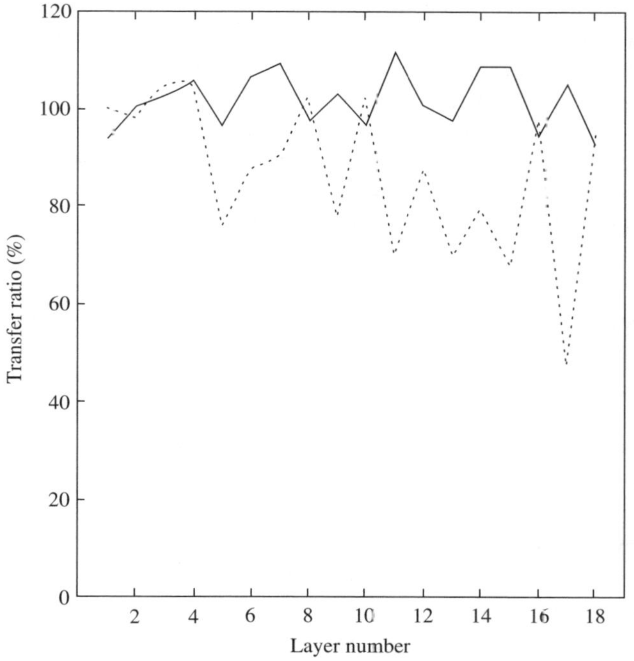

Figure 10.14 *Variation of transfer ratio with layer number for the deposition of arachidic acid on a subphase of 2.10^{-4} M $CdCl_2$ at pH = 5.7, π = 25 mN/m, and T = 25° C. Solid line: 64 mm/min; dotted line: 64 mm/min, and held under water for 10 min*

divalent cation is available, at deposition velocities large enough to perturb the electrical double layer and with short time exposure under the liquid. The role of the $pH = pK_a$ and the presence of a divalent cation, although not well understood, indicate a certain amount of a crystal-like structure in LB films. Atomic force microscopy images of LB films deposited on all kinds of solid surfaces (Zasadinsky *et al.*, 1994) show that the main requirement for long-range order in the alkyl chains is to have an underlying headgroup to headgroup interface. This fact may also explain why transition from Y- to X-type depositions occurs only at or after the second cycle (Figure 10.12). Stripe-shaped ridges on films of metal salt arachidates have been explained on the basis of ordered patterns of ionized or salt-forming carboxylic headgroups and non-ionized headgroups (Sigiyama *et al.*, 1998).

10.7 Conclusions

The experiments described in this chapter and the concepts put forward to explain these experiments are an attempt to understand the mechanics of LB depositions, a large and fascinating scientific problem. There should be no doubt that flow patterns and the relative movement of the air–water interface determine feasibility of LB deposition. The Marangoni-like effects generated by the stretching of the film on the interface help to create a region for high transfer ratios by controlling, within certain limits, the velocity of the interface. One must remember though that one-molecule-wide films cannot support stresses substantial enough to change qualitatively the flow patterns near the moving contact line. However, electrical double layers on the film at the air–water interface and similar double layers surrounding the films deposited on the solid substrate create attraction/repulsion effects that determine contact angles and flow patterns.

The movement of the solid substrate and the movement of the film at the air–water interface due to the pressure-controlling barriers create disturbances in the electrical double layers that allow deposition of multiple LB films. This is a novel way to look at the LB technique and puts multilayer depositions under a totally new light. Electrical double layers can be modified in a number of chemical, electrical, and mechanical ways to enhance depositions, opening a wide range of alternatives for the development of multilayer deposition techniques.

There are many questions remaining and many puzzling, unexplained effects such as the effect of cation size and valence on film stability. These questions point to the need to develop a better understanding of the crystal-like structure of LB films, the role of molecular and structural forces in creating these structures, and the nature and stability of electrical double layers subject to mechanical perturbations in the underlying subphase.

There is also a growing need to develop characterization techniques that can be applied, *in situ*, during deposition and would allow determination of film structure, charge, and electrical double-layer characteristics. The transfer ratio is a crude and sometimes misleading method of characterization but unfortunately, in this aspect, we have not advanced far past Katherine Blodgett's primitive experiments. Brewster angle microscopy (BAM), atomic force microscopy (AFM), and attenuated total reflectance infrared spectroscopy (ATR) are some of the emerging techniques that may help to bridge this gap as they become standard instrumentation associated with LB troughs.

The applications of LB films are far reaching. Highly ordered LB films have many interesting properties, but more important than their technological applications is the fact that these films hold the key to nature's molecular world where order is essential to function.

10.8 Summary

The transfer onto the surface of a solid substrate of successive monolayers of divalent soaps compressed on the surface of water in a Langmuir trough was described by Blodgett (Blodgett, 1935). A Langmuir trough is a container with moving barriers for manipulation of a film of an amphiphilic compound at the air–water interface. Films are deposited on solid substrata moved up and down, out and into the water subphase. The term Langmuir–Blodgett (LB) technique is currently used to denote the deposition of monolayers by transfer from the air–water interface onto a solid surface.

Single, i.e. one-layer, LB films show a remarkable ordered structure. The precise thickness of mono-molecular assemblies and the degree of control over their molecular architecture have firmly established LB films in ultrathin film technology as an essential building block of micro- and nano-technologies.

Acknowledgements

The research work that provided the foundations of this article was supported by grant CTS-0002150 of the National Science Foundation.

References

Aveyard R., Binks B.P., Fletcher P.D.I. and Ye X. 1992. Dynamic contact angles and deposition efficiency for transfer of docosanoic acid onto mica from $CdCl_2$ subphases as a function of pH, Thin Solid Films, 210, 36–38.

Aveyard R., Binks B.P., Fletcher P.D.I. and Ye X. 1995. Contact angles and transfer ratios measured during the Langmuir–Blodgett deposition of docosanoic acid onto mica from $CdCl_2$ subphases, Colloid Surf., A: Physicochem. Eng. Aspects, 94, 279–289.

Barraud A., Perrot H., Billard V., Martelet C. and Therasse J. 1993. Study of immunoglobulin G thin layers obtained by the LB method: application to immunosensors, Biosens. Bioelectron., 8, 39–48.

Berg J.C. (Ed.). 1993. Wettability. Surfactant Science Series. Marcel Dekker, New York.

Bikerman J.J. 1939. On the formation and structure of multilayers, Proc. R. Soc. Lond. A Math. Phys. Sci., 170, 130–144.

Blake T.D. and Ruschak K.J. 1997. Wetting: static and dynamic contact lines. In, Liquid Film Coating. Scientific Principles and their Technological Implications, Kistler S.F. and Schweizer P.M. (Eds.). Chapman & Hall, New York, pp. 63–97.

Blodgett K.B. 1935. Films built by depositing successive monomolecular layers on a solid surface, J. Am. Chem. Soc., 57, 1007–1022.

Bowden M.J. and Thompson L.F. 1979. Resists for fine line lithography, Solid State Technol., 22(5), 72–82.

Cerro R.L. 2003. Moving contact lines and Langmuir–Blodgett film deposition, J. Colloid Interface Sci., 257, 276–283.
Churaev N.V. 1995. The relation between colloid stability and wetting, J. Colloid Interface Sci., 172, 479–484.
Colic M. and Miller J.D. 2000. The significance of interfacial water structure in colloidal systems – dynamic aspects. In, Interfacial Dynamics, Vol. 88, Kallay N. (Ed.), Marcel Dekker, New York.
Derjaguin B.V., Churaev N.V. and Muller V.M. 1987. Surface Forces. Plenum Press, New York.
Diaz E. and Cerro R.L. 2003. Flow Bifurcation during Removal of a Solid from a Liquid Pool. AIChE Annual Meeting, San Francisco, CA.
Diaz M.E. and Cerro R.L. 2004. Transition from split streamline to dip-coating during Langmuir–Blodgett film deposition, Thin Solid Films, 460, 274–278.
Dussan V E.B. 1979. On the spreading of liquids on solid surfaces: static and dynamic contact angles, Annu. Rev. Fluid Mech., 11, 371–400.
Edwards D.A., Brernner H. and Wasan D.T. 1991. Interfacial Transport Processes and Rheology. Butterworth-Heinemann, Boston.
Elliot D.J., Furlong D.N. and Grieser F. 1999. Formation of CdS and HgS nanoparticles in LB films, Colloid. Surface., 155, 101–110.
Evans S.D., Sanassy P. and Ulman A. 1994. Mixed alkanethiolate monolayers as substrates for studying the Langmuir deposition process, Thin Solid Films, 243, 325–329.
Fuentes J., Savage M. and Cerro R.L. 2005. On the effect of molecular forces on moving contact lines, J. Fluid Mech. (submitted).
Gaines G.L. 1966. Insoluble Monolayers at the Gas Liquid Interfaces. John Wiley & Sons, New York.
Gaines G.L. 1977. Contact angles during monolayer deposition, J. Colloid Interface Sci., 59, 438–446.
de Gennes P.G. 1986. Deposition of Langmuir–Blodgett layers, Colloid Polym. Sci., 264, 463–465.
de Gennes P.G., Hua X. and Levinson P. 1990. Dynamics of wetting: local contact angles, J. Fluid Mech., 212, 55–63.
Grahame D.C. 1947. The electrical double layer and the theory of electrocapillarity, Chem. Rev., 41, 441–501.
Gu Y. and Li D. 2000. The potential of glass surface in contact with aqueous solutions, J. Colloid Interface Sci., 226, 328–339.
Gutoff E.B. and Kendrick C.E. 1982. Dynamic contact angles, AIChE J., 28, 459–466.
Honig E.P. 1973. Langmuir–Blodgett deposition ratios, J. Colloid Interface Sci., 45, 92–102.
Huh C. and Scriven L.E. 1971. Hydrodynamic model of the steady movement of a solid/liquid/fluid contact line, J. Colloid Interface Sci., 35, 85–101.
Hunter R.J., Ottewill R.H. and Rowell R.L. 1981. Zeta Potential in Colloid Science. Academic Press, London.
Israelachvili J.N. 1985. Intermolecular and Surface Forces. Academic Press, London.
Jameson G.J. and Garcia del Cerro M.C. 1976. Theory for the equilibrium contact angle between a gas, a liquid, and a solid, Trans. Faraday Soc., 72, 883–895.
Kanicky J.R., Poniatowski A.F., Mehta N.R. and Shah D.O. 2000. Cooperativity among molecules at interfaces in relation to various technological processes: effect of chain length on the pKa of fatty acid salt solutions, Langmuir, 16, 172–177.
Keller D.J., Korb J.P. and McConnell H.M. 1987. Theory of shape transitions in two-dimensional phospholipid domains, J. Phys. Chem., 91, 6417–6422.
Kruyt H.R. (Ed.). 1952. Colloid Science. Elsevier Publishing, Amsterdam.
Miller C.A. and Ruckenstein E. 1974. The origin of flow during wetting of solids, J. Colloid Interface Sci., 48, 368–373.
Miyamoto K. and Scriven L.E. 1982. Breakdown of Air Film Entrainment by Liquid Coated on Web. AIChE Annual Meeting.

Overbeek J.T.G. 1952. Electrochemistry of the double layer. In, Colloid Science, Kruyt H.R. (Ed.), Vol. 1. Elsevier Publishing, pp. 115–193.

Peng J.B., Abraham B.M., Dutta P. and Ketterson J.B. 1985. Contact angle of lead stearate-covered water on mica during the deposition of LB assemblies, Thin Solid Films, 134, 187–193.

Peterson I.R. 1996. Langmuir–Blodgett Techniques. Marcel Dekker, New York.

Peterson I.R., Russell G.J. and Roberts G.G. 1983. A new model for the deposition of w-tricosenoic acid LB film layers, Thin Solid Films, 109, 371–378.

Petrov J.G. 1986. Dependence of the maximum speed of wetting on the interactions in the three-phase contact zone, Colloids Surf., 17, 283–294.

Petrov J.G., Kuhn H. and Mobius D. 1980. Three-phase contact line motion in the deposition of spread monolayers, J. Colloid Interface Sci., 73, 66–75.

Petrov J.G. and Petrov P.G. 1998. Molecular hydrodynamic description of Langmuir–Blodgett deposition, Langmuir, 14, 2490–2496.

Roberts G.G. (Ed.) 1990. Langmuir–Blodgett Films. Plenum Press, New York.

Roth K.M., Rajeev N.D., Dabke B., Gryko D.T., Clausen C., Lindsey J.S., Bocian D.F. and Kuhr W.G. 2000. Molecular approach toward information storage based on the redox properties of porphyrins in self-assembled monolayers, J. Vac. Sci. Technol. B Microelectron. Process. Phenomena, 18, 2359–2364.

Russel W.B., Saville D.A and Schowalter W. R.. 1989. Colloidal Dispersions. Cambridge University Press, London.

Sanassy P. and Evans S.D. 1993. Mixed alkanethiol monolayers on gold surfaces: substrates for Langmuir–Blodgett film deposition, Langmuir, 9, 1024–1027.

Savelski M.J., Shetty S.A., Kolb W.B. and Cerro R.L. 1995. Flow patterns associated with the steady movement of a solid/liquid/fluid contact line, J. Colloid Interface Sci., 176, 117–127.

Schwartz D.K. 1997. Langmuir–Blodgett film structure, Surf. Sci. Rep., 27, 241–334.

Seto J., Nagai T., Ishimoto C. and Watanabe H. 1985. Frictional properties of magnetic media coated with Langmuir–Blodgett films, Thin Solid Films, 134, 101–108.

Sigiyama N., Shimizu A., Nakamura M., Nakagawa Y., Nagasawa Y. and Ishida H. 1998. Molecular-scale structures of Langmuir–Blodgett films of fatty acids observed by atomic force microscopy (II) – cation dependence, Thin Solid Films, 331, 170–175.

Srinivasan M.P., Higgins B.G., Stroeve P. and Kowel S.T. 1988. Entrainment of aqueous subphase in Langmuir–Blodgett films, Thin Solid Films, 159, 191–205.

Tien H.T., Barish R.H., Gu L.-Q. and Ottova A.L. 1998. Supported bilayer lipid membranes as ion and molecular probes, Anal. Sci., 14, 3–18.

Usui S. and Healy T.W. 2002. Zeta potential of stearic acid monolayer at the air–aqueous solution interface, J. Colloid Interface Sci., 250, 371–378.

Zasadinsky J.A., Viswanathan R., Madsden L., Garnaes J. and Schwartz D.K. 1994. Langmuir–Blodgett films, Science, 263, 1726–1733.

Zhang L.Y. and Srinivasan M.P. 2001. Hydrodynamics of subphase entrainment during Langmuir–Blodgett deposition, Colloids Surf., 193, 15–33.

11

Advances in Logic-Based Optimization Approaches to Process Integration and Supply Chain Management

Ignacio E. Grossmann

11.1 Introduction

The objective of this chapter is to provide an overview of new developments in discrete/continuous optimization with applications to process integration and supply chain management problems. The emphasis is on logic-based optimization which is becoming a new promising area in process systems engineering.

Discrete/continuous optimization problems, when represented in algebraic form, correspond to mixed-integer optimization problems that have the following general form:

$$\begin{array}{ll} \min & Z = f(x,y) \\ \text{s.t.} & h(x,y) = 0 \\ & g(x,y) \leq 0 \\ & x \in X, y \in \{0,1\}^m \end{array} \qquad \text{(MIP)}$$

where $f(x,y)$ is the objective function (e.g. cost), $h(x,y) = 0$ are the equations that describe the performance of the system (material balances, production rates), and $g(x,y) \leq 0$ are inequalities that define the specifications or constraints for feasible plans and schedules. The variables x are continuous and generally correspond to state variables, while y are the discrete variables, which generally are restricted to take 0–1 values to

Chemical Engineering Edited by Miguel A. Galán and Eva Martin del Valle

define for instance the assignments of equipment and sequencing of tasks. Problem (MIP) corresponds to a mixed-integer nonlinear program (MINLP) when any of the functions involved are nonlinear. If all functions are linear, it corresponds to a mixed-integer linear program (MILP). If there are no 0–1 variables, the problem (MIP) reduces to a nonlinear program (NLP) or linear program (LP) depending on whether or not the functions are linear.

It should be noted that (MIP) problems, and their special cases, may be regarded as steady-state models. Hence, one important extension is the case of dynamic models, which in the case of discrete time models gives rise to multiperiod optimization problems, while for the case of continuous time it gives rise to optimal control problems that contain differential-algebraic equation (DAE) models.

Mathematical programming (MP), and optimization in general, has found extensive use in process systems engineering. A major reason for this is that in these problems there are often many alternative solutions, and hence, it is often not easy to find the optimal solution. Furthermore, in many cases the economics is such that finding the optimum solution translates into large savings. Therefore, there might be a large economic penalty to just sticking to suboptimal solutions. In summary, optimization has become a major technology that helps companies to remain competitive.

Applications in process integration (process design and synthesis) have been dominated by NLP and MINLP models due to the need for the explicit handling of performance equations, although simpler targeting models in process synthesis can give rise to LP and MILP problems. An extensive review of optimization models for process integration was outlined by Grossmann *et al.* (1999). In contrast, supply chain management problems tend to be dominated by linear models, LP and MILP, for planning and scheduling (refer to Grossmann *et al.*, 2002 for a review). Finally, global optimization has concentrated more on design than on operations problems, since nonconvexities in the design problems are more likely to yield suboptimal solutions since the corresponding bounds for the variables are rather loose in these problems. It is also worth noting that all of these applications have been facilitated not only by progress in optimization algorithms, but also by the advent of modeling techniques (Williams, 1985) and systems such as GAMS (Brooke *et al.*, 1998), AMPL (Fourer *et al.*, 1992), and AIMMS (Bisschop and Entriken, 1993).

In the next section we describe new developments in discrete/continuous logic-based optimization. We provide an overview of generalized disjunctive programming (GDP) and its relation with MINLP. We describe several algorithms for GDP that include branch and bound (BB), decomposition and mixed-integer reformulations. We also describe recent developments for cutting plane techniques, global optimization of nonconvex GDP problems, and constraint programming (CP). Several examples are presented to illustrate the capabilities of these methods.

11.2 Logic-Based Discrete and Continuous Optimization

11.2.1 Review of Mixed-Integer Optimization

The conventional way of modeling discrete/continuous optimization problems has been through the use of 0–1 and continuous variables, and algebraic equations and inequalities.

For the case of linear functions, this model corresponds to a MILP model, which has the following general form:

$$\begin{aligned} \min \quad & Z = a^{\mathrm{T}} y + b^{\mathrm{T}} x \\ \text{s.t.} \quad & Ay + Bx \le d \qquad \text{(MILP)} \\ & x \in R^n,\ y \in \{0,1\}^m \end{aligned}$$

In problem (MILP) the variables x are continuous, and y are discrete variables, which generally are binary variables. As is well known, problem (MILP) is NP-hard. Nevertheless, an interesting theoretical result is that it is possible to transform it into an LP with the convexification procedures proposed by Lovász and Schrijver, (1991), Sherali and Adams, (1990), and Balas *et al.* (1993). These procedures consist of sequentially lifting the original relaxed $x-y$ space into higher dimension and projecting it back to the original space so as to yield, after a finite number of steps, the integer convex hull. Since the transformations have exponential complexity, they are only of theoretical interest, although they can be used as a basis for deriving cutting planes (e.g. lift and project method by Balas *et al.*, 1993).

As for the solution of problem (MILP), it should be noted that this problem becomes an LP problem when the binary variables are relaxed as continuous variables, $0 \le y \le 1$. The most common solution algorithms for problem (MILP) are LP-based branch and bound (BB) methods, which are enumeration methods that solve LP subproblems at each node of the search tree. This technique was initially conceived by Land and Doig, (1960), Balas, (1965), and later formalized by Dakin, (1965). Cutting plane techniques, which were initially proposed by Gomory, (1958), and consist of successively generating valid inequalities that are added to the relaxed LP, have received renewed interest through the works of Crowder *et al.* (1983), Van Roy and Wolsey, (1986), and especially the lift and project method of Balas *et al.* (1993). A recent review of branch and cut methods is outlined by Johnson *et al.* (2000). Finally, Benders decomposition (Benders, 1962) is another technique for solving MILPs in which the problem is successively decomposed into LP subproblems for fixed 0–1 and a master problem for updating the binary variables.

Software for MILP solver includes OSL, CPLEX, and XPRESS which use the LP-based BB algorithm combined with cutting plane techniques. MILP models and solution algorithms have been developed and applied successfully to many industrial problems (e.g. Kallrath, 2000).

For the case of nonlinear functions, the discrete/continuous optimization problem is given by the MINLP model:

$$\begin{aligned} \min \quad & Z = f(x,y) \\ \text{s.t.} \quad & g(x,y) \le 0 \\ & x \in X,\ y \in Y \qquad \text{(MINLP)} \\ & X = \{x \mid x \in R^n,\ x^{\mathrm{L}} \le x \le x^{\mathrm{U}},\ Bx \le b\} \\ & Y = \{y \mid y \in \{0,1\}^m,\ Ay \le a\} \end{aligned}$$

where $f(x,y)$ and $g(x,y)$ are assumed to be convex, differentiable, and bounded over X and Y. The set X is generally assumed to be a compact convex set, and the discrete set Y is a polyhedral of integer points. Usually, in most applications it is assumed that $f(x,y)$ and $g(x,y)$ are linear in the binary variables y.

A recent review of MINLP solution algorithms is outlined by Grossmann (2002). Algorithms for the solution of problem (MINLP) include the BB method, which is a direct extension of the linear case of MILPs (Gupta and Ravindran, 1985; Borchers and Mitchell, 1994; Leyffer, 2001). The branch-and-cut method by Stubbs and Mehrotra (1999), which corresponds to a generalization of the lift and project cuts by Balas *et al.* (1993), adds cutting planes to the NLP subproblems in the search tree. Generalized Benders decomposition (GBD) (Geoffrion, 1972) is an extension of Benders decomposition and consists of solving an alternating sequence of NLP (fixed binary variables) and aggregated MILP master problems that yield lower bounds. The outer-approximation (OA) method (Duran and Grossmann, 1986; Yuan *et al.*, 1988; Fletcher and Leyffer, 1994) also consists of solving NLP subproblems and MILP master problems. However, OA uses accumulated function linearizations which act as linear supports for convex functions, and yield stronger lower bounds than GBD that uses accumulated Lagrangean functions that are parametric in the binary variables. The LP-NLP-based BB method by Quesada and Grossmann (1992) integrates LP and NLP subproblems of the OA method in one search tree, where the NLP subproblem is solved if a new integer solution is found and the linearization is added to all the open nodes. Finally the extended cutting plane (ECP) method by Westerlund and Pettersson (1995) is based on an extension of Kelley's (1960) cutting plane method for convex NLPs. The ECP method also solves successively an MILP master problem but it does not solve NLP subproblems as it simply adds successive linearizations at each iteration.

11.3 Generalized Disjunctive Programming

Given the difficulties in the modeling and scaling of mixed-integer problems, the following major approaches based on logic-based techniques have emerged: generalized disjunctive programming (GDP) (Raman and Grossmann, 1994); mixed logic linear programming (MLLP) (Hooker and Osorio, 1999); and constraint programming (CP) (Hentenryck, 1989) The motivations for this logic-based modeling have been to facilitate the modeling, reduce the combinatorial search effort, and improve the handling of the nonlinearities. In this chapter we will mostly concentrate on GDP. A general review of logic-based optimization is outlined by Hooker (2000).

GDP (Raman and Grossmann, 1994) is an extension of disjunctive programming (Balas, 1979) that provides an alternate way of modeling (MILP) and (MINLP) problems. The general formulation of a (GDP) is as follows:

$$\begin{aligned}
\min \quad & Z = \sum_{k \in K} c_k + f(x) \\
\text{s.t.} \quad & g(x) \le 0 \\
& \bigvee_{j \in J_k} \begin{bmatrix} Y_{jk} \\ h_{jk}(x) \le 0 \\ c_k = \gamma_{jk} \end{bmatrix}, \quad k \in K \\
& \Omega(Y) = \text{True} \\
& x \in R^n, \; c \in R^m, \; Y \in \{\text{true, false}\}^m
\end{aligned} \qquad \text{(GDP)}$$

where Y_{jk} are the Boolean variables that decide whether a term j in a disjunction $k \in K$ is true or false, and x are continuous variables. The objective function involves the term $f(x)$ for the continuous variables and the charges c_k that depend on the discrete choices in each disjunction $k \in K$. The constraints $g(x) \le 0$ hold regardless of the discrete choice, and $h_{jk}(x) \le 0$ are conditional constraints that hold when Y_{jk} is true in the jth term of the kth disjunction. The cost variables c_k correspond to the fixed charges, and are equal to γ_{jk} if the Boolean variable Y_{jk} is true. $\Omega(Y)$ are logical relations for the Boolean variables expressed as propositional logic.

It should be noted that problem (GDP) can be reformulated as an MINLP problem by replacing the Boolean variables with binary variables y_{jk},

$$\begin{aligned} \min \quad & Z = \sum_{k\in K}\sum_{j\in J_k} \gamma_{jk} y_{jk} + f(x) \\ \text{s.t.} \quad & g(x) \le 0 \\ & h_{jk}(x) \le M_{jk}(1 - y_{jk}), \ j \in J_k, \ k \in K \\ & \sum_{j\in J_k} y_{jk} = 1, \ k \in K \\ & Ay \le a \\ & 0 \le x \le x^{\mathrm{U}}, \ y_{jk} \in \{0,1\}, \ j \in J_k, \ k \in K \end{aligned} \qquad \text{(BM)}$$

where the disjunctions are replaced by 'big-M' constraints which involve a parameter M_{jk} and binary variables y_{jk}. The propositional logic statements $\Omega(Y) = \text{True}$ are replaced by the linear constraints $Ay \le a$ as described by Williams (1985) and Raman and Grossmann (1991). Here we assume that x is a non-negative variable with finite upper bound x^{U}. An important issue in model (BM) is how to specify a valid value for the big-M parameter M_{jk}. If the value is too small, then feasible points may be cut off. If M_{jk} is too large, then the continuous relaxation might be too loose yielding poor lower bounds. Therefore, finding the smallest valid value for M_{jk} is the desired selection. For linear constraints, one can use the upper and lower bound of the variable x to calculate the maximum value of each constraint, which then can be used to calculate a valid value of M_{jk}. For nonlinear constraints one can in principle maximize each constraint over the feasible region, which is a non-trivial calculation.

11.3.1 Convex Hull Relaxation of Disjunction

Lee and Grossmann (2000) have derived the convex hull relaxation of problem (GDP). The basic idea is as follows. Consider a disjunction $k \in K$ that has convex constraints,

$$\bigvee_{j\in J_k} \begin{bmatrix} Y_{jk} \\ h_{jk}(x) \le 0 \\ c = \gamma_{jk} \end{bmatrix} \quad 0 \le x \le x^{\mathrm{U}}, \ c \ge 0 \qquad \text{(DP)}$$

where $h_{jk}(x)$ are assumed to be convex and bounded over x. The convex hull relaxation of disjunction (DP) (Stubbs and Mehrotra, 1999) is given as follows:

$$\begin{aligned} & x = \sum_{j\in J_k} \nu^{jk}, \quad c = \sum_{j\in J} \lambda_{jk}\gamma_{jk} \\ & 0 \le \nu^{jk} \le \lambda_{jk} x^{\mathrm{U}}_{jk}, \ j \in J_k \\ & \sum_{j\in J_k} \lambda_{jk} = 1, \ 0 \le \lambda_{jk} \le 1, \ j \in J_k \\ & \lambda_{jk} h_{jk}(\nu^{jk}/\lambda_{jk}) \le 0, \ j \in J_k \\ & x, \ c, \ \nu^{jk} \ge 0, \ j \in J_k \end{aligned} \qquad \text{(CH)}$$

where ν^{jk} are disaggregated variables that are assigned to each term of the disjunction $k \in K$, and λ_{jk} are the weight factors that determine the feasibility of the disjunctive term. Note that when λ_{jk} is 1, then the jth term in the kth disjunction is enforced and the other terms are ignored. The constraints $\lambda_{jk}h_{jk}(\nu^{jk}/\lambda_{jk})$ are convex if $h_{jk}(x)$ is convex as discussed by Hiriart-Urruty and Lemaréchal (1993, p. 160). A formal proof is provided by Stubbs and Mehrotra (1999). Note that the convex hull (CH) reduces to the result by Balas (1985) if the constraints are linear. On the basis of the convex hull relaxation (CH), Lee and Grossmann (2000) proposed the following convex relaxation program of (GDP).

$$
\begin{aligned}
\min \quad & Z^{\mathrm{L}} = \sum_{k\in K}\sum_{j\in J_k} \gamma_{jk}\lambda_{jk} + f(x) \\
\text{s.t.} \quad & g(x) \le 0 \\
& x = \sum_{j\in J_k} \nu^{jk}, \quad \sum_{j\in J_k} \lambda_{jk} = 1, \ k \in K \\
& 0 \le \nu^{jk} \le \lambda_{jk} x^{\mathrm{U}}_{jk}, \quad j \in J_k, \ k \in K \\
& \lambda_{jk} h_{jk}(\nu^{jk}/\lambda_{jk}) \le 0, \quad j \in J_k, \ k \in K \\
& A\lambda \le a \\
& 0 \le x, \ \nu^{jk} \le x^{\mathrm{U}}, \ 0 \le \lambda_{jk} \le 1, \ j \in J_k, \ k \in K
\end{aligned}
\qquad \text{(CRP)}
$$

where x^{u} is a valid upper bound for x and ν. For computational reasons, the nonlinear inequality is written as $\lambda_{jk}h_{jk}(\nu_{jk}/(\lambda_{jk}+\varepsilon)) \le 0$ where ε is a small tolerance. Note that the number of constraints and variables increases in (CRP) compared with problem (GDP). Problem (CRP) has a unique optimal solution and it yields a valid lower bound to the optimal solution of problem (GDP) (Lee and Grossmann, 2000). Problem (CRP) can also be regarded as a generalization of the relaxation proposed by Ceria and Soares (1999) for a special form of problem (GDP). Grossmann and Lee (2003) proved that problem (CRP) has the useful property that the lower bound is greater than or equal to the lower bound predicted from the relaxation of problem (BM).

11.4 Solution Algorithms for GDP

11.4.1 Branch and Bound

For the linear case of problem (GDP), Beaumont (1991) proposed a BB method which directly branches on the constraints of the disjunctions where no logic constraints are involved. Also for the linear case Raman and Grossmann (1994) developed a BB method which solves the (GDP) problem in hybrid form, by exploiting the tight relaxation of the disjunctions and the tightness of the well-behaved mixed-integer constraints. There are also BB methods for solving problem (GDP). In particular, a disjunctive BB method can be developed that directly branches on the term in a disjunction using the convex hull relaxation (CRP) as a basic subproblem (Lee and Grossmann, 2000). Problem (CRP) is solved at the root node of the search tree. The branching rule is to select the least infeasible term in a disjunction first. Next, we consider a dichotomy where we fix the value $\lambda_{jk} = 1$ for the disjunctive term that is closest to being satisfied, and consider on the other hand the convex hull of the remaining terms ($\lambda_{jk} = 0$).

When all the decision variables λ_{jk} are fixed, problem (CRP) yields an upper bound to problem (GDP). The search is terminated when the lower and the upper bounds are the same. The algorithm has finite convergence since the number of the terms in the disjunction is finite. Also, since the nonlinear functions are convex, each subproblem has a unique optimal solution, and hence the bounds are rigorous.

11.4.2 Reformulation and Cutting Planes

Another approach for solving a linear GDP is to replace the disjunctions either by big-M constraints or by the convex hull of each disjunction (Balas, 1985; Raman and Grossmann, 1994). For the nonlinear case, a similar way of solving the problem (GDP) is to reformulate it into the MINLP by restricting the variables λ_{jk} in problem (CRP) to 0–1 values. Alternatively, to avoid introducing a potentially large number of variables and constraints, the GDP might also be reformulated as the MINLP problem (BM) by using big-M parameters, although this leads to a weaker relaxation (Grossmann and Lee, 2003). One can then apply standard MINLP solution algorithms (i.e. BB, OA, GBD, and ECP).

To strengthen the lower bounds one can derive cutting planes using the convex hull relaxation (CRP). To generate a cutting plane, the following 2-norm separation problem (SP), a convex QP, is solved:

$$
\begin{aligned}
\min \quad & \phi(x) = (x - x_R^{BM,n})^T (x - x_R^{BM,n}) \\
\text{s.t.} \quad & g(x) \le 0 \\
& x = \sum_{i \in D_k} \nu_{ik}, \quad k \in K \\
& y_{ik} h_{ik}(\nu_{ik}/y_{ik}) \le 0, \quad i \in D_k, \quad k \in K \qquad \text{(SP)} \\
& \sum_{i \in D_k} y_{ik} = 1, \quad k \in K \\
& Ay \le a \\
& x, \nu_{ik} \in R^n, \quad 0 \le y_{ik} \le 1
\end{aligned}
$$

where $x_R^{BM,n}$ is the solution of problem (BM) with relaxed $0 \le y_{ik} \le 1$. Problem (SP) yields a solution point x^* which belongs to the convex hull of the disjunction and is closest to the relaxation solution $x_R^{BM,n}$. The most violated cutting plane is then given by

$$
(x^* - x_R^{BM,n})^T (x - x^*) \ge 0 \qquad \text{(CP1)}
$$

The cutting plane in (CP1) is a valid inequality for problem (GDP). Problem (BM) is modified by adding the cutting plane (CP1) as follows:

$$
\begin{aligned}
\min \quad & Z = \sum_{k \in K} \sum_{i \in D_k} \gamma_{ik} y_{ik} + f(x) \\
\text{s.t.} \quad & g(x) \le 0 \\
& h_{ik}(x) \le M_{ik}(1 - y_{ik}), \quad i \in D_k, \quad k \in K \\
& \sum_{i \in D_k} y_{ik} = 1, \quad k \in K \qquad \text{(CP)} \\
& Ay \le a \\
& \beta^T x \le b \\
& x \in R^n, \quad 0 \le y_{ik} \le 1
\end{aligned}
$$

where $\beta^T x \le b$ is the cutting plane (CP1). Since we add a valid inequality to problem (BM), the lower bound obtained from problem (CP) is generally tighter than before adding the cutting plane.

This procedure for generating the cutting plane can be used by solving the separation problem (SP) only at the root node. It can also be used to strengthen the MINLP problem (BM) before applying methods such as OA, GBD, and ECP. It is also interesting to note that cutting planes can be derived in the (x, y) space, especially when the objective function has binary variables y.

Another application of the cutting plane is to determine if the convex hull formulation yields a good relaxation of a disjunction. If the value of $\|x^* - x_{\mathrm{R}}^{\mathrm{BM},n}\|$ is large, then it is an indication that this is the case. A small difference between x^* and $x_{\mathrm{R}}^{\mathrm{BM},n}$ would indicate that it might be better to simply use the big-M relaxation. It should also be noted that Sawaya and Grossmann (2004) have recently developed the cutting plane method for linear GDP problems using the 1, 2, and ∞ norms, and relying on the theory of subgradient optimization.

11.4.3 GDP Decomposition Methods

Türkay and Grossmann (1996) have proposed logic-based OA and GBD algorithms for problem (GDP) by decomposition into NLP and MILP subproblems. For fixed values of the Boolean variables, $Y_{jk} = \text{true}$ and $Y_{ik} = \text{false}$ for $j \neq i$, the corresponding NLP subproblem is derived from (GDP) as follows:

$$
\begin{aligned}
\min \quad & Z = \sum_{k \in K} c_k + f(x) \\
\text{s.t.} \quad & g(x) \leq 0 \\
& \left.\begin{array}{l} h_{jk}(x) \leq 0 \\ c_k = \gamma_{jk} \end{array}\right\} \text{ for } Y_{jk} = \text{true}, \ j \in J_k, \ k \in K \qquad \text{(NLPD)} \\
& \left.\begin{array}{l} B^i x = 0 \\ c_k = 0 \end{array}\right\} \text{ for } Y_{ik} = \text{false}, \ i \in J_k, \ k \in K \\
& x \in R^n, \ c \in R^m
\end{aligned}
$$

For every disjunction k only the constraints corresponding to the Boolean variable Y_{jk} that is true are enforced. Also, fixed charges γ_{jk} are applied to these terms. After K subproblems (NLPD) are solved, sets of linearizations $l = 1, \ldots, K$ are generated for subsets of terms $L_{jk} = \{l | Y_{jk}^l = \text{true}\}$, then one can define the following disjunctive OA master problem:

$$
\begin{aligned}
\min \quad & Z = \sum_{k \in K} c_k + \alpha \\
\text{s.t.} \quad & \left.\begin{array}{l} \alpha \geq f(x^l) + \nabla f(x^l)^{\mathrm{T}}(x - x^l) \\ g(x^l) + \nabla g(x^l)^{\mathrm{T}}(x - x^l) \leq 0 \end{array}\right\} \quad l = 1, \ldots, L \\
& \begin{bmatrix} Y_{jk} \\ h_{jk}(x^l) + \nabla h_{jk}(x^l)^{\mathrm{T}}(x - x^l) \leq 0, l \in L_{jk} \\ c_k = \gamma_{jk} \end{bmatrix} \vee \begin{bmatrix} \neg Y_{jk} \\ B^k x = 0 \\ c_k = 0 \end{bmatrix}, \ k \in K \qquad \text{(MGDP)} \\
& \Omega(Y) = \text{True} \\
& x \in R^n, \ c \in R^m, \ Y \in \{\text{true}, \text{false}\}^m
\end{aligned}
$$

Before solving the MILP master problem, it is necessary to solve various subproblems (NLPD) in order to produce at least one linear approximation of each of the terms in the disjunctions. As shown by Türkay and Grossmann (1996), selecting the smallest number of subproblems amounts to the solution of a set covering problem. In the context of flowsheet synthesis problems, another way of generating the linearizations in (MGDP) is by starting with an initial flowsheet and optimizing the remaining subsystems as in the modeling/decomposition strategy (Kocis and Grossmann, 1987).

Problem (MGDP) can be solved by the methods described by Beaumont (1991), Raman and Grossmann (1994), and Hooker and Osorio (1999). For the case of process networks, Türkay and Grossmann (1996) have shown that if the convex hull representation of the disjunctions in (MGDP) is used, then assuming $B_k = I$ and converting the logic relations $\neg(Y)$ into the inequalities $Ay \leq a$ leads to the MILP reformulation of (NLPD) which can be solved with OA. Türkay and Grossmann (1996) have also shown that while a logic-based generalized Benders method (Geoffrion, 1972) cannot be derived as in the case of the OA algorithm, one can exploit the property for MINLP problems that performing one Benders iteration (Türkay and Grossmann, 1996) on the MILP master problem of the OA algorithm is equivalent to generating a generalized Benders cut. Therefore, a logic-based version of the generalized Benders method performs one Benders iteration on the MILP master problem. Also, slack variables can be introduced to problem (MGDP) to reduce the effect of nonconvexity as in the augmented-penalty MILP master problem (Viswanathan and Grossmann, 1990).

11.4.4 Hybrid GDP/MINLP

Vecchietti and Grossmann (1999) have proposed a hybrid formulation of the GDP and algebraic MINLP models. It involves disjunctions and mixed-integer constraints as follows:

$$
\begin{aligned}
\min \quad & Z = \sum_{k \in K} c_k + f(x) + d^{\mathrm{T}} y \\
\text{s.t.} \quad & g(x) \leq 0 \\
& r(x) + Dy \leq 0 \\
& Ay \leq a \\
& \bigvee_{j \in J_k} \begin{bmatrix} Y_{jk} \\ h_{jk}(x) \leq 0 \\ c_k = \gamma_{jk} \end{bmatrix}, \quad k \in K \\
& \Omega(Y) = \text{True} \\
& x \in R^n, \;\; c \in R^m, \;\; y \in \{0,1\}^q, \;\; Y \in \{\text{true, false}\}^m
\end{aligned}
\qquad \text{(PH)}
$$

where x and c are continuous variables and Y and y are discrete variables. Problem (PH) can reduce to a GDP or to an MINLP, depending on the absence and presence of the mixed-integer constraints and disjunctions and logic propositions. Thus, problem (PH) provides the flexibility of modeling an optimization problem as a GDP, an MINLP, or a hybrid model, making it possible to exploit the advantage of each model.

An extension of the logic-based OA algorithm for solving problem (PH) has been implemented in LOGMIP, a computer code based on GAMS (Vecchietti and Grossmann, 1999). This algorithm decomposes problem (PH) into two subproblems, the NLP and the MILP master problems. With fixed discrete variables, the NLP subproblem is solved.

Then at the solution point of the NLP subproblem, the nonlinear constraints are linearized and the disjunction is relaxed by convex hull to build a master MILP subproblem which will yield a new discrete choice of (y, Y) for the next iteration.

11.5 Global Optimization Algorithm of Nonconvex GDP

In the preceding sections of this chapter we assumed convexity in the nonlinear functions. However, in many applications nonlinearites give rise to nonconvex functions that may yield local solutions, not guaranteeing the global optimality. Global optimization of nonconvex programs has received increased attention due to their practical importance. Most of the deterministic global optimization algorithms are based on the spatial BB algorithm (Horst and Tuy, 1996), which divides the feasible region of continuous variables and compares lower bound and upper bound for fathoming each subregion. The one that contains the optimal solution is found by eliminating subregions that are proved not to contain the optimal solution.

For nonconvex NLP problems, Quesada and Grossmann (1995) proposed a spatial BB algorithm for concave separable, linear fractional, and bilinear programs using linear and nonlinear underestimating functions (McCormick, 1976). For nonconvex MINLP, Ryoo and Sahinidis (1995) and later Tawarmalani and Sahinidis (2002) developed BARON, which branches on the continuous and discrete variables with bounds reduction method. Adjiman *et al.* (1997, 2000) proposed the SMIN-αBB and GMIN-αBB algorithms for twice-differentiable nonconvex MINLPs. Using a valid convex underestimation of general functions as well as for special functions, Adjiman *et al.* (1996) developed the αBB method which branches on both the continuous and discrete variables according to specific options. The branch-and-contract method (Zamora and Grossmann, 1999) has bilinear, linear fractional, and concave separable functions in the continuous variables and binary variables, uses bound contraction, and applies the OA algorithm at each node of the tree. Kesavan and Barton (2000b) developed a generalized branch-and-cut (GBC) algorithm, and showed that their earlier decomposition algorithm (Kesavan and Barton, 2000a) is a specific instance of the GBC algorithm with a set of heuristics. Smith and Pantelides (1997) proposed a reformulation method combined with a spatial BB algorithm for nonconvex MINLP and NLP, which is implemented in the gPROMS modeling system.

11.5.1 GDP Global Optimization Algorithms

We briefly describe two global optimization algorithms. The first was proposed by Lee and Grossmann (2001) and is for the case when the problem (GDP) involves bilinear, linear fractional, and concave separable functions. First, these nonconvex functions of continuous variables are relaxed by replacing them with underestimating convex functions (McCormick, 1976; Quesada and Grossmann, 1995). Next, the convex hull of each nonlinear disjunction is constructed to build a convex NLP problem (CRP). At the first step, an upper bound is obtained by solving the nonconvex MINLP reformulation (BM) with the OA algorithm. This upper bound is then used for the bound contraction. The feasible region of continuous variables is contracted with an optimization subproblem that incorporates the valid underestimators and the upper bound value and that minimizes or maximizes each variable in turn. The tightened convex GDP problem is then solved in the first level of a two-level BB algorithm, in which a discrete BB search is performed on the disjunctions to predict lower bounds. In the second level, a spatial BB method is used to solve nonconvex NLP problems for updating the upper bound. The algorithm exploits the convex hull relaxation for the discrete search and the fact that the spatial BB is restricted to fixed discrete variables in order to predict tight lower bounds.

The second algorithm is by Bergamini *et al.* (2004); it does not require spatial BB searches as it uses piecewise linear approximations. The algorithm considers the logic-based OA algorithm (Türkay and Grossmann, 1996) and is based on constructing a master problem that is a valid bounding representation of the original problem, and on solving the NLP subproblems to global optimality. The functions are assumed to be sums of convex, bilinear, and concave terms. To maintain rigorously the bounding properties of the MILP master problem, linear under- and overestimators for bilinear and concave terms are constructed over a grid with the property of having zero gap in the finite set of points. The set of these approximation points is defined over subdomains defined by bounds of variables and solution points of the previous NLP subproblems. For bilinear terms, the convex envelope by McCormick is used. Disjunctions are used to formulate the convex envelope in each subdomain, and the convex hull of these disjunctions is used to provide the tightest relaxation. It should be noted that binary variables are needed for the discrete choice of the corresponding subdomains. Linear fractional functions are treated similarly. Piecewise linear subestimations replace the concave terms.

The solution of the NLP subproblems to global optimality can be performed by fixing the topology variables in the MILP and by successively refining the grid of the piecewise linear approximations. Alternatively, a general-purpose NLP algorithm for global optimization (e.g. BARON code by Tawarmalani and Sahinidis, 2002) can be used. It should be noted that the NLP subproblems are reduced problems, involving only continuous variables related to a process with fixed structure. This allows the tightening of the variable bounds, and therefore reduces the computational cost of solving it to global optimality.

11.6 Constraint Programming and Hybrid MILP/CP Methods

In order to overcome difficulties in modeling and scalability of MP models, a trend has emerged to combine MP with symbolic logic reasoning into the quantitative. Among these attempts, one of the more promising approaches has been the development of CP, which has proved to be particularly effective in scheduling applications. CP is essentially based on the idea that inference methods can accelerate the search for a solution.

CP (Hentenryck, 1989; Hooker, 2000) is a relatively new modeling and solution paradigm that was originally developed to solve feasibility problems, but it has been extended to solve optimization problems as well. CP is very expressive as continuous integers as well as Boolean variables are permitted and moreover, variables can be indexed by other variables. Constraints can be expressed in algebraic form (e.g. $h(x) \leq 0$), as disjunctions (e.g. $[A1x \leq b1] \vee [A2x \leq b2]$), or as conditional logic statements (e.g. if $g(x) \leq 0$ then $r(x) \leq 0$). In addition, the language can support special implicit functions such as the *all different* $(x1, x2, \ldots, xn)$ constraint for assigning different values to the integer variables $x1, x2, \ldots, xn$. The language consists of C++ procedures, although the recent trend has been to provide higher-level languages such as OPL. Other commercial CP software packages include ILOG Solver (ILOG, 1999), CHIP (Dincbas *et al.*, 1988), and ECLiPSe (Wallace *et al.*, 1997).

Optimization problems in CP are solved as constraint satisfaction problems (CSP), where we have a set of variables, a set of possible values for each variable (domain), and a set of constraints among the variables. The question to be answered is as follows: Is there an assignment of values to variables that satisfies all constraints? The solution of CP models is based on performing constraint propagation at each node by reducing the domains of the variables. If an empty domain is found the node is pruned. Branching is performed whenever a domain of an integer, binary or Boolean variable has more than one element, or when the bounds of the domain of a continuous variable do not lie within a tolerance. Whenever a solution is found, or a domain of a variable is reduced, new constraints are added. The search terminates when no further nodes must be examined. The effectiveness of CP depends on the propagation mechanism behind constraints. Thus, even though many constructs and constraints are available, not all of them have efficient propagation mechanisms. For some problems, such as scheduling, propagation mechanisms have been proven to be very effective. Some of the most common propagation rules for scheduling are the 'time-table' constraint (Le Pape, 1998), the 'disjunctive-constraint' propagation (Baptiste and Le Pape, 1996; Smith and Cheng, 1993), the 'edge-finding' (Nuijten, 1994; Caseau and Laburthe, 1994), and the 'not-first, not-last' (Baptiste and Le Pape, 1996).

Since MILP and CP approaches appear to have complementary strengths, in order to solve difficult problems that are not effectively solved by either of the two, several researchers have proposed models that integrate the two paradigms. The integration between MILP and CP can be achieved in two ways (Hooker, 2002; Hentenryck, 2002):

(1) By combining MILP and CP constraints into one hybrid model. In this case a hybrid algorithm that integrates constraint propagation with linear programming in a single search tree is also needed for the solution of the model (e.g. Heipcke, 1999; Rodosek *et al.*, 1999).
(2) By decomposing the original problem into two subproblems: one MILP and one CP subproblem. Each model is solved separately and information obtained while solving one subproblem is used for the solution of the other subproblem (Jain and Grossmann, 2001; Bockmayr and Pisaruk, 2003).

Maravelias and Grossmann (2004) have recently developed a hybrid MILP/CP method for the continuous time state-task-network (STN) model in which different objectives such as profit maximization, cost minimization, and makespan minimization can be handled. The proposed method relies on an MILP model that represents an aggregate of the original MILP model. This method has been shown to produce order of magnitude reductions in CPU times compared to standalone MILP or CP models.

11.7 Examples in Process Integration

11.7.1 Synthesis of Separation System

This problem, a joint collaboration with BP (Lee *et al.*, 2003), deals with the synthesis of a separation system of an ethylene plant in which the mixture to be separated includes

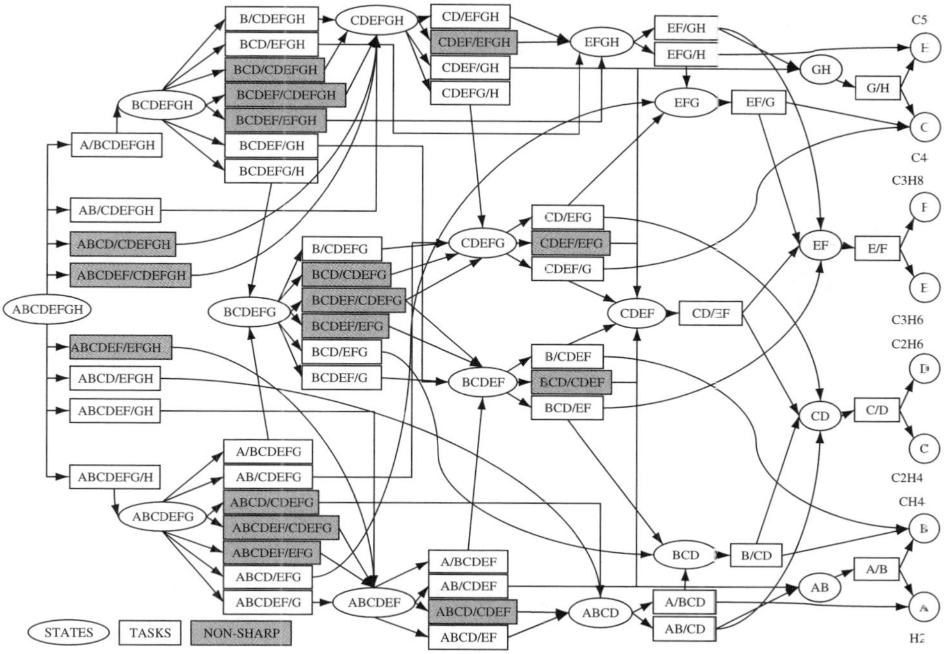

Figure 11.1 *Superstructure of separation of ethylene plant*

hydrogen, methane, ethane, ethylene, propane, propylene, and C4s, C5s, and C6s. For each potential separation task a number of separation technologies such as dephlegmators, membranes, PSA, physical and chemical absorption were considered in addition to the standard distillation columns and cold boxes. The superstructure of this problem, which includes 53 separation tasks, is shown in Figure 11.1. This problem was formulated as a GDP problem and reformulated as an MINLP by applying both big-M and convex hull transformations. The problem involved 5800 0–1 variables, 24 500 continuous variables, and 52 700 constraints, and was solved with GAMS DICOPT (CONOPT2/CPLEX) in 3 h of CPU time on a Pentium III machine. Compared to the base-case design the optimal flowsheet that is shown in Figure 11.2 included a dephlegmator and a physical absorber, and one less distillation column, achieving a $20 million reduction in the cost, largely from reduced refrigeration.

11.7.2 Retrofit Planning Problem

In this problem it is assumed that an existing process network is given where each process can possibly be retrofitted for improvements such as higher yield, increased capacity, and reduced energy consumption. Given limited capital investments to make process improvements and cost estimations over a given time horizon, the problem consists of identifying those modifications that yield the highest economic improvement in terms of economic potential, which is defined as the income from product sales minus the cost of raw materials, energy, and process modifications. Sawaya and Grossmann (2004) have developed a GDP model for this problem, which is a modification of work by Jackson and Grossmann (2002).

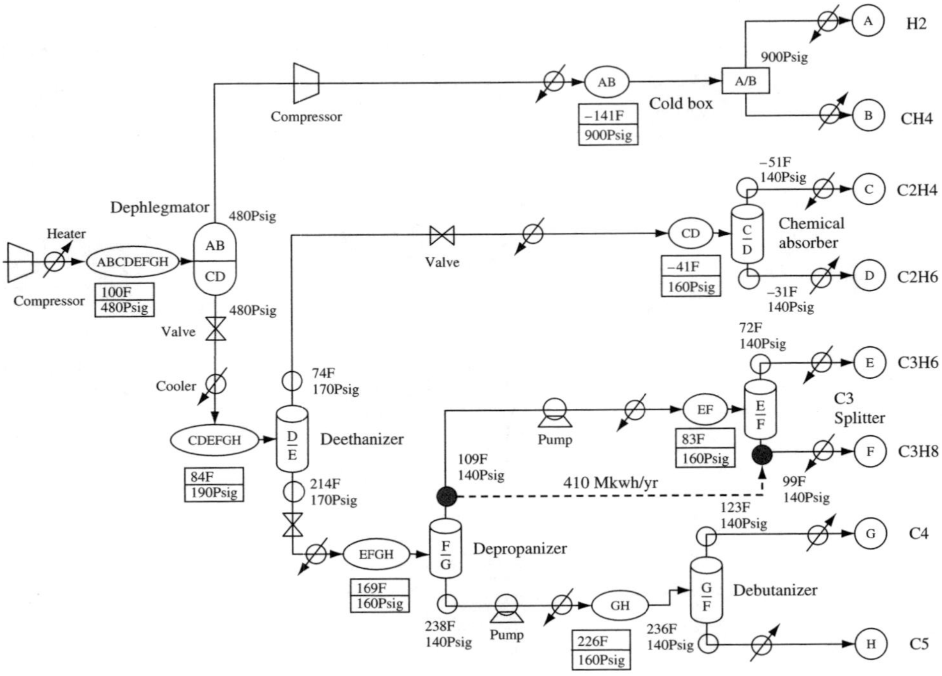

Figure 11.2 *Optimal structure of ethylene plant*

For a 10-process instance (Figure 11.3) that involves the production of products (G,H,I,J,K,L,M) from raw materials (A,B,C,D,E), the 4-period MILP model was formulated with the big-M and convex hull reformulations. The former involved 320 0–1 variables, 377 continuous variables, and 1957 constraints; the latter involved 320 0–1 variables, 1097 continuous variables, and 2505 constraints. The big-M model was solved in 1913 s and 1 607 486 nodes, while the latter required only 5.8 s and 2155 nodes. This reduction was achieved because the convex hull formulation had a gap of only 7.6% versus the 60.3% gap of the big-M model. It should be noted that with 120 cuts the gap in the big-M model reduced to only 7.9%, with which the MILP was solved in a total of 68 s, of which 22 were for the cut generation.

11.7.3 Wastewater Treatment Network

This example corresponds to a synthesis problem of a distributed wastewater multicomponent network, which is taken from Galan and Grossmann (1998). Given a set of process liquid streams with known composition, a set of technologies for the removal of pollutants, and a set of mixers and splitters, the objective is to find the interconnections of the technologies and their flowrates to meet the specified discharge composition of pollutant at minimum total cost. Discrete choices involve deciding what equipment to use for each treatment unit. Figure 11.4 shows the superstructure of a specific example with three contaminants and three choices of separation technologies per contaminant. Lee and Grossmann (2001) formulated the problem as a GDP model that involves 9 Boolean

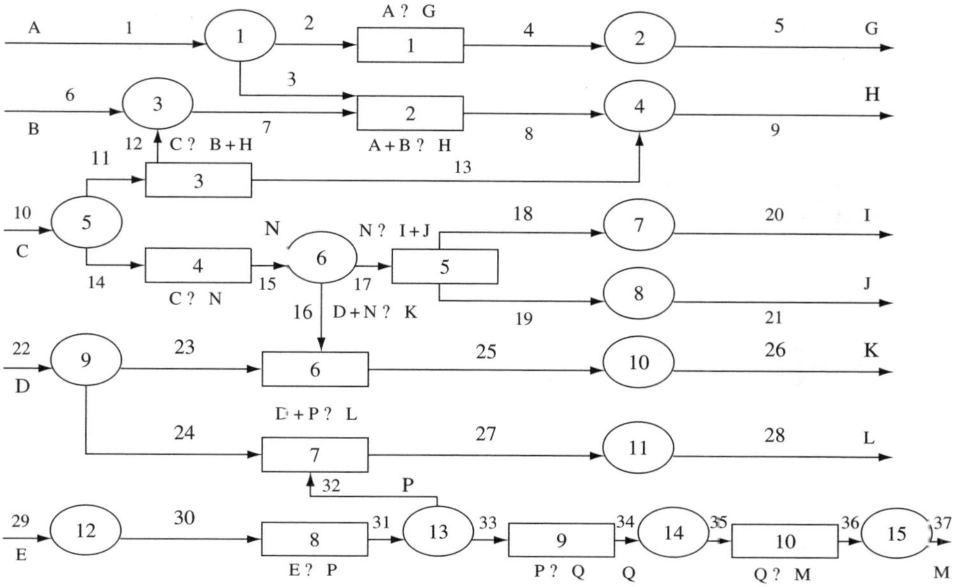

Figure 11.3 *Process network for retrofit planning*

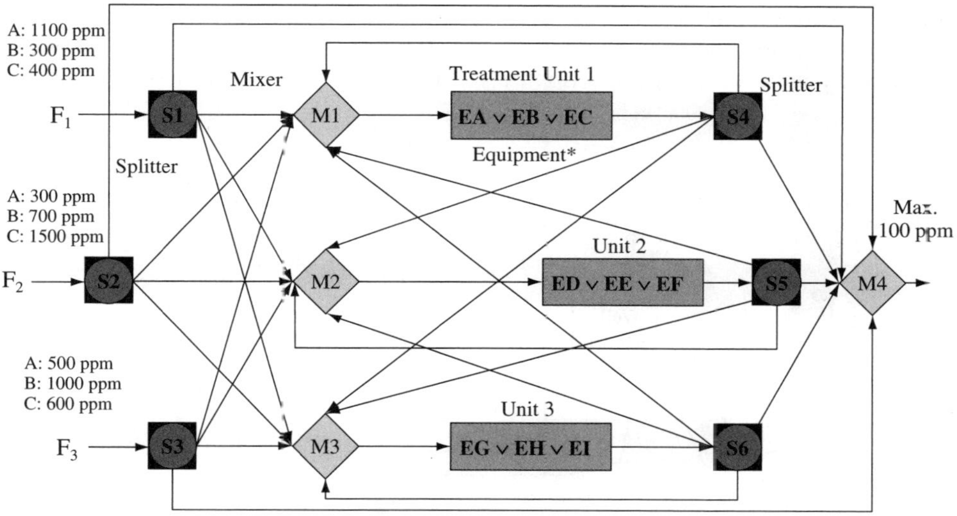

Figure 11.4 *Superstructure water treatment plant*

variables, 237 continuous variables, and 281 constraints. The two-level BB method by Lee and Grossmann (2003) required about 5 min of CPU time, while the method by Bergamini *et al.* (2004) required less than 2 min. The optimal solution with a cost of 1 692 583 \$/year is shown in Figure 11.5.

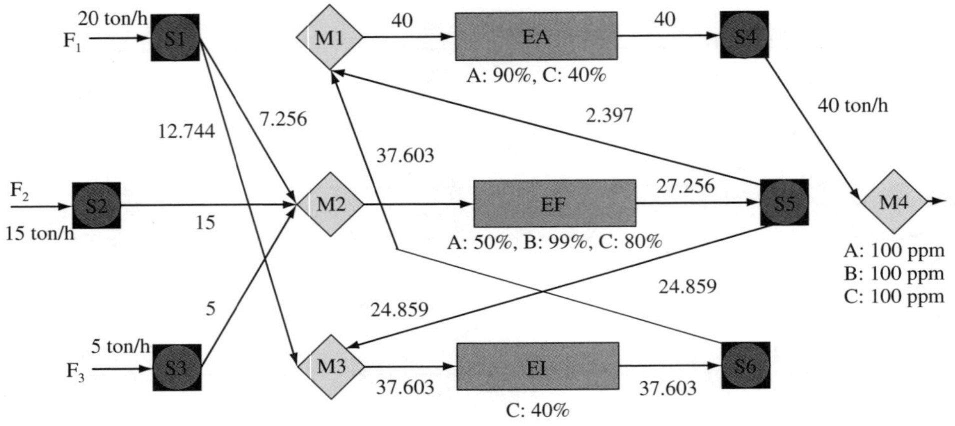

Figure 11.5 *Optimal wastewater treatment plant*

11.8 Examples in Supply Chain Management

11.8.1 Hydrocarbon Field Infrastructure Planning

In this example we consider the design, planning, and scheduling of an offshore oilfield infrastructure over a planning horizon of 6 years divided into 24 quarterly periods where decisions need to be made (Van den Heever and Grossmann, 2000). The infrastructure under consideration consists of 1 production platform (PP), 2 well platforms (WP), and 25 wells and connecting pipelines (Figure 11.6). Each oilfield (F) consists of a number

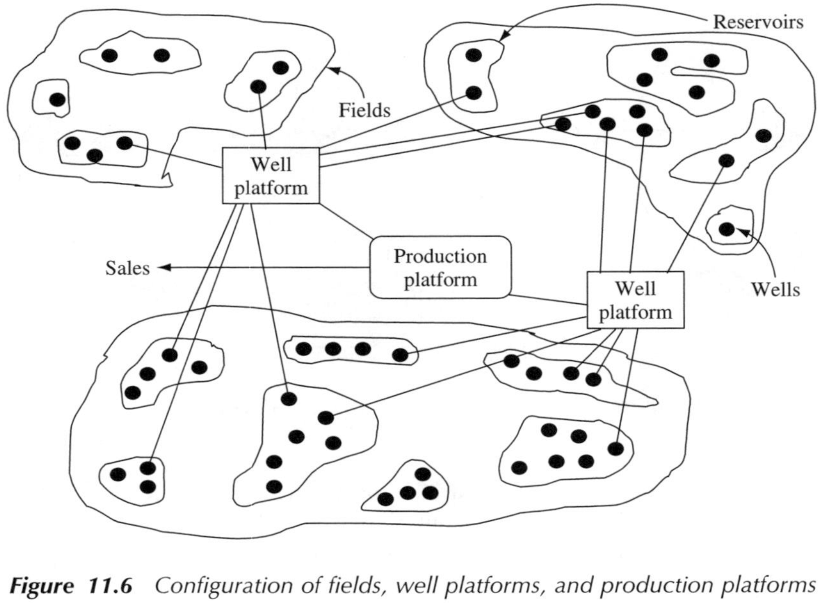

Figure 11.6 *Configuration of fields, well platforms, and production platforms*

of reservoirs (R), while each reservoir in turn contains a number of potential locations for wells (W) to be drilled. Design decisions involve the capacities of the PPs and WPs, as well as decisions regarding which WPs to install over the whole operating horizon. Planning decisions involve the production profiles in each period, as well as decisions regarding when to install PPs and WPs included in the design, while scheduling decisions involve the selection and timing of drilling of the wells. This leads to an MINLP model with 9744 constraints, 5953 continuous variables, and 700 0–1 variables. An attempt to solve this model with a commercial package such as GAMS (Brooke *et al.*, 1998) (using DICOPT (Viswanathan and Grossmann, 1990)) with CPLEX 6.6 (ILOG, 2001) for the MILPs and CONOPT2 (Drud, 1992) for the NLPs on an HP 9000/C110 workstation results in a solution time of 19 386 CPU seconds.

To overcome this long solution time, Van den Heever and Grossmann (2000) developed an iterative aggregation/disaggregation algorithm which solved the model in 1423 CPU seconds. This algorithm combines the concepts of bilevel decomposition, time aggregation, and logic-based methods. The application of this method led to an order of magnitude reduction in solution time, and produced an optimal net present value of \$68 million. Figure 11.7 shows the total oil production over the 6-year horizon, while Table 11.1 shows the optimal investment plan obtained. Note that only 9 of the 25 wells were chosen in the end. This solution resulted in savings in the order of millions of dollars compared to the heuristic method used in the oilfield industry that specifies almost all the wells being drilled.

11.8.2 Supply Chain Problem

The example problem in Figure 11.8, a multisite continuous production facility for multiple products, was solved by Bok *et al.* (2000) over a 7-day horizon to illustrate the performance of their model in three cases: (1) no intermittent deliveries of raw materials without product changeovers; (2) intermittent deliveries without changeovers; and (3) intermittent deliveries with changeovers. It is obvious that case 3 is the most rigorous and detailed. Cases 1 and 2 can be obtained by relaxing the discrete nature of case 3. In the case of intermittent deliveries, the minimum time interval between successive

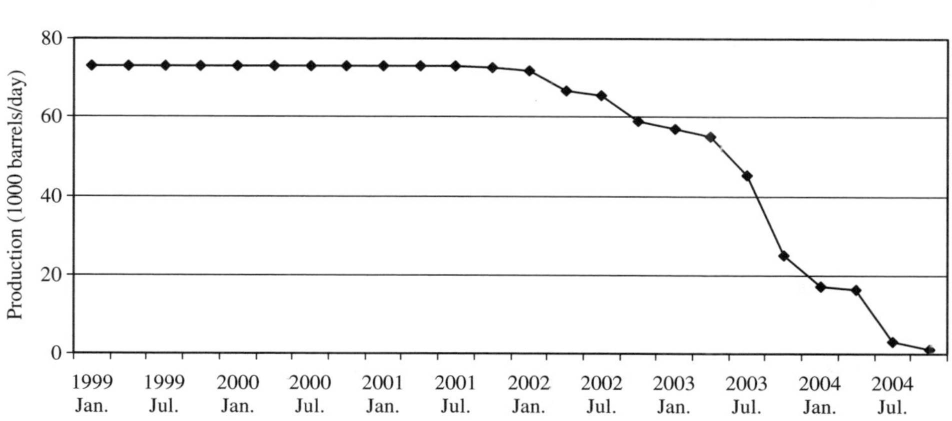

Figure 11.7 *Production profile over 6-year horizon*

Table 11.1 *The optimal investment plan*

Item		Period invested
PP		Jan 1999
WP1		Jan 1999
Reservoir	*Well*	
2	4	Jan 1999
3	1	Jan 1999
5	3	Jan 1999
4	2	Apr 1999
7	1	Jul 1999
6	2	Oct 1999
1	2	Jan 2000
9	2	Jan 2000
10	1	Jan 2000

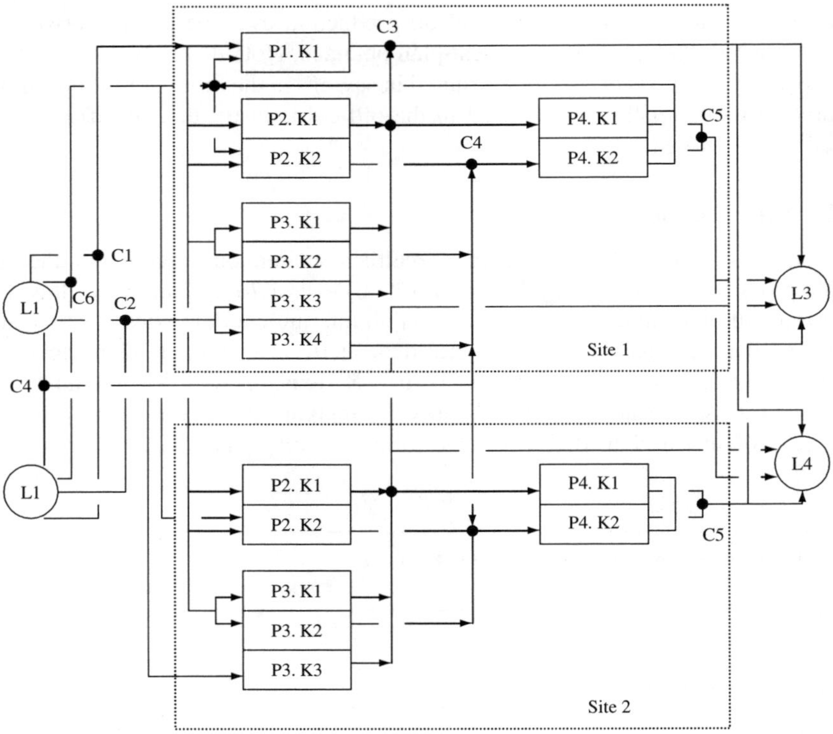

Figure 11.8 *Multisite production facility for supply chain problem*

deliveries is assumed to be 2 days regardless of the chemicals or the sites. The problem was modeled using the GAMS modeling language and solved in the full space using the CPLEX solver on an HP 9000/7000. The optimization results for this example are as follows. In case 1, 2034 variables and 1665 constraints are required and no 0–1 variable is needed because there are no intermittent deliveries nor any changeovers. The problem was solved in only 2 s CPU time. The more rigorous the model (cases 2 or 3), the larger the number of 0–1 variables required. This in turn results in more computation time for the optimization. Case 2 for changeovers required 224 0–1 variables, 1810 continuous variables, and 1665 constraints solving in 4 min CPU. Case 3, which includes changeovers and intermittent supplies, involved 392 0–1 variables, 1642 continuous variables, and 1665 constraints solving in 8 min CPU. Figure 11.9 shows the optimization results for case 3 that considers the intermittent deliveries and changeovers. Bok *et al.* (2000) proposed a decomposition algorithm in order to be able to solve larger problems.

11.8.3 Scheduling of Batch Plants

Consider the batch scheduling problem that is given through the STN shown in Figure 11.10, which is an extension of the work by Papageorgiou and Pantelides (1996).

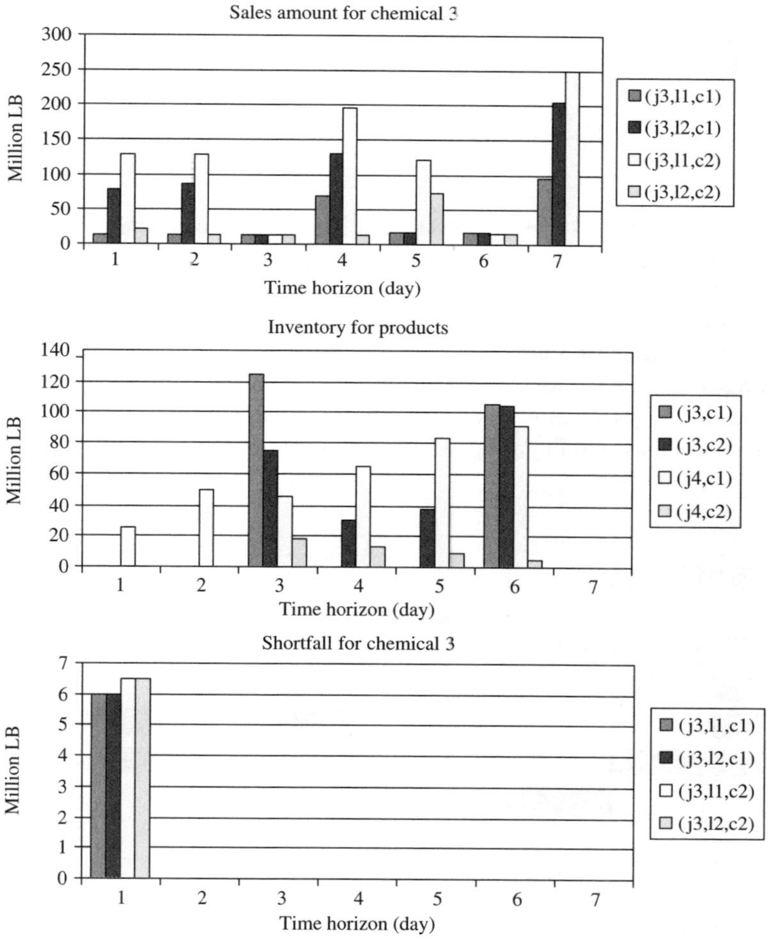

Figure 11.9 *Results for sales, inventory, and shortfalls in supply chain problem*

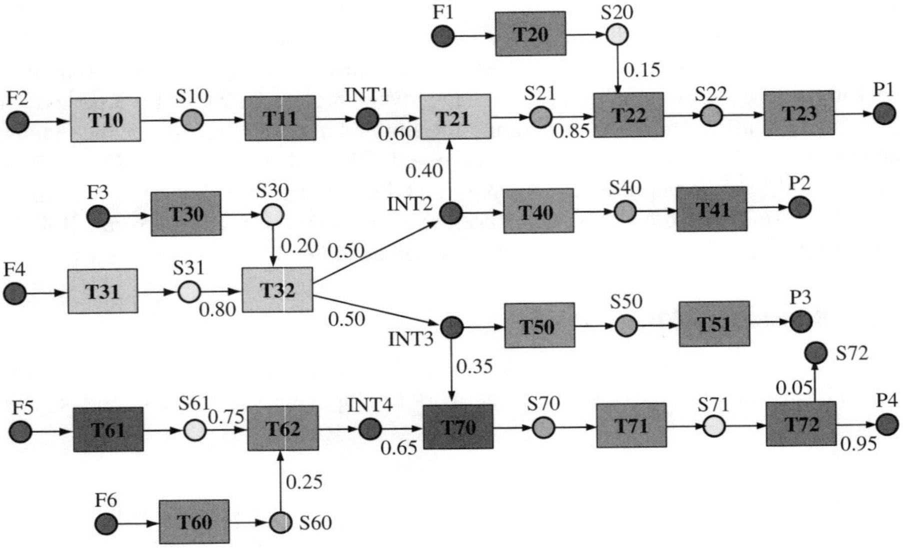

Figure 11.10 State-task-network example

The STN consists of 27 states, 19 tasks, and has 8 equipment units available for the processing. The objective is to find a schedule that produces 5 tons each for products P1, P2, P3, and P4. The problem was originally modeled with the continuous-time MILP by Maravelias and Grossmann (2003) involving around 400 0–1 variables, 4000 continuous variables, and 6000 constraints. Not even a feasible solution to this problem could be found with CPLEX 7.5 after 10 h. In contrast, the hybrid MILP/CP model required only 2 s and five major iterations between the MILP and CP subproblems! Note that the optimal schedule shown in Figure 11.11 is guaranteed to be the global optimum solution.

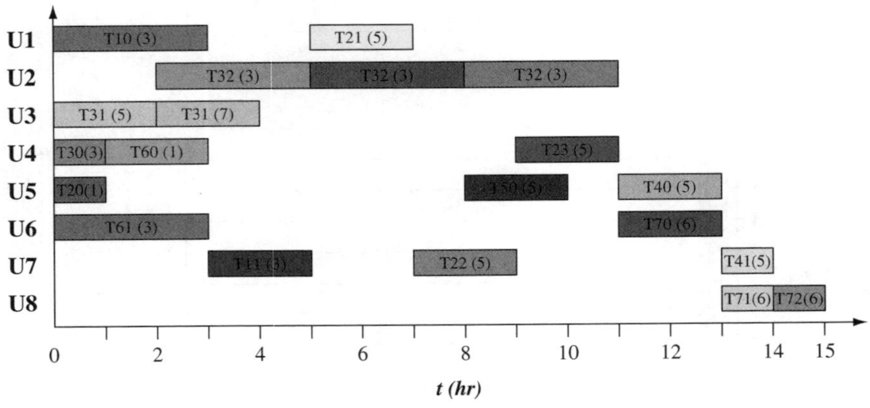

Figure 11.11 Optimal schedule

11.9 Conclusions

Mixed-integer optimization techniques (MILP, MINLP) have proved to be essential in modeling process integration problems (process synthesis and design) as well as supply chain management problems (planning and scheduling). The former tends to be largely nonlinear, while the latter tends to be linear. Major barriers that have been encountered with these techniques are modeling, scaling, and nonconvexities. It is the first two issues that have motivated logic-based optimization as a way of facilitating the modeling of discrete/continuous problems, and of reducing the combinatorial search space. The GDP formulation has shown to be effective in terms of providing a qualitative/quantitative framework for modeling, and an approach that yields tighter relaxations through the convex hull formulation. It was also shown that global optimization algorithms can be developed for GDP models and solved in reasonable time for modest-sized problem. Finally, the recent emergence of CP offers an alternative approach for handling logic in discrete scheduling problems. Here the development of hybrid methods for scheduling seems to be particularly promising for achieving order of magnitude reductions in the computations. The power and scope of the techniques were demonstrated on a variety of process integration and supply chain management problems.

Acknowledgments

The author gratefully acknowledges financial support from the National Science Foundation under Grant ACI-0121497 and from the industrial members of the Center for Advanced Process Decision-Making (CAPD) at Carnegie Mellon University.

References

Adjiman C.S., Androulakis I.P., Maranas C.D. and Floudas C.A. 1996. A global optimization method, αBB, for process design, Comput. Chem. Eng., 20(Suppl.), S419–S424.

Adjiman C.S., Androulakis I.P. and Floudas C.A. 1997. Global optimization of MINLP problems in process synthesis and design, Comput. Chem. Eng., 21(Suppl.), S445–S450.

Adjiman C.S., Androulakis I.P. and Floudas C.A. 2000. Global optimization of mixed-integer nonlinear problems, AIChE J., 46(9), 1769–1797.

Balas E. 1965. An additive algorithm for solving linear programs with zero-one variables, Oper. Res., 13, 517–546.

Balas E. 1979. Disjunctive programming, Ann. Discrete Math., 5, 3–51.

Balas E. 1985. Disjunctive programming and a hierarchy of relaxations for discrete optimization problems, SIAM J. Alg. Discrete Methods, 6, 466–486.

Balas E., Ceria S. and Cornuejols G. 1993. A lift-and-project cutting plane algorithm for mixed 0-1 programs, Math. Program., 58, 295–324.

Baptiste P. and Le Pape C. 1996. Disjunctive constraints for manufacturing scheduling: Principles and extensions, Int. J. Comput. Integr. Manufacturing, 9(4), 306–341.

Beaumont N. 1991. An algorithm for disjunctive programs, Eur. J. Oper. Res., 48, 362–371.

Benders J.F. 1962. Partitioning procedures for solving mixed variables programming problems, Numerische Mathematik, 4, 238–252.

Bergamini M.L., Aguirre P. and Grossmann I.E. 2004. Logic Based Outer Approximation for Global Optimization of Synthesis of Process Networks, submitted for publication.

Bisschop J. and Entriken R. 1993. AIMMS: The Modeling System. Paragon Decision Technology.

Bockmayr A. and Pisaruk N. 2003. Detecting infeasibility and generating cuts for MIP using CP. In, Proceedings of CP-AI-OR 2003, 24–34.

Bok J.-K., Grossmann I.E. and Park S. 2000. Supply chain optimization in continuous flexible process networks, I&EC Res., 39, 1279–1290.

Borchers B. and Mitchell J.E. 1994. An improved branch and bound algorithm for mixed integer nonlinear programming, Comput. Oper. Res., 21, 359–367.

Brooke A., Kendrick D., Meeraus A. and Raman R. 1998. GAMS Language Guide. GAMS Development Corporation, Washington, DC.

Caseau Y. and Laburthe F. 1994. Improved CLP scheduling with task intervals. In, Proceedings of 11th International Conference on Logic Programming.

Ceria S. and Soares J. 1999. Convex programming for disjunctive optimization, Math. Program., 86(3), 595–614.

Crowder H.P., Johnson E.L. and Padberg M.W. 1983. Solving large-scale zero-one linear programming problems, Oper. Res., 31, 803–834.

Dakin R.J. 1965. A tree search algorithm for mixed integer programming problems, Comput. J., 8, 250–255.

Dincbas M., Van Hentenryck P., Simonis H., Aggoun A., Graf T. and Berthier F. 1988. The constraint logic programming language CHIP. In, FGCS-88: Proceedings of International Conference on Fifth Generation Computer Systems, Tokyo, pp. 693–702.

Duran M.A. and Grossmann I.E. 1986. An outer-approximation algorithm for a class of mixed-integer nonlinear programs, Math. Program., 36, 307–339.

Drud A.S. 1992. CONOPT – A Large-Scale GRG Code, ORSA J. Comput., 6, 207–216.

Fletcher R. and Leyffer S. 1994. Solving mixed nonlinear programs by outer approximation, Math. Program., 66(3), 327–349.

Fourer R., Gay D.M. and Kernighan B.W. 1992. AMPL: A Modeling Language for Mathematical Programming. Duxbury Press, Belmont, CA.

Galan B. and Grossmann I.E. 1998. Optimal design of distributed wastewater treatment networks, Ind. Eng. Chem. Res., 37, 4036–4048.

Geoffrion A.M. 1972. Generalized Benders decomposition, J. Optimization Theory Appl., 10(4), 237–260.

Gomory R.E. 1958. Outline of an algorithm for integer solutions to linear programs, Bull. Am. Math. Soc., 64, 275–278.

Grossmann I.E. 2002. Review of nonlinear mixed-integer and disjunctive programming techniques for process systems engineering, J. Optimization Eng., 3, 227–252.

Grossmann I.E. and Lee S. 2003. Generalized disjunctive programming: nonlinear convex hull relaxation and algorithms, Comput. Optimization Appl., 26, 83–100.

Grossmann I.E., Caballero J.P. and Yeomans H. 1999. Mathematical programming approaches to the synthesis of chemical process systems, Korean J. Chem. Eng., 16(4), 407–426.

Grossmann I.E., van den Heever S.A. and Harjunkoski I. 2002. Discrete optimization methods and their role in the integration of planning and scheduling, AIChE Symposium Series No. 326, Vol. 98, pp. 150–168.

Gupta O.K. and Ravindran V. 1985. Branch and bound experiments in convex nonlinear integer programming, Manag. Sci., 31(12), 1533–1546.

Heipcke S. 1999. An example of integrating constraint programming and mathematical programming, Electron. Notes Discrete Math., 1(1).

Hentenryck P.V. 1989. Constraint Satisfaction in Logic Programming. MIT Press, Cambridge, MA.

Hentenryck P.V. 2002. Constraint and integer programming in OPL, INFORMS J. Comput., 14(4), 345–372.

Hiriart-Urruty J. and Lemaréchal C. 1993. Convex Analysis and Minimization Algorithms. Springer-Verlag, Berlin.

Hooker J. 2000. Logic Based Methods for Optimization: Combining Optimization and Constraint Satisfaction. John Wiley and Sons Inc., New York.

Hooker J.N. 2002. Logic, optimization, and constraint programming, INFORMS J. Comput., 4(4), 295–321.

Hooker J.N. and Osorio M.A. 1999. Mixed logical/linear programming, Discrete Appl. Math., 96–97(1–3), 395–442.

Horst R. and Tuy H. 1996. Global Optimization: Deterministic Approaches, 3rd edition. Springer-Verlag, Berlin.

ILOG. 1999. ILOG OPL Studio 2.1., User's Manual. ILOG.

Jackson J.R. and Grossmann I.E. 2002. High level optimization model for the retrofit planning of process networks, I&EC Res., 41, 3762–3770.

Jain V. and Grossmann I.E. 2001. Algorithms for hybrid MILP/CP model for a class of optimization problems, INFORMS J. Comput., 13, 258–276.

Johnson E.L., Nemhauser G.L. and Savelsbergh M.W.P. 2000. Progress in linear programming based branch-and-bound algorithms: an exposition, INFORMS J. Comput., 12, 2–23.

Kallrath J. 2000. Mixed integer optimization in the chemical process industry: experience, potential and future, Trans. I. Chem. E, 78(Part A), 809–822.

Kesavan P. and Barton P.I. 2000a. Decomposition algorithms for nonconvex mixed-integer nonlinear programs, Am. Inst. Chem. Eng. Symp. Ser., 96(323), 458–461.

Kesavan P. and Barton P.I. 2000b. Generalized branch-and-cut framework for mixed-integer nonlinear optimization problems, Comput. Chem. Eng., 24, 1361–1366.

Kelley J.E. Jr. 1960. The cutting-plane method for solving convex programs, J. SIAM, 8, 703.

Kocis G.R. and Grossmann I.E. 1987. Relaxation strategy for the structural optimization of process flowsheets, Ind. Eng. Chem. Res., 26, 1869.

Land A.H. and Doig A.G. 1960. An automatic method for solving discrete programming problems, Econometrica, 28, 497–520.

Lee S. and Grossmann I.E. 2000. New algorithms for nonlinear generalized disjunctive programming, Comput. Chem. Eng., 24, 2125–2141.

Lee S. and Grossmann I.E. 2001. A global optimization algorithm for nonconvex generalized disjunctive programming and applications to process systems, Comput. Chem. Eng., 25, 1675–1697.

Lee S. and Grossmann I.E. 2003. Global optimization of nonlinear generalized disjunctive programming with bilinear equality constraints: Applications to process networks, Comput. Chem. Eng., 27, 1557–1575.

Lee S., Logsdon J.S., Foral M.J. and Grossmann I.E. 2003. Superstructure optimization of the olefin separation process. In, Proceedings ESCAPE-13, Kraslawski A. and Turunen I. (Eds.), pp. 191–196.

Le Pape C. 1998. Implementation of resource constraints in ILOG SCHEDULE: A library for the development of constrained-based scheduling systems, Intell. Syst. Eng., 3(2), 55–66.

Leyffer S. 2001. Integrating SQP and branch-and-bound for mixed integer nonlinear programming, Comput. Optimization Appl., 18, 295–309.

Lovász L. and Schrijver A. 1991. Cones of matrices and set-functions and 0-1 optimization, SIAM J. Optimization, 12, 166–190.

Maravelias C.T. and Grossmann I.E. 2003. A new general continuous-time state task network formulation for short term, scheduling of multipurpose batch plants, I&EC Res., 42, 3056–3074.

Maravelias C.T. and Grossmann I.E. 2004. A hybrid MILP/CP decomposition approach for the continuous time scheduling of multipurpose batch plants, Comput. Chem. Eng. (in press).

McCormick G.P. 1976. Computability of global solutions to factorable nonconvex programs: part I – convex underestimating problems, Math. Program., 10, 147–175.

Nemhauser G.L. and Wolsey L.A. 1988. Integer and Combinatorial Optimization. Wiley-Interscience, New York.

Nuijten W.P.M. 1994. Time and resource constrained scheduling: A constraint satisfaction approach. Ph.D. Thesis, Eindhoven University of Technology.

ILOG 2001. ILOG OPL Studio 3.5: The optimization Language, ILOG Inc.

Papageorgiou L.G. and Pantelides C.C. 1996. Optimal campaign planning/scheduling of multipurpose batch/semicontinuous plants. 2. Mathematical decomposition approach, Ind. Eng. Chem. Res., 35, 510–529.

Quesada I. and Grossmann I.E. 1992. An LP/NLP based branch and bound algorithm for convex MINLP optimization problems, Comput. Chem. Engng., 16(10/11), 937–947.

Quesada I. and Grossmann I.E. 1995. A global optimization algorithm for linear fractional and bilinear programs, J. Global Optimization, 6(1), 39–76.

Raman R. and Grossmann I.E. 1991. Relation between MILP modelling and logical inference for chemical process synthesis, Comput. Chem. Eng., 15(2), 73–84.

Raman R. and Grossmann I.E. 1994. Modelling and computational techniques for logic based integer programming, Comput. Chem. Eng., 18(7), 563–578.

Rodosek R., Wallace M.G. and Hajian M.T. 1999. A new approach to integrating mixed integer programming and constraint logic programming, Ann. Oper. Res., 86, 63–87.

Ryoo H.S. and Sahinidis N.V. 1995. Global optimization of nonconvex NLPs and MINLPs with applications in process design, Comput. Chem. Eng., 19(5), 551–566.

Sawaya N.W. and Grossmann I.E. 2004. A cutting plane method for solving linear-generalized disjunctive programming problems (submitted).

Sherali H.D. and Adams W.P. 1990. A hierarchy of relaxations between the continuous and convex hull representations for zero-one programming problems, SIAM J. Discrete Math., 3(3), 411–430.

Smith E.M.B. and Pantelides C.C. 1997. Global optimization of nonconvex NLPs and MINLPs with applications in process design, Comput. Chem. Eng., 21(1001), S791–S796.

Smith S.F. and Cheng C-C. 1993. Slack-based heuristics for constrained satisfaction scheduling. Proceedings of 11th National Conference on Artificial Intelligence.

Stubbs R. and Mehrotra S. 1999. A branch-and-cut method for 0-1 mixed convex programming, Mathematical Programming, 86(3), 515–532.

Tawarmalani M. and Sahinidis N.V. 2002. Convexification and global optimization in continuous and mixed-integer nonlinear programming: Theory, algorithms, software, and applications. In, Nonconvex Optimization and Its Applications Series, Vol. 65. Kluwer Academic Publishers, Dordrecht.

Türkay M. and Grossmann I.E. 1996. Logic-based MINLP algorithms for the optimal synthesis of process networks, Comput. Chem. Eng., 20(8), 959–978.

Van den Heever S.A. and Grossmann I.E. 2000. An iterative aggregation/disaggregation approach for the solution of a mixed integer nonlinear oilfield infrastructure planning model, I&EC Res., 39, 1955–1971.

Van Roy T.J. and Wolsey L.A. 1986. Valid inequalities for mixed 0-1 programs, Discrete Appl. Math., 14, 199–213.

Vecchietti A. and Grossmann I.E. 1999. LOGMIP: a disjunctive 0-1 nonlinear optimizer for process systems models, Comput. Chem. Eng., 23, 555–565.

Viswanathan J. and Grossmann I.E. 1990. A combined penalty function and outer-approximation method for MINLP optimization, Comput. Chem. Eng., 14, 769.

Wallace M., Novello S. and Schimpf J. 1997. ECLiPSe: a platform for constraint logic programming, ICL Syst. J., 12(1), 159–200.

Westerlund T. and Pettersson F. 1995. An extended cutting plane method for solving convex MINLP problems, Comput. Chem. Eng., 19(Suppl.), S131–S136.

Williams H.P. 1985. Mathematical Building in Mathematical Programming. John Wiley, Chichester.

Yuan X., Zhang S., Piboleau L. and Domenech S. 1988. Une Methode d'optimization Nonlineare en Variables Mixtes pour la Conception de Porcedes, Rairo Recherche Operationnele, 22, 331.

Zamora J.M. and Grossmann I.E. 1999. A branch and bound algorithm for problems with concave univariate, bilinear and linear fractional terms, J. Global Optim., 14(3), 217–249.

12

Integration of Process Systems Engineering and Business Decision Making Tools: Financial Risk Management and Other Emerging Procedures

Miguel J. Bagajewicz

12.1 Introduction

Economics has always been part of engineering, so talking about its integration in our discipline seems rather odd. Moreover, many companies, especially those dealing with risky projects, employ advanced financial tools in their decision making. For example, the oil industry is relatively more sophisticated than other industries in its supply chain management tools and the associated finances. However, these are not widespread tools that all engineers employ and certainly not tools that are used in education or in academic papers. Indeed, only certain aspects of the tools that economists and financiers use, namely a few profitability measures, are fully integrated into our education and engineering academic circles. This has started to change in recent years, but many tools are still completely out of sight for mainstream chemical engineers.

This is not a review article, so not all the work that has been published on the matter will be cited or discussed. Rather, the intention is to discuss some of the more relevant and pressing issues and provide some direction for future work. It is also an article that targets engineers as the audience.

Chemical Engineering Edited by Miguel A. Galán and Eva Martin del Valle

As a motivating example, one could start with the following statement of a typical process design problem.

> 'Design a plant to produce chemical X, with capacity Y'

For a long time, this is the problem that many capstone design classes used to propose to students to solve (and some still do). This is fairly well known. The answer is a flowsheet, optimized to follow certain economics criteria, with a given cash flow profile (costs and prices are given and are often considered fixed throughout time), from which a net present value and a rate of return is obtained. In the 1980s, environmental considerations started to be added, but these were mostly used as constraints that in the end usually increased costs. It is only recently that engineers started to talk about green engineering and sustainability, but in most cases, these are still considered as constraints of the above design problem, not as valid objectives.

Reality is, however, more complex than the assumptions used for the above problem: raw materials change quality and availability, demands may be lower than expected, and products may require different specifications through time, all of which should be accomplished with one plant. So in the 1980s engineers proposed to solve the following flexible design problem.

> 'Design a flexible plant to produce chemical X, with capacity Y, capable of working in the given ranges of raw materials availability and quality and product specifications'

While the problem was a challenge to the community, it hardly incorporated any new economic considerations. The next step was to include uncertainty. Thus the revised version was:

> 'Design a plant to produce chemical X, taking into account uncertain raw materials and product prices, process parameters, raw material availability and product demand, given the forecasts, and determine when the plant should be built as well as what expansions are needed'
>
> Substitute 'plant' with 'network of processes' or 'product' and you have supply chain problems or product engineering.

This has been the typical problem of the 1990s. However, very few industries have embraced the tools and the procedures and only a few schools teach it at the undergraduate level. As noted, this was later extended to networks of plants and supply chains, a subject that is still somewhat foreign in undergraduate chemical engineering education. Notice first that the fixed capacity requirement and the flexibility ranges are no longer included. The engineer is expected to determine the right capacity and the level of flexibility that is appropriate for the design. In doing so, this maximizes expectation of profit. *The profit measures (net present value or rate of return), however, did not change and are the same as the one engineers have been using for years.* The novelty is the planning aspect and the incorporation of uncertainty.

In addition, one has to realize that not all design projects are alike. Some are constrained by spending and some not, some are performed to comply with regulations and do not necessarily target profit, but rather cost. Thus, the type of economic analysis and the associated tools change.

Where do the engineers go from here? This chapter addresses some answers to this question. Many of the most obvious pending questions/issues, some of which intersect with others being already explored, are as follows:

- What is the financial risk involved in a project?
- What is the project impact on of the financial status of the company that is considering this project, namely indicators such as,
  - liquidity ratios (assets to liabilities)?
  - cash position, debts, etc?
  - short-, medium-, and long-term shareholder value, or in the case of a private company the dividends, among others?
- How does the size of the company in relation to the capital involved in the project shape the decision maker's attitudes? It is not the same type of decision making one makes when belonging to a big corporation rather than to a medium-size private company, even when the financial indicators are similar. In other words, the question at large is how the project impacts on company market value.
- How the decision making related to the project can reflect the strategic plans of the company and, most importantly, vice versa, that is, *how to take into account the strategic plan of the company at the level of project or investment decision making*?
- Can 'here and now' decisions and design parameters be managed in relation to targets or aspiration levels for the different indicators listed above?
- Can short- and long-term contracts and options be factored in at the time of the decision making, not afterwards as control actions, to increase profitability?
- When should projects be based on taking equity and be undertaken with no increment in profit because they are instrumental for other projects?
- Can one plan to alter the exogenous parameters, like prices, demands, to affect the expected profit and/or the aforementioned indicators?
- Should one consider advertisement as part of the decision making, or product presentation (form, color, etc.), that is, the psychology of the user?
- Should sociology/psychology/advertisement/etc. be incorporated into the decision making by modeling the different decisions vis-à-vis the possible response of the market? In other words, should one start considering the market demand as susceptible to being shaped, rather than using it as simple forecasted data?

The answer to the above list of questions (which is by no means exhaustive) is slowly and strongly emerging. The latest Eighth International Symposium on Process Systems Engineering held in China (PSE 2003) had many of the above issues as the central theme, but there is substantial earlier pioneering work. For example, in an article mostly devoted to prepare us for the information technology (IT) age (now in full development), and its impact on corporate management, Robertson *et al.* (1995) examined the lack of proper communication in the corporate flow loop (Figure 12.1). They argue that the four major components of this loop (manufacturing, procuring, managing, and marketing) operate almost as separate entities with minimal data sharing.

Notwithstanding the lack of data sharing, which will (or is being) corrected, the real issue is that the different elements of the loop also start to share the same goals and methods, as Bunch and Iles (1998) argued. In many of the examples illustrated later in this chapter, decisions at the level of manufacturing, like for example the scheduling of operations, are influenced by the company's cash position and are related to pricing, etc. This involves marketing, procuring, and manufacturing in the solution of the problem. Corporate management, procuring, and marketing should also work together to solve investment problems, etc. This is the nature of the challenge and the core of

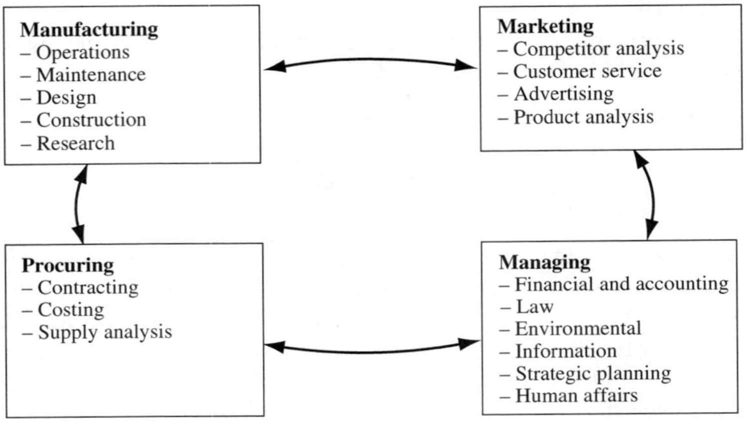

Figure 12.1 *Corporate information loop (following Robertson* et al.*, 1995)*

our analysis: chemical engineering methods and procedures, which were mostly related to manufacturing, are now increasingly involving/including the other components of the corporate loop in integrated models.

12.2 Project Evaluation as Chemical Engineers Know It

Engineers have all been likely to be educated with some exposure to the classic book on process design authored by Peters and Timmerhaus, which was recently updated (Peters *et al.*, 2003). Even in this last update, the part dealing with economics contains mostly the same chapter on profitability as earlier editions with small changes. Other available textbooks do not depart from this recipe. The recommended measures of profitability are:

- Internal rate of return
- Pay out time
- Net present value (NPV)
- Discounted cash flow rate of return.

For the most part, these methods consider that the plant is build at some known point, the time at which the whole capital investment is used, and that profits are somehow predictable throughout the time horizon. The methods respond to a project evaluation paradigm that was crafted years ago in the era when computers were not powerful enough and/or even available, and when uncertainty in modeling was manageable only for small problems.

Extensions of these measures to uncertain future conditions have been made, especially in the form of expected net present value. Another problem with all measures is the uncertainty of how long the plant will be in operation, at what point preventive maintenance will be intensified, or when some revamps will take place. In the old days, all these difficulties were ignored because of the inability or, actually, the lack of knowledge of how to handle uncertainty beyond a simple and reduced set of scenarios. In other words, the model was simplified for two reasons: an engineer should be able to do calculations and uncertainty was too complex to handle. The excuse is not valid anymore.

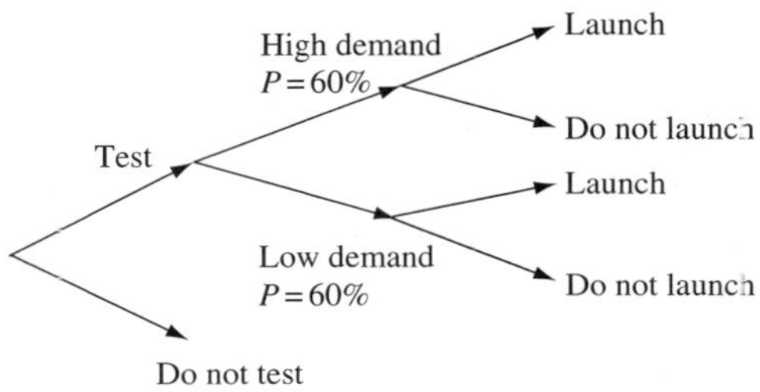

Figure 12.2 *Example of a decision tree*

Despite the aforementioned general tendency, uncertainty in project evaluations has been handled for years in various forms. Various branches of engineering still use decision models, trees, and payoff tables (McCray, 1975; Riggs, 1968; Gregory, 1988; Schuyler, 2001). Decision trees are good tools as long as decisions are discrete (e.g. to build a plant or not, to delay construction or not, etc.). A typical tree for investment decision making is illustrated next. Consider a company trying to decide if it wants to invest 5 million to test a product in the market, and if a test is positive, invest 50 million, or skip the test. A decision tree for this case is shown in Figure 12.2. In this decision tree, two types of nodes typically exist, those associated with decisions (test or not test) and the outcomes or external conditions (high/low demand), to which probabilities are associated.

Thus, to build a decision tree, one needs to *explicitly* enumerate all possible scenarios and the responses (decisions) to such scenarios. However, 'for some problems, . . . , a combinatorial explosion of branches makes calculations cumbersome or impractical' (Schuyler, 2001). One way that this problem is ameliorated (but not solved) is by introducing Monte Carlo simulations at each node of the decision tree. However, this does not address the problem of having to build the tree in the first place. In addition, trees are appropriate for the case where discrete decisions are made. Continuous decisions like for example the size of the investment, or more specifically, the size of a production plant, cannot be easily fit into decision trees without discretizing.

A separate paragraph needs to be devoted to dynamic programming (Bellman, 1957; Denardo, 1982). This technique is devoted to solving sequential decision making processes. It has been applied to resource allocation, inventory management, routing in networks, production control, etc. In many aspects this technique is equivalent to two-/multi-stage stochastic programming, with the added benefits that under certain conditions, some properties of the solutions (optimality conditions) are known and are helpful for the solution procedure. In fact, under certain conditions, one can obtain the solution recursively, moving backwards from the last node to the first. The technique can be applied to problems under uncertainty. Recently, there has been a revival of the usage of this technique in chemical engineering literature due fundamentally to the recent work of Professor Westerberg (Cheng *et al.*, 2003, 2004).

By the late 1980s the engineering community had started to introduce two-stage stochastic programming (Birge and Louveaux, 1997) in problems like planning, scheduling, etc. (Liu and Sahinidis, 1996; Iyer and Grossmann, 1998a; Gupta and Maranas, 2000, and many others). Two-stage stochastic programming is briefly outlined next using linear functions for simplicity. The dynamic programming approach is outlined briefly later.

12.2.1 Two-Stage Stochastic Programming

Two features characterize these problems: the uncertainty in the problem data and the sequence of decisions. Several model parameters, especially those related to future events, are considered random variables with a certain probability distribution. In turn, some decisions are taken at the planning stage, that is, before the uncertainty is revealed, while a number of other decisions can only be made after the uncertain data become known. The first decisions are called *first-stage decisions* and the decisions made after the uncertainty is unveiled are called *second-stage* or *recourse decisions*, and the corresponding period is called the second stage. Typically, first-stage decisions are structural and most of the time related to capital investment at the beginning of the project, while the second-stage decisions are often operational. However, some structural decisions corresponding to a future time can be considered as second-stage decisions. This kind of situation is formulated through the so-called multi-stage models, which are a natural extension of the two-stage case. Among the two-stage stochastic models, the expected value of the cost (or profit) resulting from optimally adapting the plan according to the realizations of uncertain parameters is referred to as the *recourse function*. Thus, a problem is said to have *complete recourse* if the recourse cost (or profit) for every possible uncertainty realization remains finite, independently of the nature of the first-stage decisions. In turn, if this statement is true only for the set of feasible first-stage decisions, the problem is said to have *relatively complete recourse* (Birge and Louveaux, 1997). This condition means that for every feasible first-stage decision, there is a way of adapting the plan to the realization of uncertain parameters. The following literature covers the technique in more detail: Infanger (1994), Kall and Wallace (1994), Higle and Sen (1996), Birge and Louveaux (1997), Marti and Kall (1998), and Uryasev and Pardalos (2001). In addition, Pistikopoulos and Ierapetritou (1995), Cheung and Powell (2000), Iyer and Grossmann (1998b), and Verweij *et al.* (2001) discuss solution techniques for these problems.

The general extensive form of a two-stage mixed-integer linear stochastic problem for a finite number of scenarios can be written as follows (Birge and Louveaux, 1997): Model SP:

$$\text{Max} \quad E[\text{Profit}] = \sum_{s \in S} p_s q_s^{\mathrm{T}} y_s - c^{\mathrm{T}} x \tag{12.1}$$

s.t.

$$Ax = b \tag{12.2}$$

$$T_s x + W y_s = h_s \quad \forall s \in S \tag{12.3}$$

$$x \geq 0 \;\; x \in X; \quad y_s \geq 0 \;\; \forall s \in S \tag{12.4}$$

In the above model, x represents the first-stage mixed-integer decision variables and y_s are the second-stage variables corresponding to scenario s, which has occurrence probability p_s. The objective function is composed of the expectation of the profit generated from operations minus the cost of first-stage decisions (capital investment). The uncertain parameters in this model appear in the coefficients q_s, the technology matrix T_s, and in the independent term h_s. When W, the recourse matrix, is deterministic the problem is called to be of fixed *recourse*. Cases where W is not fixed are found for example in portfolio optimization when the interest rates are uncertain (Dupacova and Römisch, 1998).

It is worth noticing that decision trees are in fact a particular case of two-stage programming. In other words, one can code through rules (mathematical in this case) the same decisions one makes in the tree explicitly, but in two-stage programming, one can also add logical constraints, if-the-else rules, etc., so there is no need for explicit enumeration of all options.

Aside from the issue of the plant life and the possible future upgrades, which complicate the modeling, there is yet another very important difficulty with these methods: the models are isolated from considering the size of the company, the health of its finances, even the temporary lack of liquidity or the abundance thereof as was pointed out above. Take, for example, the simple question: Should the project be started this year, next year, or two years down the road? The answer relies on forecasting of course, and the choice can be modeled using current two-stage stochastic programming methods, but maximizing the above measures is not proper most of the time, as the answer is not the same if the project is undertaken by a big corporation or a small company.

One important point to make is that before any treatment of risk or uncertainty, a solid deterministic model needs to be developed.

Summarizing: Chemical engineers have understood uncertainty and flexibility and have incorporated it within a two (multi)-stage process decision optimization models. In doing so, chemical engineers are not embracing the use of decision trees, which, as claimed, are a particular case of the former. Integration of financial indicators other than financial risk as well as strategic planning as a whole has barely started.

12.3 Project Evaluation the Way Economists and Financiers Practice It

One learns from books on financial management (Keown *et al.*, 2002, Smart *et al.*, 2004) that maximization of shareholder wealth, that is, maximization of the price of the existing common stock, is the real goal of a firm, not just maximization of profit as engineers are trained to think. Some alternative form of maximizing dividends should be substituted if the company is non-publicly owned. They claim that such a goal also benefits society because 'scarce resources will be directed to their most productive use by businesses competing to create wealth'. Finance management also teaches that several other issues are of importance for that goal including, among others:

- risk management, that is, its eventual reduction;
- risk diversification, that is, risky projects can be combined with other less risky ones in a balanced portfolio;
- cash flow management includes borrowing, raising investor's money, and also buying and selling securities;
- liquidity of the firm (ratio of assets to liabilities) and available cash, which affects investment and operating decisions.

To deal with risk, mostly they measure it using variability (or volatility), which is incorrect in almost all engineering project cases, as is explained later. They diversify by adding stocks to the portfolio.

12.3.1 Profit Maximization

Capital budgeting, the process through which the company analyzes future cash inflows and outflows, is performed using concepts that are extensions of the tools engineers know.

The firm *cost of capital*, which is the hurdle rate that an investment must achieve before it increases shareholder value, is one key aspect of these decisions that engineers have overlooked. Such cost of capital is measured typically by the firm's weighted average cost of capital (WACC) rate k_{WACC}. For a firm that uses only debt and common equity to finance its projects, this rate is given by:

$$k_{\text{WACC}} = [\text{After tax cost of debt}] \times \omega + [\text{Cost of equity}](1 - \omega) \tag{12.5}$$

where ω is the portion of debt that one is financing, the cost of debt is that rate paid for borrowed money, and the cost of equity is the rate that shareholders expect to get from the cash retained in the business and used for this project. The latter rate is larger than the former, of course.

In practice k_{WACC} is more complex to calculate because there are several debts incurred at different times and they require common equity as well as preferred equity. In addition, new capital may be raised through new stock offerings. Finally, one is faced with the problem of calculating a return of a project that has multiple decisions at different times, with uneven and uncertain revenues. Clearly, this simple formula needs some expansion, to add the complexities of projects containing multiple first- and second-stage decisions through time.

Financial management also suggests the alternative that the appropriate discount rate to evaluate the NPV of a project is the weighted average cost of capital, based on one important assumption that the risk profile of the firm is constant over time. In addition, this is true only when the project carries the same risk as the whole firm. When that is not true, which is most of the time, finance management has more elaborate answers, like managerial decisions that 'shape' the risk.

They also manage projects for *market value added* (MVA). The free cash flow model provides the *firm value*:

$$\text{Firm value} = \sum_i \frac{\text{Free cash flow}(i)}{(1 + k_{\text{WACC}})^i} + \frac{\text{Terminal value}}{(1 + k_{\text{WACC}})^n} \tag{12.6}$$

where the summation is extended over the period of n periods of planning. This expression uses k_{WACC} and refers to the whole company. The *firm value* is used to get the *market value added* of the investment.

$$\text{Market value added} = \text{Firm value} - \text{Investment} \tag{12.7}$$

which is a formula very similar to the *net present value* that engineers use for projects. In fact, the only difference is the value of the hurdle rate. Because the formula is a measure of the total wealth created by a firm at a given time, extending over a long time horizon, financial experts recommend the use of a shorter term measure, the *economic value added* (EVA) for period t.

$$\text{EVA}_t = \text{Net profit}_t - k_{\text{WACC}}[\text{Invested capital}] \tag{12.8}$$

where the net profit is computed after taxes. Thus, the MVA is the present value of all future EVA. Quite clearly, finance experts warn, managing for an increased EVA at any given time may lead to a non-optimal MVA.

In turn, the shareholder value can be obtained as follows: The *firm value* is the sum of *debt value* plus *equity value*. Then, if one knows the long-term interest-bearing liabilities, one has the *debt value*. Then one can obtain the shareholder value,

$$\text{Shareholder value} = \frac{\text{Equity value}}{\text{Number of shares}} = \frac{\text{Firm value} - \text{Debt value}}{\text{Number of shares}} \tag{12.9}$$

In principle, as noted above, the shareholder value is what one wants to maximize. This is what is true for the whole company and therefore implies one has to consider all projects at the same time. Thus, one can write

$$\text{Shareholder value} = \frac{\text{Equity value}}{\text{Number of shares}} = \frac{\sum_{\mathrm{p}} [\text{Firm value}_{\mathrm{p}}(x_{\mathrm{p}}) - \text{Debt value}_{\mathrm{p}}(x_{\mathrm{p}})]}{\text{Number of shares}} \tag{12.10}$$

where the summation is extended over different projects that the firm is pursuing or considering pursuing and x_{p} is the vector of first-stage ('here and now') decisions to be made. Thus, if the projects are generating similar equity value, no simplification is possible and decisions have to be made simultaneously for all projects. Hopefully, procedures that will do this interactively, that is, change the decisions of all projects at the same time, will be developed.

However, which shareholder value does one want to the maximize? The one corresponding to the next quarter company report, or a combination of shareholder values at different points in the future? In other words, is there such a thing as an optimal investment and operating strategy/path? This looks like an optimal control problem!

And then, there is the dividend policy. Is it possible that this should be decided *together with* and not *independently from* the specific project first-stage variables?

The 'here and now' decisions (x_{p}) involve several technical choices of the processes themselves (catalysts, technologies, etc.) which require detailed modeling and also some other 'value drivers', like advertisements to increase sales, alliances to penetrate markets, investment in R&D, company acquisition, cost-control programs, inventory control, control of the customer paying cycles (a longer list is given by Keown *et al.*, 2002). Most of these 'knobs and controls' are called second-stage ('wait and see') decisions, but many are also first-stage decisions.

The literature on strategic planning (Hax and Majluf, 1984) has models that deal directly with shareholder value. They use different models (market to book values, profitability matrices, etc.) to obtain corporate market value, which take into account the company reinvestment policy, dividend payments, etc. One cannot help also mentioning some classic and highly mathematical models from game theory and other analytical approaches, some of which are discussed elegantly by Debreu (1959) and Danthine and Donaldson (2002).

A brief glance at the literature tells us that economists are not yet so keen on using two-stage stochastic models. They understand, of course, the concept of options in projects, but many are still 'locked' to the use of point measures like NPV and decision trees (De Reyck *et al.*, 2001).

Finally, some of the financial ratios that are waiting to be embraced by engineering models are:

- Liquidity ratios
  - Current ratio = current assets/current liabilities
  - Acid test or quick ratio = (current asset inventories)/current liabilities
  - Average collection period: accounts receivable/daily credit sales
  - Accounts receivable turnover = credit sales/accounts receivable
  - Inventory turnover = costs of goods sold/inventory
- Operating profitability ratios
  - Operating income return on investment = income/total assets
  - Operating profit margin = income/sales
  - Total assert turnover = sales/ total assets
  - Accounts receivable turnover = sales/accounts receivable
  - Fixed assets turnover = sales/net fixed assets
- Financial ratios
  - Debt ratio = total debt/total assets
  - Times interest earned = operating income/interest expense
  - Return on equity = net income/common equity

While all these indicators focus on different aspects of the enterprise, they should be at least used as constraints in engineering models.

It is therefore imperative that engineers incorporate these measures and objectives in project evaluation, *when and if*, of course, *decisions at the technical level have an impact on the outcome*. In other words, how much of the project is financed by equity is a decision to make together with the technical decisions about size and timing of every project and the technical decisions of the project itself, such as the selection of technologies, catalysts, etc. This last aspect is what makes the integration a must!

12.3.2 Risk Management

The other major component influencing business decisions is risk. First, one needs to distinguish business risk from financial risk.

Business risk is measured by the non-dimensional ratio of variability (standard deviation) to expected profit before taxes and interest (Keown *et al.*, 2002; Smart *et al.*, 2004). Therefore, the same variability associated with a larger profit represents less business risk. Thus, one can use this ratio to compare two investments, but when it comes to managing risk for one investment, the objective seems to be the usual: maximize profit and reduce variability. As will be discussed later in greater detail, these are conflicting goals. Measures to reduce business risk include product diversification, reduction of fixed costs, managing competition, etc. More specifically, the change in product price and fixed costs is studied through the degree of operating leverage (DOL) defined in various forms, one being the ratio of revenue before fixed costs to earnings before interest and taxes (EBIT).

Engineers have not yet caught up in relating these concepts with their models. As usual, the mix includes some second-stage decisions, but most of them are first-stage 'here and now' decisions. Modeling through two-stage stochastic programming and including technical decisions in this modeling is the right answer. Some aspects of this modeling are discussed below.

Financial risk is in some cases defined as the 'additional variability in earnings ... and the additional chance of insolvency ... caused by the use of financial leverage' (Keown *et al.*, 2002). In turn, the financial leverage is the amount of assets of the firm being financed by securities bearing a fixed or limited rate of return. Thus, the degree of financial leverage (DFL) is defined as the ratio of EBIT to the difference of EBIT and the total interest expense I, that is,

$$\mathrm{DFL} = \frac{\mathrm{EBIT}(x)}{\mathrm{EBIT}(x) - I(x)} \qquad (12.11)$$

In other words, business and financial risk differ fundamentally in that one considers interest paid and the other does not. Both are considered related to variability. As is shown later, the claim is that this is the wrong concept to use in many cases.

Another very popular definition of risk is through the risk premium or *beta*. This is defined as the slope of the curve that gives market returns as a function of S&P 500 Index returns; in other words, comparing how the investment compares with the market. The concept of 'beta' (the slope of the curve) is part of the capital asset pricing model (CAPM) proposed by Lintner (1969) and Sharpe (1970), which intends to incorporate risk into valuation of portfolios and it can also be viewed as the increase in expected return in exchange for a given increase in variance. However, this concept seems to apply to building stock portfolios more than to technical projects within a company.

Financial risk is also assessed through point measures like risk-adjusted return of capital (RAROC), risk-adjusted net present value (RPV), Sharpe ratio (Sharpe, 1966). It is unclear if these point measures are proper ways of assessing risk, much less managing it, in engineering projects. This point is expanded below.

Economists also consider risk as 'multidimensional' (Dahl *et al.*, 1993). They have coined names for a variety of risks. Some of these, applied mostly to stocks, bonds, and other purely financial instruments, are market risk (related to the CAPM model and the above described parameter 'beta'), volatility risk (applied to options, primarily), currency risk, credit risk, liquidity risk, residual risk, inventory risk, etc.

The managing of net working capital is used by finance experts to manage risk. The working capital is the total assets of the firm that can be converted to cash in a one-year period. In turn, the net working capital is the difference between assets and liabilities. Thus increasing the net working capital reduces the chance of low liquidity (lack of cash or ability to convert assets into cash to pay bills in time). This is considered as short-term risk. Several strategies are suggested to maintain an appropriate level of working capital (Finnerty, 1993).

A separate consideration needs to be made for inventory, which in principle is used to be able to uncouple procuring from manufacturing and sales. In this regard it is mostly considered as a risk hedging strategy that increases costs. Finally, contracts, especially option contracts and futures, are other risk hedging tools.

Recently, risk started to be defined in terms of another point measure introduced by J.P. Morgan, *value at risk* or VaR (Jorion, 2000). This is defined as the difference between the expected profit and the profit corresponding to 5% cumulative probability. Many other 'mean-risk' models use measures such as tail value at risk, weighted mean

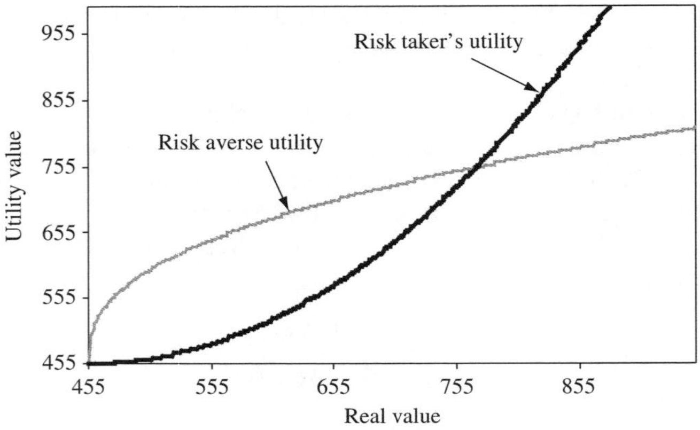

Figure 12.3 *Utility functions*

deviation from a quantile, and the tail Gini mean difference (reviewed by Ogryczak and Ruszcynski, 2002), to name a few.

More advanced material (Berger, 1980; Gregory, 1988; Danthine and Donaldson, 2002) proposes the use of *expected utility theory* to assess risks. This theory proposes to assign a value (different from money) to each economic outcome. Figure 12.3 illustrates the utility function of a risk-averse decision maker, who values (in relative terms) small outcomes more than large outcomes. It also shows the utility function of a risk taker who places more value in higher outcomes. In most cases, the utility curve is constructed in a somehow arbitrary manner, that is, taking two extreme outcomes and assigning a value of 0 to the less valued and the value of 1 to the most valued one. Then there are procedures that pick intermediate outcomes and assign a value to them until the curve is constructed.

This theory leads to the definition of loss functions as the negative utility values, which are used to define and manipulate risk (Berger, 1980). To do this, a decision rule must be defined. Thus, risk is defined as the expected loss for that particular decision rule. This, in turn, leads to the comparison of decision rules. Engineering literature contains some reference to this theory. As is discussed below, expected utility has a lot of potential as a decision making tool. All that is needed is to start putting it in the context of the emerging two-stage stochastic modeling.

Some important things, one learns from the review of basic financing are the following:

1) The majority of the tools proposed are deterministic, although some can be extended to expectations on profit distributions and therefore decision trees are presented as advanced material in introductory finance books. Quite clearly, one would benefit from using two-stage stochastic programming instead.
2) Risk is considered a univariate numerical measure like variability or value at risk (VaR), which is the difference between the project expected outcome and the profit corresponding to (typically) 5% cumulative probability. Opportunities at high profit levels are rarely discussed or considered.
3) Financiers only know how to evaluate a project. They can manipulate it on the financial side, but they cannot manipulate it in its technological details because they need

engineering expertise for that. This is the Achilles heel of their activity. Engineers, in turn, cannot easily take into account the complexity of finances. Both need each other more than ever.

12.4 Latest Progress of Chemical Engineering Models

Decision making is an old branch of management sciences, a discipline that has always had some overlap with engineering, especially industrial engineering. Some classical books on the subject (Riggs, 1968; Gregory, 1988; Bellman, 1957) review some of the different techniques, namely:

- resource allocation (assignment, transportation);
- scheduling (man–machine charts, Gantt charts, critical path scheduling, etc.);
- dynamic programming (Bellman, 1957; Denardo, 1982);
- risk (reviewed in more detail in the next section) through the use of decision trees, regret tables, and utility theory.

Notwithstanding the value of all these techniques, the new emerging procedures rely heavily on two-stage stochastic programming and some revival of dynamic programming. It is argued here that several techniques, like decision trees and utility theory, are special cases of two-stage stochastic programming. Others claim the same when advocating the dynamic programming approach (Cheng *et al.*, 2003, 2004). They proposed to model decision making as a multiobjective Markov decision process.

For example, in recent years, the integration of batch plant scheduling with economic activities belonging to procuring and marketing has been pioneered by the books by Puigjaner *et al.* (1999, 2000). These contain full chapters on financial management in batch plants where something similar to the corporate information loop (Figure 12.1), as viewed by engineers and economists, is discussed. These authors discuss the notion of enterprise wide resource management systems (ERM), one step above enterprise resource planning (ERP). They outline the cycle of operations involving cash flow and working capital, the management of liquidity, the relationships to business planning, etc. as it relates mostly to batch plants. They even direct attention to the role of pricing theory and discuss the intertwining of these concepts with existing batch plant scheduling models. These summary descriptions of the role of cash and finances in the context of batch plants are the seeds of the mathematical models that have been proposed afterwards. Extensive work was also performed by many other authors in a variety of journal articles. A partial (clearly incomplete) list of recent work directly related to the integration of process systems engineering and economic/financial tools is the following:

- Investment planning (Sahinidis *et al.*, 1989; Liu and Sahinidis, 1996; McDonald and Karimi, 1997; Bok *et al.*, 1998; Iyer and Grossmann, 1998a; Ahmed and Sahinidis, 2000a, Cheng *et al.*, 2003, 2004).
- Operations planning (Ierapetritou *et al.*, 1994; Ierapetritou and Pistikopoulos, 1994; Pistikopoulos and Ierapetritou, 1995; Iyer and Grossmann, 1998a; Lee and Malone, 2001; Lin *et al.*, 2002; Mendez *et al.*, 2000; McDonald, 2002; Maravelias and Grossmann, 2003; Jackson and Grossmann, 2003; Mendez and Cerdá, 2003).

- Refinery operations planning (Shah, 1996; Lee *et al.*, 1996; Zhang *et al.*, 2001; Pinto *et al.*, 2000; Wenkai *et al.*, 2002; Julka *et al.*, 2002b; Jia *et al.*, 2003; Joly and Pinto, 2003; Reddy *et al.*, 2004; Lababidi *et al.*, 2004; Moro and Pinto, 2004).
- Design of batch plants under uncertainty (Subrahmanyan *et al.*, 1994; Petkov and Maranas, 1997).
- Integration of batch plant scheduling and planning and cash management models (Badell *et al.*, 2004; Badell and Puigjaner, 1998, 2001a,b; Romero *et al.*, 2003a,b).
- Integration of batch scheduling with pricing models (Guillén *et al.*, 2003a).
- Integration of batch plant scheduling and customer satisfaction goals (Guillén *et al.*, 2003b).
- Technology selection and management of R&D (Ahmed and Sahinidis, 2000b; Subramanian *et al.*, 2000).
- Supply chain design and operations (Wilkinson *et al.*, 1996; Shah, 1998; Bok *et al.*, 2000, Perea-Lopez *et al.*, 2000; Bose and Pekny, 2000; Gupta and Maranas, 2000; Gupta *et. al.*, 2000; Tsiakis *et al.*, 2001; Julka *et al.*, 2002a,b; Singhvi and Shenoy, 2002; Perea-Lopez *et al.*, 2003; Mele *et al.*, 2003; Espuña *et al.*, 2003; Neiro and Pinto, 2003).
- Agent-based process systems engineering (Julka *et al.*, 2002a,b; Siirola *et al.*, 2003).
- Financial risk through the use of a variety of approaches and in several applications (Applequist *et al.*, 2000; Gupta and Maranas, 2003a; Mele *et al.*, 2003; Barbaro and Bagajewicz, 2003, 2004a,b; Wendt *et al.*, 2002; Orcun *et al.*, 2002).
- New product development (Schmidt and Grossmann, 1996; Blau and Sinclair, 2001; Blau *et al.*, 2000).
- Product portfolios in the pharmaceutical industry (Rotstein *et al.*, 1999).
- Options trading and real options (Rogers *et al.*, 2002, 2003; Gupta and Maranas, 2003b, 2004).
- Transfer prices in supply chain (Gjerdrum *et al.*, 2001).
- Oil drilling (Iyer *et al.*, 1998; Van den Heever *et al.*, 2000, 2001; Van den Heever and Grossmann, 2000; Ortiz-Gómez *et al.*, 2002).
- Supply chain in the pharmaceutical industry (Papageorgiou *et al.*, 2001; Levis and Papageorgiou, 2003).
- Process synthesis using value added as an objective function (Umeda, 2004). This chapter revisits dynamic programming approaches.

The rest of this chapter concentrates on discussing some aspects of the integration that have received attention by engineers, namely,

- financial risk
- effect of inventories
- regular, future, and option contracts
- budgeting
- pricing
- consumer satisfaction.

Some work is also mentioned that calls for the integration of finances and other disciplines with key ideas of product engineering and the chemical supply chain.

12.5 Financial Risk Management

12.5.1 Definition of Risk

There are various definitions of risk in the engineering literature, most of them rooted in the finance field, of course.

A good measure of risk has to take into account different risk preferences and therefore one may encounter different measures for different applications or attitudes toward risk. The second property that a risk measure should have is that when it focuses on particular outcomes, like low profit that are to be averted, one would like also to have information about the rest of the profit distribution. Particularly, when one compares one project to another, one would like to see what it is that one loses in other portions of the spectrum as compared to what one gains by averting risk.

Some of these alternative measures that have been proposed are now reviewed:

- *Variability*: That is, standard deviation of the profit distribution. This is the most common assumption used in the non-specialized financial literature, where investment portfolios (stocks primarily) are considered. Mulvey *et al.* (1995) introduced the concept of *robustness* as the property of a solution for which the objective value for any realized scenario remains 'close' to the expected objective value over all possible scenarios and used the variance of the cost as a 'measure' of the robustness of the plan, i.e. less variance corresponds to higher robustness. It is obvious that the smaller the variability, the less negative deviation from the mean. But it also implies smaller variability on the optimistic side. Thus, either the distribution is symmetric (or this is assumed) or one does not care about the optimistic side. This is the specific assumption of stock portfolio optimization, but it is known not to be correct for other type of investments, especially multi-year ones (Smart *et al.*, 2004). Thus, the use of variability as a measure of risk is being slowly displaced by engineers (not necessarily by the finance community) in favor of other measures. Nonetheless, it is still being used. Tan (2002), for example, provides means to reduce variability by using capacity options in manufacturing. It has the added disadvantage that it is nonlinear.
- *Cumulative probability for a given aspiration level*: This is the correct way of defining risk when one wants to reduce its measure to a single number because unlike variance it deals with the pessimistic side of the distribution only. Consider a project defined by x. Risk is then defined by

$$\mathrm{Risk}(x, \Omega) = \mathrm{P}\{\mathrm{Profit}(x) \leq \Omega\} \tag{12.12}$$

  where Profit(x) is the actual profit, showed in Figure 12.4 as the shaded area. This definition has been used by the petroleum industry for years (McCray, 1975). In the process systems literature this definition was used by Rodera and Bagajewicz (2000), Barbaro and Bagajewicz (2003, 2004a), and Gupta and Maranas (2003a). Figure 12.5 depicts a cumulative distribution curve, which also represents risk as a function of all aspiration levels. This is the preferred representation because, as is discussed later, one can best manage risk using it.
- *Downside risk*: This measure, introduced by Eppen *et al.* (1989) in the framework of capacity planning for the automobile industry, is an alternative and useful way of

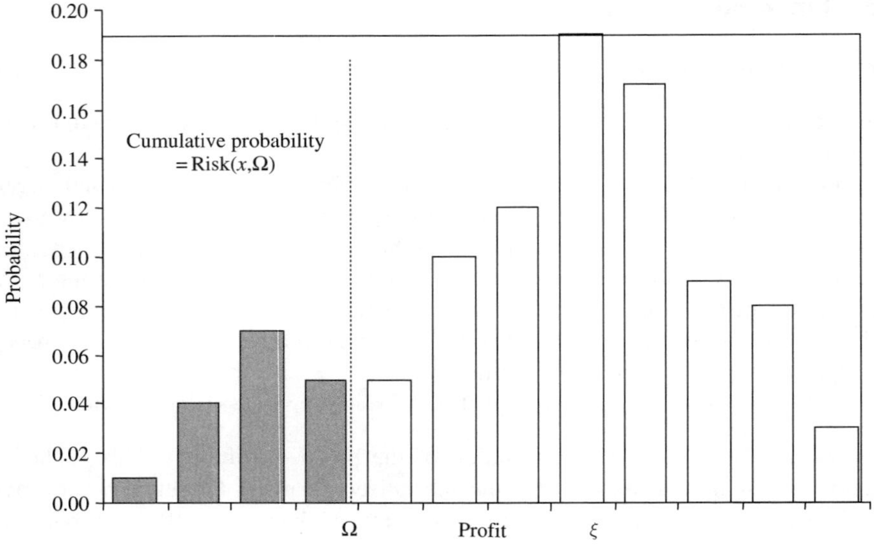

Figure 12.4 *Definition of risk. Discrete case*

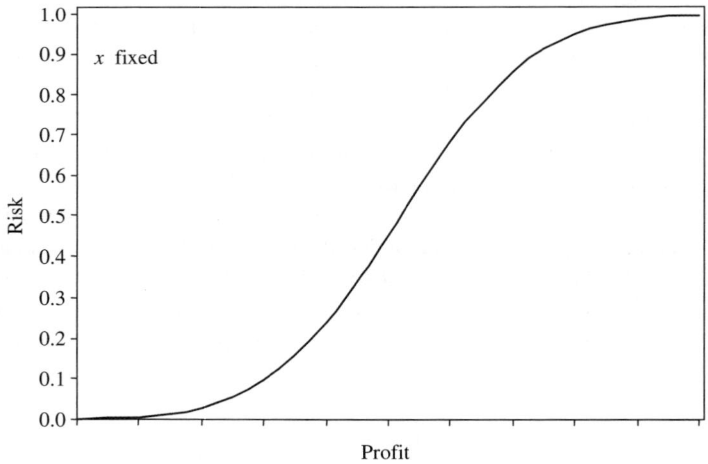

Figure 12.5 *Risk curve, continuous case*

measuring risk using the concept of currency. Consider the positive deviation from a profit target Ω for design x, $\delta(x, \Omega)$, defined as follows:

$$\delta(x, \Omega) = \begin{cases} \Omega - \text{Profit}(x) & \text{If Profit}(x) < \Omega \\ 0 & \text{Otherwise} \end{cases} \tag{12.13}$$

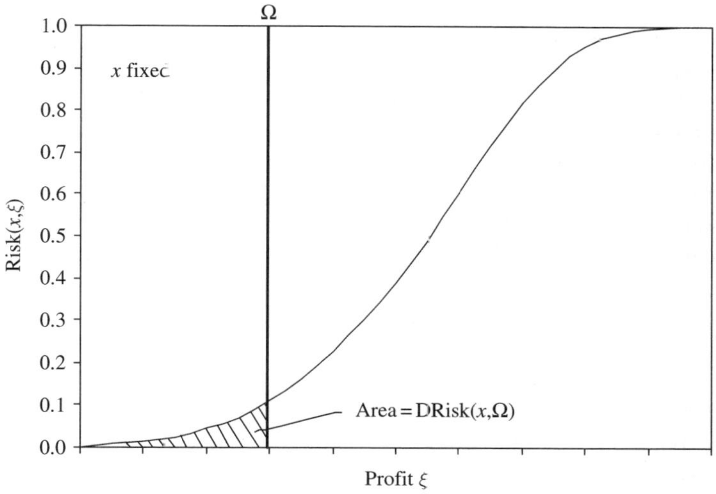

Figure 12.6 *Interpretation of downside risk*

Downside risk is then defined as the expectation of $\delta(x, \Omega)$, that is, $\text{DRisk}(x, \Omega) = E[\delta(x, \Omega)]$. This form has been very useful computationally to identify process alternatives with lower risk, as is discussed below. Barbaro and Bagajewicz (2003, 2004a) proved that downside risk is just an integral of the risk curve, as shown in Figure 12.6. Moreover, they proved that downside risk is not monotone with risk, that is, two designs can have the same risk for some aspiration level, but different downside risk. Moreover, projects with higher risk than others can exhibit lower downside risk. Therefore, minimizing one does not imply minimizing the other. However, this measure has several computational advantages and was used to generate solutions where risk is managed using goal programming (Barbaro and Bagajewicz, 2003, 2004a). Gupta and Maranas (2003a) discuss these measures (risk and downside risk) as well.

- *Upper partial mean*: This was proposed by Ahmed and Sahinidis (1998). It is defined as the expectation of positive deviation from the mean, that is, $\text{UPM}(x) = E[\Delta(x)]$, where $\Delta(x)$ is defined the same way as $\delta(x, \Omega)$, but using $E[\text{Profit}(x)]$ instead. In other words, the UPM is defined as the expectation of the positive deviation of the second-stage profit. The UPM is a linear and asymmetric index since only profits that are below the expected value are measured. However, in the context of risk management at the design stage, this measure cannot be used because it can underestimate the second-stage profit by not choosing optimal second-stage policies. Indeed, because of the way the UPM is defined, a solution may falsely reduce its variability just by not choosing optimal second-stage decisions. This is discussed in detail by Takriti and Ahmed (2003), who present sufficient conditions for a measure of a robust optimization to assure that the solutions are optimal (i.e. not stochastically dominated by others). For these reasons, downside risk is preferred, simply because the expectation of positive deviation is done with respect to a fixed target (Ω) and not the changing profit expectation.
- *Value at risk* (VaR): Discussed in detail by Jorion, 2000, this was introduced by J.P. Morgan (Guldimann, 2000) and is defined as the expected loss for a certain

confidence level usually set at 5% (Linsmeier and Pearson, 2000). A more general definition of VaR is given by the difference between the mean value of the profit and the profit value corresponding to the p-quantile. For instance, a portfolio that has a normal profit distribution with zero mean and variance σ, VaR is given by $z_p\sigma$ where z_p is the number of standard deviations corresponding to the p-quantile of the profit distribution. Most of the uses of VaR are concentrated on applications where the profit probability distribution is assumed to follow a known symmetric distribution (usually the normal) so that it can be calculated analytically. The relationship between VaR and Risk is generalized as follows (Barbaro and Bagajewicz, 2004a):

$$\mathrm{VaR}(x, p_\Omega) = \mathrm{E}[\mathrm{Profit}(x)] - \mathrm{Risk}^{-1}(x, p_\Omega) \tag{12.14}$$

where p_Ω is the confidence level related to profit Ω, that is, $p_\Omega = \mathrm{Risk}(x, \Omega)$. Notice that VaR requires the computation of the inverse function of Risk. Moreover, since Risk is a monotonically increasing function of Ω, one can see from equation 12.14 that VaR is a monotonically decreasing function of p_Ω.

While computing VaR as a post-optimization measure of risk is a simple task and does not require any assumptions on the profit distribution, it poses some difficulties when one attempts to use it in design models that manage risk. Given its computational shortcomings, VaR is only convenient to use as a risk indicator because of its popularity in finance circles.

Finally, sometimes the risks of low liquidity measured by the cash flow at risk (CFAR) are more important than the *value at risk* (Shimko, 1998).

Companies that operate with risky projects identify VaR or similar measures directly with potential liability, and they would hold this amount of cash through the life of a project, or part of it.

- *Downside expected profit* (DEP): For a confidence level p_Ω (Barbaro and Bagajewicz, 2004a), this is defined formally as the expectation of profit below a target corresponding to a certain level of risk p_Ω, that is, $\mathrm{DEP}(x, p_\Omega) = E[\gamma(x, \Omega)]$, where

$$\gamma(x, \Omega) = \begin{cases} \mathrm{Profit}(x) & \text{If } \mathrm{Profit}(x) \le \Omega \\ 0 & \text{Otherwise} \end{cases} \tag{12.15}$$

and $\Omega = \mathrm{Risk}^{-1}(x, p_\Omega)$. Plotting DEP as a function of the risk is revealing because at low risk values some feasible solutions may exhibit larger risk adjusted present value. The relationship between DEP, risk, and downside risk is

$$\mathrm{DEP}(x, p_\Omega) = \int_{-\infty}^{\Omega} \xi f(x, \xi)\mathrm{d}\xi = \Omega\mathrm{Risk}(x, \Omega) - \mathrm{DRisk}(x, \Omega) \tag{12.16}$$

where $f(x, \xi)$ is the profit distribution.

- *Regret analysis* (Riggs, 1968): This is an old tool from decision theory that has been used in a variety of ways to assess and manage risk (Sengupta, 1972; Modiano, 1987). Its use as a constraint in the context of optimization under uncertainty and aiming at the managing of financial risk has been suggested by Ierapetritou and Pistikopoulos (1994). The traditional way of doing regret analysis requires the presence of a table of profits for different designs under all possible scenarios. One way to generate such a table is to use the sampling average algorithm (Verweij *et al.*, 2001) to solve a deterministic design, scheduling and/or planning model for several scenarios, one at a time or a certain number at a time, to obtain several designs (characterized by first-stage variables). The next step is to fix these first-stage variables to the values obtained and solve the model to obtain the profit of that design under every other scenario. The different criteria to choose the preferred solution are as follows:

- The *maximum average* criterion states that one should choose the design that performs best as an average for all scenarios. This is equivalent to choosing the solution with best ENPV.
- The *maximax* criterion suggests choosing the design that has the highest profit value in the profit table. This represents an *optimistic* decision in which all the bad scenarios are ignored in favor of a single good scenario.
- The *maximin* criterion states that the design that performs best under the worst conditions is chosen. This is equivalent to identifying the worst-case value (minimum over all scenarios) for each design and choosing the design with the best worst-case value (or the maximum–minimum).

Aseeri and Bagajewicz (2004) showed that none of these strategies can guarantee the identification of the best risk-reduced solutions, although in many instances they can be used to identify promising and good solutions. For example, Bonfill *et al.* (2004) used the maximization of the worst case as a means to obtain solutions that reduce risk at low expectations.

- *Chance constraints* (Charnes and Cooper, 1959): In essence, chance expressions are not 'other' than risk, as defined above, but usually applied to outcomes other than cost or profit. Vice versa, financial risk can be thought of as a chance expression applied to profit. Many authors (Orcun *et al.*, 2002; Wendt *et al.*, 2002) use chance expressions by evaluating the probability that a design or a system can meet a certain uncertain parameter. Typical chance constraints have been used in scheduling of plant operations to assess the probabilities of meeting certain levels of demand. Aseeri and Bagajewicz (2004) showed that this approach is less efficient than straight risk curve analysis and is in fact a special case of it. For example, a chance constraint for the production, e.g. Production $\leq$ Demand, should be replaced by Production $\leq F^{-1}(1-\alpha)$, where F is the cumulative distribution for the demand and α is the chosen confidence level. But a model with these types of constraints is just *one* instance of a sampling algorithm. Thus, the approach of using chance constraints is a subset of the sampling average algorithm discussed above.
- *The Sharpe ratio* (Sharpe, 1966): This is given by the expected excess return of investment over a risk-free return divided by the volatility, that is,

$$S = \frac{r - rf}{\sigma} \tag{12.17}$$

  where r and rf are the expected return and the risk-free return, respectively, and σ is the volatility and can be used directly to assess risk in investments (Shimko, 1997).
- *Risk-adjusted return on capital* (RAROC): This is the quotient of the difference between the expected profit of the project adjusted by risk and the capital (or value) at risk of an equivalent investment and the value at risk. This value is a multiple of the

Sharpe ratio in portfolio optimization, although this assertion is only valid for symmetric distributions. This particular measure has not been used in two-stage stochastic engineering models to manage risk. This is not preferred because, as explained below, it is better to depart from single valued measures looking at the whole risk curve behavior instead.

- *Certainty equivalent approach* (Keown *et al.*, 2002): In this approach a certainty equivalent is defined. This equivalent is the amount of cash required with certainty to make the decision maker indifferent between this sum and a particular uncertain or risky sum. This allows a new definition of net present value by replacing the uncertain cash flows by their certain equivalent and discounting them using a risk-free interest rate.
- *Risk premium*: Applequist *et al.* (2000) suggest benchmarking new investments against the historical risk premium mark. Thus, they propose a two-objective problem, where the expected net present value and the risk premium are both maximized. The technique relies on using the variance as a measure of variability and therefore it penalizes/rewards scenarios at both sides of the mean equally, which is the same limitation that is discussed above.
- *Risk-adjusted NPV* (RPV) (Keown *et al.*, 2002): This is defined as the net present value calculated using a risk-adjusted rate of return instead of the normal return rate required to approve a project. However, Shimko (2001) suggests a slightly different definition where the value of a project is made up of two parts, one forms the 'not at risk' part, discounted using the risk-free return rate, and the other forms the part 'at risk' discounted at the fully loaded cash plus risk cost.
- *Real option valuation* (ROV): Recently, Gupta and Maranas (2004) revisited a real-option-based concept to project evaluation and risk management. This framework provides an entirely different approach to NPV-based models. The method relies on the arbitrage-free pricing principle and risk neutral valuation. Reconciliation between this approach and the above-described risk definitions is warranted.
- *Other advanced theories*: Risk evaluation and its management continue to be an object of research. For example, Jia and Dyer (1995) propose a method to weigh risk (defined through the variance and assuming symmetry) against value. These models are consistent with expected utility theory. More generally, some define risk as just the probability of an adverse economic event and associate these adverse effects to something other than pessimistic profit levels (Blau and Sinclair, 2001). For example, Blau *et al.* (2000) when analyzing drug development define risk as the probability of having more candidates in the pipeline than available resources, which would result in delays in product launching. While all these are valid risk analysis, they are nonetheless, simplifications that one needs to remember one is doing. The ultimate risk analysis stems from the financial risk curve based on profit of the whole enterprise. This is explored in more detail later in the chapter.

Fortunately, computers and tools to handle uncertainty and risk are widely available these days: @Risk (Palisade *http://www.palisade.com*), Crystal Ball (Decision Engineering, *http://www.crystalball.com*); Risk Analyzer (Macro Systems, *http://www.macrosysinc.com/*), Risk + (C/S Solutions, *http://www.cs-solutions.com*) among many others. In other efforts, Byrd and Chung (1998) prepared a program for DOE to assess risk in petroleum exploration. They use decision trees. There are some Excel templates used in chemical engineering classes (O'Donnel *et al.*, 2002). Therefore, there is no excuse anymore for not obtaining the expected net present value or other profitability measures and performing risk analysis by using these tools. Reports available from the web pages cited above indicate that the use of these tools is becoming popular. Its teaching in senior chemical engineering classes should be encouraged. All these Excel-based programs require that one builds the model, as in two-stage programming. Therefore, it is unclear how far one can go with these Excel-based modeling methods versus the use of two-stage stochastic programming.

Conclusions

- The use of variance should be avoided because it incorporates information from the upside, when in fact one is targeting the downside profit.
- Point measures (VaR, RAROC, beta, etc.) are useful but incomplete. They do not depict what is taking place in the upside profit region and can lead to wrong conclusions.
- Regret analysis is potentially misleading and therefore should be used with caution.
- Chance values on specific constraints are weaker indicators of risk.
- The direct use of the probabilistic definition of risk (given by the cumulative distribution curve) or the closely related concept of downside risk as a means of assessing risk is recommended.

12.5.2 Risk Management at the Design Stage

Most of the strategies devoted to managing risk in projects at the design stage target variability. One very popular tool is known as 'six-sigma' (Pande and Holpp, 2001). Companies also make use of 'failure mode effects analysis' (Stamatis, 2003), which is a procedure originated at NASA in which potential failures are analyzed and measures to prevent it are discussed.

To manage risk while using two-stage stochastic models, one can use a constraint, restricting variability, risk itself, downside risk, VaR, etc., or incorporate chance constraints as well as regret functions as done by Ierapetritou and Pistikopoulos (1994). Constraints including variability are nonlinear and, as discussed above, are not favored anymore. Others have not been attempted (VaR). Next, constraints using risk and downside risk for two-stage stochastic programming are discussed.

Since uncertainty in the two-stage formulation is represented through a finite number of independent and mutually exclusive scenarios, a risk constraint can be written as follows:

$$\text{Risk}(x, \Omega) = \sum_{s \in S} p_s z_s(x, \Omega) \leq R_\Omega \tag{12.18}$$

where z_s is a new binary variable defined for each scenario as follows:

$$z_s(x, \Omega) = \begin{cases} 1 & \text{If } q_s^{\mathrm{T}} y_s > \Omega \\ 0 & \text{Otherwise} \end{cases} \quad \forall s \in S \tag{12.19}$$

and R_Ω is the desired maximum risk at the aspiration level Ω. A constraint to manage downside risk can be written in a similar fashion as follows:

$$\text{DRisk}(x, \Omega) = \sum_{s \in S} p_s \delta_s(x, \Omega) \leq \text{DR}_\Omega \tag{12.20}$$

where $\delta_s(x, \Omega)$ is defined as in equation 12.13 for each scenario and DR_Ω is the upper bound of downside risk. Note that both expressions are linear. The former includes binary variables, while the latter does not. Since binary variables add computational burden, Barbaro and Bagajewicz (2004a) preferred and suggested the use of downside risk.

Thus, this representation of risk is favored and variability, upper partial mean, regret functions, chance constraints, VaR, and the risk premium are disregarded.

Now, adding the constraints is easy, but picking the aspiration levels is not. In fact, Barbaro and Bagajewicz (2003, 2004a) have suggested that the conceptual scheme is multiobjective in nature. Indeed, one wants to minimize risk at various aspiration levels at the same time as one wants to maximize the expected profit, which is equivalent to pushing the curve to the right. All this is summarized in Figure 12.7.

The (intuitive) fact that lowering the risk at low expectations is somehow incompatible with maximizing profit was formally proven in the engineering literature by Barbaro and Bagajewicz (2004a). In fact, the different solutions one can obtain using the multiobjective approach are depicted in Figure 12.8. Indeed, if only one objective at low aspirations is used (Ω_1), then the risk curve (curve 2) is lower than the one corresponding to maximum profit (SP). A similar thing can be said for curve 3, which corresponds to minimizing the risk at high aspiration levels. Curve 4 corresponds to an intermediate balanced answer. In all cases, one finds that the curves intersect the maximum profit solution (SP) at some point (they are not stochastically dominated by it) and they have (naturally) a lower expected profit.

To obtain all these curves, Barbaro and Bagajewicz (2004a) proposed to solve several goal programming problems penalizing downside risk with different weights, thus obtaining a spectrum of solutions from which the decision maker could choose. They also discuss the numerical problems associated with this technique. Gupta and Maranas (2003a) also suggested the use of this definition of risk, but have not pursued the idea so

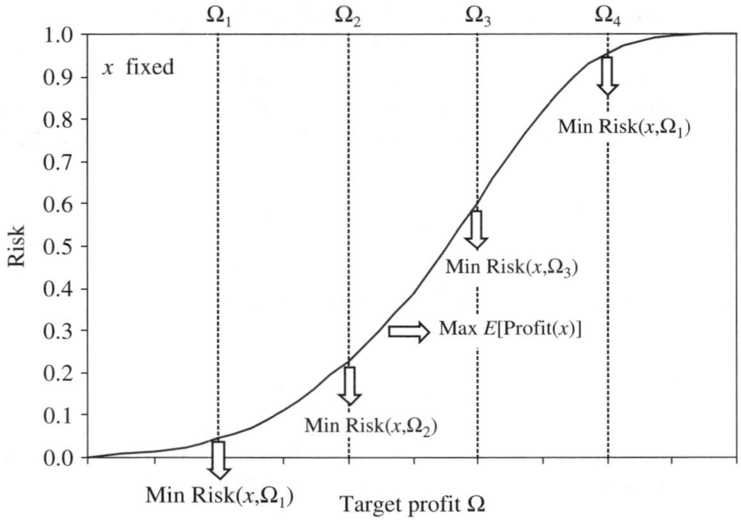

Figure 12.7 *Multiobjective approach for risk management*

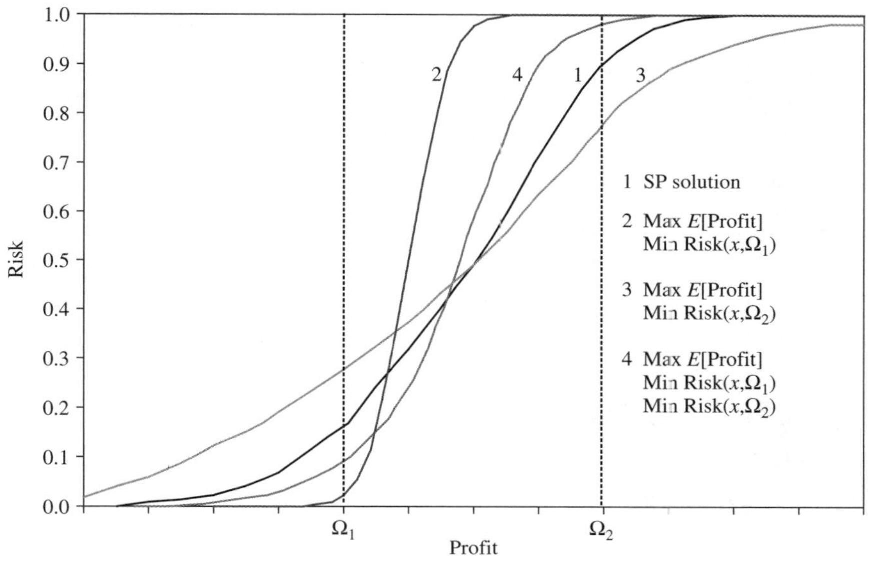

Figure 12.8 *Spectrum of solutions obtainable using a multiobjective approach for risk management*

far. Bonfill *et al.* (2004) also showed that maximizing the worst-case scenario outcome renders a single curve (not a spectrum) that has lower risk at low expectations. Conceivably, one can maximize the best-case scenario and obtain the optimistic curve as in case 3 (Figure 12.8).

In practice, after trying this approach in several problems, the technique was proven computationally cumbersome for some cases (too many scenarios were needed to get smooth risk curves) and the determination of a 'complete' (or at least representative) risk curve spectrum elusive, because too many aspiration levels need to be tried.

To ameliorate the computational burden of goal programming, an alternative way of decomposing the problem and generating a set of solutions was proposed (Aseeri and Bagajewicz, 2004). This decomposition procedure, which is a simple version of the sampling average algorithm (Verweij *et al.*, 2001), now follows:

1) Solve the full problem for each of the n_s scenarios at a time obtaining a solution (x_s, y_s). The values of the first-stage variables x_s obtained are kept as representative of the 'design' variables for this scenario to be used in step c).
2) Use the profit of these n_s solutions to construct a (fictitious) risk curve. This curve is an upper bound to the problem.
3) Solve the full problem for all n_s scenarios, n_s times, fixing the first-stage variables x_s obtained in step a) in each case. This provides a set of n_s solutions $(x_s, y_{s1}, y_{s2}, \ldots, y_{sn_s})$ that constitute the spectrum of solutions.
4) Identify the curve with largest expected profit and determine the gap between this curve and the one for the upper bound.
5) A (not so useful) lower bound curve can be identified by taking the largest value of all curves for each aspiration level.

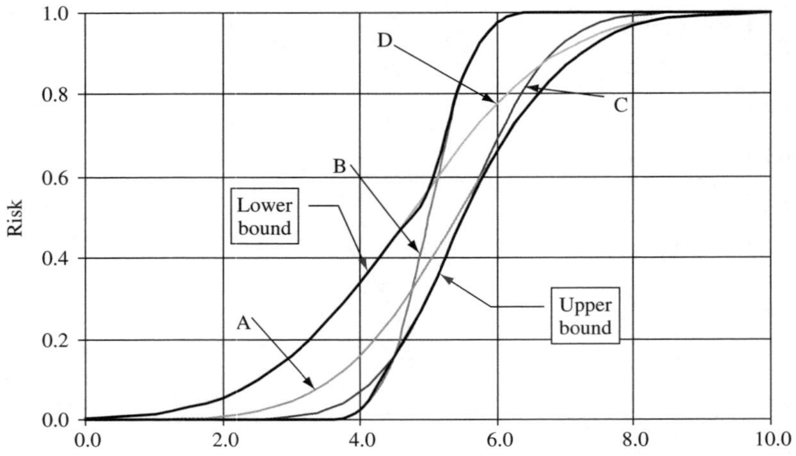

Figure 12.9 *Upper bound curve and spectrum of solutions*

The assumption is that given a sufficiently large number of scenarios, one will be able to capture all possible (or significant) solutions, thus generating the entire spectrum. Figure 12.9 illustrates the procedure for four curves (A, B, C, and D). Design A contributes to the upside of the upper bound risk curve, while design B contributes to the downside of it. The middle portion of the upper bound risk curve is the contribution of design C. The lower bound risk curve is contributed from two designs B in the upside and D in the downside. One final warning needs to be added: upper bounds can be constructed only if the problems can be solved to rigorous global optimality.

12.5.3 Automatic Risk Evaluation and New Measures

All widely used measures of risk are related to the downside portion of the risk curve. In striving to minimize risk at low expectations, they rarely look at what happens on the upside. In other words, a risk averse decision maker will prefer curve 2 (Figure 12.8), while a risk taker will prefer curve 3. In reality, no decision maker is completely risk averse or completely risk taker. Therefore, some compromise like the one offered by curve 4 needs to be identified. Thus, some objective measure that will help identify this compromise is needed. If such a measure is constructed, the evaluation can be automated so that a decision maker does not have to consider and compare a large number of curves visually. Aseeri *et al.* (2004) discussed some measures and proposed others such as:

- *Opportunity value* (or *upside potential*), which is defined the same way as VaR but on the upside. OV and VaR are illustrated in Figure 12.10 where two projects are compared, one with expected profit of 3 (arbitrary units) and the other of 3.4. The former has a VaR of 0.75, while the latter has a VaR of 1.75. Conversely, the upside potential of these two projects is 0.75 and 3.075, respectively. Considering a reduction in VaR without looking at the change in OV can lead to solutions that are too risk averse.
- The ratio of OV to VaR, which can be used in conjunction with the expected profit to sort solutions out.

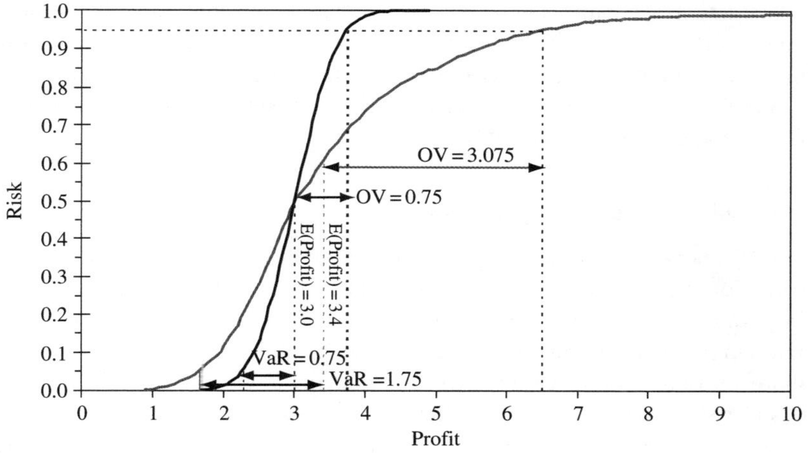

Figure 12.10 *Opportunity value (OV) or upside potential vs. VaR*

- The *risk area ratio* (RAR), which is defined as the quotient of the areas between a solution and the maximum expected profit solution (SP). More specifically, it is given by the ratio of the *opportunity area* (O_Area), enclosed by the two curves above their intersection, to the *risk area* (R_Area), enclosed by the two curves below their intersection (Figure 12.11):

$$\mathrm{RAR} = \frac{\mathrm{O\_Area}}{\mathrm{R\_Area}} \tag{12.21}$$

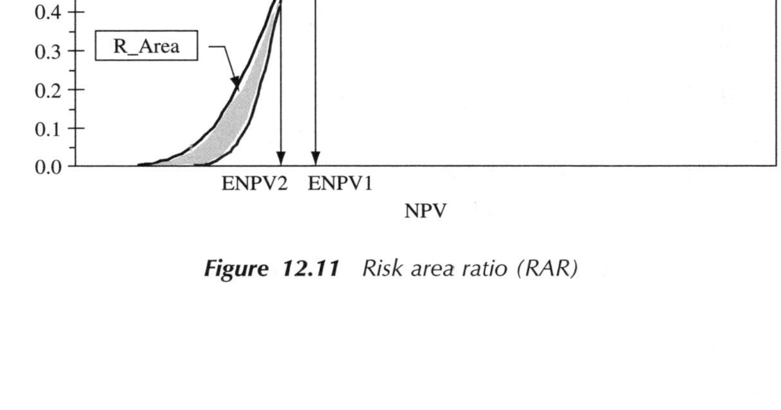

Figure 12.11 *Risk area ratio (RAR)*

- By construction, the ratio cannot be smaller than 1, but the closer this ratio is to 1, the better is the compromise between upside and downside profit. Note also that this is only true if the second curve is minimizing risk in the downside region. If risk on the *upside* is to be minimized, then the relation is reversed (i.e. O_Area is below the intersection and R_Area is above it).

12.5.4 Use of Expected Utility Theory

As discussed above, expected utility can be reconciled with the two-stage stochastic framework. For example, if one uses the nonlinear coordinate transformation of real value into utility value given by the utility function (Figure 12.3), one can modify the view of the risk curve, as shown in Figure 12.12. If such utility function can be constructed based more on quantitative relations to shareholder value, then one does not need to perform any risk management at all. One could speculate that it suffices to maximize utility value, but only if one has identified the ultimate objective function associated with the company's optimum financial path. It is worth noting that anything less, like the net present value, which can be considered a utility function too, will require the analysis of different curves before a final choice is made.

12.5.5 Markov Decision Models and Dynamic Programming

This approach, recently suggested by Cheng *et al.* (2003), proposes to rely on a Markov decision process, modeling the design/production decisions at each epoch of the process as a two-stage stochastic program. The Markov decision process used is similar in nature to a multi-stage stochastic programming where structural decisions are also considered as possible recourse actions. Their solution procedure relies on dynamic programming techniques and is applicable only if the problems are separable and monotone. In addition, they propose to depart from single-objective paradigms, and use a multiobjective approach rightfully claiming that cost is not necessarily the only objective and that other objectives are usually also important, like social consequences, environmental impact, process sustainability. Among these other objectives, they include risk (measured by

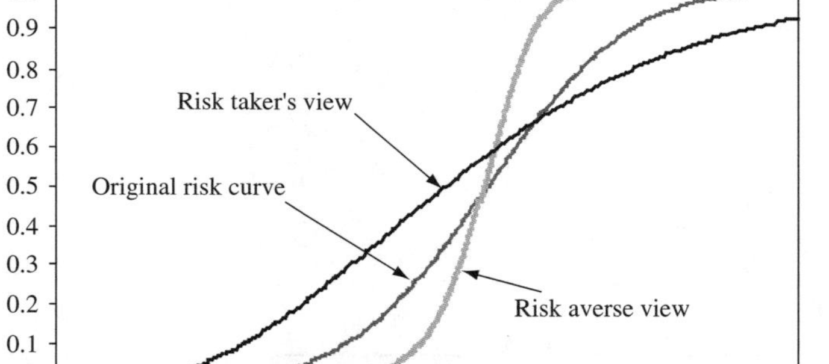

Figure 12.12 Risk curves based on utility functions

downside risk, as introduced by Eppen *et al.*, 1989) and, under the assumption that decision makers are risk averse, they claim it should be minimized. Aside from the fact that some level of risk could be tolerable at low profit aspirations in order to get larger gains at higher ones, thus promoting a risk-taking attitude, this assumption has some important additional limitations. Since downside risk is not only a function of the first-stage decisions but also of the aspiration or target profit level, minimizing downside risk at one level does not imply its minimization at another. Moreover, minimizing downside risk does not necessarily lead to minimizing financial risk for the specified target. Thus, treating financial risk as a single objective presents some limitations, and it is proposed that risk be managed over an entire range of aspiration levels as discussed above. This may present some problems for the dynamic programming approach making two-stage programming more appealing.

Conclusions

- Models with chance or regret constraints are less efficient because they can only generate a subset of the spectrum solutions at best.
- The big difference between this engineering view and that of the economists is that they rely on point measures because they consider risk as the behavior of the distribution at low profit values, while the engineers try to strike a balance at all profit levels.
- Risk management can be best performed by the generation of the spectrum of solutions followed by the identification of the more desirable solutions, as opposed to penalization of stochastic solutions using any measure, including risk directly.
- Such spectrum can be obtained using goal programming, worst-case and best-case scenario maximization, and/or by a decomposition procedure based on the sampling average algorithm.
- The screening of solutions can be best made by looking at the area ratio.
- The use of utility functions, if they can be constructed in direct relation to shareholder value, would eliminate or reduce the need for risk management because the utility function already contains it.

12.5.6 Case Studies

Gas commercialization in Asia. Aseeri and Bagajewicz (2004) considered the problem of investing in the distribution and use of gas in the region. Transportation through pipelines (whenever possible), LNG, and CNG ships was considered. The use of GTL technologies to produce gasoline, ammonia, and methanol were also considered. Many producers (Australia, Indonesia, Iran, Kazakhstan, Malaysia, Qatar, and Russia) and buyers in the region (Japan, China, India, South Korea, and Thailand) were considered. The scope of the project extends from the year 2005 to 2030 and the capital investment was limited. The planning model maximized the expected net present value and used the structure of classical planning models under uncertainty (Sahinidis *et al.*, 1989) and the risk analysis was performed using risk curve generation using the decomposition procedure based on the sample averaging method explained above. The solution to the problem included the number of ships that need to be purchased in each period of time, the number, location, and corresponding capacity of the plants to be built and the countries whose demand is to be partially (or fully) satisfied.

For an investment limit of 3 billion dollars in the first time period and 2 billion dollars in the third time period with the other four time periods having no investments allowed,

Table 12.1 *Results for stochastic model (200 scenarios). Gas commercialization in Asia*

Time period	FCI	Processing facilities				Transportation to:				
		Indo (GTL)				China		Thai		Avrg. ships
		Cap	Flow	Feed	Ships	Ships	Flow	Ships	Flow	
T1	3.00	–	–	–	–	–	–	–	–	–
T2	–	4.43	4.25	283.1	5.0	1.12	0.76	3.88	3.48	5.00
T3	1.90	4.43	4.43	295.5	5.0	–	–	4.94	4.43	4.94
T4	–	7.18	7.09	472.6	8.0	0.44	0.30	7.56	6.79	8.00
T5	–	7.18	7.18	479.0	8.0	–	–	8.00	7.18	8.00
T6	–	7.18	7.18	479.0	8.0	–	–	8.00	7.18	8.00

Note: Capacities and flow are in million tons per year and feed gas flow is in billion standard cubic feet per year.

the model gave the results shown in Table 12.1. The first part of the table (processing facilities) shows the existing (available) capacities. The fixed capital investment (FCI) appears on the time period prior to capacity increases because of construction time (4 years). The required gas feed amounts are indicated on the 'Feed' column in billion SCF/year. Also the numbers of ships available for transportation are indicated in the 'Ships' column.

Thus, a GTL plant should be built in Indonesia in the first time period with a capacity of 4.43 million tons/year and five ships are to be built/purchased for the transportation of the GTL product. An expansion in the third time period to increase the capacity to 7.18 million tons/year as well as the purchase of three additional ships is suggested. The second part of the table (transportation) shows the number of ships that are assigned to transport products to different markets as well as the yearly flow of transported products (fractional ships should be understood as fractions of the year that each ship is allotted to a certain route). Not all the investment is utilized in the third period, which is explained by the fact that increased capacity leads to the need for more ships, money for which is not available.

When downside risk at 3.5 billion dollars is penalized, a design that reduces risk and does not have a large effect on ENPV was obtained. The design obtained is illustrated in Table 12.2. This result also suggests a GTL process, but at another supplier location

Table 12.2 *Results for stochastic model (200 scenarios) with downside risk at \$B 3.5 minimized. Gas commercialization in Asia*

Time period	FCI	Processing facilities				Transportation to:				
		Mala (GTL)				China		Thai		Avrg. ships
		Cap	Flow	Feed	Ships	Ships	Flow	Ships	Flow	
T1	3.00	–	–	–	–	–	–	–	–	–
T2	–	4.57	4.47	297.9	4.0	1.16	0.98	2.79	3.49	3.95
T3	1.89	4.57	4.57	304.9	4.0	–	–	3.66	4.57	3.66
T4	–	7.49	7.32	488.2	6.0	0.42	0.35	5.58	6.97	6.00
T5	–	7.49	7.49	499.6	6.0	–	–	6.00	7.49	6.00
T6	–	7.49	7.49	499.6	6.0	–	–	6.00	7.49	6.00

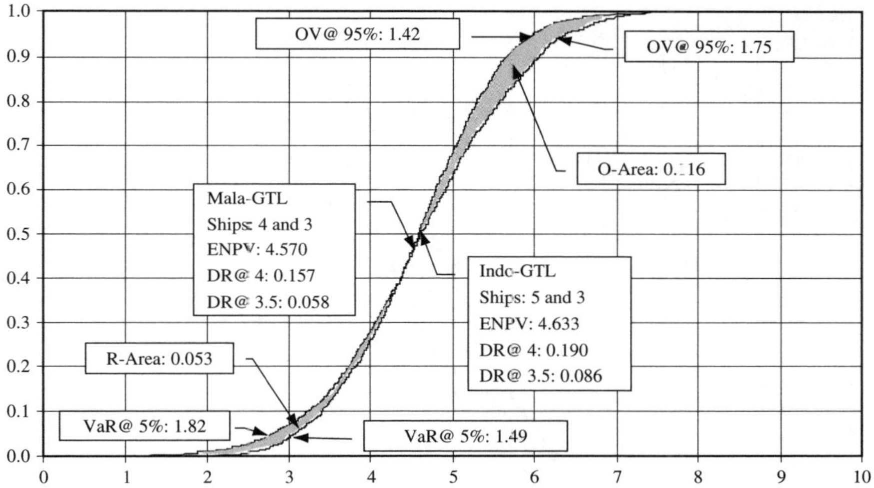

Figure 12.13 *Comparison of risk curves for gas commercialization in Asia*

(Malaysia). Investment in Malaysia manages to reduce risk over that in Indonesia due to the lower volatility of natural gas prices in Malaysia. Figure 12.13 compares the risk curves and shows values of VaR and OV. Table 12.3 compares the risk indicators more closely. The VaR reduces from 18.1% but the OV (UP) reduces 18.9% and the risk area ratio (RAR) is equal to 2.2. This means that the loss in opportunity is more than twice the gain in risk reduction. The application of the decomposition procedure rendered similar solutions to those obtained using the full stochastic model. The use of regret analysis in this case produced similar but slightly less profitable answers.

Figure 12.14 shows the upper and lower bound risk curves as well as the solution that maximizes ENPV and the one that minimizes risk. It was noticed during the construction of the lower bound risk curve that 89.4% of its points were mainly contributed by one single bad design. When this design was excluded, a tighter and more practical lower bound risk curve was obtained, which is the one depicted.

Offshore oil drilling. Aseeri *et al.* (2004) considered the problem of scheduling the drilling of wells in offshore reservoirs and planning their production using a basic model similar to that of Van den Heever *et al.* (2000). Uncertainties in reservoir parameters (productivity index) and oil price were considered. In addition, budgeting constraints tracing cash flow and debts were added. One field consisting of three reservoirs was assumed (Figure 12.15). In each reservoir two wells can be drilled for which estimates

Table 12.3 *Value at risk for the alternative solutions. Gas commercialization in Asia*

Solution	VaR(5%)	UP(95%)	Risk @ 3.5	DRisk @ 3.5
NGC	1.82	1.75	14.4%	0.086
NGC-DR	1.49	1.42	12.0%	0.058

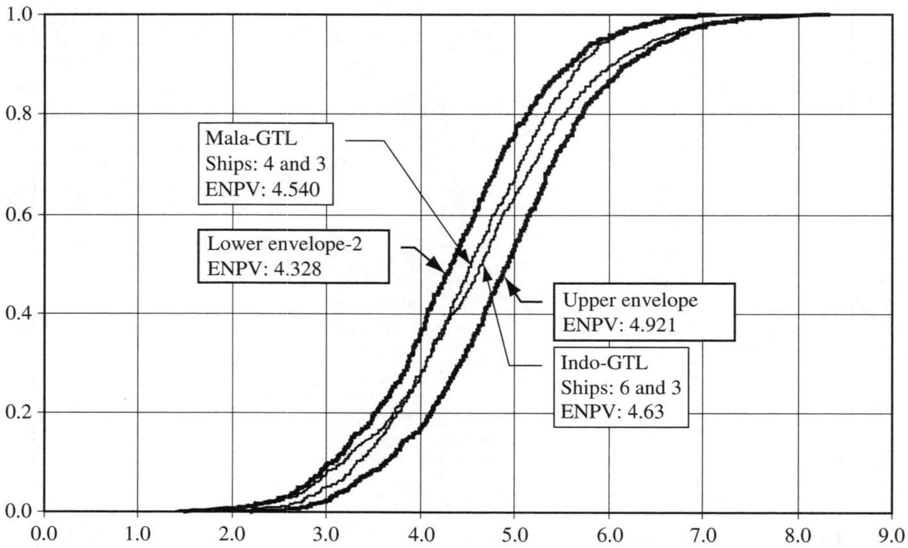

Figure 12.14 *Upper and lower bound risk curves for gas commercialization in Asia*

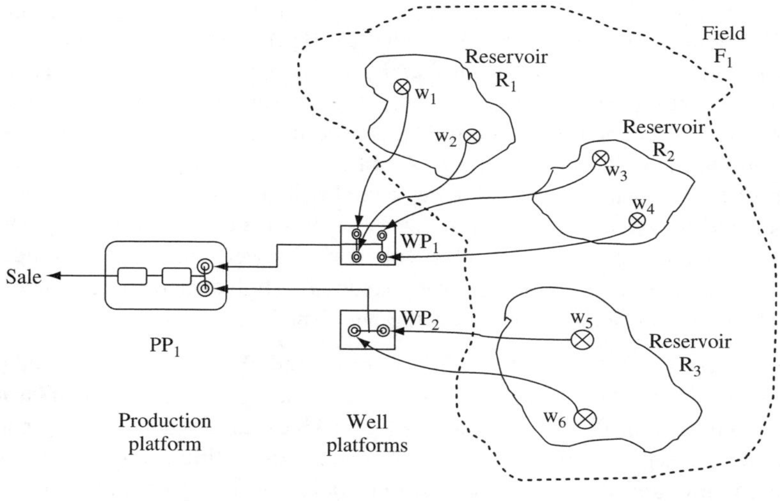

Figure 12.15 *Offshore drilling superstructure*

of the drilling cost as well as the expected productivity index are assumed to be known. The wells in reservoirs R_1 and R_2 can be connected to a well platform WP_1 and the wells in reservoir R_3 can be connected to well platform WP_2. Both well platforms are to be connected to a production platform in which crude oil is processed to separate gas from oil and then oil is sent to customers. The objective of this problem is to maximize the net present value of the project. The decision variables in the model are reservoir choice,

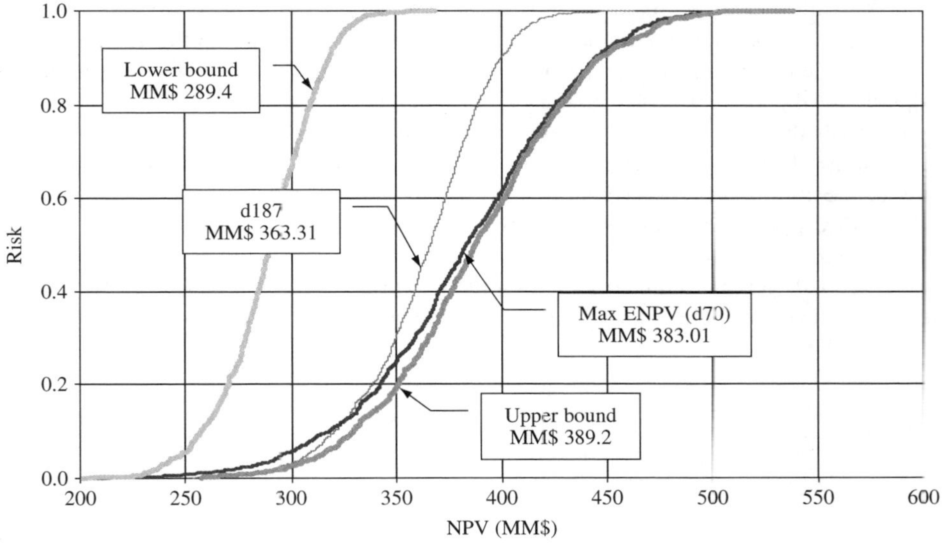

Figure 12.16 *Solutions and bounds for the offshore drilling case study*

candidate well sites, capacities of well and production platforms, and fluid production rates from wells. The problem is solved for a 6-year planning horizon with quarterly time periods (24 time periods).

Applying the decomposition procedure described above, the solutions and the corresponding bounds shown in Figure 12.16 were obtained. The gap between the optimal solution and the upper bound is less than 1.6%. The production rates and reservoir pressure profiles are, of course, different. The maximum profit solution opens production of wells w6, w5, none, w3, w4, w2, w1 in months three through nine, respectively. The alternative less risky design opens production of wells w3, w4, w1, w2, none, w6, w5 in the same months. Platforms are built in one time period before the wells are opened.

In the less risky solution, VaR reduced from 87.12 to 55.39 or 36.4% and OV reduced from 78.81 to 45.19 or 42.7%. The resulting RAR is 16.4. This is an indication of how significant the reduction in opportunity is compared to the small reduction in risk.

Design of water networks. Koppol and Bagajewicz (2003) considered the problem of designing water utilization networks. Water is used in many operations, mainly washing, or as direct steam in process plants. Water is put in contact with organic phases from which the contaminants are extracted. Such water utilization systems consist of networks of water reuse and partial regeneration, aimed at the reduction of cost. A review article by Bagajewicz (2000) offers a detailed description of the different reuse and regeneration schemes that have been proposed, as well as the variety of solution procedures proposed. In addition, Koppol *et al.* (2003) discuss zero liquid discharge cycles.

The problem consists of determining a network of interconnections of water streams among the processes so that the expected cost is minimized while processes receive water of adequate quality, with or without change of flows. Thus, one is allowed to reuse wastewaters from other processes, diluting it with fresh water if there is a need for it and eventually placing a treatment process in between uses. The uncertainty in this case comes mainly from the contaminant loads that the water will pick up in each process.

One single-contaminant example involving six water using processes and solved by Koppol and Bagajewicz (2003) assumes 20% uncertainty for the contaminant loads and capital costs that are comparable to reductions in operation cost achieved by using reuse connections. The effects of financial risk considerations are illustrated by showing two results, one minimizing costs (Figure 12.17) and the other minimizing risk (Figure 12.18).

The corresponding risk curves are depicted in Figure 12.19. Note that the risk curves are inverted because this problem pursues minimization of cost and not maximization of profit. Second, the minimum cost solution reduces operating costs (consumes 107.5 ton/h of fresh water), while the risk reduced solution reduces the capital cost of interconnection

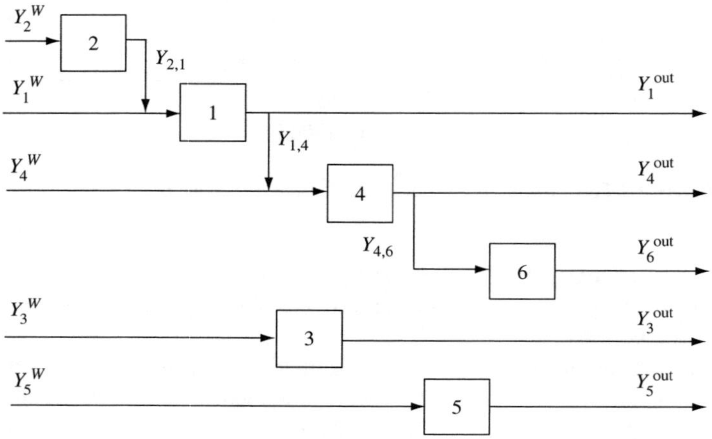

Figure 12.17 *Minimum expected cost water network*

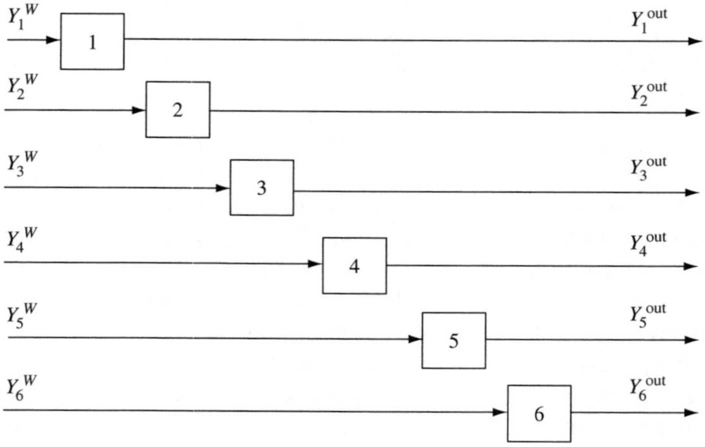

Figure 12.18 *Minimum risk water network*

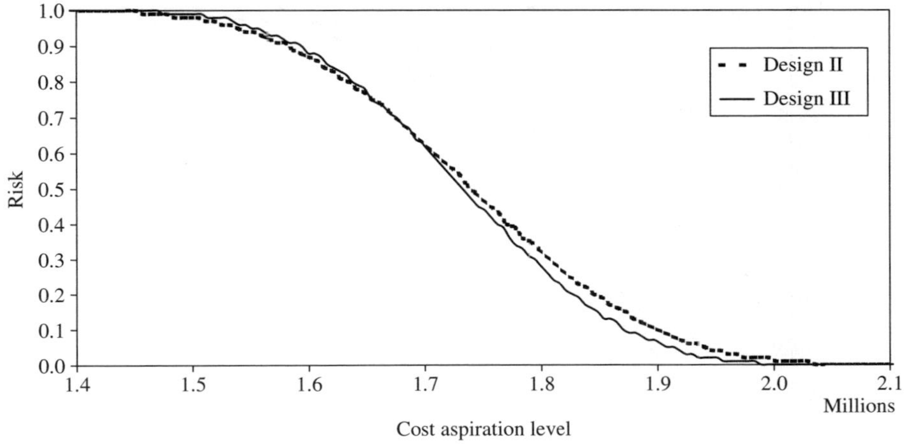

Figure 12.19 *Risk curves for water networks*

piping and has a larger freshwater consumption (134.9 ton/h). The latter has a slightly higher cost (1.74 vs. 1.73 million dollars/year). One important thing to point out from these solutions is that the less risky solution makes no reuse of spent water, that is, it says one should not build any network of interconnections

Planning of the retrofit of heat exchanger networks. Barbaro and Bagajewicz (2003) considered the problem of adding area and heat exchangers between different plants to help save energy in the total site. This is a typical retrofit problem, with the exception that they also address how the placement of new units should be scheduled through time. The uncertainty considered is in the price of energy. The first-stage decisions are the schedule of additions of exchangers and the second-stage decisions consist of the energy consumption of the different units. The possibilities of reducing throughput because of the lack of installed capacity are taken into account. Results on small-scale examples show that financial risk considerations motivate changes in the decision making.

12.6 Effect of Inventories on Financial Risk

It is common accepted knowledge that inventory hedges from price, availability, and demand variations, and their impact on the profitability of the operations. It is also known that maintaining such inventory has a cost, both capital and operative. That risk is automatically reduced is not necessarily true unless risk is managed specifically as is briefly shown next. Contrary to the assumption that operating at zero inventory (produce to order) always increases profit, it will be also shown that inventories do not represent a reduction in expected profit.

Barbaro and Bagajewicz (2004b) showed how the hedging effect of inventories can be better appreciated through the analysis of the risk curves. They presented an extension of the deterministic mixed-integer linear programming formulation introduced by Sahinidis *et al.* (1989) for planning under uncertainty. The model considers keeping inventories

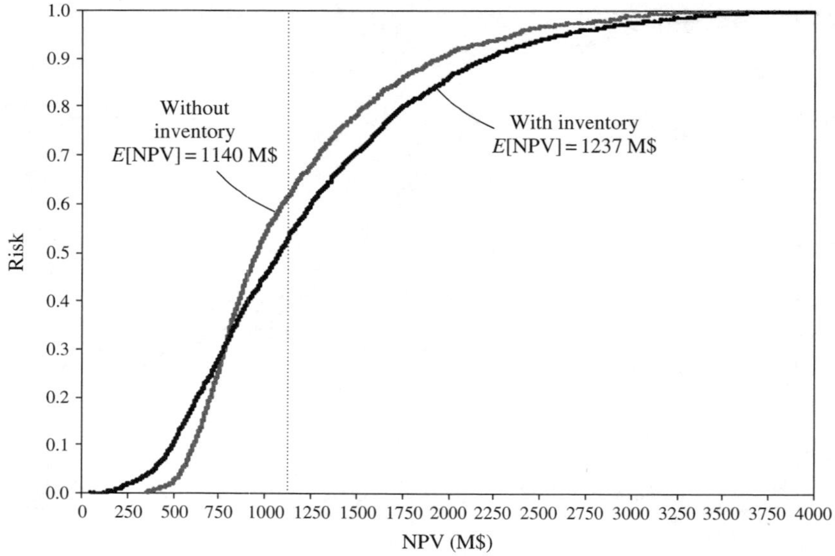

Figure 12.20 *Solutions of investment planning in process networks (with and without inventory)*

of raw materials, products, and intermediate commodities when uncertain prices and demands are considered. Details of the planning solution are omitted here, concentrating on the analysis of the risk curves. Figure 12.20 compares the solutions with and without the use of inventory. It is apparent from the figure that:

- the solution that makes use of inventory has higher expected profit, which is contrary to existing perceptions;
- risk exposure at low aspiration levels is higher when inventory is considered.

In turn, Figure 12.21 shows the spectrum of solutions obtained using downside risk through goal programming. In this spectrum, several solutions that reduce risk even compared to the solution not using inventory can be found. Thus:

- Solutions do not increase opportunities for high profit. The risk area ratio is expected to be large, more of a reason to watch the curve and not rely on point measure indicators.
- The usual perception that inventory helps reduce risk is confirmed, *but it requires risk to be specifically managed.*

Interestingly, many articles devoted to inventory risk, especially in management science, consider variance as a measure of risk (Gaur and Seshadri, 2004) and proceed consequently. While there are many other intricacies behind the relationship between inventories, risk hedging, and expected profit that engineers have not yet grasped, the use of variance constitutes the first head-on collision between both approaches.

In addition, by focusing on how external factors (product demand, prices, etc.) translate into changes in a company's assets through a two-stage stochastic programming approach, the decision maker can manage risk and also uncover several strategic options such as capacity integration (Gupta *et al.*, 2000).

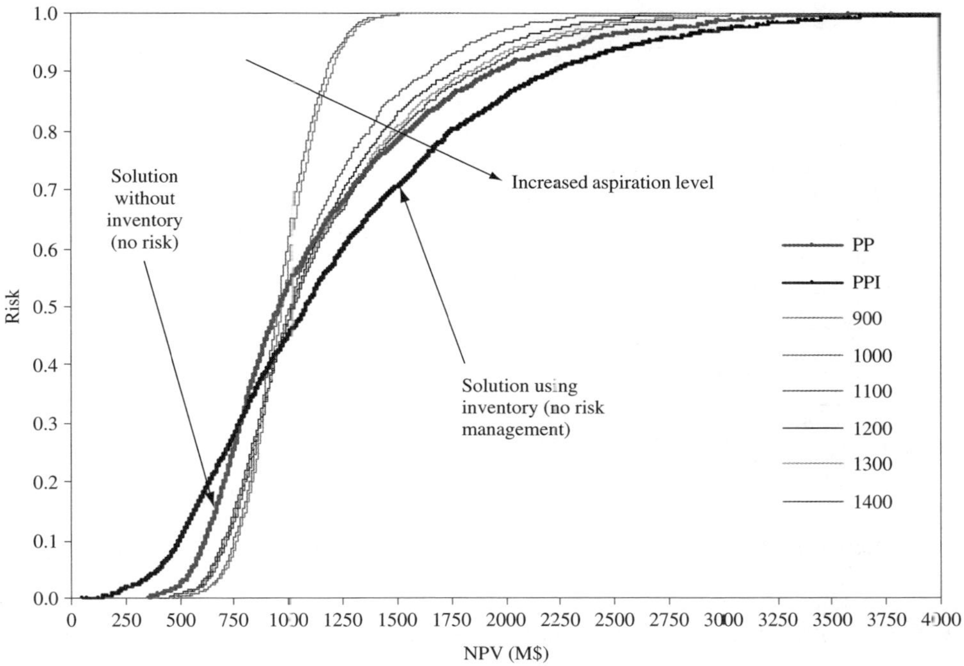

Figure 12.21 *Spectrum of solutions of investment planning in process networks (with and without inventory)*

12.7 Effect of Contracts and Regulations in Project Planning

12.7.1 Regular Fixed Contracts

A contract is a binding agreement which obligates the seller to provide the specified product and obligates the buyer to pay for it under specific terms and conditions. One method of managing the risk when prices are uncertain is to use long-term fixed-price contracts especially with raw material suppliers but also with consumers downstream of the supply chain. However, the risk arising if the spot market price for natural gas turns out to be, in average, less than the fixed contract price cannot be avoided (Derivatives and Risk Management, EIA, 2002). This is addressed below by option contracts.

Aseeri and Bagajewicz (2004) illustrated the effects of contracts using the problem of commercializing gas in Asia (outlined earlier). Natural gas prices were assumed to have fixed prices at the supplier location at their mean values. The risk curves are shown in Figure 12.22. We summarize the results as follows:

- For this case, the difference of expected profit is very small. Actually the plan that uses contracts has slightly higher profit (0.6%), but it is unclear if this is a real gain or just a numerical effect. The solution with contracts chooses the same locations as the one without contracts but with different capacities.

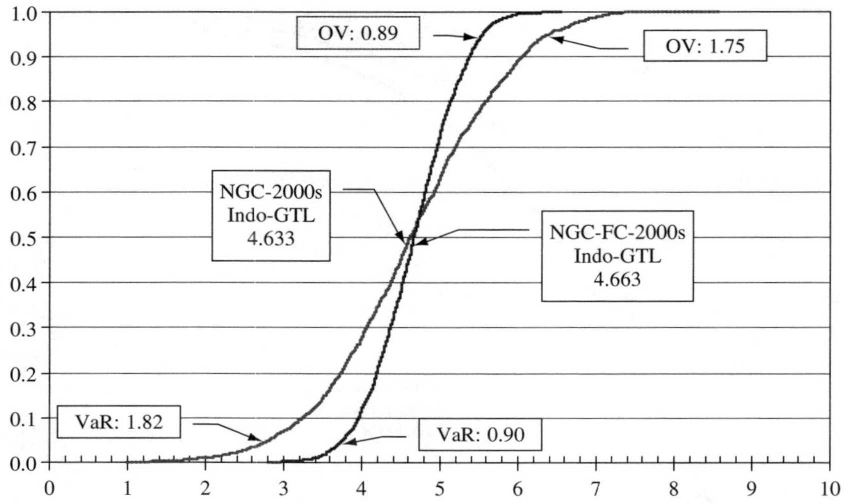

Figure 12.22 *Effect of fixed price contracts on gas commercialization in Asia*

- Risk is substantially reduced (about 50% reduction in VaR), but OV also reduces by roughly the same amount. Thus, contracts have a hedging effect from bad scenarios, but also prevent high profit materializing in optimistic scenarios.
- Contracts have a larger risk reduction effect compared to plain risk management without using them. This can be seen by comparing with Figure 12.14.

12.7.2 Effect of Option Contracts

Futures and *option* contracts are often referred as *derivatives* (Hull, 1995). A futures contract is an agreement to buy or sell an asset at a certain time in the future for a certain price. In turn, there are two basic kinds of option contracts: calls and puts. A call option gives the holder the right to buy an asset by a certain date and for a certain price. On the other hand, a put option gives the holder the right to sell an asset by a certain date and for a certain price. These contracts are traded daily in many exchanges such as the Chicago Board of Trade (CBOT), the Chicago Mercantile Exchange (CMB), the New York Futures Exchange (NFE), and the New York Mercantile Exchange (NYMEX) among others. These derivatives are agreed and the option holder party pays a premium (option cost) to gain the privilege of exercising his/her options. It consists of two components, an intrinsic value and a time value. The intrinsic value is measured as the difference between the strike price and the market price. In the case of gas commercialization in Asia the market price is the mean expected price of gas. If the two are equal then the intrinsic value is 0. The time value is the extra amount which the option buyer is willing to pay to reduce the risk that the price may become worse than the mean values during the time of the option. The time value is affected by two elements: the length of the time period for the option and the anticipated volatility of prices during that time (SCORE, 1998).

Barbaro and Bagajewicz (2004b) introduced specific constraints that can be used in the context of two-stage stochastic investment planning models. Similarly to fixed contracts, the usual assumption that option contracts hedge risk automatically at low profit levels is not always true. Specific risk management is required.

Aseeri and Bagajewicz (2004) showed that risk curve analysis can be also used to determine the right premium to pay depending on what side of the negotiation one is. Figure 12.23 shows the risk curves for the results of a stochastic model run using different premium costs. We notice that with a premium unit cost of 2% of the mean value, the option contract shifts the risk curve substantially to the right; that is, it considerably increases the profit at almost all scenarios. The results with 4%, 6%, and 8% could be acceptable to the supplier since they have a significant chance of success. Any price greater than 8% is not attractive to the buyer. They also run the model penalizing downside risk, showing that, risk can be managed as well. In fact, option contracts can produce a 38% reduction of VaR with a small reduction in risk area ratio (RAR), much smaller than the case of fixed contracts (although these last ones reduce VaR by 50%), all at the same value of expected profit. Thus,

- option contracts do not automatically reduce risk and require risk management;
- they are excellent tools to reduce risk at low profit expectations, maintaining upside potential (UP or OV).

Rogers *et al.* (2002, 2003) discuss the use of real options in pharmaceutical R&D projects, Gupta and Maranas (2003b) discuss the use of emission option contracts in the technology selection for pollution abatement, and Rico-Ramirez *et al.* (2003) use real options in batch distillation. Gupta and Maranas (2003b) recognize that variance cannot be used for risk management because of its symmetric nature, but we should give credit to Ahmed and Sahinidis (1998) for pointing this out first. Finally, the finance community has proposed means of managing risk through real options (Dixit and Pindyck, 1994; Trigeorgis, 1999).

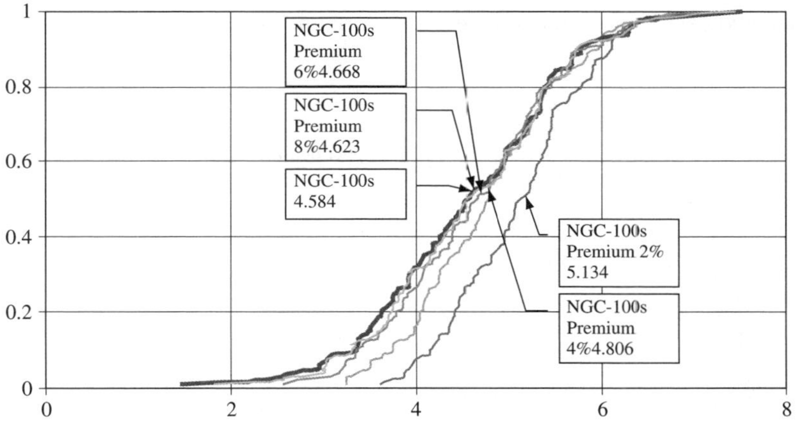

Figure 12.23 *Effect of derivative premium on gas commercialization in Asia*

12.7.3 Effect of Regulations

In a recent article Oh and Karimi (2004) explore the effect of regulations (bilateral and multilateral international trade agreements, import tariffs, corporate taxes in different countries, etc.) on the capacity expansion problem. They point out that barring the work of Papageorgiou *et al.* (2001), who explore the effect of corporate taxes in the optimization of a supply chain for a pharmaceutical industry, very little attempt has been made to incorporate other regulatory issues in the capacity expansion problem. However, they point to other attempts in location-allocation problems and in production-distribution problems (Cohen *et al.*, 1989; Arntzen *et al.*, 1995; Goetschalckx *et al.*, 2002), which include tariffs, duty drawbacks, local content rules, etc. for a multinational corporation.

12.8 Integration of Operations Planning and Budgeting

Cash flow needs to be managed at the first stage, not at the second stage. This is contained in the pioneering work of Badell and Puigjaner (1998), Badell and Puigjaner (2001a,b), Romero *et al.* (2003a,b), and Badell *et al.* (2004). In these articles, the group of Professor Puigjaner establishes the links between procuring, financial management, and manufacturing; that is, proposes the use of models that break the walls existing between these three entities. Basically, deterministic cash flow is considered at the same time as scheduling of operations, batch plants in this case. Also, the articles provide some background on the literature of cash management models and the need to use and integrate them. More specifically, Romero *et al.* (2003a) propose the merging of scheduling and planning with cash management models (Charnes *et al.*, 1959; Robichek *et al.*, 1965; Orgler, 1970; Srinivasan, 1986). Their model shows that the profit increases by considering these activities together because procuring does not buy expensive raw materials too early in the process. The integrated approach, instead, proposes to rearrange the schedule to accumulate some cash to buy these expensive raw materials. The root of the difference is not only better cash management, but also a departure from Miller and Orr's (1966) model (Figure 12.24), where it is recommended to borrow or buy securities, only when a lower or upper bound is reached. A flat profile, like the one in Figure 12.25,

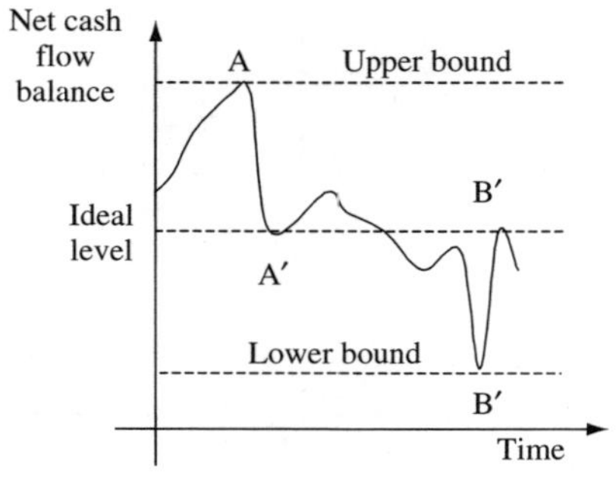

Figure 12.24 *Miller and Orr's cash management model*

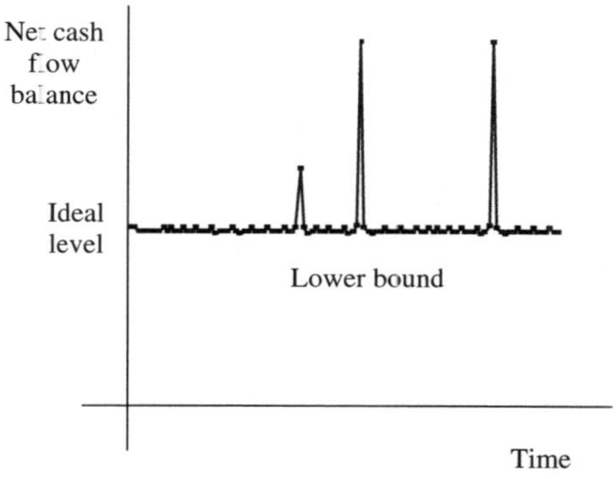

Figure 12.25 *Ideal cash flow model*

is desirable, and achievable only with the integrated model. In this flat profile, one only sees spikes due, for example, to the short span between the inflow of cash and the buying of securities. As one can see, no downward spikes are observed because the outflow of money can be planned.

Romero *et al.* (2003a) also present preliminary work where the above ideas are also extended to consider uncertainty and financial risk. Some sort of budgeting, although not full cash management, was considered by Van den Heever *et al.* (2000) where royalty payments are taken into account in the offshore drilling problem. Although royalties were not included, full cash management with uncertainty was considered by Aseeri *et al.* (2004) for the same offshore oil drilling problem, results of which have been outlined above. Van den Heever and Grossmann (2000) proposed an aggregation/disaggregation method to solve the problem. The same group of researchers added tax and royalty calculations to the problem, which increased its numerical complexity (Van den Heever *et al.*, 2000) and studied the use of big-M constraints as well as disjunctive programming. Finally, Van den Heever *et al.* (2001) proposed a Lagrangean decomposition procedure.

12.9 Integration of Operations Planning and Pricing

Guillén *et al.* (2003a) suggested that pricing policies considered in an integrated manner with scheduling decisions (integration of manufacturing and marketing) increase profit. They discuss the existing models for pricing and point out that the supply curve, which is dependent on manufacturing costs, can be altered. In other words, altering the production schedule should and indeed does have an effect on the fixed costs per unit used in existing classical pricing models (Dorward, 1987; Mas-Collel *et al.*, 1995). They assumed an iterative model in which product prices are first fixed to obtain a schedule, which in turn can be used again in an iterative manner to obtain new product costs which enable the calculation of new processes. They showed that this model does not always converge. Moreover, an alternative model in which process *and* production schedules are obtained simultaneously is proposed. The integrated model, they show, produces different schedules and prices and allows larger profits. Indeed, for a case of three products, the

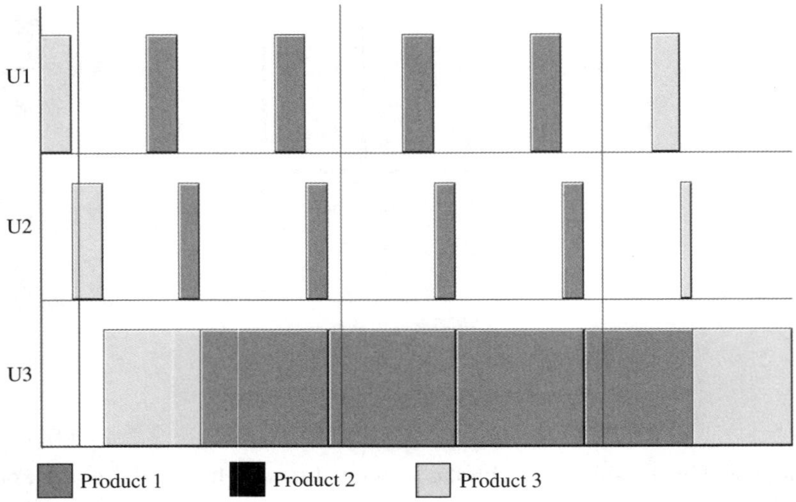

Figure 12.26 *Gantt chart using the iterative method for scheduling and pricing*

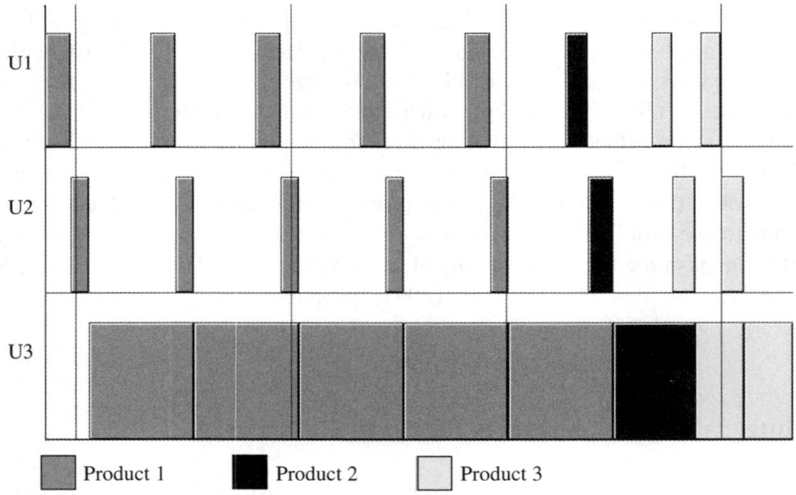

Figure 12.27 *Gantt chart using the integrated model for scheduling and pricing*

iterative model renders the Gantt diagram of Figure 12.26 and the integrated model renders the one in Figure 12.27, both choosing different processes and the latter having almost twice the profit of the former. In these diagrams, the corresponding product produced in each stage is shown on top of each batch.

In addition, Guillén *et al.* consider uncertainty in the demand–price relation parameters. Thus, they build a stochastic model, in which processes are first-stage decisions, not parameters as is common in batch scheduling models, and sales are second-stage variables. The model renders different schedules and prices (Figure 12.28). The resulting schedule

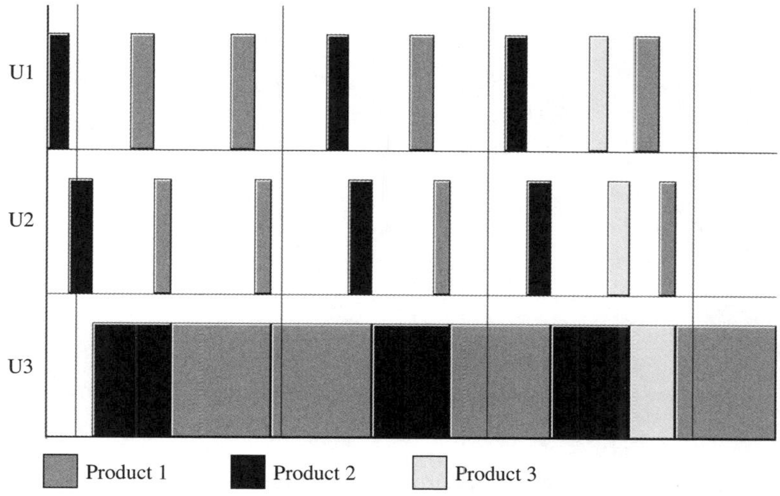

Figure 12.28 Gantt chart for the stochastic case for scheduling and pricing

implies a mixed product campaign and not a single product campaign as occurred in the deterministic case. This schedule seems to be more robust.

In order to show the capability of the proposed formulation of risk management, the problem is modified so as to reduce the risk associated at low targets. The Gantt chart corresponding to one solution with lower risk and consequently lower expected profit is shown in Figure 12.29. The risk curves of both the stochastic solution (SP) and the one with lower risk are shown in Figure 12.30. The risk at low expectations (profits under a target of $6500) was reduced to 0.

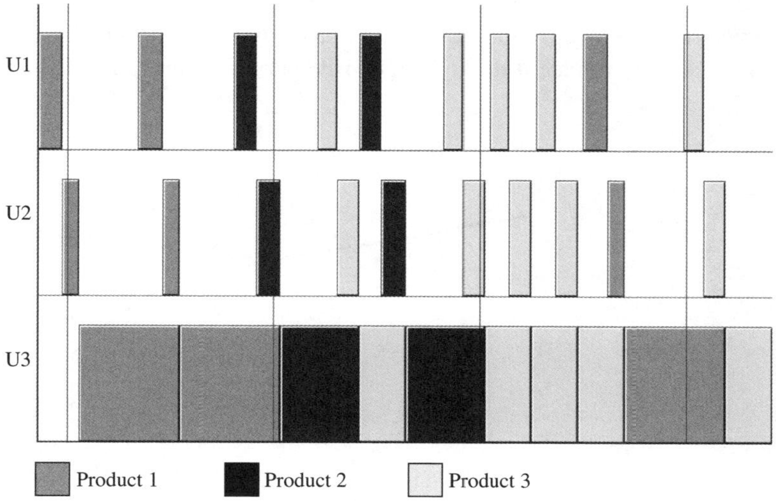

Figure 12.29 Gantt chart for the risk managing solution for scheduling and pricing

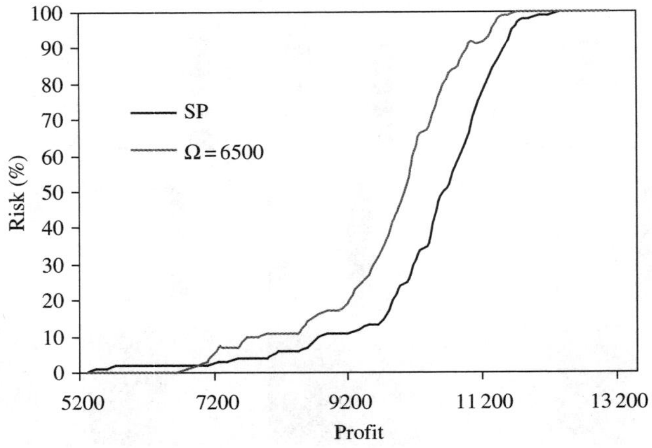

Figure 12.30 *Risk curves for scheduling and pricing*

12.10 Consumer Satisfaction

Consumer satisfaction was considered in conjunction with supply chain management by Tsiakis *et al.* (2001). They measure it using the quotient between sales and demand. Guillén *et al.* (2003b) study a similar problem: a supply chain with warehouses, markets, and distribution centers. Instead of constraining the customer satisfaction, they assume a multiobjective model and construct a Pareto surface. They also solve a stochastic model where demands are uncertain, the opening of plants/warehouses is first-stage variables and the sales are second-stage variables. The two resulting Pareto curves are different (Figures 12.31 and 12.32) revealing the need for stochastic models. Moreover, Guillén *et al.*, define customer satisfaction risk and discuss the interrelation between three different objectives: NPV, customer satisfaction, and financial risk. Finally, they define compounded risk and evaluate it through composite curves (Figure 12.33).

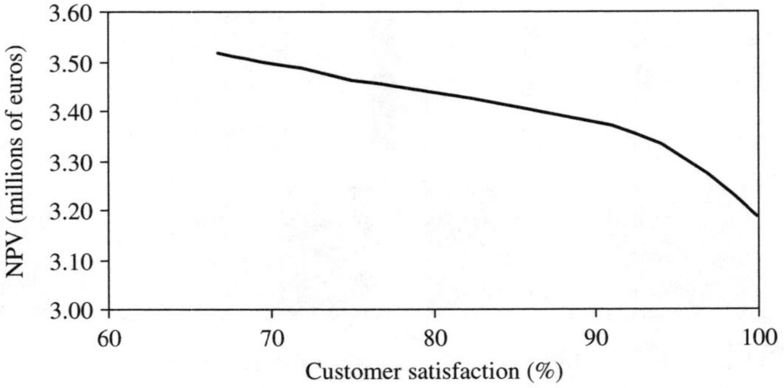

Figure 12.31 *Deterministic Pareto curve: profit–customer satisfaction*

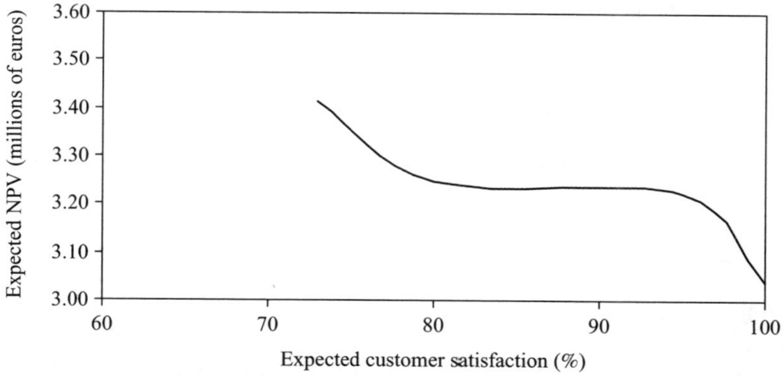

Figure 12.32 Stochastic Pareto curve: profit–customer satisfaction

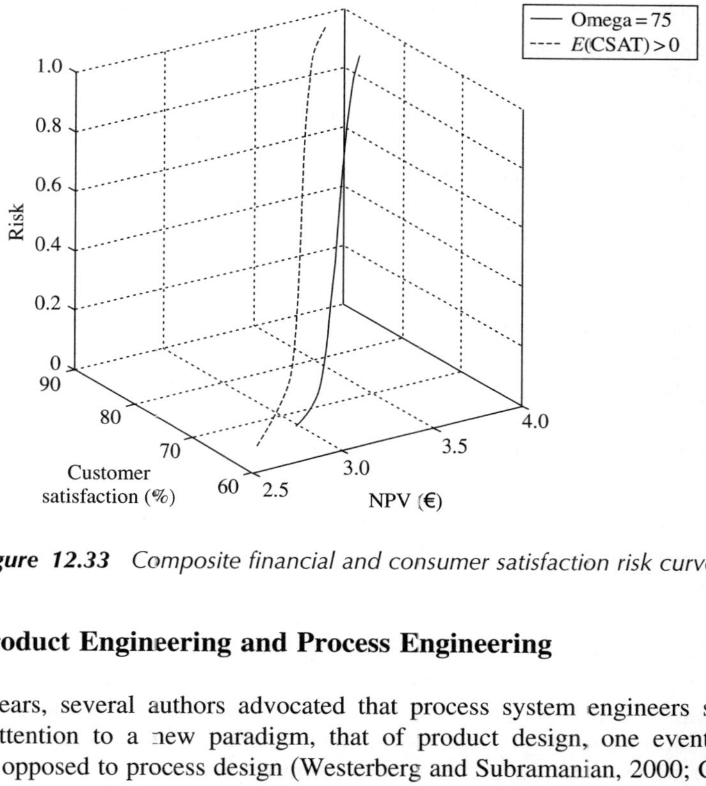

Figure 12.33 Composite financial and consumer satisfaction risk curves

12.11 Product Engineering and Process Engineering

In recent years, several authors advocated that process system engineers should pay increased attention to a new paradigm, that of product design, one eventually (and apparently) opposed to process design (Westerberg and Subramanian, 2000; Cussler and Moggridge, 2001). The suggestion, although implicit, is that process engineering is a mature field, while product design is a relatively virgin field, at least virgin from the use of tools and methods of the PSE community. One good example of efforts following the suggested path is the article by Wibowo and Ng (2001), who analyze the issues associated with the fabrication of creams and pastes. We already mentioned in the preceding text several articles dealing with drug development, which are in fact product design.

Table 12.4 Process design versus product design (taken from Cussler, 2003)

Process design	Product design
1. Batch vs. continuous	1. Customer need
2. Input/Output	2. Idea generation
3. Recycles	3. Selection
4. Separation/Heat	4. Manufacture

Following the same trend, Cussler (2003) illustrates some of the differences between the two paradigms (Table 12.4). To do the comparison he uses the so-called conceptual design paradigm (Douglas, 1988), which is very similar to the onion model (Smith and Linnhoff, 1988; Smith, 1995), and is widely used. This is the order that many books on process and product design follow (Seider *et al.*, 2004). There are of course other approaches to process design, such as the reducible superstructure approach (best represented by the book of Biegler *et al.*, 1997). Table 12.4, nonetheless, has a few interesting features.

1) It makes product design the center.
2) It suggests *ad hoc* idea generation and selection steps that presumably vary from product to product, for which a systematic search is not available or has to be constructed case by case (Cussler and Moggridge, 2001). We have seen examples on searches driven by functionalities (refrigerants, drugs, etc.) like those performed by Camarda and Maranas (1999) and Sahinidis and Tawarlamani (2000) and also described by Achenie *et al.* (2002).
3) In fact, some who advocate product design (Cussler and Moggridge, 2001) only call it chemical product design, which rules out mechanical and electronic and electromechanical devices, etc. In fact, it is only a matter of time until this expands to all products. Evans (2003), for example, has recently emphasized the upcoming integration between the process industries as providers of commodities and the discrete industries the providers of package goods, devices, appliances, automobiles, etc.
4) Its title suggests that these are somehow opposite and to an extent mutually exclusive activities.

Recently, Stephanopoulos (2003) reemphasized the idea of product design and suggested that manufacturing is indeed migrating from process (commodity)-centric to product-centric, all this judged by the performance of the companies in the stock market. He suggests that a company should maximize value-addition through the supply chain and that while the process-centered industry focuses on commodities, the product-centered industry focuses on identification of customer needs as a driving force. He asks whether 'Process Systems Engineering is prepared to engineer (design and manufacture) products, or should someone else do it?' One must wonder whether he refers to the role of chemical engineers in all end-user products, not only chemical products.

In parallel to this push for extending the borders of chemical process systems engineering to areas that have been traditionally the reserve of industrial engineers, a new concept of supply chemical chain was recently discussed in detail by Grossmann and Westerberg (2000), and in the United States National Academies report by Breslow and Tirrell (2003). The chemical supply chain extends from the molecule level to the whole enterprise. Breslow and Tirrell (2003) suggest that:

> Another important aspect in the modeling and optimization of the chemical supply chain is the description of the dynamics of the information and material flow through the chain. This will require a better understanding of the integration of R&D, process design, process operation, **and business logistics**. The challenge will be to develop quantitative models that can be used to better coordinate and optimize the chemical enterprise. Progress will be facilitated by new advances in information technology, particularly through advances in the Internet and by new methods and mathematical concepts. Advances in computer technology will play a central role. Fulfilling the goal of effectively integrating the various functions (R&D, design, production, logistics) in the chemical supply chain will help to **better meet customer demands**, and effectively adapt in the new world of e-commerce. Concepts related to planning, scheduling, and control that have not been widely adopted by chemical engineers should play a prominent role in the modeling part of this problem. Concepts and tools of computer science and operations research will play an even greater role in terms of impacting the implementation of solutions for this problem. (The bold was added)

The report cites the need for 'integration of several parts of the chemical supply chain', which 'will give rise to a number of challenges, such as multi-scale modeling for molecular dynamics, integration of planning, scheduling and control (including Internet based), and integration of measurements, control, and information systems', but falls short of discussing the full integration with economics management and business.

It is here suggested that:

1) Process engineering, which is mistakenly associated with the production of commodities, is an integral part of product design. Product design cannot exist without process design. So there is no antagonism of any sort.
2) When the chemical supply chain is considered in the context of process design, one realizes that the chemical supply chain contains many of the elements of product design. Indeed, it deals with the chemistry, the selection, the manufacturing, and the supply chain of entire enterprises, single- or multi-company, as in the alliances that Stephanopoulos (2003) suggested, that deliver the product to the customer.

Thus, product design, the newly proposed paradigm, can be constructed by putting marketing (idea generation and other tools) in front of the chemical supply chain, recognizing that process systems engineering is an essential tool used in the chemical supply chain and putting the customer and the market at both ends of the supply chain, upfront as an object of study for its needs and potential responses, and at the other end as the entity that shapes the demand and provides the feedback (Figure 12.34). Or perhaps, the chemical supply chain box needs to be broken apart into smaller pieces, each interacting with the rest in different forms.

Some elements addressing how one might address all these decision making processes are described by Pekny (2002) who explores the role of different algorithm architectures in large-scale engineering problems.

We notice that in Figure 12.34 the interactions are in both directions, suggesting that there is no sequential approach, as Cussler would suggest, but rather the notion that, since all activities influence each other, we ought to consider them simultaneously. We have also added some dotted line boxes to indicate that integrated models already exist between the different segments. Noticeably, the upfront identification of customer needs is not yet integrated with the rest of the activities. There is therefore a need for modeling in this area.

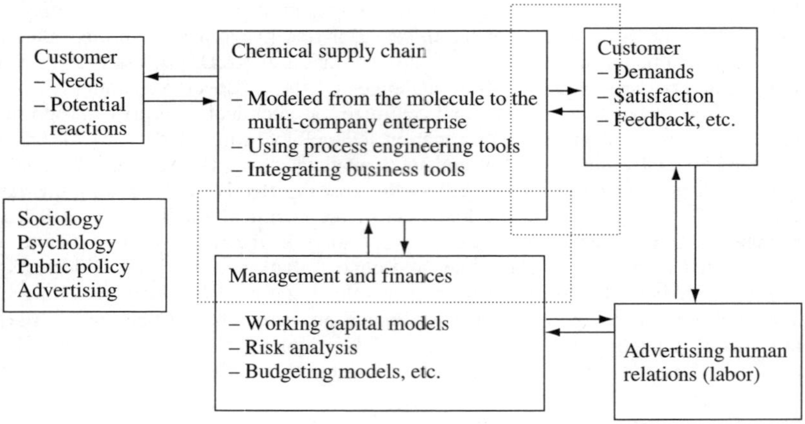

Figure 12.34 *Product development and delivery supply chain*

12.12 Retrofit

While most engineering assumes grass-roots activities, the issue of revamping the existing infrastructure of a company is sometimes even more challenging and it certainly involves integration with finances. Specifically, we leave for some future discussion the retrofit activity of the existing industry.

12.13 The Environment

Recently, there has been much discussion about sustainability and the global life cycle assessment of processes and products (Nebel and Wright, 2002). While some authors have started to incorporate this as a constraint in process design, others have opted to leave sustainability (or some equivalent measure) as an alternative objective to be considered together with profitability and perhaps financial risk in Pareto surfaces (Cheng *et al.*, 2003). Grossmann (2003) discusses some of what he calls 'timid' efforts by the process systems engineering community to assess sustainability (Marquardt *et al.*, 2000) and there is already work on industrial ecology performed by chemical engineers (Bakshi, 2000; Bakshi and Fiskel, 2003). Batterham (2003) provides an insightful analysis of sustainability emphasizing the fact that sustainability is no longer a *constraint* that comes from regulations but is becoming a genuine objective of corporations in such a way that 'both companies and society can benefit'. Time will tell how genuine these efforts are. We claim here that sustainability is both a constraint and an alternative objective, and leave the analysis for future work.

12.14 Conclusions

This is not a review of the large number of developments concerning the integration of business tools are examined with process and product engineering. Rather, some tools are examined that help the integration, which has been recently proposed and used. The article focuses on financial risk and proposes that the proper paradigm is through handling of risk curves, especially if they are immersed in a two- (or multi-) stage stochastic model. It was also proposed to extend the connections to shareholder value.

Full integration of several disciplines – management, finances, industrial engineering, and chemical engineering, among others – is slowly taking place. The result is a 'beginning to end' modeling of the product research/development and delivery supply chain. In turn, more interactions with other disciplines (public policy, psychology, etc.) will come. At some point, with powerful computers and adequate modeling, one can dream of the whole process being fully integrated into one single model. Then, one can start to ask whether, with so much access to information and with so many tools to respond optimally, we will reach a state where competition will cease to have a meaning.

Acknowledgements

After we presented our first article on risk management at the 1999 AIChE Meeting (Rodera and Bagajewicz, 2000), we got busy developing the theory (Barbaro and Bagajewicz, 2003, 2004a,b), and than worked on applications of that theory to several problems. Some of these experiences are summarized in this chapter. I thank deeply my students who were instrumental in helping me articulate the vision. A fruitful sabbatical stay with the group of Dr Lluis Puigjaner at the Polytechnic of Catalunya (UPC) taught me invaluable lessons through many passionately argued ideas and discussions with students and professors.

My thanks also go to the following persons who read my original manuscript and were instrumental in improving it: Arthur Westerberg and Ignacio Grossmann (Carnegie Mellon University), Anshuman Gupta and Costas Maranas (Pennsylvania State University), Frank Zhu and Gavin Towler (UOP), Larry Evans (Aspentech), Jeffrey Siirola (Eastman), and Jesus Salas (University of Oklahoma).

Finally, I thank the editors of this book for the financial support and the opportunity to think out loud.

References

Achenie L.E.K., Gani R. and Venkatasubramanian V. 2002 (Eds.). Computer Aided Molecular Design: Theory and Practice. Elsevier Publishers.

Ahmed S. and Sahinidis N.V. 1998. Robust process planning under uncertainty, Ind. Eng. Chem. Res., 37(5), 1883–1892.

Ahmed S. and Sahinidis N.V. 2000a. Analytical investigations of the process planning problem, Comput. Chem. Eng., 23(11–12), 1605–1621.

Ahmed S. and Sahinidis N.V. 2000b. Selection, acquisition and allocation of manufacturing technology in a multi-product environment, Manage. Sci. (submitted).

Applequist G., Pekny J.F. and Reklaitis G.V. 2000. Risk and uncertainty in managing chemical manufacturing supply chains, Comput. Chem. Eng., 24(9/10), 2211.

Arntzen B., Brown G., Harrison T. and Trafton L. 1995. Global supply chain management at digital equipment corporation, Interfaces, 25, 69.

Aseeri A., Gorman P. and Bagajewicz M. 2004. Financial risk management in offshore oil infrastructure planning and scheduling, Ind. Eng. Chem. Res. 43(12), 3063–3072 (special issue honoring George Gavalas).

Aseeri A. and Bagajewicz M. 2004. New measures and procedures to manage financial risk with applications to the planning of gas commercialization in Asia, Comput. Chem. Eng., 28(12), 2791–2821.

Badell M. and Puigjaner L.A. 1998. New conceptual approach for enterprise resource management systems. In, FOCAPO AIChE Symposium Series, New York, Pekny J.F. and Blau G.E. (Eds.), 94 (320) 217.

Badell M. and Puigjaner L. 2001a. Discover a powerful tool for scheduling in ERM systems, Hydrocarbon Process., 80(3), 160.

Badell M. and Puigjaner L. 2001b. Advanced enterprise resource management systems, Comput. Chem. Eng., 25, 517.

Badell M., Romero J., Huertas R. and Puigjaner L. 2004. Planning, scheduling and budgeting value-added chains, Comput. Chem. Eng., 28, 45–61.

Bagajewicz M. 2000. A review of recent design procedures for water networks in refineries and process plants, Comput. Chem. Eng., 24(9), 2093–2115.

Bagajewicz M. and Barbaro A.F. 2003. Financial risk management in the planning of energy recovery in the total site, Ind. Eng. Chem. Res., 42(21), 5239–5248.

Bakshi B.R. 2000. A thermodynamic framework for ecologically conscious process systems engineering, Comput. Chem. Eng., 24, 1767–1773.

Bakshi B.R. and Fiskel J. 2003. The quest for sustainability: challenges for process systems engineering, AIChE J., 49(6), 1350–1358.

Barbaro A. and Bagajewicz M. 2003. Financial risk management in planning under uncertainty. FOCAPO 2003 (Foundations of Computer Aided Process Operations), Coral Springs, FL, USA, January 2003.

Barbaro A.F. and Bagajewicz M. 2004a. Managing financial risk in planning under uncertainty, AIChE J., 50(5), 963–989.

Barbaro A.F. and Bagajewicz M. 2004b. Use of inventory and option contracts to hedge financial risk in planning under uncertainty, AIChE J., 50(5), 990–998.

Batterham R.J. 2003. Ten years of sustainability: where do we go from here?, Chem. Eng. Sci., 58, 2167–2179.

Bellman R.E. 1957. Dynamic Programming. Princeton University Press.

Berger J.O. 1980. Statistical Decision Theory. Springer-Verlag, New York.

Biegler L., Grossmann I.E. and Westerberg A.W. 1997. Systematic Methods of Chemical Process Design. Prentice Hall, New Jersey.

Birge J.R. and Louveaux F. 1997. Introduction to Stochastic Programming. Springer, New York.

Blau G., Mehta B., Bose S., Pekny J., Sinclair G., Keunker K. and Bunch P. 2000. Risk management in the development of new products in highly regulated industries, Comput. Chem. Eng., 24, 659–664.

Blau G.E. and Sinclair G. 2001. Dealing with uncertainty in new product development, CEP, 97(6), 80–83.

Bok J., Lee H. and Park S. 1998. Robust investment model for long range capacity expansion of chemical processing networks under uncertain demand forecast scenarios, Comput. Chem. Eng., 22, 1037–1050.

Bok J., Grossmann I.E. and Park S. 2000. Supply chain optimization in continuous flexible process networks, Ind. Eng. Chem. Res., 39, 1279–1290.

Bonfill A., Bagajewicz M., Espuña A. and Puigjaner L. 2004. Risk management in scheduling of batch plants under uncertain market demand, Ind. Eng. Chem. Res., 43(9), 2150–2159.

Bose S. and Pekny J.F. 2000. A model predictive framework for planning and scheduling problems: a case study of consumer goods supply chain, Comput. Chem. Eng., 24(2–7), 329–335.

Breslow R. and Tirrell M.V. (Co-Chairs). 2003. Beyond the molecular frontier. challenges for chemistry and chemical engineering. Committee on Challenges for the Chemical Sciences in the 21st Century. The United States National Academies Press.

Bunch P.R. and Iles C.M. 1998. Integration of planning and scheduling systems with manufacturing business processes, Annual AIChE Meeting, Miami, FL, paper 235g.

Byrd L.R. and Chung F.T.-H. 1998. Risk analysis and decision-making software package user's manual. Prepared for the US DOE, National Petroleum Technology Office.

Camarda K.V. and Maranas C.D. 1999. Optimization in polymer design using connectivity indices, Ind. Eng. Chem. Res., 38, 1884–1892.

Charnes A. and Cooper W.W. 1959. Chance-constrained programming, Manag. Sci., 6, 73.

Charnes A., Cooper W.W. and Miller M.H. 1959. Application of linear programming to financial budgeting and the costing of funds. In, The Management of Corporate Capital, Solomon E. (Ed.). Free Press of Glencoe, Illinois.

Cheng, L., Subrahmanian E. and Westerberg A.W. 2003. Design and planning under uncertainty: issues on problem formulation and solution, Comput. Chem. Eng., 27(6), 781–801.

Cheng, L., Subrahmanian E. and Westerberg A.W. 2004. Multi-objective decisions on capacity planning and production-inventory control under uncertainty, Ind. Eng. Chem. Res., 43, 2192–2208.

Cheung R.K.M. and Powel W.B. 2000. Shape – a stochastic hybrid approximation procedure for two-stage stochastic programs, Oper. Res., 48(1), 73–79.

Cohen M.A., Fisher M. and Jaikumar R. 1989. International manufacturing and distribution networks: a normative model framework. In, Managing International Manufacturing, Ferdows K. (Ed.). Amsterdam, North Holland, p. 67.

Cussler E.L. 2003. Chemical Product Design and Engineering (Plenary Talk). AIChE Annual Conference, San Francisco. Paper 430a, November 2003.

Cussler E.L. and Moggridge G.D. 2001. Chemical Product Design, 1st edition. Cambridge University Press.

Dahl H., Meeraus A. and Zenios S. 1993. Some financial optimization models: I Risk management. In, Financial Optimization, Zenios S.A. (Eds.). Proceedings of a Conference held at the Wharton School, University of Pennsylvania, Philadelphia, USA.

Danthine J.P. and Donaldson J.B. 2002. Intermediate Financial Theory. Prentice Hall, New Jersey.

De Reyck B., Degraeve Z. and Vandenborre R. 2001. Project Options Valuation with Net Present Value and Decision Tree Analysis. Working Paper, London Business School. Presented at 2001 INFORMS. INFORMS Conference, Miami Beach, November.

Debreu G. 1959. Theory of Value. An Axiomatic Analysis of Economic Equilibrium. John Wiley, New York.

Denardo E.V. 1982. Dynamic Programming: Models and Applications. Prentice Hall, New Jersey.

Dixit A.K. and Pindyck R.S. 1994. Investment under Uncertainty. Princeton University Press, Princeton, NJ.

Dorward N. 1987. The Price Decision: Economic Theory and Business Practice. Harper & Row, London.

Douglas J.M. 1988. Conceptual Design of Chemical Processes. McGraw Hill, New York.

Dupacova J. and Römisch W. 1998. Quantitative stability for scenario-based stochastic programs, Prague Stochastics 98, S. 119–124, JCMF, Praha.

EIA. 2002. Derivatives and Risk Management in the Petroleum, Natural Gas, and Electricity Industries. Energy Information Administration, October 2002. Available at http://www.eia.doe.gov.

Eppen G.D, Martin R.K. and Schrage L. 1989. A scenario approach to capacity planning. Oper. Res., 37, 517–527.

Espuña A., Rodrigues M.T., Gimeno L. and Puigjaner L. 2003. A holistic framework for supply chain management. Proceedings of ESCAPE 13. Lappeenranta, Finland, 1–4 June.

Evans L. 2003. Achieving Operational Excellence with Information Technology. AICHE Spring Meeting, New Orleans, March.

Finnerty J.E. 1993. Planning Cash Flow. AMACON (American Management Association), New York.

Gaur V. and Seshadri S. 2004. Hedging Inventory Risk through Market Instruments. Working Paper at the Stern School of Business, New York University.

Gjerdrum J., Shah N. and Papageorgiou L.G. 2001. Transfer prices for multienterprise supply chain optimization, Ind. Eng. Chem. Res., 40, 1650–1660.

Goetschalckx M., Vidal C.J. and Dogan K. 2002. Modeling and design of global logistics systems: a review of integrated strategic and tactical models and design algorithms, Eur. J. Oper. Res., 143, 1.

Gregory G. 1988. Decision Analysis. Plenum Press, New York.

Grossmann I.E. 2003. Challenges in the new millennium: product discovery and design, enterprise and supply chain optimization, global life cycle assessment. PSE 2003. 8th International Symposium on Process Systems Engineering, China.

Grossmann I.E. and Westerberg A.W. 2000. Research challenges in process systems engineering. AIChE J., 46, 1700–1703.

Guillén G., Bagajewicz M., Sequeira S.E., Tona R., Espuña A. and Puigjaner L. 2003a. Integrating pricing policies and financial risk management into scheduling of batch plants. PSE 2003. 8th International Symposium on Process Systems Engineering, China, June 2003.

Guillén G., Mele F., Bagajewicz M., Espuña A. and Puigjaner L. 2003b. Management of financial and consumer satisfaction risks in supply chain design. Proceedings of ESCAPE 13. Lappeenranta, Finland, 1–4 June 2003.

Guldimann T. 2000. The story of risk metrics, Risk, 13(1), 56–58.

Gupta A. and Maranas C.D. 2000. A two-stage modeling and solution framework for multisite midterm planning under demand uncertainty, Ind. Eng. Chem. Res., 39, 3799–3813.

Gupta A. and Maranas C. 2003a. Managing demand uncertainty in supply chain planning, Comput. Chem. Eng., 27, 1219–1227.

Gupta A. and Maranas C. 2003b. Market-based pollution abatement strategies: risk management using emission option contracts, Ind. Eng. Chem. Res., 42, 802–810.

Gupta A. and Maranas C. 2004. Real-options-based planning strategies under uncertainty, Ind. Eng. Chem. Res., 43(14), 3870–3878.

Gupta A., Maranas C.D. and McDonald C.M. 2000. Midterm supply chain planning under demand uncertainty: customer demand satisfaction and inventory management, Comput. Chem. Eng., 24, 2613–2621.

Hax A.C. and Majluf N.S. 1984. Strategic Management: An Integrated Perspective. Prentice Hall, New Jersey.

Higle J.L. and Sen S. 1996. Stochastic Decomposition. A Statistical Method for Large Scale Stochastic Linear Programming. Kluwer Academic Publishers, Norwell, MA.

Hull J. 1995. Introduction to Futures and Options Markets. Prentice Hall, Englewood Cliffs, NJ.

Ierapetritou M.G. and Pistikopoulos E.N. 1994. Simultaneous incorporation of flexibility and economic risk in operational planning under uncertainty, Comput. Chem. Eng., 18(3), 163–189.

Ierapetritou M.G., Pistikopoulos E.N. and Floudas C.A. 1994. Operational planning under uncertainty, Comput. Chem. Eng., 18(Suppl.), S553–S557.

Infanger G. 1994. Planning under Uncertainty: Solving Large-Scale Stochastic Linear Programs. Boyd and Fraser, Danvers, MA.

Iyer R.R. and Grossmann I.E. 1998a. A bilevel decomposition algorithm for long-range planning of process networks, Ind. Eng. Chem. Res., 37, 474–481.

Iyer R.R. and Grossmann I.E. 1998b. Synthesis of operational planning of utility systems for multiperiod operation, Comput. Chem. Eng., 22, 979–993.

Iyer R.R., Grossmann I.E., Vasantharajan S. and Cullick A.S. 1998. Optimal planning and scheduling of offshore oil field infrastructure investment and operations, Ind. Eng. Chem. Res., 37, 1380–1397.

Jackson J.R. and Grossmann I.E. 2003. Temporal decomposition scheme for nonlinear multisite production planning and distribution models, Ind. Eng. Chem. Res., 42, 3045–3055.
Jia J. and Dyer J.S. 1995. Risk-Value Theory. Working Paper, Graduate School of Business, University of Texas at Austin. Presented at INFORMS Conference, Los Angeles.
Jia Z., Ierapetritou M. and Kelly J.D. 2003. Refinery short-term scheduling using continuous time formulation: crude oil operations, Ind. Eng. Chem. Res., 42, 3085–3097.
Joly M. and Pinto J.M. 2003. Mixed-integer programming techniques for the scheduling of fuel oil and asphalt production, Trans. IChemE, 81(Part A) 427–447.
Jorion P. 2000. Value at Risk. The New Benchmark for Managing Financial Risk, 2nd edition. McGraw Hill, New York.
Julka N., Srinivasan R. and Karimi I. 2002a. Agent-based supply chain management – 1: framework, Comput. Chem. Eng., 26, 1755–1769.
Julka N., Karimi I. and Srinivasan R. 2002b. Agent-based supply chain management – 2: a refinery application, Comput. Chem. Eng., 26, 1771–1781.
Kall P. and Wallace S.W. 1994. Stochastic Programming. John Wiley & Sons, Chichester.
Keown A.J., Martin J.D., Petty J.W. and Scott D.F. 2002. Financial Management: Principles and Applications, 9th edition. Prentice Hall, New Jersey.
Koppol A. and Bagajewicz M. 2003. Financial risk management in the design of water utilization systems in process plants, Ind. Eng. Chem. Res., 42(21), 5249–5255.
Koppol A., Bagajewicz M., Dericks B.J. and Savelski M.J. 2003. On zero water discharge solutions in the process industry, Adv. Environ. Res., 8(2), 151–171.
Lababidi H.M.S., Ahmed M.A., Alatiqi I.M. and Al-Enzi A.F. 2004. Optimizing the supply chain of a petrochemical company under uncertain operating and economic conditions, Ind. Eng. Chem. Res., 43(1), 63–73.
Lee Y.G. and Malone M.F. 2001. A general treatment of uncertainties in batch process planning, Ind. Eng. Chem. Res., 40, 1507–1515.
Lee H., Pinto J.M., Grossmann I.E. and Park S. 1996. Mixed integer linear programming model for refinery short-term scheduling of crude oil unloading with inventory management, Ind. Eng. Chem. Res., 35, 1630–1641.
Levis A.A. and Papageorgiou L.G. 2003. Multisite capacity planning for the pharmaceutical industry using mathematical programming. Proceedings of the ESCAPE 13 Meeting, Lappeenranta, Finland, 1–4 June 2003.
Lin X., Floudas C., Modi S. and Juhasz N. 2002. Continuous time optimization approach for medium-range production scheduling of a multiproduct batch plant, Ind. Eng. Chem. Res., 41, 3884–3906.
Linsmeier T.J. and Pearson N.D. 2000. Value at risk, Financ. Anal. J., 56(2), 47–68.
Lintner J. 1969. The valuation of risk assets and the selection of risky investments in stock portfolios and capital budgets, Rev. Econ. Stat., 51(2), 222–224.
Liu M.L. and Sahinidis N.V. 1996. Optimization in process planning under uncertainty, Ind. Eng. Chem. Res., 35, 4154–4165.
Maravelias C.T. and Grossmann I.E. 2003. A new continuous-time state task network formulations for short term scheduling of multipurpose batch plants. Proceedings of the ESCAPE 13 Meeting, Lappeenranta, Finland, 1–4 June 2003.
Marquardt W.L., Von Wedel L. and Bayer B. 2000. Perspectives on lifecycle process modeling, AIChE Symp. Ser., 96(323), 192–214.
Marti K. and Kall P. 1998 (Eds.). Stochastic Programming Methods and Technical Applications, LNEMS, Vol. 458. Springer-Verlag, Berlin.
Mas-Collel A., Whinston M. and Green J. 1995. Microeconomics Theory. University Press, Oxford.
McCray A.W. 1975. Petroleum Evaluations and Economic Decisions. Prentice Hall, New Jersey.
McDonald C. 2002. Earnings optimized planning for business decision making in global supply chains. 9th Mediterranean Congress of Chemical Engineering, Barcelona, Spain. Lecture A. November 2002.

McDonald C.M. and Karimi I.A. 1997. Planning and scheduling of parallel semi-continuous processes 1. Production planning, Ind. Eng. Chem. Res., 36, 2691–2700.

Mele F., Bagajewicz M., Espuña A. and Puigjaner L. 2003. Financial risk control in a discrete event supply chain. Proceedings of the ESCAPE 13 Meeting, Lappeenranta, Finland, 1–4 June.

Mendez C.A. and Cerdá J. 2003. Dynamic scheduling in multiproduct batch plants, Comput. Chem. Eng., 27, 1247–1259.

Mendez C.A., Henning G.P. and Cerdá J. 2000. Optimal scheduling of batch plants satisfying multiple product orders with different due dates, Comput. Chem. Eng., 24(9–10), 2223–2245.

Miller M.H. and Orr R. 1966. A model of the demand for money by firms, Q. J. Econ., 80(3), 413.

Modiano E.M. 1987. Derived demand and capacity planning under uncertainty, Oper. Res., 35, 185–197.

Moro L.F.L. and Pinto J.M. 2004. Mixed-integer programming approach for short-term crude oil scheduling, Ind. Eng. Chem. Res., 43(1), 85–94.

Mulvey J.M., Vanderbei R.J. and Zenios S.A. 1995. Robust optimization of large-scale systems, Oper. Res., 43, 264–281.

Nebel B.J. and Wright R.T. 2002. Environmental Science: Toward a Sustainable Future. Prentice Hall, New Jersey.

Neiro S.M.S. and Pinto J.M. 2003. Supply Chain Optimization of Petroleum Refinery Complexes. FOCAPO 2003. (Foundations of Computer Aided Process Operations). Coral Springs, FL, USA, January.

O'Donnel B., Hickner M.A. and Barna B.A. 2002. Economic risk analysis. Using analytical and Monte Carlo techniques, Chemical Engineering Education, Spring.

Ogryczak W. and Ruszcynski A. 2002. Dual stochastic dominance and related mean-risk models, Siam J. Optim., 13(1), 60–78.

Oh H.-C. and Karimi I.A. 2004. Regulatory factors and capacity expansion planning in global chemical supply chains, Ind. Eng. Chem. Res., 43, 3364–3380.

Orcun S., Joglekar G. and Clark S. 1996. Scheduling of batch processes with operational uncertainties, Comp. Chem. Eng., 20, Suppl., S1191–S1196.

Orcun S., Joglekar G. and Clark S. 2002. An iterative optimization-simulation approach to account for yield variability and decentralized risk parameterization, AIChE Annual Meeting, Indianapolis, Indiana, paper 266b.

Orgler Y.E. 1970. Cash Management. Wadsworth Pub., California.

Ortiz-Gómez A., Rico-Ramírez V. and Hernández-Castro S. 2002. Mixed integer multiperiod model for the planning of oilfield production, Comp. Chem. Eng., 26, 703–714.

Pande P.S. and Holpp L. 2001. What is Six Sigma? McGraw Hill, New York.

Papageorgiou L.G., Rotstein G.E. and Shah N. 2001. Strategic supply chain optimization for the pharmaceutical industries, Ind. Eng. Chem. Res., 40, 275–286.

Pekny J. 2002. Algorithm architectures to support large-scale process systems engineering applications involving combinatorics, uncertainty, and risk management, Comp. Chem. Eng., 26, 239–267.

Perea-Lopez E., Grossmann I.E. and Ydstie B.E. 2000. Dynamic modeling and decentralized control of supply chains, Ind. Eng. Chem. Res., 40(15), 3369–3383.

Perea-Lopez E., Ydstie B.E. and Grossmann I.E. 2003. A model predictive control strategy for supply chain optimization, Comp. Chem. Eng., 27, 1201–1218.

Peters M.S., Timmerhaus K.D. and West R.E. 2003. Plant Design and Economics for Chemical Engineers, 5th edition. McGraw Hill, New York.

Petkov S.B. and Maranas C.D. 1997. Multiperiod planning and scheduling of multiproduct batch plants under demand uncertainty, Ind. Eng. Chem. Res., 36, 4864–4881.

Pinto J.M., Joly M. and Moro L.F.L. 2000a. Planning and scheduling models for refinery operations, Comp. Chem. Eng., 24, 2259–2276.

Pistikopoulos E.N. and Ierapetritou M.G. 1995. Novel approach for optimal process design under uncertainty, Comput. Chem. Eng., 19(10), 1089–1110.

PSE 2003. 8th International Symposium on Process Systems Engineering (Computer Aided Chemical Engineering, Volume 15A & 15B), Bingzhe Chen and Art westerberg (Eds.), Computer Aided Chemical Engineering Series, Elsevier.

Puigjaner L., Espuña A., Graells M. and Badell M. 2000. Advanced Concepts on Batch Processes Integration and Resource Conservation Economics. UPC Press, ISBN 84-930526-7-1.

Puigjaner L., Dovì V., Espuña A., Graells M., Badell M. and Maga L. 1999. Fundamentals of Process Integration and Environmental Economics. UPC Press Barcelona, Spain.

Reddy P.C.P., Karimi I.A. and Srinivasan R. 2004. A novel solution approach for optimizing crude oil operations, AIChE J., 50(6), 1177–1197.

Rico-Ramirez V., Diwekar U.M. and Morel B. 2003. Real option theory from finance to batch distillation, Comput. Chem. Eng., 27, 1867–1882.

Riggs J.L. 1968. Economic decision models for engineers and managers. McGraw Hill, NY.

Robertson J.L., Subrahmanian E., Thomas M.E. and Westerberg A.W. 1995. Management of the design process: the impact of information modeling. In, Biegler L.T. and Doherty M.F. (Eds.), Fourth International Conference on Foundations of Computer Aided Process Design Conference, AIChE Symposium Series No. 304, 91, CACHE Corp. and AIChE, 154–165.

Robichek A.A., Teichroew D. and Jones J.M. 1965. Optimal short-term financing decision, Manag. Sci., 12, 1.

Rodera H. and Bagajewicz M. 2000. Risk assessment in process planning under uncertainty, AIChE Annual Meeting, Los Angeles.

Rogers M.J., Gupta A. and Maranas C.D. 2002. Real options based analysis of optimal pharmaceutical R&D portfolios, Ind. Eng. Chem. Res., 41, 6607–6620.

Rogers M.J., Gupta A. and Maranas C.D. 2003. Risk Management in Real Options Based Pharmaceutical Portfolio Planning. FOCAPO, 241–244.

Romero J., Badell M., Bagajewicz M. and Puigjaner L. 2003a. Risk management in integrated budgeting-scheduling models for the batch industry, PSE 2003, 8th International Symposium on Process Systems Engineering, China.

Romero J., Badell M., Bagajewicz M. and Puigjaner L. 2003b. Integrating budgeting models into scheduling and planning models for the chemical batch industry, Ind. Eng. Chem. Res., 42(24), 6125–6134.

Rotstein G.E., Papageorgiou L.G., Shah N., Murphy D.C. and Mustafa R. 1999. A product portfolio in the pharmaceutical industry, Comput. Chem. Eng., S23, S883–S886.

Sahinidis N.V., Grossmann I.E., Fornari R.E. and Chathrathi M. 1989. Optimization model for long range planning in the chemical industry, Comput. Chem. Eng., 13, 1049–1063.

Sahinidis N.V. and Tawarlamani M. 2000. Applications of global optimization to process molecular design, Comput. Chem. Eng., 24, 2157–2169.

Schmidt C.W. and Grossmann I.E. 1996. Optimization models for the scheduling of testing tasks in new product development, Ind. Eng. Chem. Res., 35(10), 3498–3510.

Schuyler J. 2001. Risk and Decision Analysis in Projects, 2nd edition, Project Management Institute, Newton Square, PA.

SCORE. 1998. Stock options reference educator. Education and Training Department, Hong Kong Exchanges and Clearing Limited. http://www.hkex.com.hk/TOD/SCORE/english/, version 1.0.

Seider W.D., Seader J.D. and Lewin D.R. 2004. Product and Process Design Principles. John Wiley, New York.

Sengupta J.K. 1972. Stochastic Programming: Methods and Applications. North Holland, Amsterdam.

Shah N. 1996. Mathematical programming techniques for crude oil scheduling, Comp. Chem. Eng., Suppl., S1227–S1232.

Shah N. 1998. Single and multisite planning and scheduling: current status and future challenges. In, Pekny J. and Blau G. (Eds.), Proceedings of the Conference in Foundations of Computer Aided Process Operations, AIChE Symposium Series, No 320, CACHE Publications.

Sharpe W.F. 1966. Mutual fund performance, J. Bus., January, 119–138.
Sharpe W.F. 1970. Portfolio Theory and Capital Markets. McGraw Hill, New York.
Shimko D. 1997. See Sharpe or be flat, Risk, July, 29.
Shimko D. 1998. Cash before value, Risk, July, 45.
Shimko D. 2001. NPV no more: RPV for risk-based valuation. Available at www.ercm.com.
Siirola J.D., Hauan S. and Westerberg A.W. 2003. Toward agent-based process systems engineering: proposed framework and application to non-convex optimization, Comp. Chem. Eng, 27, 1801–1811.
Singhvi A. and Shenoy U.V. 2002. Aggregate planning in supply chains by pinch analysis, Trans. IChemE, part A, September.
Smart S.B, Megginson W.L. and Gitman L.J. 2004. Corporate Finance. Thomson-South Western, Ohio.
Smith R. and Linnhoff B. 1988. The design of separators in the context of overall processes, Trans. IChemE, ChERD, 66, 195.
Smith R. 1995. Chemical Process Design. McGraw Hill, New York.
Srinivasan V. 1986. Deterministic cash flow management, Omega, 14(2), 145–166.
Stamatis D.H. 2003. Failure Mode and Effect Analysis: FMEA from Theory to Execution. American Society for Quality, Milwaukee, WI.
Stephanopoulos G. 2003. Invention and innovation in a product-centered chemical industry: general trends and a case study. AIChE Annual Conference (Nov). 55th Institute Lecture, San Francisco.
Subrahmanyam S., Pekny J. and Reklaitis G.V. 1994. Design of batch chemical plants under market uncertainty, Ind. Eng. Chem. Res., 33, 2688–2701.
Subramanian D., Pekny J. and Reklaitis G.V. 2000. A simulation–optimization framework for addressing combinatorial and stochastic aspects on an R&D pipeline management problem, Comp. Chem. Eng., 24, 1005–1011.
Tan B. 2002. Managing manufacturing risks by using capacity options, J. Oper. Res. Soc., 53, 232–242.
Takriti S. and Ahmed S. 2003. On robust optimization of two-stage systems, Math. Program., 99(1), 109–126.
Trigeorgis L. 1999. Real Options. Managerial Flexibility and Strategy in Resource Allocation. MIT Press, Cambridge, MA.
Tsiakis P., Shah N. and Pantelides C.C. 2001. Design of multi-echelon supply chain networks under demand uncertainty, Ind. Eng. Chem. Res., 40, 3585–3604.
Umeda T. 2004. A conceptual framework for the process system synthesis and design congruent with corporate strategy, Ind. Eng. Chem. Res., 43(14), 3827–3837.
Uryasev S., Panos M. and Pardalos. 2001 (Eds.). Stochastic Optimization: Algorithm and Applications. Kluwer Academic Publisher, Norwell, MA.
Van den Heever S. and Grossmann I.E. 2000. An iterative aggregation/disaggregation approach for the solution of the mixed integer nonlinear oilfield infrastructure planning model, Ind. Eng. Chem. Res., 39(6), 1955.
Van den Heever S., Grossmann I.E., Vasantharajan S. and Edwards K. 2000. Integrating complex economic objectives with the design and planning of offshore oilfield infrastructures, Comput. Chem. Eng., 24, 1049–1055.
Van den Heever S., Grossmann I.E., Vasantharajan S. and Edwards K. 2001. A Lagrangean decomposition heuristic for the design and planning of offshore field infrastructures with complex economic objectives, Ind. Eng. Chem. Res., 40, 2857–2875.
Verweij B., Ahmed S., Kleywegt A.J., Nemhauser G. and Shapiro A. 2001. The sample average approximation method applied to stochastic routing problems: a computational study, Comput. Appl. Optim., 24, 289–333.
Wendt M., Li P. and Wozny G. 2002. Nonlinear chance constrained process optimization under uncertainty, Ind. Eng. Chem. Res., 41, 3621–3629.

Wenkai L., Hui C. and Hua B. 2002. Scheduling crude oils unloading, storage and processing, Ind. Eng. Chem. Res., 41, 6723–6734.

Westerberg A.W. and Subramanian E. 2000. Product design, Comp. Chem. Eng., 24(2–7), 959–966.

Wibowo C. and Ka M.Ng. 2001. Product-oriented process synthesis and development: creams and pastes, AIChE J., 47(12), 2746–2767.

Wilkinson S.J., Cortier A., Shah N. and Pantelides C.C. 1996. Integrated production and distribution scheduling on a European wide basis, Comp. Chem. Eng., 20, S1275–S1280.

Zhang J., Zhu X.X. and Towler G.P. 2001. A level-by-level debottlenecking approach in refinery operation, Ind. Eng. Chem. Res., 40, 1528–1540.

Index